Advances in Polar Ecology

Volume 6

Series Editor
Dieter Piepenburg
Helmholtz Centre for Polar and Marine Research, Alfred Wegener Institute,
Bremerhaven, Bremen, Germany

Advances in Polar Ecology

In recent years, the polar regions have received increased scientific and public interest. Both the Arctic and Antarctic have been recognized as key regions in the regulation of the global climate and polar ecosystems have been identified to be particularly susceptible to the ongoing environmental changes. Consequently, the international efforts in polar research have been enhanced considerably, and a wealth of new findings is being produced at a growing rate by the international community of polar researchers. The aim of the book series Advances in Polar Ecology is to foster the progress in the scientific knowledge about the polar and sub-polar regions of both hemispheres by contributing to the fast and wide-ranging dissemination of novel scientific information gained from recent studies of sea, freshwater and land biota among polar researchers, environmental managers and policy makers. Advances in Polar Ecology's broad ecology-oriented scope encompasses environmental and biological research of both recent and past ecosystems. The Series offers an outlet to publish contributions (monographs, edited works, conference proceedings, etc.) addressing scientific topics that need more comprehensive and in-depth analysis than the length of typical journal articles allow for. These include, but are not limited to, thorough accounts of the current status of active research areas of wide importance that outline promising perspectives for future research. An international editorial board oversees the rigorous peer review that all contributions will be subjected to.

More information about this series at https://link.springer.com/bookseries/10290

Eugene G. Morozov • Mikhail V. Flint
Vassily A. Spiridonov
Editors

Antarctic Peninsula Region of the Southern Ocean

Oceanography and Ecology

Springer

Editors
Eugene G. Morozov
Shirshov Institute of Oceanology
Moscow, Russia

Mikhail V. Flint
Shirshov Institute of Oceanology
Moscow, Russia

Vassily A. Spiridonov
Shirshov Institute of Oceanology
Moscow, Russia

ISSN 2468-5712 ISSN 2468-5720 (electronic)
Advances in Polar Ecology
ISBN 978-3-030-78926-8 ISBN 978-3-030-78927-5 (eBook)
https://doi.org/10.1007/978-3-030-78927-5

This Springer imprint is published by the registered company Springer Nature Switzerland AG
The registered company address is: Gewerbestrasse 11, 6330 Cham, Switzerland

Abstract

This book is dedicated to the recent research in the Atlantic sector of the Southern Ocean. The book includes descriptions and analyses of the most important features of the Antarctica Peninsula region.

The book is based on results of a Russian expedition in the region of the Antarctic Peninsula as well as on a review of an earlier research in the region. The main goal of the research was to study the physical properties and ecology of this key region of the Southern Ocean. Data analysis is supplemented with numerical modeling of the circulation in the adjacent region of the Southern Ocean and research on rogue waves. The main objects of the study were the properties of seawater, currents, ecosystem, and biological communities in the Powell Basin of the northwestern Weddell Sea, the Bransfield Strait, and the Antarctic Sound.

The Antarctic marine ecosystem is unique due to its history, which is millions of years long, its remarkable biodiversity, and the high productivity of basic trophic levels of the ecosystem. This ecosystem is efficiently used by organisms such as seabirds, seals, and whales that feed on Antarctic krill.

A CTD/LADCP survey was conducted in the Powell Basin in the northern Weddell Sea. Temperature and salinity profiles at each station were supplemented with measurements of optical and chemical properties of seawater and plankton catches. The survey covered the hydrological front between the water of the Weddell Sea and very cold water in the western part of the sea above a sea bottom elevation covered with floating ice and icebergs.

The Powell Basin is a link between the ecosystems of the Weddell Sea and the Scotia Sea, providing for the accumulation and redistribution of krill transported with the waters of the western branch of the Weddell Cycle. Studies of the water column structure, optical characteristics, and chlorophyll fluorescence revealed a significant variety of types of the vertical structure of the upper layer (in particular, a very non-uniform distribution of meltwater and phytoplankton distribution). A significant proportion of the waters resulting from the melting of sea ice and icebergs is associated with an increased concentration of phytoplankton. An abundance of juvenile krill was found in the western part of the Powell Basin, which is transported with the water and drifting ice from the southern parts of the Weddell Sea.

Benthic studies conducted in the Powell Basin and its surrounding submarine elevations revealed the existence of highly diverse benthic communities. Their composition of macrotaxones and the dominance of particular trophic groups is related to the depth and oceanographic conditions. For the first time in the study area, bottom sampling was combined with video surveillance and photography.

Acknowledgments

The expedition onboard the research vessel *Akademik Mstislav Keldysh* was supported by the State Task of the Ministry of Higher Education and Science of Russia. The expedition was financed under state tasks: Shirshov Institute of Oceanology 0128-2019-0008; Il'ichev Pacific Oceanological Institute FWMM-2019-0007; Zhirmunsky National Scientific Center of Marine Biology, Far East Branch, Russian Academy of Sciences 0208-2019-0013; Marine Hydrophysical Institute 0555-2019-0003; Institute of Biology of Southern Seas AAAA-A19-119100290162-0; Institute of Problems of Marine Technology 13.1902.21.0012; and Severtsov Institute of Ecology and Evolution 0089-2019-0021.

Analysis and data processing were supported by grants of the Russian Foundation for Basic Research nos. 20-08-00246, 19-57-60001.

The editors and authors of the book are grateful to numerous colleagues who participated in the field work and helped to perform observations in the ocean. We are also grateful to the crew and captain of the vessel L.V. Sazonov and members of the crew V.V. Nedbailov, V.V. Kvakin, S.V. Abramov, Yu.K. Kovyrshin, V.V. Ivanov, A.V. Goskov, L.A. Salimgaraeva, N.N. Malashkin, V.I. Kokitko, N.P. Matviy, D.O. Khaldin, S.V. Artyushenko, and L.P. Kozlova. We thank our colleagues who made a great contribution to the organization of the expedition: A.V. Sokov, director of the Institute, V.P. Tereshchenkov, head of the Scientific Fleet Department; and the Atlantic Fleet Base headed by A.V. Mysherina with the assistance of V.M. Pyatakov and S.Yu. Lukashin.

Eugene Morozov

Mikhail Flint

Vassily Spiridonov

Introduction

This book is an overview of the research performed by Russian scientists in the region of the Antarctic Peninsula, Powell Basin, Drake Passage, Bransfield Strait, and Antarctic Sound in January–March 2020. Field works are supplemented with theoretical research and numerical modeling. The studies were performed during cruise 79 of the R/V *Akademik Mstislav Keldysh* aimed at investigation of ecosystems in the Atlantic sector of Antarctica. During this cruise, studies on physical oceanography, biology, and ecology of the region were carried out within a multidisciplinary research performed by cooperation of research institutes of Russia: Shirshov Institute of Oceanology, Russian Academy of Sciences; Il'ichev Pacific Oceanological Institute, Far East Branch, Russian Academy of Sciences; Zhirmunsky National Scientific Center of Marine Biology, Far East Branch, Russian Academy of Sciences; Institute of Problems of Marine Technology, Far East Branch, Russian Academy of Sciences; Kovalevsky Institute of Biology of Southern Seas, Russian Academy of Sciences; Marine Hydrophysical Institute, Russian Academy of Sciences; Severtsov Institute of Ecology and Evolution, Russian Academy of Sciences; Lavrentiev Institute of Hydrodynamics, Siberian Branch of the Russian Academy of Sciences, and also representatives of the Russian Federal Research Institute of Fisheries and Oceanography, Russian Federal Fishery Agency; Geographical Faculty of the Moscow State University; and Southern Research Center, Russian Academy of Sciences. Processing of satellite information was carried out by the Aerocosmos Research Institute.

Field works in the Antarctic region on cruise 79 of the R/V Akademik Mstislav Keldysh (January 16–March 1, 2020) were carried out in the year of the 200th anniversary of the discovery of Antarctica by the Russian expedition on the *Vostok* and *Mirny* sloops. This was the first Russian Antarctic expedition held in 1819–1821, headed by F.F. Bellingshausen and M.P. Lazarev. Their expedition was aimed to reach the Southern Ocean and find land in the proximity of the South Pole. We also give credit to the First Soviet Antarctic Expedition headed by M.M. Somov and V.G. Kort. The expedition lasted from 1955 to 1957. In 1982–1983, another Antarctic expedition was organized by the Shirshov Institute of Oceanology.

The main studies in the 2020 expedition in the Southern Ocean were focused on identifying the mechanisms of the formation of biological productivity in the confluence zone of the Antarctic Circumpolar Current and the Weddell Sea waters, while assessing the diversity of pelagic and bottom biota and structural and functional features of pelagic and bottom ecosystems. The studies at the stations included hydrophysical, hydrochemical, and optical measurements, measurements of primary and bacterial production, and plankton catches. Materials on the bottom fauna and photographs of the bottom surface were obtained using a submersible vehicle.

During the cruise, data were collected that characterize the environment state, and plankton communities including krill, ichthyoplankton, macronecton, bottom fauna, and spatiotemporal variations in the most important components of the ecosystem.

The expedition in 2019–2020 aimed at studying the Antarctic ecosystems in the key regions of transport and interaction of the water masses in the Atlantic sector of Antarctica. It was organized in accordance with the objectives for the study of the World Ocean formulated in "Strategy for Development of Russian Federation Activity in Antarctica up to 2020 and the More Distant Future." The investigations were focused on obtaining new results needed for a comprehensive assessment of the state of Antarctic ecosystems, their vulnerability to impacts of fisheries, the impact of current climate change on Antarctic ecosystems, assessing the role and place of Antarctica in the global climatic changes, and studying the climate-forming role of the waters and ice of the Southern Ocean and their influence on the global natural processes.

The study presented in this book is based on recent observations, analysis of historical data, and literature reviews. Field studies analyzed in the book are supplemented by theoretical and numerical studies and reviews of previous oceanographic research projects.

A total of 104 scientists from 10 institutes of the Russian Academy of Sciences as well as from the Moscow State University and VNIRO institute took part in the research. During the field works in the Southern Ocean, 123 oceanographic stations were occupied with hydrophysical, hydrochemical, and hydrobiological observations. A map of the oceanographic stations in the cruise is shown in Fig. 1 and a photo of a rogue wave in the region is shown in Fig. 2.

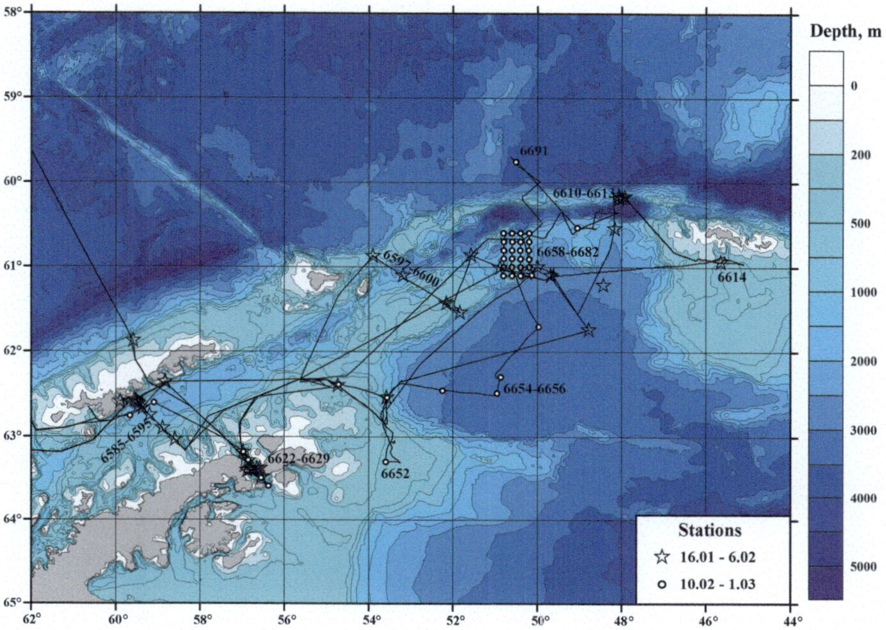

Fig. 1 Stations during cruise 79 of the R/V *Akademik Mstislav Keldysh*

Fig. 2 Photo of a rogue wave in the Drake Passage (A.V. Grigoriev)

Contents

Part I
Atmosphere-Ocean Interaction and Physical Oceanography

Chapter 1
Geostrophic and Wind-Driven Components of the Antarctic Circumpolar Current

Nikolay A. Diansky, Varvara V. Bagatinskaya, Anatoly V. Gusev, and Eugene G. Morozov

Abstract The contributions of wind and thermohaline factors to the formation of the mean climatic structure of the Antarctic Circumpolar Current (ACC) were studied using numerical simulations with the INMOM ocean general circulation model. The goal of our research was to separate the wind and thermohaline ACC components. The simulations were carried out for summer (February) and winter (August) conditions. The ACC structure of three jets was revealed by numerical simulations using the diagnosis-adjustment approach to the EN4 climatic data of observations. Wind circulation in the surface layer turns to the left with respect to the direction of wind stress (southern hemisphere) according to Ekman's theory. Due to the stronger winds in winter, the response of drift velocities is stronger than in summer. It is shown that, in spite of strong winds over the Southern Ocean, the thermohaline factor of circulation is generally much stronger than the drift factor. Nevertheless, the contribution of the wind component to the increase in the zonal velocity in the Drake Passage near Antarctica can be as high as 15–20% of the thermohaline velocity in the upper layer. Wind contributes to a decrease in the mean dynamic topogra-

N. A. Diansky (✉)
Lomonosov Moscow State University, Moscow, Russia

Zubov State Oceanographic Institute, Moscow, Russia

Marchuk Institute of Numerical Mathematics, Russian Academy of Sciences, Moscow, Russia

V. V. Bagatinskaya
Lomonosov Moscow State University, Moscow, Russia

A. V. Gusev
Zubov State Oceanographic Institute, Moscow, Russia

Marchuk Institute of Numerical Mathematics, Russian Academy of Sciences, Moscow, Russia

Shirshov Institute of Oceanology, Russian Academy of Sciences, Moscow, Russia

E. G. Morozov
Shirshov Institute of Oceanology, Russian Academy of Sciences, Moscow, Russia

E. G. Morozov et al. (eds.), *Antarctic Peninsula Region of the Southern Ocean*, Advances in Polar Ecology 6, https://doi.org/10.1007/978-3-030-78927-5_1

phy (MDT) by more than 6 cm from the open ocean to the Antarctic coast. The MDT decrease caused by thermohaline factors over this distance exceeds 2 m, which agrees well with the finding that the contribution of thermohaline factors to the ACC dynamics significantly exceeds the contribution of wind. However, winds are subjected to seasonal variations. The effect of wind on the formation of the barotropic stream function is more pronounced than on the MDT. Thermohaline transport in the Drake Passage is almost the same (~110 Sv) in winter and summer. Due to wind forcing, the total ACC transport around Antarctica increases on average by 9–11 Sv in February and by 12–17 Sv in August. In the Drake Passage, the wind transport increases less than over the entire ACC: by 8 Sv in summer and by 12 Sv in winter. Thermohaline factors comprise a significantly greater contribution to the ACC dynamics than wind forcing. A three-jet structure of the ACC was revealed based on simulations using the INMOM model and EN4 data. It is shown that this three-jet structure of the ACC is of a thermohaline nature.

Keywords Antarctic Circumpolar Current · Wind contribution · Thermohaline factor · Numerical model · Barotropic stream function

1.1 Introduction

The Antarctic Circumpolar Current (ACC) flows along the entire circumference around Antarctica. The ACC is the only current linking three major oceans. There is no land connecting the Antarctic continent with the other continents; therefore, this current is circumpolar. This is the most powerful current on the globe. According to the recent investigations, its volume transport varies over a wide range; for example, 80 Sv is reported in Pfister et al. (2014), 100 Sv in Stammer et al. (2018), 130 Sv in Zika et al. (2012), 144 Sv in Koshlyakov et al. (2013), 150 Sv in our recent paper (Farneti et al. 2015), and 173 Sv in Donohue et al. (2016) (1 Sv = million m^3/s). The ACC blocks meridional transport and heat exchange between the Antarctic continent and warm waters of the ocean north of the current. The ACC forms two large gyres between the current and continent: the Ross and Weddell gyres.

The current includes three main fronts: The Subantarctic Front (SAF), the Polar Front (PF), and the Southern ACC (SACC) Front. The multi-jet structure of the ACC was described in Burkov (1994) and Orsi et al. (1995). These jets were identified from the data of hydrographic stations. However, the structure of the ACC is more complex.

When satellite altimetry was introduced into ocean science in the last decade of the twentieth century, a number of works appeared based on the satellite data (Sokolov and Rintoul 2007). They revealed that the ACC includes a larger number of jets. Sokolov and Rintoul (2007, 2009) suggested a nine-jet ACC structure. Tarakanov and Gritsenko (2018) revealed that the ACC structure includes an even greater number of jets. They identified 12 simultaneously existing jets.

The ACC is formed predominantly due to strong westerly winds and sea surface fluxes of heat and fresh water caused by ice melting in Antarctica. Eventually, the ACC appeared strongly related to the zonal thermohaline stratification of the Southern Ocean, which is formed under long-term wind forcing and heat and freshwater fluxes; it is also steered by the bottom topography and coastline. For millions of years the thermohaline stratification has been adjusting to the nonuniform heating of the Earth and wind forcing at the surface. The ACC balances the momentum imparted by the winds. The momentum caused by wind induces northward motion due to the Coriolis force. This northward Ekman transport is balanced by a deep pressure-driven southward flow and eddy transport across the current over the entire ocean depth, which is also directed southward (Koshlyakov and Tarakanov 2011). Hence, upwelling of deep waters appears, which are transformed to surface waters with a resulting northward flow.

Therefore, the ACC structure is rather complex, and it would be interesting to find out the relative contributions of wind and gradient components to the ACC intensity. The goal of this research is to quantitively estimate the geostrophic and wind-driven components of the ACC. Numerical simulation is a good tool for this. The ocean general circulation model (OGCM) INMOM (Institute of Numerical Mathematics Ocean Model) (Moshonkin et al. 2018) is used in this research. The INMOM has already been used in the prognostic mode for simulating Antarctic Bottom Water flows (Frey et al. 2019) as well as for investigation of the Southern Ocean in the scope of the Coordinated Ocean-ice Reference Experiments, Phase II (CORE-II) (Downes et al. 2015; Farneti et al. 2015). The results of simulation with the INMOM appeared comparable with the leading world OGCMs. In this research, to study the ACC structure, the INMOM is used in the diagnosis mode with application of the diagnosis-adjustment method (Sarkisyan and Sündermann 2009).

1.2 Description of the INMOM Ocean Global Circulation Model and Scenarios of Experiments

The INMOM belongs to the class of σ-coordinate ocean models (Griffies 2006), in which an isobathic coordinate is introduced as

$$\sigma = (z+\zeta)/(H+\zeta), \tag{1.1}$$

where z is an ordinary vertical coordinate referenced from the ocean undisturbed state and directed downward, H is the depth of the basin from the undisturbed surface, and ζ is the deviation of sea surface height (SSH) from its undisturbed state, which is positive when the free surface is plane $z = 0$. The prognostic variables are horizontal velocity components, potential temperature, salinity, and SSH. The horizontal pressure gradient components are written in the Jacobian form with localized pressure function that allows us to reduce the error related to nonlinear density dependence on pressure. The lateral viscosity operator is written in the null-trace tensor form that results in zero dissipation in the case of solid rotation. The lateral diffusion operator for temperature and salinity is implemented as mixing along isopycnal surfaces using the semi-Lagrangian algorithm, which is important for correct reproduction of thermohaline structure by long-term simulations. Coefficients of vertical turbulent viscosity and diffusion are computed according to Mellor and Yamada (1982) with modifications in Galperin et al. (1988) and Allen et al. (1995). A full description of the INMOM and principles of its numerical implementation is presented in Moshonkin et al. (2018).

The INMOM is parallelized to run efficiently on massively parallel computing systems using the domain decomposition method. This method divides the computational domain into rectangular blocks so that each processor can compute independently. The computational domain is divided only in the horizontal dimensions leaving the vertical dimension unchanged due to a fixed number of vertical levels for each horizontal cell of the σ-coordinate model. The message passing interface (MPI) is used as the main technique for parallelization. The MPI-IO is used for parallel implementation of input and output in the INMOM, which allows processors to read and write files asynchronously.

The conditions at the open boundary $\sigma = 0$ are set as fluxes of fresh water, heat, salt, and momentum at the sea surface. The real freshwater flux is used to account for the basin water content change, which allows us to avoid setting artificial salt fluxes over open water. In some cases (as done in this research), the Dirichlet conditions can be set for temperature and/or salinity, i.e., the prescribed sea surface temperature (SST) and salinity (SSS).

A sea ice model, embedded into the INMOM for better simulations at high latitudes, consists of three components: thermodynamics (Yakovlev 2009), transport (Smolarkiewicz 1984), and elastic-viscous-plastic rheology (Hunke 2001).

The INMOM version for the World Ocean used in this investigation is constructed in a curvilinear orthogonal coordinate system by means of conformal transformation of the geographic system. To avoid the problem of grid convergence near the North Geographical Pole, the "north" pole on the model grid is located at point 90°E, 60°N, while the model "south" pole coincides with the South Geographical Pole. The spatial resolution is 0.25° × 0.25° in horizontal coordinates (1440 × 720 grid nodes) and 33 σ-levels nonuniformly distributed in depth. The time step for integration of model equations was set as 5 min to provide solution stability. The model bottom topography was constructed from the GEBCO data (Becker et al. 2009).

The circulation simulation over the considered climatic period was performed using the A.S. Sarkisyan's "diagnosis-adjustment" method (Sarkisyan and Sündermann 2009). The rationale for this method is the two-stage mutual adjustment between the initially prescribed thermohaline ocean state and simulated velocity of currents. These stages are as follows.

The first stage, referred to as "diagnosis," consists in "freezing" the climatic thermohaline ocean state by prescribing it according to the observed data. The global fields of temperature and salinity from the observational data objective analysis EN4.2.1 (Good et al. 2013) (with corrections (Gouretski and Reseghetti 2010)) were used for simulations. The EN4 was developed at the MetOffice (UK) as a product of the two European projects: Enhanced Ocean Data Assimilation and Climate Prediction (ENACT) and ENSEMBLES (Ingleby and Huddleston 2007). The initial conditions for simulations were prepared as EN4.2.1 data averaged over the period 1993–2012, separately for February and August in order to separately estimate strongly different summer and winter ocean states. February and August are chosen as the most typical months of oceanographic summer and winter, respectively.

The wind stress only is set as external forcing computed by using the datasets from the CORE database designed for experiments with OGCMs (Large and Yeager 2009). To restore the chosen ocean dynamics climatic states, the 1993–2009 climatic mean wind stress was computed for February and August. This averaging period is shorter than for temperature and salinity data EN4.2.1 because the time span of the CORE data is limited to 2009.

Since the thermohaline state does not change over time during the simulation, the sea ice model is not used in this formulation. The heat and fresh water fluxes are not needed for integration over time in OGCM, while wind stress is assumed as transferred directly from the atmosphere to the ocean. The latter assumption is justified because sea ice drift velocity is in good agreement with sea surface velocity, and, in fact, sea ice is just a part of the ocean's upper layer. Thus, the diagnosis stage is needed to produce the circulation based on the observed climatic thermohaline state.

The second stage referred to as "adjustment" consists of a further model run with computation of all the oceanic characteristics (including thermohaline ones) starting from the data obtained at the diagnosis stage. In addition to wind stress for velocity equations, the boundary conditions are required for temperature and salinity equations. The simplest method for this is setting Dirichlet conditions. It allows us to avoid the necessity to use a sea ice model and compute sea surface fluxes, which could be incorrect due to using the time averaged data for principally nonlinear processes. The adjustment state is needed to adjust the original thermohaline state to the model conditions: resolution, topography, circulation, etc., as shown in (Sarkisyan and Sündermann (2009) and Frey et al. (2019).

Thus, in our approach, the model is used only for restoring the full ocean dynamics including the gradient and wind components of circulation. Using the INMOM, all the components of ocean dynamics are reconstructed by the prescribed temperature, salinity, and wind stress: SSH, circulation, meridional overturning circulation stream function, and others. In this approach, the OGCM role is secondary because it is just an instrument for reconstructing circulation based on the observed thermohaline state. One should note that restoring the gradient component of circulation by this approach is similar to the "dynamical" method. However, using the complete OGCM enables one to avoid the problem of setting the "reference surface," which is the weakest point of the "dynamical" method, because the OGCM restores the total 3D pattern of currents from surface to bottom, taking wind-induced currents into account.

The following set of data was prepared for the purposes of this research. The model run in the diagnosis mode continued for 30 days using the prescribed temperature and salinity from EN4. During this time, the model adjusts to the given density fields at middle and high latitudes, and it does not make much sense to continue further simulations (Sarkisyan and Sündermann 2009).

It should be noted that, except for the narrow near-equatorial zone, ocean dynamics is in quasi-geostrophic state, i.e., density defines the structure of gradient geostrophic currents. Since ocean dynamics is quickly (several inertial periods) adjusted to the quasi-geostrophic equilibrium, the circulation can be restored during the short time of the diagnosis run as long as 1 month. Simulations in the adjustment mode continued for 60 days. Two months is usually enough for the adjustment of thermohaline state (Sarkisyan and Sündermann 2009). If we continue the further simulations, the model will change the initial thermohaline fields obtained from the observational data in accordance with its model climate, which depends on the parameters of the model and, first of all, on its spatial resolution.

Diagnostic and adjustment simulations were performed for February and August conditions. In the ocean, the typical winter and summer months are February and August. Due to the thermal inertia of the ocean, the annual variation lags behind relative to the calendar by about 1 month (Levitus et al. 1994). Thus, consideration of the climatic conditions in February and August makes it possible to study the ACC separately for the characteristic winter and summer conditions.

Two pairs of numerical experiments separately for February and August were carried out to reveal the roles of gradient and wind-driven currents in the formation of the ACC dynamics. All experiments were performed in the diagnosis mode with fixed temperature and salinity fields from the previous adjustment simulations. The duration of each run was 30 days. During the first run for each pair (February, August), the diagnostic simulations were performed, setting the just described CORE-based wind stress for each month, while in the second run, the wind stress was zeroed.

When performing simulations in the adjustment mode, the previously described sea surface Dirichlet conditions were set for temperature and salinity. Since the climatic mean (i.e., smoothed) SST, and SSS fields were used, the simulated ACC dynamics turned out insufficiently developed. As a result, the zonal velocity and the meridional sea surface height slope from Antarctica to middle latitudes were also underestimated. To minimize this effect, while repeating the adjustment simulations, a compensation term was added to the right-hand side of the meridional velocity equation at each time step in the latitudinal zone from Antarctica to 45°S (the latitudes where SSH is close to zero), which is proportional to the difference between the two zonal mean SSHs: the first one is estimated from the data in Fig. 1.3 from Armitage et al. (2018), and the second one is simulated with the INMOM. This allowed us to obtain the ACC dynamic characteristics close to real ones.

The simulation results were averaged over the last 5 days of the runs to analyze them. Averaging is needed since there are inertial oscillations and gravity waves in the solution (needing to be suppressed), which is caused by the not yet stabilized near-equatorial circulation. We use the global ocean model in our simulations. The output of results was set at the horizontal model grid and 33 standard oceanographic depth levels. The simulations showed that this INMOM version adequately reproduces the climatic mean observed structure of the general hydrophysical fields by the proposed scenario at middle and high latitudes (Diansky and Bagatinsky 2019; Bagatinsky and Diansky 2021). Subsequently, we present the comparative analysis of the results of research.

1.3 Geostrophic and Wind-Driven Parts of the Antarctic Circumpolar Current

The main goal of this research is to study the contributions of wind and thermohaline properties to the formation of the ACC climatic mean structure. In general, the formation of ocean circulation occurs due to all the factors affecting the ocean. However, purely formally, the ocean circulation can be decomposed into two main components, gradient and wind-driven ones based on the diagnostic simulations proposed here using the OGCM. The fixed density field forms a gradient component, while the wind forcing forms a drift component. If wind forcing is absent in the proposed formulation, then the thermohaline gradient circulation caused by the fixed density gradients would be purely reproduced in the OGCM. However, to reproduce the wind-driven circulation, we cannot turn off the thermohaline part of the ocean circulation using an OGCM with a uniform density distribution. This is because the formation of drift circulation is closely related to the depth distribution of the parameterized vertical turbulent viscosity coefficient, which depends not only on the wind stress but also on the density distribution over depth. Thus, it is extremely difficult to simulate the wind-driven circulation in a homogeneous ocean that would correspond to the real one. Therefore, it is easier to get a correct estimate by subtracting the thermohaline part from the total ocean circulation simulated without a wind contribution as done in this work.

Figure 1.1a, b shows the structures of summer (February) and winter (August) wind stresses, which are in good agreement with the results in Yu et al. (2020). The figures indicate that in winter, due to the higher atmospheric activity, the westerly winds over the ACC associated with the southern Ferrel cell (Huang and McElroy 2014) increase significantly. This is especially notable in the Indian Ocean sector, where the wind stress in winter almost doubles. In summer, the wind distribution along the ACC is more uniform. Predominantly easterly winds are manifested in the circle south of ~65°÷63°S except for the Drake Passage due to the existence of the meridional polar cell in the atmospheric large-scale circulation over Antarctica (Qian et al. 2016). These winds are also stronger in winter, due to the intensification of katabatic winds over Antarctica (Yu et al. 2020), which amplify the surface winds of the Antarctic meridional cell.

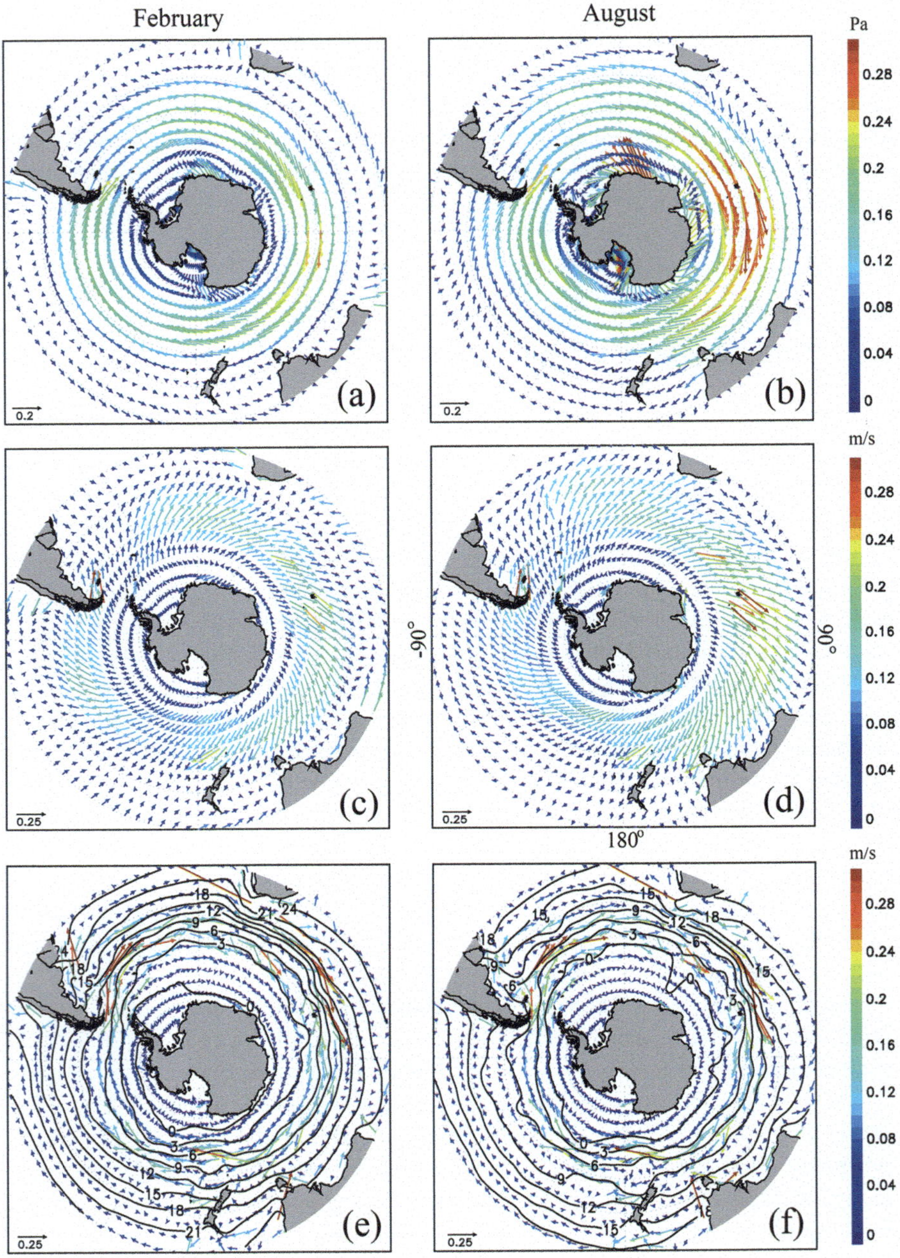

Fig. 1.1 Mean wind stress in 1993–2009 (vectors in color, Pa) calculated from the CORE v2 data for February conditions (**a**); the same, but for the climatic August conditions (**b**); difference between the total and thermohaline surface current velocities (vectors in color, m/s) for February conditions (**c**); the same, but for the climatic August conditions (**d**); thermohaline surface current velocities (vectors in color, m/s) and temperature (black contour lines, °C) based on the experimental results without taking into account wind stress for February conditions (**e**); the same, but for the climatic August conditions (**f**)

Wind circulation contributes to surface drift currents according to Ekman's theory with a turn of ~45° to the left (southern hemisphere) relative to the wind stress. This is shown in Fig. 1.1c, d, which presents the wind-driven surface velocities obtained by subtracting the gradient currents from the total ones. Due to stronger winds in winter, the response to drift velocities in this season is generally also stronger than in summer, which is especially notable in the Indian Ocean sector, because here the wind stress significantly increases in winter compared to summer.

In general, these figures indicate that near the surface wind-driven and thermohaline zonal velocities are comparable in magnitude. However, especially in the Drake Passage, the gradient zonal velocities in the ACC are several times higher (in the Indian Ocean sector, almost by one order of magnitude) than the wind-driven ones. This suggests that the thermohaline factor of the ACC formation significantly prevails over the wind-driven one, despite rather strong winds over the Southern Ocean. However, we repeat that this is the formal result, since the formation of the ACC itself is associated with both factors. It happens because the wind-driven velocities are turned by ~45° to the north; hence, their zonal component is approximately two times smaller than the velocity magnitude, while thermohaline velocities are directed almost along the latitude circles.

To assess the contributions of gradient and wind-driven currents to the ACC dynamics, we make a comparative analysis of the simulations in the two cases: a complete formulation and without wind forcing. Only thermohaline circulation is simulated in the experiments without wind. Contour lines in Fig. 1.2 show the distribution of SSH deviations from the undisturbed state in February and August. Further, we refer to these climatic mean SSH deviations as the mean dynamic topography (MDT) since it is formed by the actual ocean dynamics (Armitage et al. 2018; Maximenko et al. 2014).

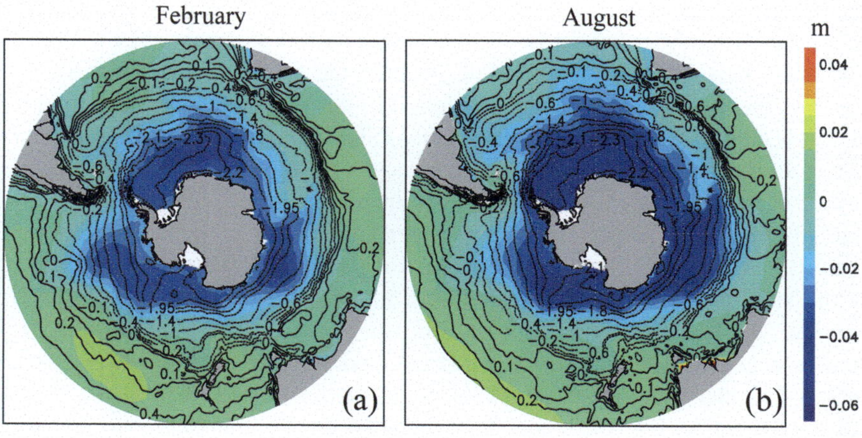

Fig. 1.2 Thermohaline part of the MDT in the Southern Ocean (contour lines, m) and the wind part of MDT defined as the difference between full and thermohaline SSH (color scale, m) for the climatic February conditions (**a**); the same, but for the climatic August conditions (**b**). Dotted lines indicate negative differences, and solid lines indicate positive ones

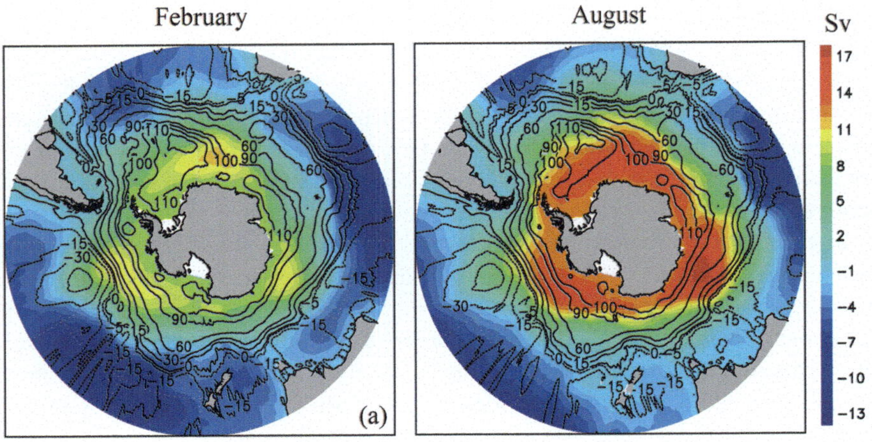

Fig. 1.3 Thermohaline barotropic stream function transport in the Southern Ocean (contour lines, Sv) and the difference between the total and thermohaline stream function transports (color scale, Sv) for the climatic February conditions (**a**); the same, but for August (**b**)

If we add thermohaline SSH and wind variations shown in Fig. 1.2, then the total MDT simulated using the OGCM INMOM would be in good agreement with Armitage et al. (2018) Fig. 1.3a) that was calculated by the geodetic method using radar altimetry data from the CryoSat-2 (CS-2) mission (Wingham et al. 2006) relative to the Earth's gravitational field according to the GOCO05c (GOCE) model (Fecher et al. 2017).

It is worth noting here that its restoration is possible only through the ocean dynamics. The MDT is a basic surface indicator of the ocean circulation in a wide variety of oceanographic applications. It can be calculated using two different approaches: oceanic and geodesic ones. The oceanic approach is based on the reconstruction of the mean ocean dynamics. In the geodesic approach, the MDT is determined by subtracting the geoid height from the mean satellite-based SSH. In this case, the relative SSH is defined referred to the World Geodetic System 1984 (WGS84) ellipsoid, which is the same as the OGCM-simulated SSH. Both methods have their drawbacks. However, as the accuracy of the observational data and accuracy of ocean-circulation reconstruction is increasing using modern models, the quality of the oceanic method for reconstructing the MDT would increase. It should be noted that at present, various semi-empirical global models are used for recovering the MDT by geodetic methods, the most advanced of which are CNES-CLS13 (Rio et al. 2014), MDOT APDRC 2014 (Maximenko et al. 2014), and DTU15MDT (Andersen and Knudsen 2016). In all these models, at the first stage, the direct method for calculating the MDT (SSH minus Geoid) is used, and at the second stage, in the CNES-CLS13 and MDOT APDRC 2014 models, the so-called dynamic-adjustment method (Sarkisyan and Sündermann 2009) is used to reconstruct the MDT from observations of temperature, salinity, and current velocities. The main disadvantage of this method, as already mentioned, is the choice of a reference surface, which is more important for the ACC because its lower boundary almost reaches the bottom. In addition, the dynamical method is unable to correctly satisfy the boundary conditions near the coast, and the drift circulation is added to it for a more accurate MDT determination. All these drawbacks are absent from the perspective OGCM-based method presented here for the MDT recovery using the OGCM. The MDT accurate calculation as a standalone problem is necessary for satellite navigation problems (Armitage et al. 2018; Maximenko et al. 2014).

If we compare the winter and summer thermohaline MDT (in the southern hemisphere) simulated by the INMOM, we can find that they are not very different. Thus, the seasonal variability of the MDT in the ACC is largely determined by the seasonal wind variability. Especially, wind can play a significant role in the SSH changes near the coast.

Taking the stronger winds and stronger cooling in winter in the Southern Ocean into account, one would expect a greater MDT decrease toward Antarctica in August (winter) as compared to February (summer). This may be due to the just-mentioned increase in the easterly winds near Antarctica in winter.

The influence of wind-driven dynamics on the latitudinal MDT change is shown in Fig. 1.2 in color. It is manifested in the wind contribution to the decreasing MDT from the open ocean basins of the Southern Ocean to the Antarctic by not more than 6 cm. At the same time, the decrease in MDT due to the thermohaline factors from the open ocean to Antarctic coast exceeds 2 m, because the contribution of the thermohaline factors to the ACC dynamics significantly exceeds the wind contribution (Fig. 1.1). However, unlike the thermohaline factor, the wind forcing is strongly subjected to seasonal changes. In winter (August), due to stronger winds, this MDT decrease is more powerful than in summer, which is manifested in more extensive areas of the MDT drop due to wind.

Contour lines in Fig. 1.3 show the distribution of the barotropic stream function (BSF) in the Southern Ocean based on the results of experiments. The BSF was defined as zero at the coast of South America. The full barotropic volume transport (sum of thermohaline and wind driven parts) in the Drake Passage reaches 122 Sv in summer and 128 Sv in winter, which is in good agreement with the most recent estimates (Zika et al. 2012; Pfister et al. 2014).

In general, there is a good agreement between the spatial distribution of the BSF and MDT both in the thermohaline and wind driven parts. This is due to the fact that in the quasi-geostrophic (the "geostrophic" approximation of currents in the ocean is carried out with good accuracy) and "rigid lid" (filtration of fast gravitational waves) approximations, the forms of the equations of barotropic circulation written out by means of BSF and MDT practically coincide (see, e.g., Sarkisyan and Sündermann (2009) or Pedlosky (1982)). It is natural to expect that the solutions of these equations with the same wind and thermohaline forcing would be close with an accuracy to a coefficient (Pedlosky 1982).

The influence of wind on the BSF formation is even more pronounced than for the MDT. The intensity of the total ACC streams due to the wind increases on average by 9–11 Sv in February and 12–17 Sv in August, which is associated with the stronger winter westerly winds in the Southern Ocean. At the same time, in summer, the greatest increase in the BSF by 12 Sv due to the wind is confined to the ACC sector south of Africa and in winter (by 18 Sv) south of Australia. It is interesting that in the Drake Passage, the wind-driven volume transport increases less than over the entire ACC: by 8 Sv in summer and by 12 Sv in winter.

Figure 1.4 shows the vertical sections of the zonal ACC velocity along 65°W in the Drake Passage based on the results of the experiment without wind forcing and with the wind contribution to its formation under the mean February and August conditions. We note that the three-jet structure of the ACC was found based on simulations using the INMOM model. It follows from Fig. 1.4 that the three-jet structure of the ACC is of thermohaline origin. The three-jet structure appeared only in the adjustment mode and was not manifested in the initial calculations in the diagnosis mode. This suggests that the EN4 climatic fields averaged over space and time do not contain this information. These jets are pronounced after joint adjustment of the model and the initial data on temperature and salinity in the adjustment mode of simulations. Their locations agree with the field data (Burkov 1994; Orsi et al. 1995; Kim and Orsi 2014).

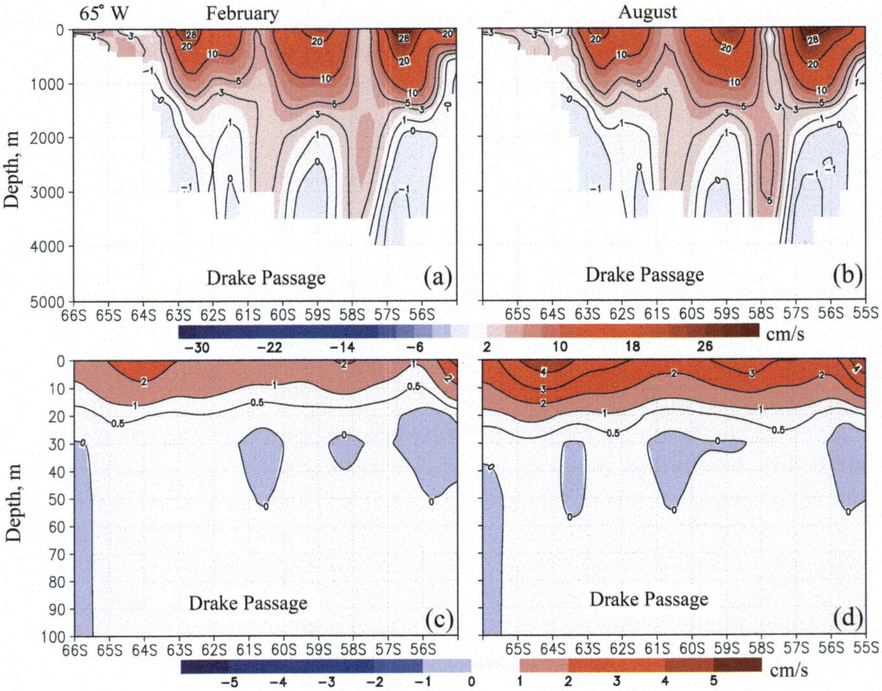

Fig. 1.4 Vertical section of the ACC thermohaline zonal velocity component (cm/s) along 65°W in the Drake Passage for the February conditions (**a**); the same but for the August conditions (**b**). Difference between the total and thermohaline zonal velocities (cm/s) in the upper layer for the February conditions (**c**); the same but for the August conditions (**d**)

There are three classical major ACC jets in the Drake Passage: the northern one is associated with the Subantarctic Front, the central one is related to the Polar Front, and the southern one is a current from the Bellingshausen Sea associated with the Southern Front of the ACC. The existence of three jets is in good agreement with the early works (Burkov 1994; Orsi et al. 1995). The recent studies reveal up to 12 individual jets (Sokolov and Rintoul 2007; Koshlyakov et al. 2013; Tarakanov and Gritsenko 2018). The resolution of the model does not allow us to distinguish a large number of jets.

The highest velocities based on our simulations are approximately 25–30 cm/s. The three jets revealed in our simulations are very close to the description given in Kim and Orsi (2014): "The central and southern jets are the most stable, and the highest velocities of currents, amounting to 35-40 cm/s according to the averaged data, were observed in the central jet. The northern stream is the most dynamic and, judging by the annual series of observations, is subjected to strong meandering and eddy formation. The currents in the strait retained their predominant northeastern direction up to a level of 3000 m, and the speed of the currents at this depth decreased to average values of 5-10 cm/s. Below 3000 m, in the Drake Passage, there was a weak westerly current with an average speed of 2-5 cm/s, associated with the transport of bottom waters from the Weddell Sea."

The underestimated velocities of the currents in the Drake Passage are explained by the fact that the climatic data on temperature and salinity used in the simulations are strongly averaged over space and time, which reduces the values of density gradients and, consequently, the gradient currents calculated from them. It is noteworthy that the velocities simulated with the INMOM using the climatic mean wind stress, temperature, and salinity data may be lower than in the case of using instantaneous data.

It follows from these figures that the contribution of the wind component in the upper layer can be almost as high as the thermohaline component. In the middle of the Drake Passage, the eastward-directed wind-induced velocity component is stronger than near the coast of Antarctica. Here, it intensifies the ACC velocity in the upper layer adding to its velocity up to ~2 cm/s in summer and up to ~4 cm/s in winter. At depths of 20–40 m, the wind-induced component decreases the ACC velocity by ~0÷1 cm/s in summer and winter. Such a multidirectional structure of wind circulation by depth can be explained by Ekman's spiral.

1.4 Discussion

In this work, we investigated the contributions of wind and thermohaline factors to the formation of the mean climatic structure of the ACC using numerical simulations with the OGCM INMOM in the diagnosis mode with application of the diagnosis-adjustment method (Sarkisyan and Sündermann 2009). It is worth noting that the formation of ACC is associated with both factors; however, wind is the original cause of the ACC existence. The suggested approach enables distinguishing two main components of the ACC: the fixed density field forms the gradient component, and the wind forcing forms the drift component. The only method to separate them is application of the ocean circulation models. If the wind effect is disabled in the proposed formulation, then the thermohaline gradient circulation would be purely reproduced by the OGCM. The drift part of the ocean circulation is determined by subtracting the thermohaline part from the total circulation.

The studies were carried out for February and August as the most typical months of the summer and winter state of the ACC. In winter, due to the higher atmospheric activity, the westerly winds over the ACC significantly increase caused by the southern Ferrel cell (Huang and McElroy 2014). This is especially notable in the sector of the Indian Ocean, where the winter wind stress almost doubles in winter. Predominantly easterly winds are manifested in the circle south of ~65°-63° S, excluding the Drake Passage due to the existence of the meridional polar cell in the atmospheric large-scale circulation over Antarctica (Qian et al. 2016) increasing in winter due to the strengthening of katabatic winds over Antarctica (Yu et al. 2020).

It is shown that according to Ekman's theory, the surface wind-driven circulation of the ACC turns by ~45° to the left with respect to the wind stress direction (Southern Hemisphere). Due to stronger winds in winter, the response of drift velocities in this season is stronger than in summer, since the wind stress in winter increases significantly compared to summer.

Sections of the zonal ACC velocity component were constructed (Fig. 1.4) along 65° W in the Drake Passage based on the mean results under February and August conditions in the experiments without wind forcing and with the wind contribution to circulation. The classical three-jet structure of the ACC was found based on the simulations using the INMOM model only in the adjustment mode.

These distributions of zonal velocity show that the average contribution of the wind component in the upper layer is equal to 15–20% of the thermohaline component. In the Drake Passage, the east-directed wind velocity component is stronger off the coast of Antarctica. A multidirectional depth structure of pure wind circulation in the ACC was revealed by the simulations: in the upper 20 m layer, it is directed to the east, and in the 20–40 m layer, to the west. This structure can be easily explained by the Ekman's spiral.

Despite the stronger winds and stronger cooling in winter in the Southern Ocean, the winter and summer MDTs simulated using the INMOM model do not differ much (Fig. 1.2). A greater drop in the level toward Antarctica in August (winter) should be expected compared to February (summer). The fact that this does not happen is due to the aforementioned intensification of the eastern winds near Antarctica in winter.

The influence of drift dynamics on the latitudinal variations in the MDT (Fig. 1.2) is manifested in the fact that, in the Southern Ocean, the wind contributes to the decrease in the MDT by not more than 6 cm in the direction from the open ocean to Antarctic coast. The MDT decrease due to thermohaline factors from the open ocean to Antarctic coast exceeds 2 m. However, unlike the thermohaline factor, the wind forcing is strongly subjected to seasonal changes. Due to stronger winds in the winter period (August), this MDT decrease is stronger than in summer, which manifests itself in wider areas of wind-induced SSH decrease.

The barotropic stream functions were calculated in winter and summer periods based on the results of the experiments. The wind contribution to the formation of the barotropic circulation in the Southern Ocean was estimated. The simulated BSF transport in the Drake Passage reaches 122 Sv in summer and 128 Sv in winter, which is slightly lower than usually assumed. An accurate estimate of the transport in the Drake Passage is still a pressing problem. For example, in Pfister et al. (2014) and Stammer et al. (2018), this transport was estimated at 80 Sv and 100 Sv, respectively. However, the estimate in Zika et al. (2012) yields 130 Sv, which is much closer to our results, and Donohue et al. (2016) report about 173 Sv.

In general, there is a correspondence (Fig. 1.3) between the spatial distributions of the BSF and MDT, which is explained by the properties of the system of equations of large-scale ocean hydrothermodynamics (Pedlosky 1982; Sarkisyan and Sündermann 2009). Almost all the features in the spatial distribution of the MDT (Fig. 1.2) are reflected in the BSF, but with greater spatial intermittency.

However, wind influence on the BSF formation is more pronounced than that on the MDT. The intensity of the total ACC transport due to the wind increases on average by 9–11 Sv in February and by 12–17 Sv in August, which is associated with stronger winter westerly winds in the Southern Ocean. In summer, the greatest BSF enhancement, by 12 Sv, is due to the wind being confined to the ACC sector south of Africa, and in winter, up to 18 Sv, south of Australia. It is interesting to note that in the Drake Passage, the wind-induced transport increases less than over the entire ACC in general: by 8 Sv in summer and by 12 Sv in winter.

The three jets revealed in our simulations are very close to the description given in Kim and Orsi (2014). These jets do not appear in diagnostic simulations but are formed when the model and initial data on temperature and salinity are jointly adjusted. This indicates that this information is not contained in the space and time mean fields of EN4. Thus, the diagnosis-adjustments method, which forms the ocean circulation through the symbiosis of observational data and a model solution, can serve as a good tool for studying ocean circulation.

It is noteworthy that the MDT is a fundamental surface for a large number of oceanographic applications and primarily for accurate determination of changes in the SSH of the World Ocean and satellite navigation (Armitage et al. 2018; Maximenko et al. 2014). The MDT recovery problem can be solved using two different approaches, namely, oceanic and geodetic. The oceanic approach is based on the reconstruction of the average dynamics of the ocean. In the geodetic approach, the MDT is determined by subtracting the geoid height from the satellite-based SSH. Both methods have their own advantages and disadvantages. However, as the accuracy of the observational data and the accuracy of reconstructing the ocean circulation using modern models increase, the accuracy of the oceanic MDT reconstruction method would undoubtedly increase.

1.5 Conclusions

The contributions of wind and thermohaline factors to the formation of the climatic mean structure of the Atlantic Circumpolar Current (ACC) were studied using numerical simulations with the ocean general circulation model (OGCM) INMOM in the diagnosis-adjustment mode (Sarkisyan and Sündermann 2009). The studies were carried out for the summer (February) and winter (August) state of the ACC.

The three-jet structure of the ACC was found based on simulations using the INMOM model. It was shown that the three-jet structure of the ACC is of the thermohaline nature. The wind-driven circulation in the surface layer of the ACC turns to the left (in the Southern Hemisphere) according to Ekman's theory with respect to the wind direction. Due to the stronger winds in winter, the response to drift velocities is stronger than in summer. Simulations show that the thermohaline factor of the ACC formation significantly prevails over the drift factor, despite rather strong winds over the Southern Ocean. Nevertheless, the distributions of the zonal velocity on the sections in the Drake Passage show that, in the upper layer, the contribution of the wind component in the upper layer can be as high as 15–20% of the thermohaline velocity. In the Drake Passage, the east-directed wind velocity component is stronger off the coast of Antarctica.

A wind-induced weakening of the ACC in the subsurface layer (20–40 m) was found, which is explained by the Ekman's spiral. The influence of drift dynamics on the latitudinal change in mean dynamic topography (MDT) is manifested in the fact that the wind contributes to a decrease in the MDT by more than 6 cm from the open ocean areas in the Southern Ocean to Antarctica.

The influence of wind on the barotropic stream function (BSF) formation is more pronounced than on the MDT. The intensity of the total ACC transport due to the wind increases on average by 9–11 Sv in February and by 12–17 Sv in August, which is associated with stronger winter westerly winds in the Southern Ocean. In the Drake Passage, the wind transport increases less than in the entire ACC: by 8 Sv in summer and by 12 Sv in winter.

Acknowledgments This work was supported by the State Assignment 0128-2019-0008 and by the Russian Foundation for Basic Research (grant no. 19-57-60001 adjustment of the INMOM model) and the Russian Science Foundation (grant no. 17-17-01295 (study of wind contributions to the Antarctic Circumpolar Current).

References

Allen JS, Newberger PA, Federiuk J (1995) Upwelling circulation on the Oregon continental shelf. Part I: Response to idealized forcing. J Phys Oceanogr 25:1843–1866

Andersen OB, Knudsen P (2016) Deriving the DTU15 global high resolution marine gravity field from satellite altimetry. In: ESA living planet symposium Prague, Czech Republic

Armitage TWK, Kwok R, Thompson AF, Cunningham G (2018) Dynamic topography and sea level anomalies of the Southern Ocean: variability and teleconnections. J Geophys Res Oceans 123:613–630. https://doi.org/10.1002/2017JC013534

Bagatinsky VA, Diansky NA (2021) Variability of the North Atlantic thermohaline circulation in different phases of the Atlantic multidecadal oscillation according to ocean objective analysis and reanalysis. Izv Atmos Ocean Phys 57(2):1–14

Becker JJ, Sandwell DT, Smith WHF, Braud J et al (2009) Global bathymetry and elevation data at 30 arc seconds resolution: SRTM30_PLUS. Mar Geod 32(4):355–371. https://doi.org/10.1080/01490410903297766

Burkov VA (1994) Antarctic jets. Oceanology 34:169–177

Diansky NA, Bagatinsky VA (2019) Thermohaline structure of waters in the North Atlantic in different phases of the Atlantic multidecadal oscillation. Izv Atmos Ocean Phys 55(6):628–639

Donohue KA, Tracey KL, Watts DR, Chidichimo MP, Chereskin TK (2016) Mean Antarctic Circumpolar Current transport measured in Drake Passage. Geophys Res Lett 43:11760–11767. https://doi.org/10.1002/2016GL070319

Downes SM, Farneti R, Uotila P, Griffies SM et al (2015) An assessment of Southern Ocean water masses and sea ice during 1988–2007 in a suite of interannual CORE-II simulations. Ocean Model 94:67–94

Farneti R, Downes SM, Griffies SM, Marsland SJ et al (2015) An assessment of Antarctic Circumpolar Current and Southern Ocean meridional overturning circulation during 1958–2007 in a suite of interannual CORE-II simulations. Ocean Model 93:84–120

Fecher T, Pail R, Gruber T, Goco Consortium (2017) GOCO05c: a new combined gravity field model based on full normal equations and regionally varying weighting. Geophys Surv 38(3):571–590. https://doi.org/10.1007/s10712-016-9406-y

Frey DI, Morozov EG, Fomin VV, Diansky NA, Tarakanov RY (2019) Regional modeling of Antarctic Bottom Water flows in the key passages of the Atlantic. J Geophys Res Oceans 124:8414–8428

Galperin B, Kantha LH, Hassid S, Rosati A (1988) A quasi-equilibrium turbulent energy model for geophysical flows. J Atmos Sci 45:55–62

Good SA, Martin MJ, Rayner NA (2013) EN4: quality controlled ocean temperature and salinity profiles and monthly objective analyses with uncertainty estimates. J Geophys Res Oceans 118:6704–6716. https://doi.org/10.1002/2013JC009067

Gouretski V, Reseghetti F (2010) On depth and temperature biases in bathythermograph data: development of a new correction scheme based on analysis of a global ocean database. Deep-Sea Res I 57:812–834

Griffies SM (2006) Some ocean model fundamentals, ocean weather forecasting: an integrated view of oceanography (Chassignet EP, Verron J, eds.), Springer, Dordrecht, 2006, pp 19–73

Huang J, McElroy MB (2014) Contributions of the Hadley and Ferrel circulations to the energetics of the atmosphere over the past 32 years. J Clim 27(7):2656–2666

Hunke EC (2001) Viscous–Plastic Sea ice dynamics with the EVP model: linearization issues. J Comput Phys 170(1):18–38

Ingleby B, Huddleston M (2007) Quality control of ocean temperature and salinity profiles – historical and real-time data. J Mar Syst 65:158–175

Kim YS, Orsi AH (2014) On the variability of Antarctic Circumpolar Current fronts inferred from 1992-2011 altimetry. J Phys Oceanogr 44(12):3054-3071

Koshlyakov MN, Gladyshev SV, Tarakanov RY, Fedorov DA (2013) Currents in the Drake Passage by the observations in October–November of 2011. Oceanology 53(1):1–12

Koshlyakov MN, Tarakanov RY (2011) Water transport across the Subantarctic front and the Global Ocean Conveyer Belt. Oceanology 51(5):721–735

Large WG, Yeager SG (2009) The global climatology of an interannually varying air–sea flux data set. Clim Dyn 33:341–364

Levitus S, Burgett R, Boyer TP (1994) World Ocean Atlas 1994, vol. 3, Salinity, NOAA Atlas NESDIS 3, U.S. Govt. Print. Off., Washington, D.C., J. 99 pp

Maximenko N, Knudsen P, Centurioni L, Andersen O, Hafner J, Melnichenko O (2014) New mean dynamic ocean topography derived from a synthesis of satellite altimeter, gravity, and scatterometer data and trajectories of Lagrangian drifters. Ocean Surface Topography Science Team Meeting. Konstanz, Germany

Mellor GL, Yamada T (1982) Development of a turbulence closure model for geophysical fluid problems. Rev Geophys Space Phys 20:851–875

Moshonkin S, Zalesny V, Gusev A (2018) Simulation of the Arctic–North Atlantic Ocean circulation with a two-equation K-omega turbulence parameterization. J Mar Sci Eng 6 № 95

Orsi AH, Whitworth Th III, Nowlin WD Jr (1995) On the meridional extent and fronts of the Antarctic Circumpolar Current. Deep-Sea Res 42(5):641–673

Pedlosky J (1982) Geophysical fluid dynamics. Springer, 624 pp

Pfister PL, Stocker TF, Rempfer J, Ritz SP (2014) Influence of the Central American seaway and Drake Passage on ocean circulation and neodymium isotopes: a model study. Paleoceanography 29:1214–1237

Qian W, Wu K, Liang H (2016) Arctic and Antarctic cells in the troposphere. Theor Appl Climatol 125:1–12. https://doi.org/10.1007/s00704-015-1485-z

Rio M-H, Mulet S, Picot N (2014) Beyond GOCE for the ocean circulation estimate: synergetic use of altimetry, gravimetry, and in situ data provides new insight into geostrophic and Ekman currents. Geophys Res Lett 41:8918–8925. https://doi.org/10.1002/2014GL061773

Sarkisyan AS, Sündermann JE (2009) Modelling ocean climate variability. Springer, Dordrecht, 374 pp. https://doi.org/10.1007/978-1-4020-9208-4

Smolarkiewicz PK (1984) A fully multidimensional positive definite advection transport algorithm with small implicit diffusion. J Comput Phys 54:325–362

Sokolov S, Rintoul SR (2007) Multiple jets of the Antarctic Circumpolar Current south of Australia. J Phys Oceanogr 37(5):1394–1412

Sokolov S, Rintoul SR (2009) The circumpolar structure and distribution of the Antarctic Circumpolar Current fronts. Part A: mean circumpolar paths. J Geophys Res 114:C11018. https://doi.org/10.1029/2008JC005108

Stammer D, Köhl A, Vlasenko A, Matei I, Lunkeit F, Schubert S (2018) A pilot climate sensitivity study using the CEN coupled adjoint model (CESAM). J Clim 31(5):2031–2056

Tarakanov RY, Gritsenko AM (2018) Jets of the Antarctic Circumpolar Current in the Drake Passage based on hydrographic section data. Oceanology 58(4):503–516

Wingham DJ, Francis CR, Baker S, Bouzinac C, Brockley D, Cullen R et al (2006) CryoSat: a mission to determine the fluctuations in Earth's land and marine ice fields. Adv Space Res 37(4):841–871. https://doi.org/10.1016/j.asr.2005.07.027

Yakovlev NG (2009) Reproduction of the large-scale state of water and sea ice in the Arctic Ocean from 1948 to 2002: Part II. The state of ice and snow cover. Izv Atmos Ocean Phys 45:478–494

Yu L, Zhong S, Sun B (2020) The climatology and trend of surface wind speed over Antarctica and the Southern Ocean and the implication to wind energy application. Atmos 11:108. https://doi.org/10.3390/atmos11010108

Zika JD, England MH, Sijp WP (2012) The ocean circulation in thermohaline coordinates. J Phys Oceanogr 42(5):708–724

Chapter 2
Multi-jet Structure of the Antarctic Circumpolar Current

Roman Yu Tarakanov

Abstract The article provides an overview of the current state of research on the jet structure of the Antarctic Circumpolar Current. Temperature criteria for the identification of individual flow jets are given. It is noted that the question of the number of jets in the ACC and its constancy in time and space remains currently open.

Keywords Antarctic Circumpolar Current · Jets · Temperature criteira of identification

The Antarctic Circumpolar Current (ACC), intensified by strong westerly winds from the mid-latitudes of the Southern Hemisphere and encircling the Antarctic continent from west to east, is the most powerful current in the World Ocean. At the same time, the only area where the ACC extends from one continental slope (South America) to another (Antarctica) is the Drake Passage (Fig. 2.1), which makes it convenient to study the ACC here. According to modern concepts, the geostrophic nature of the ACC has an equivalent barotropic structure (i.e., does not change its direction with depth) and penetrates to a considerable depth (Killworth and Hughes 2002). According to Tarakanov (2012), the ACC in the Drake Passage spreads due to orographic restrictions not deeper than 2000–2500 m in its axial part and only as deep as 1000–1500 m in the northern and southern peripheries. Note that the main barrier to the ACC is not in the Drake Passage but is a ridge south of Australia, i.e., the Australian-Antarctic Ridge with depths above the crest of about 2500 m (Fig. 2.1).

Satellite altimetry observations show that quasi-zonal large-scale ocean currents including the ACC consist of separate jets (dynamic fronts) (Berloff et al. 2009). These jets coincide with the highest horizontal gradients of density, temperature, and salinity and usually with hydrochemical parameters (Burkov 1994; Orsi et al. 1995). Hence, the distributions of the mentioned properties on the meridional

R. Yu. Tarakanov (✉)
Shirshov Institute of Oceanology, Russian Academy of Sciences, Moscow, Russia

E. G. Morozov et al. (eds.), *Antarctic Peninsula Region of the Southern Ocean*, Advances in Polar Ecology 6, https://doi.org/10.1007/978-3-030-78927-5_2

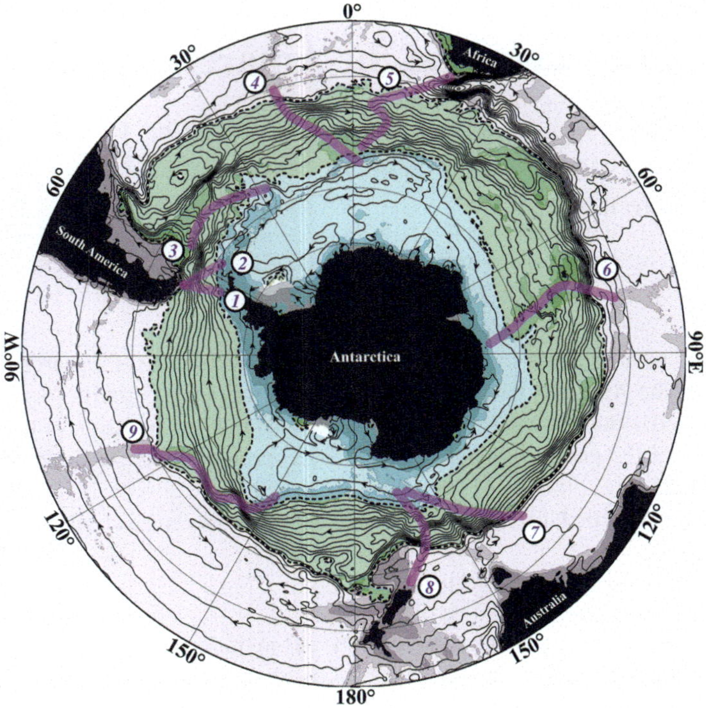

Fig. 2.1 Mean absolute dynamic topography CNES-CLS09 v.1.1 (Rio et al. 2011) in the Southern Ocean. Isolines of absolute dynamic topography (ADT) correspond to streamlines of absolute geostrophic currents on the ocean surface. The area of closed circumpolar that streamlines approximately corresponding to the ACC zone is highlighted in green. The numbers indicate the main topographic barriers on the pathway of the ACC: Hero Ridge and Phoenix Rift, Shackleton Ridge, North Scotia Ridge, South Atlantic Ridge and Bouvet Rise, Agulhas Ridge-Meteor Rise-Shona Ridge, Kerguelen Rise, Australian-Antarctic Ridge, Macquarie Mountains (Ridge), and South Pacific Rise

sections that cross the currents are characterized by a stepwise elevation of isolines from north to south, especially strong at the fronts. In addition to a dynamic front, oceanographic research often uses the concept of a hydrological front that is a surface that separates vertical structures of water masses in the ocean space. The hydrological front on the ocean surface is a line, and the dynamic front is a zone. It should be noted that hydrological and dynamic fronts may not be connected with each other, although, as a rule, the former enter the space of the latter. In accordance with the terminology adopted in Tarakanov and Gritsenko (2014a, b), the term "front" will be used in relation to hydrological fronts and "jet" or "current" in relation to dynamic fronts.

For many years, the researchers distinguished only two jets and two associated fronts in the ACC zone over the entire circumpolar circle: the Subantarctic Front (SAF) and the Polar Front (PF) located south of the SAF although in some sectors of the ACC up to four jets have been distinguished (see the review in the monograph by Sarukhanyan and Smirnov (1986)). The zone north of the SAF is called the Subantarctic Zone (SAZ), the zone between the PF and the SAF is called the South Polar Frontal Zone (SPFZ), and the zone south of the PF up to the Antarctic Slope Front that separates the waters of the open ocean from the more homogeneous waters of the Antarctic shelf is called the Antarctic Zone (AAZ), sometimes also called the South Polar Zone. For the first time, a more complex, multi-jet structure of the ACC in the entire circumpolar circle was described by Burkov (1994) based on a generalization of data from hydrological sections. He identified two additional jets inside the AAZ. Orsi et al. (1995) independently of Burkov proposed a three-jet ACC classification (Fig. 2.2) also on the basis of hydrological data, which is now considered generally accepted. They singled out the SAF, PF, and the Southern Front of the ACC (SACCF). At the same time, the Southern Front of the ACC, unlike the other two, is a purely dynamic front. In accordance with the terminology proposed in Koshlyakov et al. (2011), the SACCF and dynamic fronts associated with the PF and SAF will be called the South Antarctic (SthAC), South Polar (SPC), and Subantarctic (SAC) currents, respectively. The southern boundary of the ACC (SACCB) in this case is an exclusively hydrological front.

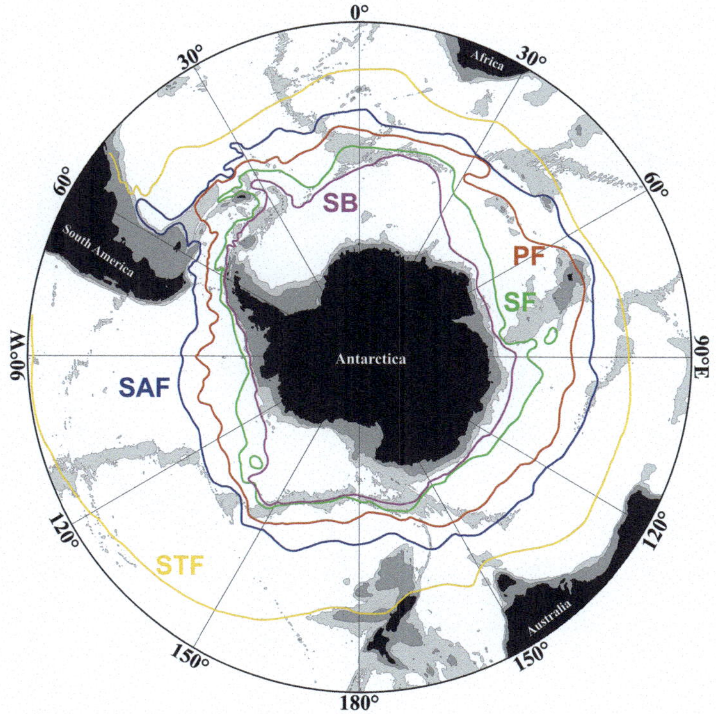

Fig. 2.2 Southern Ocean fronts after (Orsi et al. 1995). SB is southern boundary of the ACC, SF is Southern Front of the ACC, PF is Polar Front, SAF is Subantarctic Front, STF is Subtropical Front. Light and dark gray shades darken areas with depths shallower than 2500 and 1500 m, respectively

In 1990–2000, it was indicated in a number of papers on individual sectors of the Southern Ocean (e.g., Budillon and Rintoul (2003); Sokolov and Rintoul (2002, 2007); Sparrow et al. (1996)) that the ACC zone contains a significantly larger number of jets than suggested by the previously cited classification by Orsi et al. (1995). In this case, splitting and merging of separate jets were often observed, and additional jets were often interpreted only as local features of the current field. Analyzing the frontal structure of the Southern Ocean based on satellite altimetry data, trans-oceanic hydrological sections, WOCE climatology, and data from the Argo floats, Sokolov and Rintoul (2007, 2009a) proposed a more complex nine-jet ACC classification (Fig. 2.3). They identified three jets in each of the currents corresponding (in our terminology) to the SAC and SPC, and two jets to the SthAC, and also considered SACCB as an independent jet. In this case, the classical PF and SAF correspond to the northern stream of the SPC and the middle stream of the SAC (Table 2.1). All these jets were in the entire analyzed interval of altimetric

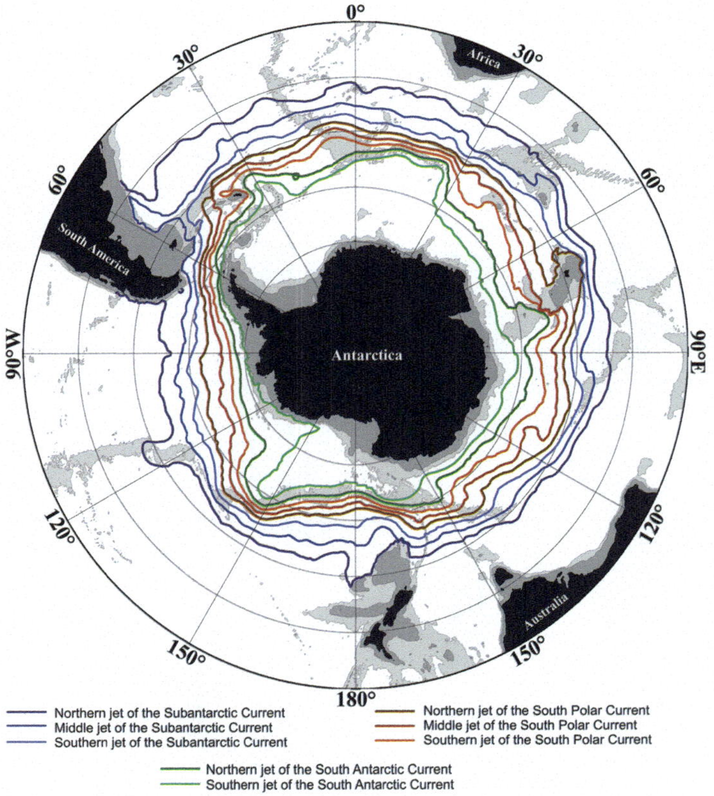

Fig. 2.3 ACC jets according to Sokolov and Rintoul (2009a) except for the jet associated with the SACCB. The names of the jets are given in accordance with the terminology proposed in Koshlyakov et al. (2011). Light and dark gray shades darken areas with depths of less than 2500 and 1500 m, respectively

Table 2.1 Criteria for determining locations of ACC fronts

ACC jets[a]	Temperature criterium	Argo based (°C) (Sokolov and Rintoul 2009a)	According to Tarakanov and Gritsenko (2014a), (°C)
SAC-northern	θ at $p = 400$ db	6.06 ± 0.79	
SAC-5	θ at $p = 400$ db		4.40–5.20
SAC-middle	θ at $p = 400$ db	4.06 ± 0.35	
SAC-3	θ at $p = 400$ db		3.10–3.23
SAC-2	θ at $p = 400$ db	2.78 ± 0.15	
SAC-southern	θ at $p = 400$ db		2.52–2.68
SPC-northern-2	θ at $p = 200$ db	1.98 ± 0.14	
SPC-northern-1	θ in the subsurface θ_{max}		2.31–2.41[b]
SPC-middle	θ in the subsurface θ_{max}	2.25 ± 0.07	
SPC-southern	θ in the subsurface θ_{max}	2.11 ± 0.09	
SthAC-northern	θ in the subsurface θ_{max}	1.93 ± 0.16	
SthAC-southern	θ in the subsurface θ_{max}	1.59 ± 0.28	

[a]Acronyms are given in according to Tarakanov and Gritsenko (2018)
[b]Based on data of survey in October–November 2008 in the middle part of the Drake Passage (Tarakanov 2014)

observations (1993–2008) and over the entire circumpolar circle were stably related to the same isohypses (isolines of absolute dynamic topography (ADT)), which are streamlines of the geostrophic current on the ocean surface. The multi-jet ACC is especially well pronounced in those regions where the current expands while passing through vast deep basins (Sokolov and Rintoul 2002, 2007, 2009a, b). In the areas of narrowing of the ACC, for example, in the Drake Passage or when crossing underwater ridges, the jets merge forming several "superjets" (Sokolov and Rintoul 2007).

Note that in the classification in Sokolov and Rintoul (2009a), the northern SAC jet is not circumpolar and partially follows the continental slope of South America, while the SACCB, with the exception of the Weddell Sea, coincides with the continental slope of Antarctica (Fig. 2.3). These results were the consequence of serious methodological errors associated with the procedures for calculating the mean ADT and the selection of individual jets. The mean ADT was calculated from the climatic dataset of temperature and salinity with a reference surface at a depth of 2500 m. If the ocean depth at a particular geographical point was shallower, then the calculation was made from the bottom. Naturally, within such a calculation procedure, a zone of increased ADT gradients will always form over the continental slope, which leads to the formation of a "false" jet over the continental slope of Antarctica and a pathway of the northern SAC jet along the continental slope of South America. Thus, the jet associated with the SACCB does not exist, and the question of the discontinuity of the northern SAC jet remains open. According to the analysis of the data of the sections in the Drake Passage (Tarakanov and Gritsenko 2018), the jet north of the middle SAC jet still exists in the region south of South America.

In the papers by Tarakanov and Gritsenko (2014a, b), an even finer jet structure of the ACC was revealed based on an integrated analysis of the hydrophysical section data south of Africa and a 19-year series of weekly digital maps of ADT in this area, built by the French agency CLS (Collecte Localisation Satellites) based on satellite altimetry observations and publicly available in the Internet (DT Global – MADT – Upd product, http://www.aviso.altimetry.fr). Here, 12 simultaneously existing jets were identified, stably tied for a significant period of time (up to 1 year) to certain isohypses. Among these jets, six jets corresponded to the SAC, four jets to the SPC, and two jets to the SthAC. The division of the SPC in the Drake Passage into four zones with different thermohaline characteristics of the near-surface layer of the ocean, but without an obvious splitting into the same number of jets, was recorded according to the hydrophysical survey data carried out in 2008 by the Shirshov Institute of Oceanology, Russian Academy of Sciences (IO RAS), in the central part of the Drake Passage (Tarakanov 2014).

The jet structure of the ACC was analyzed in Tarakanov and Gritsenko (2018) on the basis of CTD/SADCP data from two hydrophysical sections in 2010 and 2011 in the Drake Passage in light of the concept of a 12-jet structure suggested in Tarakanov and Gritsenko (2014a, b) in the sector of the Southern Ocean south of Africa. It is shown that the eight-jet classification (Sokolov and Rintoul 2009a, b) (with a decrease by one due to the absence of a jet associated with the SACCB) is insufficient to identify all the jets observed in the sections studied. In 2010, 11 zones of increased velocities (the so-called individual jets) were recorded, and in 2011 9 jets were detected. These individual jets were different combinations of 12 jets of the ACC identified by Tarakanov and Gritsenko (2018) south of Africa.

According to Ablain et al. (2017), over 25 years of satellite altimetry observations, the level of the World Ocean increased on average by 7 cm. Naturally, within such an increase, the stable timing of the cores of the ACC jets to certain isohypses led Sokolov and Rintoul (2009b) to the conclusion about the contraction of the ACC band toward Antarctica. This conclusion raises doubts, since an increase in the ocean level, for example, up to 1.5 m (i.e., more than the difference between the ADT across the ACC band), would mean that the ACC would go completely beyond the "gates" of the Drake Passage, i.e., there would be a rupture of the current by the Antarctic Peninsula. This idea taking into account the conclusions in Tarakanov and Gritsenko (2014b) about a relatively short (in time) retention of the jet cores to certain ranges of ADT requires a more detailed study of the dynamics of ADT change in the ACC band with respect to the jet structure of this flow.

A number of temperature criteria (Table 2.1, Fig. 2.4) suggested in Sokolov and Rintoul (2009a) exist for the identification of ACT jets. For the northern jet of the SPC, this criterion corresponds to the classical Botnikov (1963) criterion for the Polar Front: – 2 °C at a level of 200 db. For the jets north of the PF, the criteria are

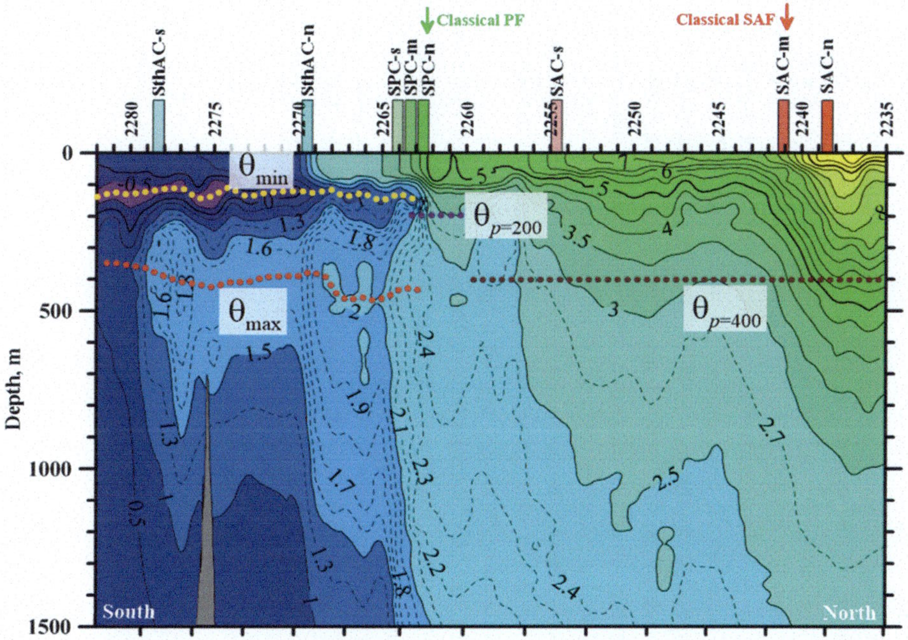

Fig. 2.4 Distribution of potential temperature θ (°C) on a conventional section crossing the ACC zone. Yellow, red, purple, and dashed brown lines show the surfaces of the minimum θ, maximum θ, θ at the 200 db level, and θ at the 400 dB level, respectively, used in Sokolov and Rintoul (2009a) to construct temperature criteria for identifying ACC jets. All these criteria, with the exception of the first, are given in Table 2.1. Colored serifs on the upper axis indicate the positions of the axes of the ACC jets in accordance with these criteria. The locations of the classical PF and PAF (Orsi et al. 1995) are also shown

constructed from the value of potential temperature at a level of 400 db, and for the jets in the south, two groups of criteria are proposed: the first is the potential temperature in the subsurface minimum layer in Winter Water, and the second is the potential temperature in the underlying layer of temperature maximum. Moreover, these criteria were based on two datasets, namely, the WOCE climatology and Argo float database. It was shown in Tarakanov (2011) that for the southern jets of the ACC, the criteria based on the properties of the Winter Water cannot be actually formulated either for the entire circumpolar circle or for its individual parts due to the strong dependence of the properties of the subsurface temperature minimum along the southern jets of the Antarctic Circumpolar Current from the local conditions of the formation of this minimum. This conclusion was then confirmed in Tarakanov (2014) and Tarakanov and Gritsenko (2018). In addition, it was shown in Tarakanov (2011) that the discrepancies in Table 2.1 between the criteria constructed according to the WOCE and Argo data are due to the fact that the first data set was calculated based on the measurements before the mid-1990s, and the Argo project was developed only in the second half of the 2000s. Thus, these discrepancies are a consequence of the time dependence of these criteria and reflect the temperature increase in the subsurface maximum layer that occurred during this period. Later, this conclusion was taken into account when analyzing the data of CTD measurements in the Drake Passage and south of Africa in 2008–2011 (Tarakanov 2014; Tarakanov and Gritsenko 2014a, 2018). Temperature criteria for the newly identified jets were formulated in Tarakanov and Gritsenko (2018).

The estimates of the ACC jet structure in the Drake Passage (Tarakanov and Gritsenko 2018) and south of Africa (Tarakanov and Gritsenko 2014a, b) suggest that a multi-jet structure with a larger number of jets than that suggested in (Sokolov and Rintoul 2009a) is common over the entire circumpolar circle, even in the areas of significant ACC narrowing such as the Drake Passage. However, the question of the number of jets in the ACC and its constancy in time and space remains currently open.

This work was supported by the RFBR grants (19-57-60001 and 18-05-00233).

References

Ablain M, Legeais JF, Prandi P et al (2017) Satellite altimetry-based sea level at global and regional scales. Surv Geophys 38(1):7–31

Berloff P, Kamenkovich I, Pedlosky J (2009) A model of multiple zonal jets in the oceans: dynamical and kinematical analysis. J Phys Oceanogr 39(11):2711–2734

Botnikov VN (1963) Geographical position of the Antarctic convergence zone in the Pacific Ocean (in Russian). Sov Antarct Exped Inf Bull Engl Transl 4(41):324–327

Budillon G, Rintoul SR (2003) Fronts and upper ocean thermal variability south of New Zealand. Antarct Sci 15(1):141–152

Burkov VA (1994) Antarctic jets. Okeanologia (Moscow) 34:169–177

Killworth PD, Hughes CW (2002) The Antarctic Circumpolar Current as a free equivalent barotropic jets. J Mar Res 60(l):19–45

Koshlyakov MN, Gladyshev SV, Tarakanov RY, Fedorov DA (2011) Currents in the Western Drake Passage according to the observations in January of 2010. Oceanology (Engl Transl) 51(2):187–198

Orsi AH, Whitworth T III, Nowlin WD Jr (1995) On the meridional extent and fronts of the Antarctic Circumpolar Current. Deep-Sea Res 42(5):641–673

Rio MH, Guinehut S, Larnicol G (2011) New CNES-CLS09 global mean dynamic topography computed from the combination of GRACE data, altimetry, and in situ measurements. J Geophys Res 116:C07018

Sarukhanyan EI, Smirnov NP (1986) Water masses and circulation of the Southern Ocean. Gidrometeoizdat, Leningrad. (in Russian)

Sokolov S, Rintoul SR (2002) Structure of Southern Ocean fronts at 140° E. J Mar Syst 37(1–3):151–184

Sokolov S, Rintoul SR (2007) Multiple jets of the Antarctic Circumpolar Current south of Australia. J Phys Oceanogr 37(5):1394–1412

Sokolov S, Rintoul SR (2009a) The circumpolar structure and distribution of the Antarctic Circumpolar Current fronts. Part A: Mean circumpolar paths. J Geophys Res 114(C11018). https://doi.org/10.1029/2008JC005108

Sokolov S, Rintoul SR (2009b) The circumpolar structure and distribution of the Antarctic Circumpolar Current fronts. Part B: Variability and relationship to sea surface height. J Geophys Res 114(C11019). https://doi.org/10.1029/2008JC005248

Sparrow MD, Heywood KJ, Brown J, Stevens DP (1996) Current structure of the South Indian Ocean. J Geophys Res 101(C3):6377–6392

Tarakanov RY (2011) Southern jets of the Antarctic Circumpolar Current in the Eastern Pacific Antarctic. Oceanology (Engl Transl) 51(4):588–598

Tarakanov RY (2012) The Scotia Sea and the Drake Passage as an orographic barrier for the Antarctic Circumpolar Current. Oceanology (Engl Transl) 52(2):157–170

Tarakanov RY (2014) Jets of the Antarctic Circumpolar Current in the central part of the Drake Passage based on the survey data in October–November of 2008. Oceanology (Engl Transl) 54(1):1–7

Tarakanov RY, Gritsenko AM (2014a) The frontal and jet structure south of Africa based on the data of SR02 transect in December of 2009. Oceanology (Engl Transl) 54(4):401–413

Tarakanov RY, Gritsenko AM (2014b) Fine-jet structure of the Antarctic Circumpolar Current south of Africa. Oceanology (Engl Transl) 54(6):677–687

Tarakanov RY, Gritsenko AM (2018) Jets of the Antarctic Circumpolar Current in the Drake Passage based on hydrographic section data. Oceanology (Engl Transl) 58(4):503–516

Chapter 3
Frontal Zone Between Relatively Warm and Cold Waters in the Northern Weddell Sea

Eugene G. Morozov, Viktor A. Krechik, Dmitry I. Frey, Alexander A. Polukhin, Vladimir A. Artemiev, Valentina V. Kasyan, Philipp V. Sapozhnikov, and Rinat Z. Mukhametianov

Abstract A strong front between cold and warm surface waters was found in the northwestern part of the Weddell Sea (Powell Basin) in February 2020. The front was manifested by sharp horizontal gradients of temperature, chemical, and optical properties. Differences in the concentrations and species of phytoplankton and zooplankton on both sides of the front were found. The front exists due to the differences in the water properties typical for the entire Weddell Sea and water from the western part of the sea, which is formed in a shallow region covered with ice and icebergs that prevent warming of the upper layer of water by the solar irradiance.

Keywords Weddell Sea · Antarctic Slope Front · Weddell Front · Differences in optical, chemical, and biological properties

3.1 Introduction

Frontal zones are widely spread in the ocean (Fedorov 1983). The processes in the frontal zones are related to mesoscale variability on spatial scales from a few kilometers to hundreds of kilometers (Monin et al. 1974). Oceanic fronts are associated with different properties of waters that are caused by the conditions of formation of water masses. Fronts are often associated with frontal rings of jet currents that carry

E. G. Morozov (✉) · V. A. Krechik · D. I. Frey · A. A. Polukhin · V. A. Artemiev
P. V. Sapozhnikov · R. Z. Mukhametianov
Shirshov Institute of Oceanology, Russian Academy of Sciences, Moscow, Russia

V. V. Kasyan
Zhirmunsky National Scientific Center of Marine Biology, Far Eastern Branch, Russian Academy of Sciences, Vladivostok, Russia

E. G. Morozov et al. (eds.), *Antarctic Peninsula Region of the Southern Ocean*, Advances in Polar Ecology 6, https://doi.org/10.1007/978-3-030-78927-5_3

water different from the ambient waters. Frontal rings separate from unstable meandering fronts of a larger scale. Frontal rings differ from the mesoscale eddies of the open ocean although these are similar phenomena (Rhines 2001; Koshlyakov 2008). Mesoscale eddies in the ocean are the dynamical analog of atmospheric weather. These eddies are energetic, time-dependent circulations about 100 km in width, found almost everywhere in the ocean (Rhines 2001). Mesoscale eddies differ from frontal eddies because mesoscale (or synoptic) eddies of the open ocean contain the same water as the surrounding water. Frontal eddies or rings transport captured water from the opposite side of the front. Isothermal, isohaline, and isopycnal surfaces in mesoscale eddies either ascend or deepen in the eddies. Hence, the shape of the θ,S-curves does not change. Despite the fact that rings or frontal eddies obey the same physical laws of the balance of pressure and Coriolis forces, the shape of the θ,S-curves differs from the curves for the surrounding waters since they carry captured water, which differs in its thermohaline, chemical, and other characteristics from the surrounding water. Orbital velocities in both eddies can be much higher than the velocities of mean currents.

Thermohaline properties of waters in the Weddell Sea are described in Orsi et al. (1993) and Schröder et al. (2002). Cold winter water occupies the upper layer up to a depth of 100–300 m. In summer, the upper layer up to 50–100 m is warmed and mixed. A layer of Warm Deep Water located below it determines the maximum temperature and salinity at intermediate depths of about 600 m. Deeper, there is a layer of Weddell Sea Deep Water (potential temperature $\theta < 2$ °C). Beneath it, there is a layer of Weddell Sea Bottom Water with a temperature below zero and colder than −0.7 °C.

The study site was located south of the frontal zone that separates the waters of the Scotia and Weddell seas; it is called the Weddell-Scotia Confluence Zone (WSCZ) (Gordon 1967; Bogdanov et al. 1969; Patterson and Sievers 1980; Muench et al. 1990). The Weddell-Scotia Confluence Zone is the boundary separating the eastward flowing waters of the northern Weddell Sea from those of the southern Scotia Sea (Patterson and Sievers 1980). The Weddell Front is located south of the study site (Fig. 3.1) based on Thompson et al. (2009). The location of the survey in the Powell Basin is shown in Fig. 3.1. The southwestern stations of the survey appeared close to the Antarctic Slope Front (ASF). This front is characterized by a ~ 20 cm s^{-1} current following the 1000-m isobath (Thompson et al. 2009; Heywood et al. 2004). This is a meandering front steered by bottom topography. It is possible for frontal eddies to separate from the front.

The goal of this research is to analyze the properties of water masses on both sides of a front found in the Powell Basin during a CTD survey on February 19–20, 2020.

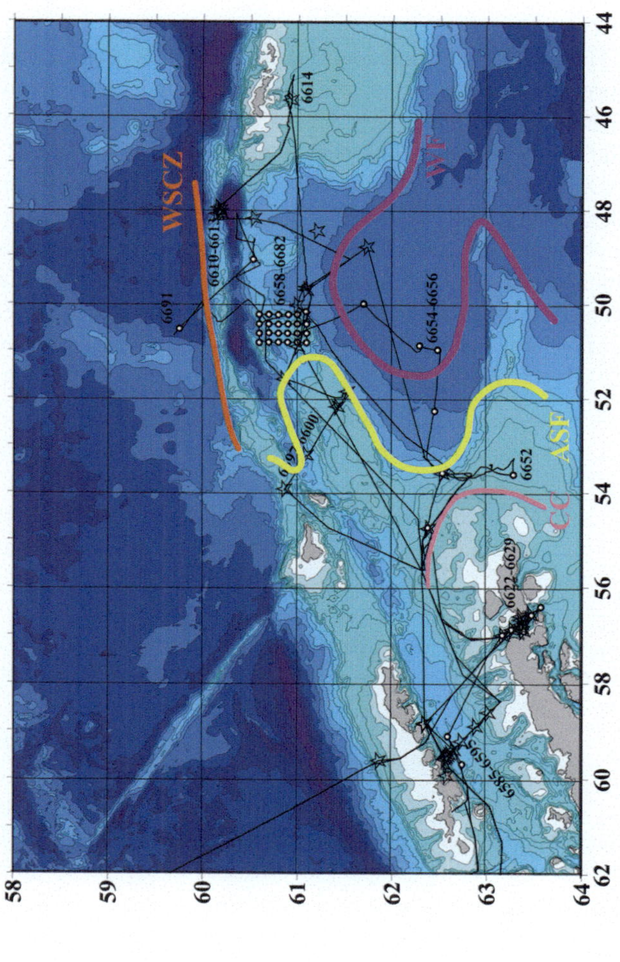

Fig. 3.1 Scheme of the stations during cruise 79 of the R/V *Akademik Mstislav Keldysh* in 2020. Stars denote stations in January and circles in February 2020. Yellow circles show the survey on February 19–20, 2020. Fronts of the Weddell Sea are indicated by colored lines: CC is coastal current, ASF is Antarctic Slope Front, WF is Weddell Front, and WSCZ is Weddell Scotia Confluence Zone

3.2 Data

The variability of physical, chemical, optical, and biological characteristics on spatial scales of about 25–30 nautical miles has been studied based on the data of a mesoscale survey of 6 × 4 stations with a size of about 30 × 24 nautical miles with a distance of 6 miles between stations. A scheme of the survey conducted during cruise 79 of the R/V *Akademik Mstislav Keldysh* on February 19–20, 2020, is shown in Fig. 3.2. The survey did not extend northward far enough to reach the Weddell-Scotia Confluence Zone.

The depth of CTD/LADCP profiling was 600 m. The survey was located so that its southwestern corner was on a seamount with a summit at a depth of about 600 m.

Profiles of temperature and salinity were measured at each station using the Sea-Bird 911 probe. The rosette with 24 5-liter Niskin bottles was also equipped with an RDI 300 kHz LADCP profiler of velocity. Bio-optical measurements were performed with a self-recording instrument PUM-200, which included a beam attenuation meter and a chlorophyll-a fluorescence meter Minitracka II (Chelsea) (light radiation at 430 nm and recording at 685 nm). Measurements of the beam attenuation coefficient were performed in the green spectrum range at a wavelength of 525 nm. Chemical properties were measured using commonly accepted methods (Stepanova et al. 2021, this book). Plankton catches were performed using Bongo and Jedy nets.

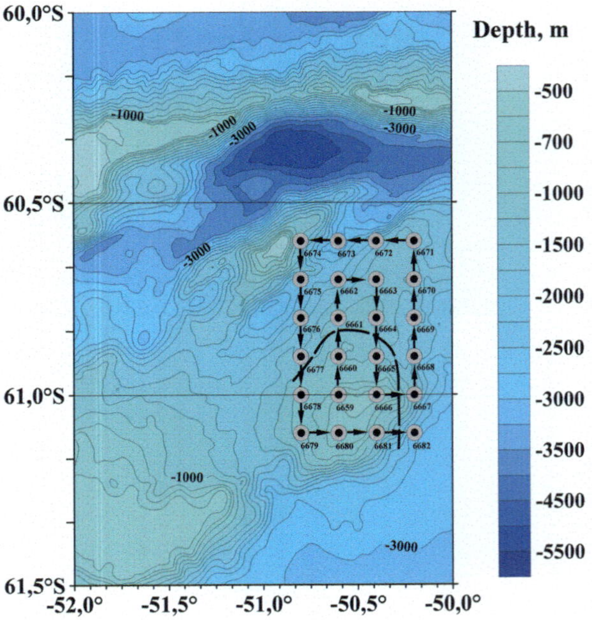

Fig. 3.2 Scheme of mesoscale survey on February 19–20, 2020. Numbers of stations are shown. The stations on the sections were made at latitudes 60°36′, 60°42′, 60°48′, 60°54′, 61°00′, and 61°06′ S and at longitudes 50°48′, 50°36′, 50°24′, and 50°12′ W. Catches using a Bongo net were carried out at stations 6659, 6662, 6665, 6669, 6671, 6674, 6676, 6679, and 6682 (white circles). The arrows show the route taken during the survey. The depth scale is in meters. The dashed black line schematically shows the front line within the survey

3.3 Results

The data collected during the survey on February 19–20, 2020, in the Powell Basin revealed the existence of a strong front between the relatively warm waters of the Weddell Sea and the very cold waters in its western part. This front exists due to the differences not only in temperature but also in other properties of water characteristic of the entire Weddell Sea (relatively warm summer waters of the Weddell Sea south of the WSCZ) and cold water from the western part of the sea, which is prevented from summer warming in shallow waters covered with ice and icebergs. The Weddell Sea water on the southern side of the WSCZ was located in the northern part of the survey, and the water from the Antarctic Peninsula shelf was in the southwestern part of the survey.

3.4 Hydrology of the Region

It was found during the survey that even on such a small scale, the water structure at various stations turned out to be completely different. In the northern part of the survey, the water structure was typical for the Weddell Sea (Orsi et al. 1993). The water structure in the northern part of the survey was similar to the structure in the entire Weddell Sea. In the upper 100–150 m, there was a warm summer surface layer with a strong vertical temperature gradient. Below it, at a depth of about 200 m, there was a summer cold intermediate layer with a subsurface temperature minimum with negative values of about −0.5 °C (Fig. 3.3). In winter, the top layer is transformed into a uniform cold layer due to the winter convection and wind mixing. Deeper than 200 m, there is a layer of Warm Deep Water with temperatures up to +0.4 °C at a depth of about 600 m; further to the bottom, the temperature decreased.

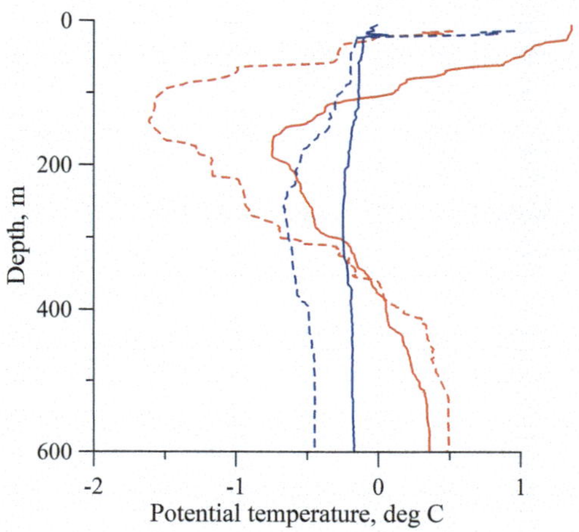

Fig. 3.3 Vertical distribution of temperature at survey stations 6660 (cold water, solid blue line) and 6673 (warm water, solid red line) and at stations near the Antarctic Slope Front 6599 (cold water, dashed blue line) and 6600 (warm water, dashed red line)

The temperature of the second type of water was negative in the entire water column (Fig. 3.3). The temperature only slightly changes with depth. This water structure is similar to that over the Antarctic Peninsula shelf and submarine elevations in the western part of the Weddell Sea west of the 50° W between latitudes 60°30′ S and 62°00′ S. The shallow depths of the sea in this region are in the range of 400–700 m. The region is covered with ice and icebergs from the southern part of the Weddell Sea. The ice prevents surface water from warming. Ice cools the water to negative temperatures, and strong internal waves above the elevation mix the entire layer of water from the surface to the bottom. These waters are called Antarctic Shelf Waters. The thickness of the mixed layer reaches 600 m. In the 2020 cruise, we could occupy stations only in the northern part of the Powell Basin. Further to the south, there were ice fields.

During the cruise on January 24–25, 2020, we made two stations southwest of the survey on February 19–10, 2020. They were located on both sides of the Antarctic Slope Front: cold station 6599 (61°25.7′ S, 52°08.4′ W) and warm station 6600 (61°31.9′ S, 51°50.4′ W). The distance between stations was 11 miles, but the water structure at both stations was completely different. Vertical distributions of potential temperatures are shown in Fig. 3.3 with dashed lines. The general structure of cold and warm profiles was similar at the Antarctic Slope Front and at the front detected during the survey, which could either be the APF or a front of a cyclonic ring of this front.

CTD casts at the stations of the survey showed that salinity increases with depth at all stations (Fig. 3.4). Stratification is more strongly determined by salinity than by temperature. At northern stations, salinity in the deep part of the water column is slightly higher than salinity in the southwestern part of the survey. Salinity in the upper layer at stations over the elevation (e.g., station 6659) was close to 33.9 PSU. Such salinities are not found in the northern section of the survey north of the front. The minimum salinity there was 34.0 PSU. The differences in temperature-salinity dependences on both sides of the front are very strong (Fig. 3.5).

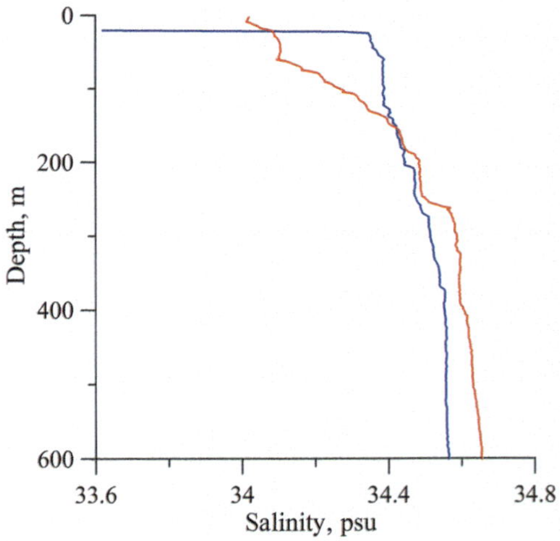

Fig. 3.4 Vertical distribution of salinity at stations 6659 (cold region, blue line) and 6671 (warm region, red line). Reproduced from Morozov et al. (2020) ©Pleiades Publishing, Ltd. Used with permission

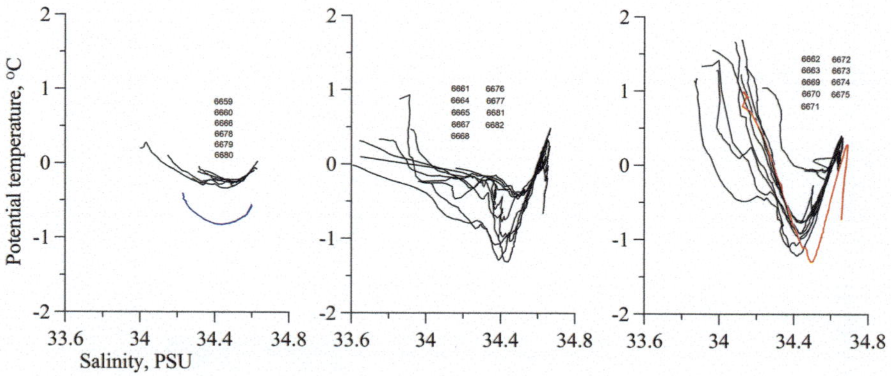

Fig. 3.5 θ, S-curves at stations in the cold region (left panel) and typical curve at the cold side of the front (blue); in the front (central panel); and in the warm region (right panel) and typical curve in the warm region (red). Numbers of stations are indicated

We performed an analysis of the regions, in which the θ,S-structure of water was similar to that found in the southwestern part of the survey. These stations were selected from the WOD18 database array (Fig. 3.6) west of the Antarctic Slope Front over shallower depths. All of these stations with cold water are located in the northwestern part of the Weddell Sea above the elevation. They are indicated with red dots in Fig. 3.6. Thus, the sources of cold water are located on the elevation in the western part of the Powell Basin in the northwestern part of the Weddell Sea at the coastal side of the Antarctic Slope Front near the Antarctic Peninsula. The water over the coastal elevation is cooled by the presence of ice and icebergs. Floating ice prevents solar radiation from heating the water in summer.

Thus, two different water structures were recorded during the mesoscale survey: warmer water with a subsurface cold layer, typical for the Weddell Sea, and other water of advective origin found at the southwestern stations of the survey that spreads to the survey region from the western shallower part of the Powell Basin (northwestern part of the Weddell Sea).

To confirm this hypothesis, let us consider the satellite altimetry data on February 19, that is, on the date of the survey (Fig. 3.7). A chart of dynamic topography is shown in Fig. 3.7. Figure 3.7 also shows the 1000-m isobath, whose location is close to the meander shape based on the satellite dynamic topography. This relationship shows the dependence of the position of the meander on the bottom topography. It was emphasized in Thompson et al. (2009) that the currents of the Antarctic Slope Front (ASF) with velocities of 20 cm s^{-1} follow the 1000-m isobath, and they are associated with cyclonic flow around the Powell Basin. Heywood et al. (2004) suggested a slightly different scheme than Thompson et al. (2009). According to their scheme, the ASF splits in two over the South Scotia Ridge. One branch of the front follows a topographically steered route out through the Weddell-Scotia Confluence into the Antarctic Circumpolar Current (ACC), and the second part forms the Weddell Front south of the Powell Basin.

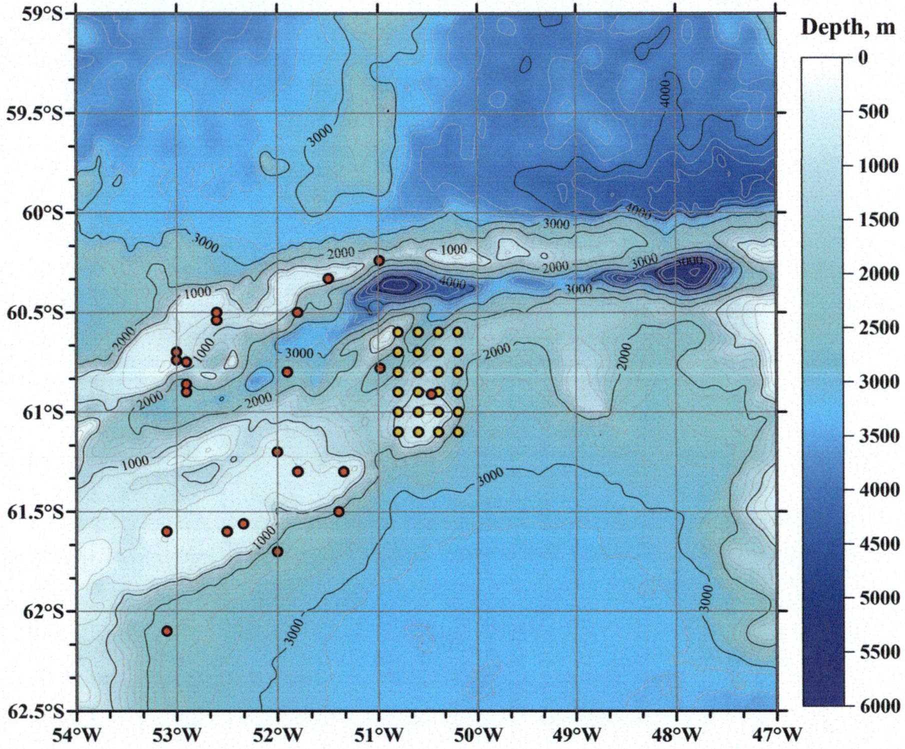

Fig. 3.6 Bathymetry of the western part of the Weddell Sea. Red dots are the positions of stations from the WOD18 database, where the structure of water masses is similar to the structure in the southwestern part of the mesoscale survey. Yellow circles indicate positions of the survey stations. Depth scale in meters is on the right. Reproduced from Morozov et al. (2020) ©Pleiades Publishing, Ltd. Used with permission

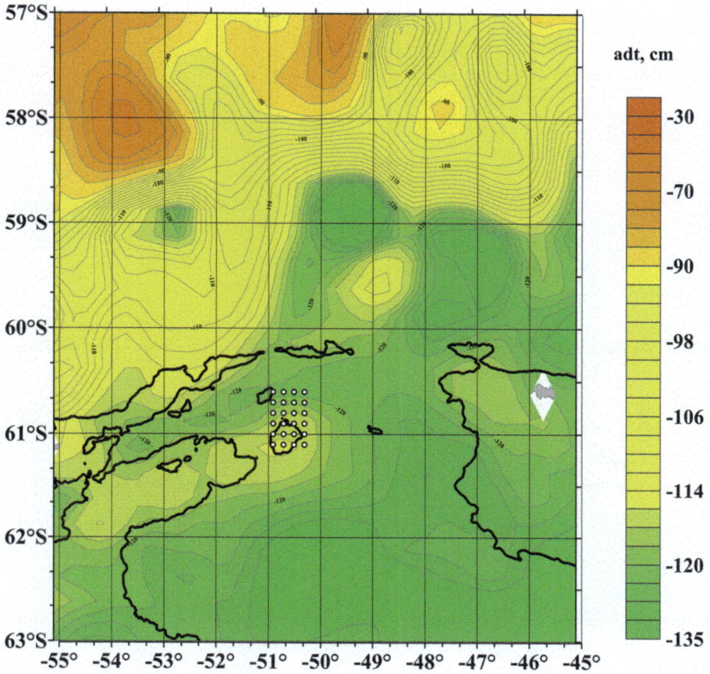

Fig. 3.7 Satellite altimetry on February 19, 2020. Absolute dynamic topography (ADT) is shown. Color scale of dynamic topography is on the right. The altimetry map shows the 1000-m isobath (thick black line) and the points of mesoscale stations (24 black circles with a white dot). The light rhombus (61°40′ S, 45°45′ W) corresponds to the South Orkney Islands. (Reproduced from Morozov et al. (2020) ©Pleiades Publishing, Ltd. Used with permission)

Waters that occupy the northern part of Powell Basin are separated with a clearly pronounced frontal zone from the waters in the Scotia Sea.

In the region of 61°30′ S, 54° W, there is a meander of the frontal zone that separates from it. It seems that this meander is formed by the influence of topography, and it is quasi-stationary. The meander moves around an elevation with depths shallower than 1000 m, which is clearly seen on the combined map of depths and altimetry. The captured water in the meander is subject to strong cooling due to the presence of large amounts of ice and icebergs in the region. When cooled, this water acquires other thermohaline properties. The easternmost part of this meander reached our mesoscale survey, bringing cold water with different properties. It is possible that this meander breaks up into several separate rings, which are located in a series over an elevation in the latitudinal zone from 61° S to 62° S. Presumably, the seamount located in the southwestern part of the mesoscale survey only partially changes the circulation in this area. This mountain is the end of the elevation, above which the meander is located.

Let us consider the spatial variability of temperature and salinity in the survey site (Fig. 3.8). Temperature and salinity in the region of the seamount are somewhat

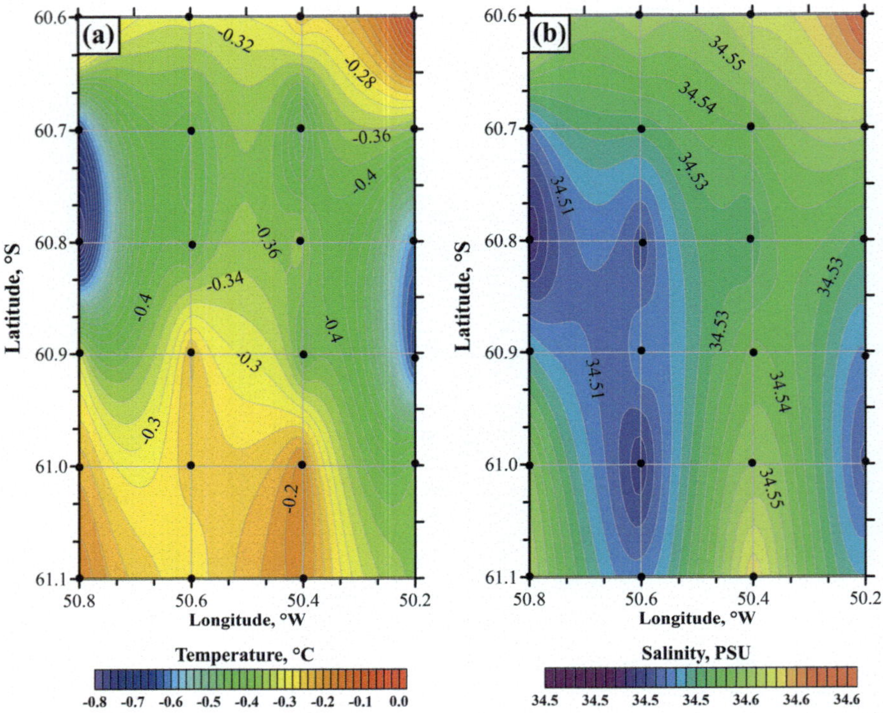

Fig. 3.8 Maps of the distribution of temperature, °C (a), and salinity, PSU (b), at a depth of 300 m during the survey on February 19–20, 2020. Stations are shown with dots. Temperature (°C) and salinity (PSU) scales are below the charts

lower than in other parts of the survey. However, as is usually the case in the ocean, such a decrease in temperature and salinity does not mean that isotherms and isohalines, and hence isopycnals, descend over the mountain under the influence of anticyclonic rotation. This effect caused by the descent of iso-surfaces is weaker than the influence of the inflow of colder and less saline water into the region of our mesoscale survey from the southwest. The effect of the presence of a seamount is greater on the formation of colder water over the elevation; this seamount is part of this elevation. The anticyclonic rotation is created by a meander located above the mountain or by a ring that has already separated from it.

A meridional section of temperature across the mesoscale survey clearly shows the differences in the water structure a short distance from the study site (Fig. 3.9). In the southern part, an almost uniform layer of cold water was found, while in the northern part, the structure of waters included a warm surface layer, a subsurface intermediate cold layer, and warmer deep water.

The velocities measured by the lowered and shipborne profilers show satisfactory agreement both with the geostrophic velocities calculated from the satellite altimetry data and between each other. The instantaneous velocities measured by the profilers were processed to remove tidal currents, estimated using the Oregon

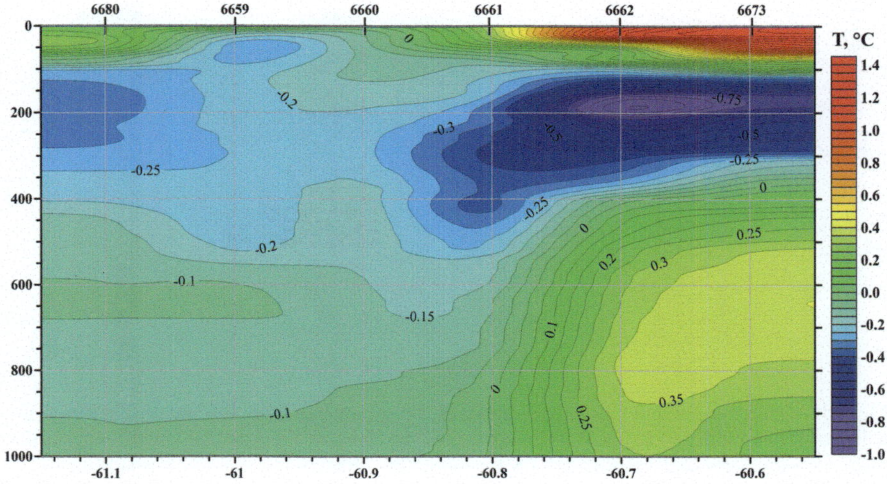

Fig. 3.9 Meridional section of temperature along 50°36′ W. Color temperature (°C) scale is on the right. Stations are marked by ticks on the top axis

State University (USA) TPXO 9 model (Egbert and Erofeeva 2002). While approaching station 6666 and after this station, the shipborne Doppler current profiler showed a high northward velocity in the entire water layer to the east of the seamount, which is typical for the eastern periphery of an anticyclonic eddy. The lowered velocity profiler also showed high northward velocities of about 50 cm/s. To the west of the seamount, high speeds were directed to the west. The current vectors of the SADCP and LADCP instruments at a depth of 400 m are shown in Fig. 3.10.

3.5 Hydrochemistry

The differences in the properties of waters north and south of the front are confirmed by the chemical determinations of the concentration of nitrates, silicates, and phosphates. The distribution of nitrates and phosphates in the surface layer (3–5 m) shows a pronounced maximum in the southwestern part of the study site at station 6659 (Fig. 3.11), which is reflected in an increase in the concentration of nutrients at the surface, as well as in a decrease in the oxygen content. This is due to the upwelling of deep waters above the elevation to the west of the mesoscale survey where the water source from the southwestern part of the survey is located. High values of nitrates and phosphates at the surface are typical for the waters south of the front and, most likely, for the waters over an elevation to the west of the mesoscale survey. This process is well known and has been repeatedly noted in other regions of the World Ocean (Doblin et al. 2016).

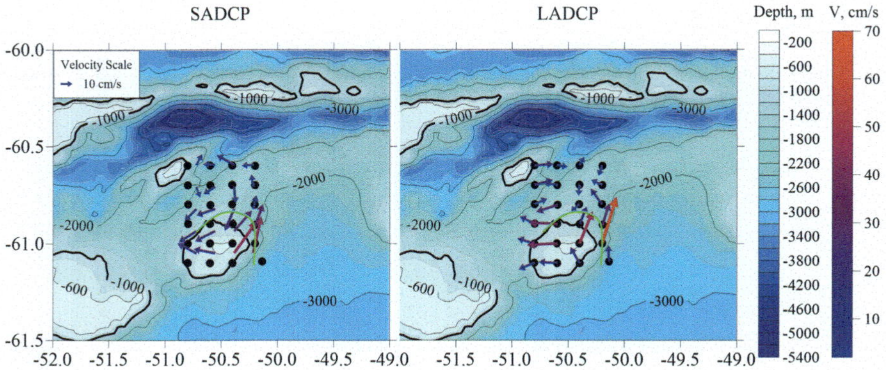

Fig. 3.10 Velocity vectors in the region of mesoscale survey based on the data of shipborne current profiler and lowered LADCP. The vectors are superimposed on the bathymetry map. Color scale of bathymetry and velocity are on the right. The map shows the 1000-m isobath (thick black line)

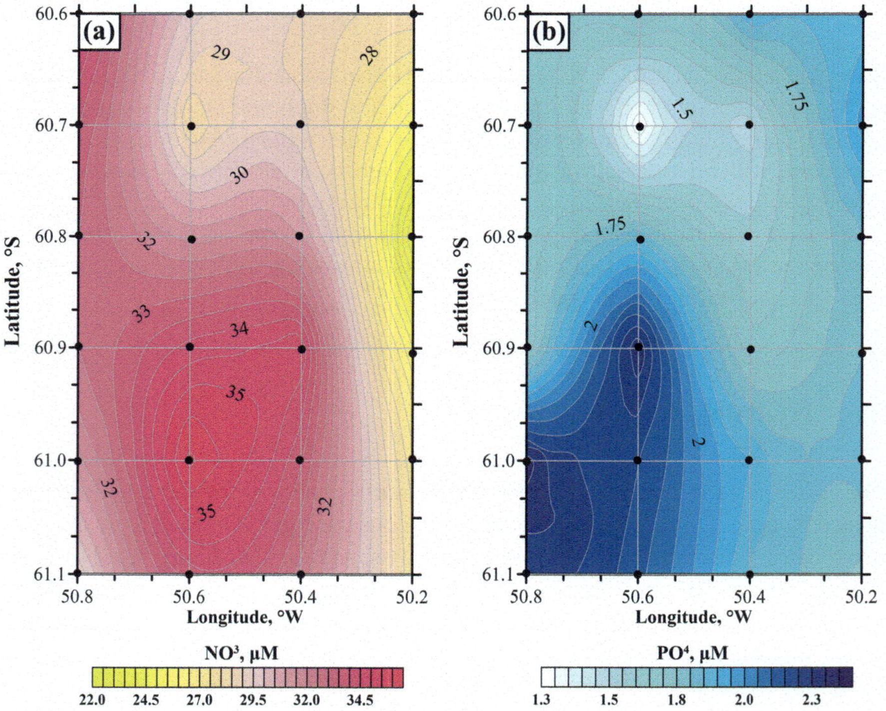

Fig. 3.11 Maps of distribution of nitrates (a) and phosphates (b) at the surface of the survey. Color scales for nitrates and phosphates (μM) are shown below. Reproduced from Morozov et al. (2020) ©Pleiades Publishing, Ltd. Used with permission

The front between the waters of the Weddell Sea and the waters above the elevation to the west of the mesoscale survey is clearly distinguished by the silicates/phosphates ratios (Fig. 3.12). The ratio of silicates to phosphates exceeds 35 at the front, which is typical for the high-latitude modification of the Weddell Sea water mass (Arzhanova and Artamonova 2014). South of the frontal zone, this ratio is smaller than 35, which indicates that these waters do not belong to the frontal zone.

We can conclude based on hydrochemical data that strong variability in the concentration of nutrients exists in the study site caused by the presence of strongly different water above the seamount (Dafner and Mordasova 1994). The advection of cold waters from the southwest and the accompanying upwelling of deep waters along the mountain slope lead to an increase in the concentration of phosphates and silicates. In the future, this can lead to the intensification of production of organic matter (Doblin et al. 2016; Ellwood, et al. 2020) and other biological processes.

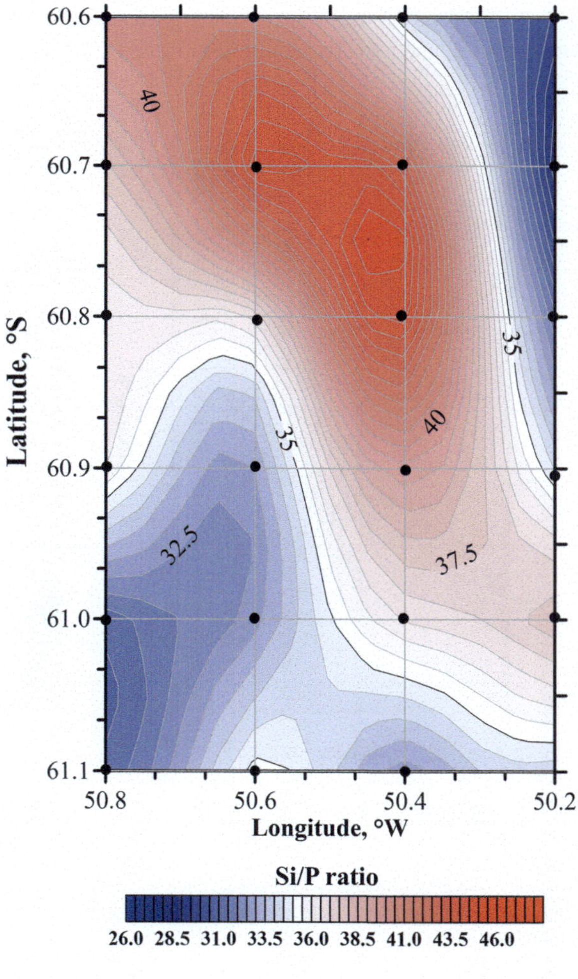

Fig. 3.12 Map showing the concentration ratio of silicates and phosphates at the surface. The ratio scale is located below. Stations are shown with dots. Reproduced from Morozov et al. (2020) ©Pleiades Publishing, Ltd. Used with permission

3.6 Optical Properties

The waters on both sides of the front are different according to their bio-optical structure in the upper layer. In the spatial distribution of the beam attenuation coefficient, the highest values were observed in the northern part of the study site in warm waters; they were as high as 0.6 m^{-1} at 525 nm (Fig. 3.13a). In the southern part of the survey, the beam attenuation coefficient did not exceed 0.3 m^{-1} at 525 nm (Fig. 3.13a). Similar difference was observed in the fluorescence distribution of chlorophyll-a: 0.5 relative units in the northern part and 0.14 r.u. in the southern part (Fig. 3.13b).

Differences in the optical characteristics are clearly pronounced in the vertical distributions of optical parameters at stations 6660 (in the southwestern part of the survey above the seamount) and 6671 (northeastern part) (Fig. 3.14).

At station 6660, the vertical profile in the thin surface layer (12–14 m) is typical for the ice melting zone. Changes in the beam attenuation coefficient and fluorescence with depth coincide with a sharp increase in salinity associated with surface freshening caused by the melting of ice. In the areas where sea ice is melting,

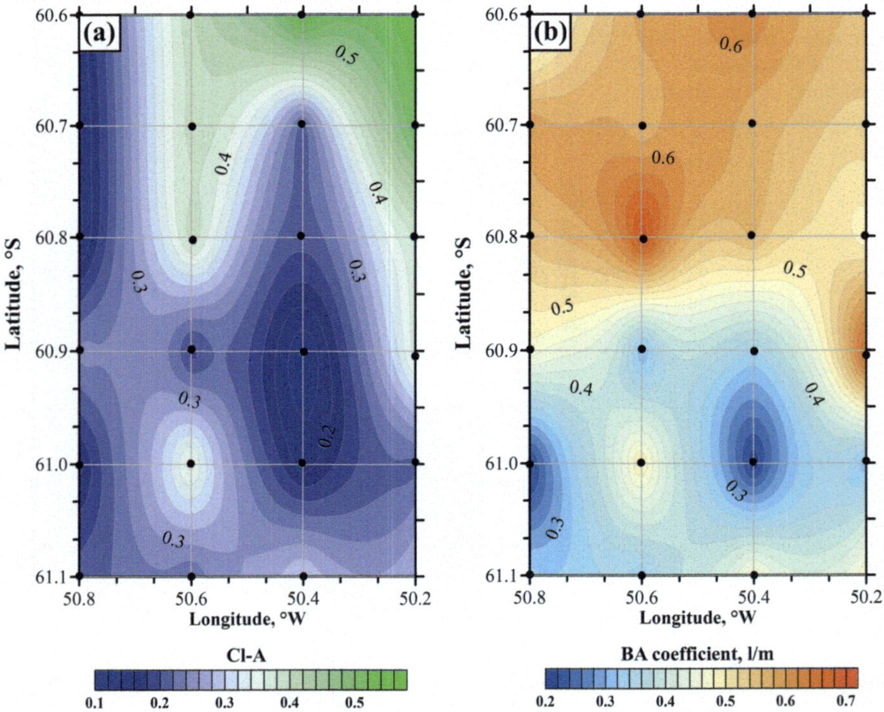

Fig. 3.13 Maps of chlorophyll-a fluorescence in relative units (**a**) beam attenuation coefficient at 525 nm (m^{-1}) (**b**) and in the surface layer. Small dots indicate stations and large dots stations 6660 and 6671. Color scales are shown below the figures

diatomic algae appear in the water and cause an increase in fluorescence and the beam attenuation coefficient. Below this zone, the temperature and fluorescence reach a quasi-permanent level; they slightly change to a depth of 200 m. The beam attenuation coefficient monotonously decreases to 0.15 m^{-1} at 525 nm.

At station 6671, the vertical distributions are strongly different. Here is an almost homogeneous upper layer of heated water ~70-m thick. The beam attenuation coefficient and fluorescence of chlorophyll-a are higher than in the southwestern part of the survey. At a depth of 25–30 m, there is a maximum of these optical properties. At a depth of 70–65 m, a sharp thermocline coincides with a sharp decrease in chlorophyll-a fluorescence and beam attenuation. The high correlation between the beam attenuation coefficient and fluorescence of chlorophyll-a indicates that nutrients dominate in the suspended matter composition (Fig. 3.14).

3.7 Phytoplankton

Observations of the state (composition, structure, and spatial organization) of phytoplankton at the stations of the mesoscale survey revealed a number of differences between communities north and south of the front. These differences affected both the set of species represented by free-floating cells and those species that form epibiontic aggregates on the surface of the chaetae (setae) of the shells of large planktonic diatoms.

South of the front, the communities were composed of large free-moving tychopelagic diatoms such as *Corethron pennatum*, *Chaetoceros criophilus*, *Ch.*

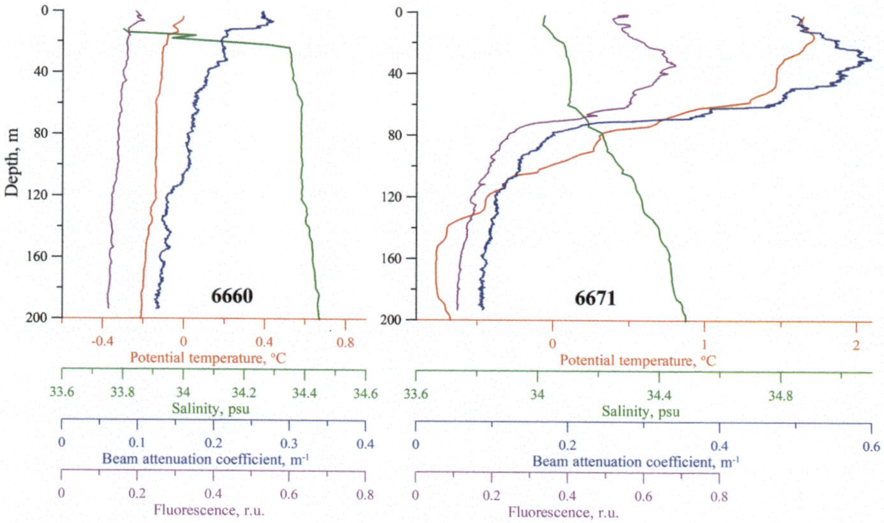

Fig. 3.14 Vertical distributions of hydrological and bio-optical parameters at stations 6660 and 6671

convolutus, *Stellarima microtrias*, and *Thalassiothrix antarctica*. All these species are common in the Antarctic waters (Kang and Fryxell 1993; Schloss et al. 2002). *Corethron pennatum* and *Ch. criophilum* were dominants in the euphotic layer. The epibiontic complexes on the surface of setae built on the basis of the matrix of colonies of heterotrophic bacteria were formed by two species of small heterotrophic flagellates and colonial species of diatoms from the ice flora (Kang et al. 1993; Lundholm and Rytter Hasle 2008; Cefarelli et al. 2010): *Fragilariopsis nana*, *F. curta*, *F. cylindrus*, occasionally *F. kerguelensis*, as well as small colonies of *Chaetoceros debilis* and *Ch. laciniosus*. *Thalassiothrix antarctica* (with a slight admixture of *C. pennatum and Ch. criophilus*) clearly dominated in the surface layer at depths of about 4 m based on the materials collected from the flow-through system. Large amounts of already mentioned flagellates and, occasionally, short ribbon-like colonies of *Fragilariopsis species (F. nana* and *F. kerguelensis*), as well as single cells of *Licmophora* spp. (Fig. 3.15) live on the matrices of sail-shaped bacterial colonies of *Th. antarctica*, which were formed on the basis of plexuses of long cells, fragments of colonies, and dead shells.

At the stations north of the front, two colonial diatom species belonging to the genus *Pseudo-nitzschia* (*P. calliantha* and *P. turgiduloides*) were the main structural elements of phytoplankton in the euphotic layer. Here they were accompanied by massive *C. pennatum* and *Ch. criophilum*, while at depths of about 4 m, the same *Th. antarctica* and *C. pennatum* cells were also common at early stages of the transition to sexual reproduction. Nevertheless, the composition of the epibiontic complexes on *Corethron* chaetae and *Th. antarctica* (in fact, looking exactly like fine bristles) has changed significantly here. Now, on the matrix of colonial bacteria, small colonies and simply groups of *Pseudo-nitzschia* cells became more abundant (Fig. 3.2). The small flagellates just mentioned became more widespread, numerous, and compact colonial settlements of small benthic species from the genera *Amphora*, *Halamphora*, and *Nitzschia* were found (a total of ten species). Mass development of *Licmophora* spp. was often observed, and the cells of *Protoperidinium*

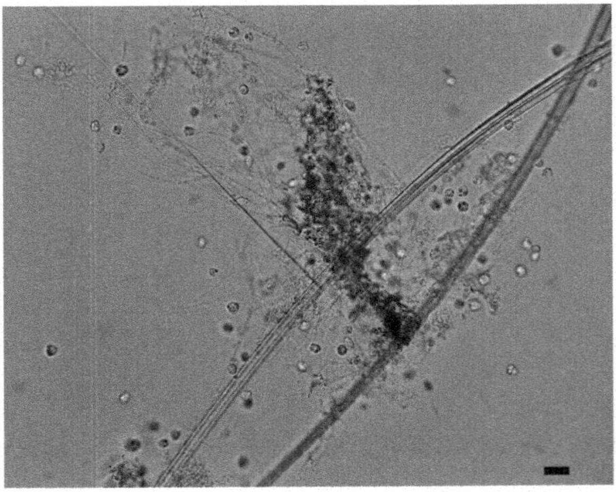

Fig. 3.15 Heterotrophic flagellates, which developed in large numbers on the surface of the matrix of colonial bacteria, as part of epibiontic complexes south of the front. Scale: 10 μm

defectum dinoflagellates were also regularly found. Colonies of *Fragilariopsis* spp. were also present, but their role in the structure of epibiontic complexes was less pronounced.

In general, one can speak about significant restructuring of phytoplankton cenoses north of the front. The changes were manifested in a general increase in the diversity of flora and the appearance of mass species belonging to the genus *Pseudo-nitzschia* in the composition of communities, as well as in strong restructuring and growth of the diversity of epibiontic complexes that form on the shells of large diatoms, in particular, on their bristle-like structures, which enable soaring of cells in the water column.

Epibiontic assemblages are of particular interest because, as our observations of the stomach contents of euphausiids have shown, they form the basis of the krill diet in these waters.

3.8 Zooplankton

Studies of zooplankton in the Powell Basin in the northern Weddell Sea have shown significant differences in the species composition, structure, and spatial distribution of communities between those south and north of the hydrological front. All stations were divided into three groups based on the cluster analysis (Fig. 3.17a).

The first group (A) was observed to the south of the front in meander waters with negative temperatures; it was characterized by the lowest abundance and species diversity of zooplankton communities (Fig. 3.17b, d). The plankton was dominated by copepods of the *Metridia*, *Oithona*, *Oncea*, and *Triconia* genera, the species characteristic of the coastal regions of the Antarctic Peninsula (Fransz 1988; Foster 1989). Deeper than 200 m, large copepods of genera of the oceanic complex, *Calanus* and *Calanoides*, and also Ostracoda were occasionally encountered. Usually, copepods of the *Calanus* and *Calanoides* genera were found at a depth of 0–250 m in summer and 500–1000 m in winter (Delgado et al. 1998), and they dominate in number in the summer (Lee et al. 2004).

The abundance and species diversity of zooplankton of the second group (B) increased in the frontal region (Fig. 3.17b, d). The appearance of a large number (over 250 ± 76 ind./m^3) of Hyperiidae and Euphausiidae eggs was noted in the warm summer surface layer (Fig. 3.16). There were fewer adult specimens of hyperiids (Hyperiidae), pteropods (*Clione*), and larvae (calyptopis stages) of euphausiids. Copepods *Paraeuchaeta antarctica* and *Rhincalanus gigas* as well as Siphonophorae (*Diphyes*) and copepods nauplii were recorded singly, but not found earlier. Large copepods (such as *Calanus propinquus*, *Calanoides acutus*), which were found south of the front in the cold layer up to 200–500 m, were found in this region above 200 m. Cold Antarctic waters were observed in the subsurface layer above 200 m in the frontal region where conditions were formed for the accumulation of large copepods.

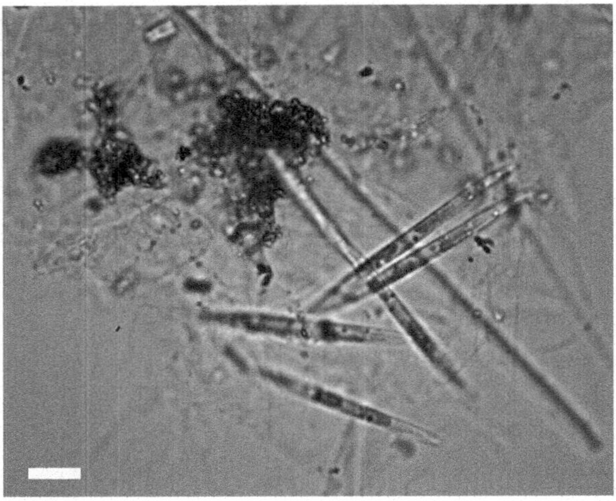

Fig. 3.16 Individual cells and a fragment of a linear colony of the diatom *Pseudo-nitzschia calliantha*, as part of an epibiontic complex formed north of the front. Scale: 10 μm

Communities (C) located north of the front in the typical waters of the Weddell Sea (Klepikov 1963) were characterized by a high abundance (more than 500 ± 103 ind./m^3) of larvae of Hyperiidae and Euphausiidae (Fig. 3.18). Hyperiidae larvae were concentrated in the Warm Deep Water layer (200–500 m), while the Euphausiidae larvae (calyptopis and furcilia stages) were concentrated in the surface warm layer (0–200 m). The krill eggs constantly sink into deeper layers, and nauplii usually hatch at a depth of 850 m or 1250–1850 m. As they rise to the surface waters, krill larvae pass through the stages of nauplii, metanauplius, calyptopis, and furcilia (Siegel and Watkins 2016). According to Makarov (1982), between December and February, up to 80% of calyptopis larvae are found at depths of 200–500 m. Later in the season, most of them are concentrated at a depth of 100–200 m, and sometimes in the upper 25-m layer. Furcilia larvae prefer depths of 200–500 m in February–March, and in April–May they accumulate in the upper 25-m layer. Krill juvenile and adult stages are consistent surface residents (0–100 m) and are thought to grow rapidly by feeding on abundant phytoplankton during the Antarctic summer (Ikeda 1985).

The highest abundance and diversity of zooplankton communities was noted in the group C, exceeding by 3–5 times those in the other studied regions (Fig. 3.17b, d). Copepods of the genera *Calanus*, *Calanoides*, *Ctenocalanus*, *Metridia*, *Oithona*, *Oncea*, *Paraeuchaeta*, *Triconia*, and *Rhincalanus* as well as Chaetognatha, Euphausiidae *(Thysanoessa)*, Pteropoda *(Clione)*, Ostrachora, Siphonophorae *(Diphyes*), and Polychaeta larvae were found in this group. The increased number of larvae in communities located north of the frontal zone indicates that this area can serve as an important feeding ground for whales, which are known to actively hunt euphausiids and hyperiids (Hempel 1985).

It is known that Euphausiidae are an important component of the macrozooplankton community in the coastal ecosystems of Antarctica (Flores et al. 2012; Driscoll et al. 2015). Krill *Euphausia superba* feeding on phytoplankton and small zooplankton serve as food for baleen whales, crabeater seals, pelagic fish, and birds.

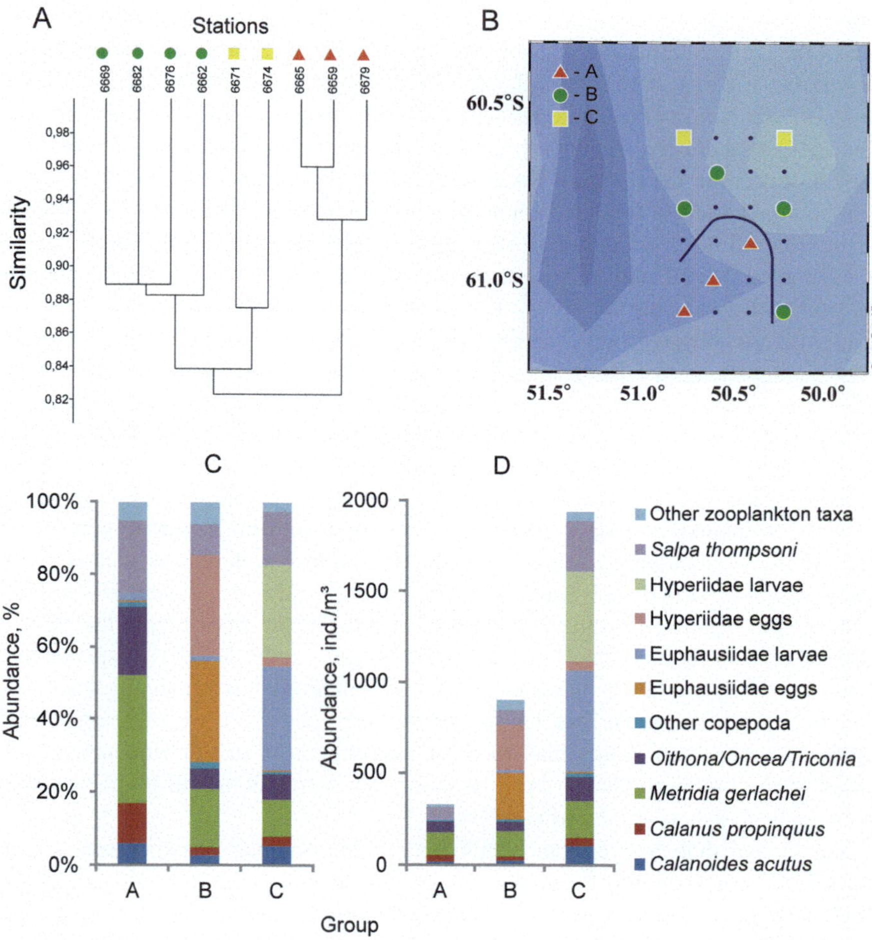

Fig. 3.17 (**a**) Results of cluster analysis based on zooplankton abundance by Bray-Curtis similarity (Clarke and Gorley 2005). Three groups (A–C) were identified. (**b**) Spatial distribution of zooplankton community groups according to the cluster analysis. (**c**) Contribution (%) of the major taxonomic groups to total zooplankton abundance. (**d**) Total abundance (ind./m^3) values for each group

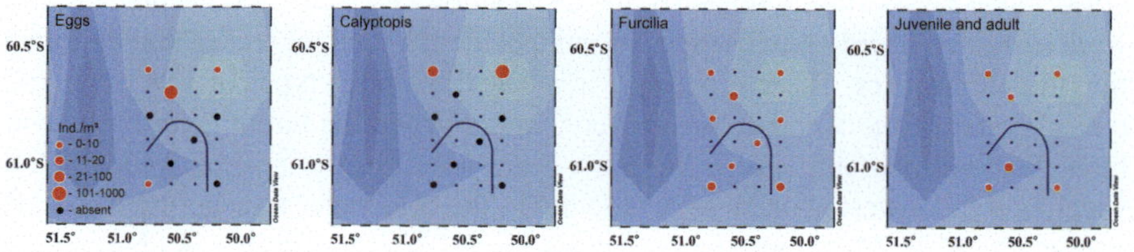

Fig. 3.18 Concentration of eggs, larval stages, and juvenile and adult krill Euphausia superba in the Powell Basin in the northern part of the Weddell Sea. The black line schematically shows the front

The habitat of Antarctic krill, *Euphausia superba*, is approximately 32 million-km^2 area of the entire Southern Ocean south of the Polar Front (Sprong and Schalk 1992; Siegel and Watkins 2016). Mackintosh (1973) more clearly defined the northern limits of distribution of krill and indicated several clusters around the continent; the Weddell-Scotia cluster, the Enderby cluster, the Kerguelen-Gaussberg cluster, two smaller clusters in the northern Ross Sea, and the Bellingshausen cluster. In addition, he noted areas of low krill abundance (e.g., 65–68°S and 30° W–20°E). Krill *E. superba* did not form dense clusters in the study site (Fig. 3.18). Single specimens of juvenile and adult *E. superba* were recorded in the communities located in the front region and to the north of it. A slight increase in concentration (up to 23 mg/m^3) was observed south of the front area. The krill was represented by individuals ranging in size from 38 to 52 mm, with an average length of 36.6 mm. Females were present at early stages IIIB and IIIA, while males were predominantly at stage IIIA. The sex ratio was 1.0: 1.3 with a slight predominance of males.

In recent decades, *Salpa thompsoni* has shown an increased prevalence at high latitudes of the Southern Ocean and is now often the dominating in the number of taxons of zooplankton in the waters of the Antarctic Peninsula. It is assumed that an increase in the abundance of this species at high-latitude waters is associated with ocean warming (Loeb and Santora 2012). In our study, the largest amount of *S. thompsoni* was noted in the communities located in the front region and to the north of it, with biomass concentrations of 0.66 g/m^3 and 1.62 g/m^3, respectively. However, the quantitative indicators of krill *E. superba* were insignificant. It is likely that this is due to the fact that salps eat up eggs and the larvae of younger-age stages of crustaceans. Salps may play an important role in the biogeochemical cycles of the Southern Ocean, but they have low correlation with higher trophic levels: pelagic fish, seabirds, whales, etc.

It is known that the Antarctic Peninsula region continues to experience sharp climate changes, including an increase in sea surface temperature (Anisimov et al. 2007) and a decrease in the extent of sea ice in winter (Stammerjohn et al. 2008). In combination with these changes in physical properties, the primary production and the structure of phytoplankton and zooplankton communities in the water column are likely to continue to change, opening up the opportunities for increasing the population density of other previously not so numerous species (Hewes et al. 2009; Loeb and Santora 2012; Voronina 1998).

Thus, the structure of zooplankton communities in the Powell Basin in the northern Weddell Sea is closely related to the distribution of water masses. Three groups were identified based on to the species composition, structure, and spatial distribution of communities. These groups are associated with two types of water masses: typical water of the Weddell Sea and cold water above the bottom elevation. The highest abundance and species diversity were noted in the typical Weddell Sea water, and the lowest abundance was found in the cold water transported from the bottom elevation. Mass reproduction of euphausiids and hyperiids was observed in the front region, in the warm summer surface layer; an accumulation of their larvae was found north of the front. Euphausiidae larvae were concentrated in the surface warm layer, while the Hyperiidae larvae were concentrated in the Warm Deep Water layer. In general, Copepoda (*Calanus propinquus*, *Calanoides acutus*, and *Metridia gerlachei*) and eggs and larvae of Hyperiidae and Euphausiidae prevailed in abundance, while Salpidae dominated in biomass. The abundance and species diversity of zooplankton increased in the northern direction across the front. Krill *E. superba* did not form dense aggregations in the study site, and the largest amount of salpa *S. thompsoni* was observed in the communities located in the front region and to the north of it. The abundance of Antarctic crustaceans krill *E. superba* is determined by a complex of processes, among which the main ones are the intensity of spawning within the breeding area, the intensity of transport of larvae by currents, and the rate of grazing by predators.

3.9 Conclusions

A strong front between cold and warm surface waters was found in the northern part of the Weddell Sea (Powell Basin) in February 2020 based on the data of a survey of 24 oceanographic stations. The differences in the water structures north and south of the front are pronounced in temperature and salinity, nutrients, and optical characteristics of the upper layer. Differences in the concentrations and species and structure of plankton communities on both sides of the front were found. Seawater properties on the cold side of the front are similar to those in the western part of the Weddell Sea, which is formed in shallow water covered with ice and icebergs. Ice prevents warming of the upper layer by solar radiation. This water is different from the warmer water in the Powell Basin. It is possible that cold rings separate from the Antarctic Slope Front and transport cold water with the properties different than that of the surrounding waters. The differences in water properties and plankton structure observed during our survey could be related to a front of a cold ring that separated from the ASF and displaced to the region of our survey.

Acknowledgments The authors thank the crew of the R/V *Akademik Mstislav Keldysh* for their hard work and assistance to collect the new data in severe Antarctic conditions. This work was supported by the State Task no. 0128-2019-0008. Field works were partly supported by the RFBR grant no. 20-08-00246.

References

Anisimov OA, Vaughan DG, Callaghan TV, Furgal T et al (2007) Polar regions (Arctic and Antarctic). In: Parry ML, Canziani OF, Palutikof JP, van der Linden PJ, Hanson CE (eds) Climate change 2007: impacts, adaptation and vulnerability. Contribution of working group II to the fourth assessment report of the intergovernmental panel on climate change. Cambridge University Press, Cambridge, pp 653–685

Arzhanova NV, Artamonova KV (2014) Hydrochemical structure of waters in the fishery areas of Antarctic krill Euphausia superba Dana. Trudy VNIRO 152:118–132

Bogdanov MA, Oradovskii SG, Solyankin EV, Khvatskii NV (1969) The frontal zone in the Scotia Sea. Oceanology 11:966–974

Cefarelli AO, Ferrario ME, Almandoz GO, Atencio AG, Akselman R, Vernet M (2010) Diversity of the diatom genus Fragilariopsis in the Argentine Sea and Antarctic waters: morphology, distribution, and abundance. Polar Biol 33:1463–1484

Clarke KR, Gorley RN (2005) PRIMER: getting started with V. 6. PRIMER-E, Plymouth

Dafner EV, Mordasova NV (1994) Influence of biotic factors on the hydrochemical structure of surface water in the Polar Frontal Zone of the Atlantic Antarctic. Mar Chem 45(1–2):137–148. https://doi.org/10.1016/0304-4203(94)90098-1

Delgado LE, Jana R, Marin VH (1998) Testing hypotheses on life-cycle models for Antarctic calanoid copepods, using qualitative, winter, zooplankton samples. Polar Biol 20:74–76. https://doi.org/10.1007/s003000050278

Doblin MA, Petrou K, Sinutok S, Seymour JR, Messer LF, Brown MV, Norman L, Everett JD, McInnes AS, Ralph PJ, Thompson PA, Hassler CS (2016) Nutrient uplift in a cyclonic eddy increases diversity, primary productivity, and iron demand of microbial communities relative to a western boundary current. PeerJ 4:e1973. https://doi.org/10.7717/peerj.1973

Driscoll RM, Reiss CS, Hentschel BT (2015) Temperature-dependent growth of Thysanoessa macrura: inter-annual and spatial variability around Elephant Island, Antarctica. Mar Ecol Prog Ser 529:49–61

Egbert GD, Erofeeva SY (2002) Efficient inverse modeling of barotropic ocean tides. J Atmos Ocean Technol 19(2):183–204. https://doi.org/10.1175/1520-0426(2002)019<0183:EIMOBO>2.0.CO;2

Ellwood MJ, Strzepek RF, Strutton PG, Trull TW, Fourquez M, Boyd PV (2020) Distinct iron cycling in a Southern Ocean eddy. Nat Commun 11:825. https://doi.org/10.1038/s41467-020-14464-0

Fedorov KN (1983) The physical nature and structure of oceanic fronts (in Russian). Hydrometizdat, Leningrad 296 pp. (1986, English translation) Springer, 333 pp

Flores H, Atkinson A, Kawaguchi S, Krafft BA, Milinevsky G, Nicol S et al (2012) Impact of climate change on Antarctic krill. Mar Ecol Prog Ser 458:1–19. https://doi.org/10.3354/meps09831

Foster BA (1989) Time and depth comparisons of sub-ice zooplankton in McMurdo Sound, Antarctica. Polar Biol 9:431–435

Fransz HG (1988) Vernal abundance, structure and development of epipelagic copepod populations of the eastern Weddell Sea (Antarctica). Polar Biol 9:107–114

Gordon AL (1967) Structures of Antarctic waters between 20°W and 170° W. In: Bushnell VC (ed) Antarctic map folio series, Folio 6. American Geographical Society, New York, 10 pp

Hempel G (1985) Antarctic marine food web. In: Siegfried WR, Condy PR, Laws RM (eds) Antarctic nutrient cycles and food webs. Springer, Berlin, pp 266–270

Hewes CD, Reiss CS, Holm-Hansen O (2009) A quantitative analysis of sources for summertime phytoplankton variability over 18 years in the South Shetland Islands (Antarctica) region. Deep-Sea Res 56:1230–1241

Heywood KJ, Naveira-Garabato AC, Stevens DP, Muench RD (2004) On the fate of the Antarctic Slope Front and the origin of the Weddell Front. J Geophys Res 109:C06021. https://doi.org/10.1029/2003JC002053

Ikeda T (1985) Life history of Antarctic krill Euphausia superba: a new look from an experimental approach. Bull Mar Sci 37(2):599–608

Kang S-H, Fryxell GA (1993) Phytoplankton in the Weddell Sea, Antarctica: composition, abundance and distribution in water-column assemblages of the marginal ice-edge zone during austral autumn. Mar Biol 116:335–348

Kang S-H, Fryxell GA, Roelke DL (1993) Fragilariopsis cylindrus compared with other species of the diatom family Bacillariaceae in Antarctic marginal ice-edge zones. Nova Hedwigia Beih 106:335–352

Klepikov VV (1963) Hydrology of the Weddell Sea. Tr Sov Antarkt Eksp (Oroceedings of the Soviet Antarctic Expedition) 17:45–93. [in Russian]

Koshlyakov MN (2008) Discovery and investigations of open ocean synoptic eddies. Izv Atmos Ocean Phys 38:682–694

Lee W, Kim S, Kang S, Bang H, Lee K, Kwak I (2004) Distribution and abundance of zooplankton in the Bransfield Strait and the western Weddell Sea during austral summer. Ocean Polar Res 26(4):607–618. https://doi.org/10.4217/OPR.2004.26.4.607

Loeb VJ, Santora JA (2012) Population dynamics of *Salpa thompsoni* near the Antarctic Peninsula: growth rates and interannual variations in reproductive activity (1993–2009). Prog Oceanogr 96(1):93–107. https://doi.org/10.1016/j.pocean.2011.11.001

Lundholm N, Rytter Hasle G (2008) Are *Fragilariopsis cylindrus* and *Fragilariopsis nana* bipolar diatoms? – morphological and molecular analyses of two sympatric species. Nova Hedwig Beih 133:231–250

Mackintosh NA (1973) Distribution of post-larval krill in the Antarctic. Discov Rep 36:95–156

Makarov RR (1982) Ontogenetic migrations in Antarctic euphausiids: "the Antarctic". Soviet Committee Antarctic Res Rep 21:118–132. (in Russian)

Monin AS, Kamenkovich VM, Kort VG (1974) Variability of the World Ocean (in Russian). Hydrometizdat, Leningrad 262 pp. Variability of the oceans (1977, English translation) Wiley. 241 pp.

Morozov EG, Frey DI, Polukhin AA, Krechik VA, Artemiev VA, Gavrikov AV, Kasyan VV, Sapozhnikov PV, Gordeeva NV, Kobylyansky SG (2020) Mesoscale variability of the ocean in the northern part of the Weddell Sea. Oceanology 60(5):573–588

Muench RD, Gunn JT, Husby DM (1990) The Weddell scotia confluence in midwinter. J Geophys Res 95(C10):18,177–18,190

Orsi AH, Nowlin WD, Whitworth T (1993) On the circulation and stratification of the Weddell Gyre. Deep-Sea Res I Oceanogr Res Pap 40(1):169–203

Patterson SL, Sievers HA (1980) The Weddell-Scotia confluence. J Phys Oceanogr 10(10):1584–1610

Rhines PB (2001) Mesoscale eddies. In: Encyclopedia of ocean sciences, 2nd edn, p 2001

Schloss IR, Ferreyra GA, Ruiz-Pino D (2002) Phytoplankton biomass in Antarctic shelf zones: a conceptual model based on Potter Cove, King George Island. J Mar Syst 36:129–143

Schröder M, Hellmera HH, Absy JM (2002) On the near-bottom variability in the northwestern Weddell Sea. Deep-Sea Res II 49:4767–4790

Siegel V, Watkins JL (2016) Distribution, biomass and demography of Antarctic krill, Euphausia superba. In: Siegel V (ed) Biology and ecology of Antarctic Krill, Advance in polar ecology series. Springer, New York, pp 21–100. https://doi.org/10.1007/978-3-319-29279-3_2

Sprong I, Schalk PH (1992) Acoustic observations on krill spring-summer migration and patchiness in the northern Weddell Sea. In: Hempel G (ed) Weddell sea ecology. Springer, New York, pp 261–268. https://doi.org/10.1007/978-3-642-77595-6_30

Stammerjohn SE, Martinson DG, Smith RC, Yuan X, Rind D (2008) Trends in Antarctic annual sea ice retreat and advance and their relation to El Nino Southern Oscillation and Southern Annular Mode variability. J Geophys Res Oceans 113:C03S90

Stepanova S, Polukhin A, Borisenko G, Chultsova A, Marina E, Popov O, Seliverstova A, Vidnichuk A, Tishchenko P (2021) Hydrochemical structure of waters in the northern Weddell Sea in austral summer 2020 (Chapter 11). In: Antarctic Peninsula region of the southern ocean. Springer, Cham

Thompson AF, Heywood KJ, Thorpe SE, Renner AHH, Trasvina A (2009) Surface circulation at the tip of the Antarctic Peninsula from drifters. J Phys Oceanogr 39:3–26. https://doi.org/10.1175/2008jpo3995.1

Voronina NM (1998) Comparative abundance and distribution of major filter-feeders in the Antarctic pelagic zone. J Mar Syst 17(1–4):375–390. https://doi.org/10.1016/S0924-7963(98)00050-5

Chapter 4
Water Masses, Currents, and Phytoplankton in the Bransfield Strait in January 2020

Eugene G. Morozov, Dmitry I. Frey, Viktor A. Krechik, Alexander A. Polukhin, and Philipp V. Sapozhnikov

Abstract Two regions exist in the Bransfield Strait: the region of the Bransfield Strait along the South Shetland Islands with Transitional Zonal Water with Bellingshausen Sea influence and the region along the northeastern coast of the Antarctic Peninsula with colder and more saline Transitional Zonal Water with Weddell Sea influence. The narrow jet stream flows along the islands and transports relatively warm water from the Bellingshausen Sea. Most of the strait is occupied by cold waters from the Weddell Sea.

Keywords Bransfield Strait · South Shetland Islands · Temperature · Salinity · Currents · Dissolved oxygen · Phytoplankton

4.1 Introduction

The Bransfield Strait is an important region of circulation in the Southern Ocean. Part of the waters of the Antarctic Circumpolar Current (ACC) flow through this strait. The relatively warm waters of the southern jets of the ACC enter the Bransfield Strait from the southwest. In the strait, they meet the cold waters from the Weddell Sea. Thus, high gradients of temperature and density are formed, which leads to the intensification of the current in the Bransfield Strait that follows along the South Shetland Islands.

As a result, two main water masses are formed in the strait: warm and relatively fresh (0.5–3.0°C and 33.1–33.9 PSU in summer) Transitional Zonal Water with Bellingshausen Sea influence (TBW) and cold and more saline Transitional Zonal Water with Weddell Sea influence (TWW) usually with negative temperatures and

E. G. Morozov (✉) · D. I. Frey · V. A. Krechik · A. A. Polukhin · P. V. Sapozhnikov
Shirshov Institute of Oceanology, Russian Academy of Sciences, Moscow, Russia

E. G. Morozov et al. (eds.), *Antarctic Peninsula Region of the Southern Ocean*, Advances in Polar Ecology 6, https://doi.org/10.1007/978-3-030-78927-5_4

salinity (34.1–34.6 PSU in summer) (Garcıa et al. 2002; Huneke et al. 2016). The deeper layers of the Bransfield Strait are occupied with Bransfield Strait Deep Water (BSDW). This water is the result of isopycnal mixing of the Weddell Sea deep waters with the winter and summer waters from the shelf areas. This water occupies the layer below WSTW (Gordon et al. 2000; Huneke et al. 2016). The upper limit of this water is usually identified as potential temperature −1.0 °C and salinity 34.5 PSU (Garcıa et al. 2002).

Moffat and Meredith (2018) write that the hydrography of Bransfield Strait is characterized by these three distinct water masses. The major part of the water volume in the strait is occupied by a mixture of shelf and deep Weddell Sea waters with temperatures below 0 °C.

A relatively stationary front is observed between two surface water masses in the strait that separates TBW and TWW in the depth range of 100–500 m. The Bransfield Current forms a narrow (up to 20 km) TBW jet, spreading to the northeast along the northern boundary of the strait (Sangrà et al. 2011; Morozov 2007; Zhou et al. 2002, 2006).

4.2 Data

In January 2020, during cruise 79 of the R/V "Akademik Mstislav Keldysh," a cross section of nine stations was occupied across the Bransfield Strait. The distance between stations varied from 1.5 to 12 miles (Fig. 4.1). Measurements of temperature and salinity were made using an SBE 911 probe; they were combined with the

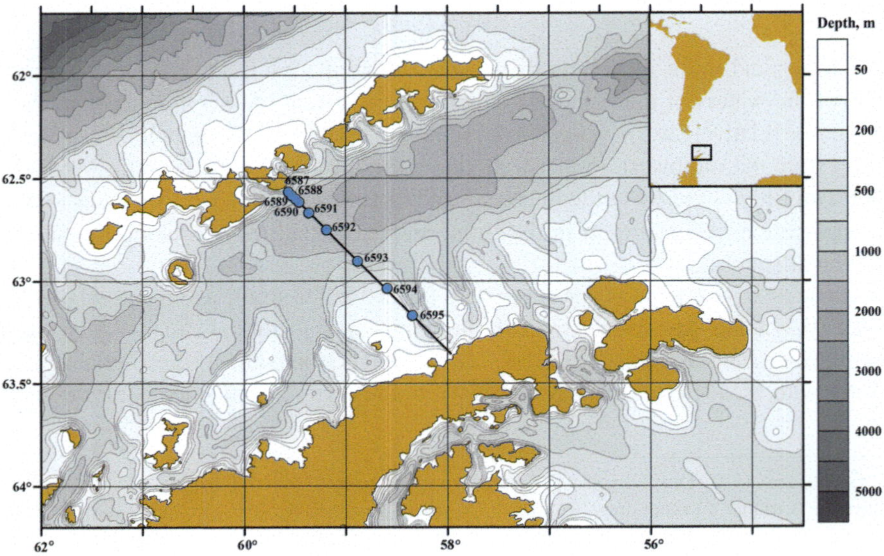

Fig. 4.1 Study site in the Bransfield Strait and the scheme of stations in 2020

measurements of currents using a Teledyne RDI Workhorse Monitor 300 kHz profiler (LADCP). During the processing, tidal currents were removed using the TPXO 9 model (Egbert and Erofeeva 2002). Hydrochemical properties were measured in water samples collected in the Niskin bottles at the stations. The length of the section from Greenwich Island (South Shetland Islands) to the shelf of the Antarctic Peninsula was 93 km, and the maximum depth is 1436 m.

Dissolved oxygen in seawater was analyzed using a modified Winkler method using a Brinkman/Dosimate-765 microburette. Analysis for alkalinity was performed by direct titration in an open cell with hydrochloric acid (0.02N) with a mixed indicator (methyl red + methylene blue).

Determination of the content in the samples of dissolved mineral phosphorus, silicates, and forms of mineral nitrogen (ammonia, nitrites, nitrates) was carried out according to the methods commonly accepted in oceanological practice.

Phytoplankton samples were taken using a modification of the Apstein net with a ring diameter of 50 cm and a cone length of 1 m, with a mill gas mesh size of 37μm. The net was lowered to a depth of 50 m and then lifted at a speed of 0.5 m/s. At each station, three samples were taken. The further processing of plankton material is described in Sapozhnikov et al. (2021) (this volume, Chap. 21).

4.3 Temperature, Salinity, and Currents

A layer of relatively warm water (1.5–1.9 °C) was found in the upper part of the water column up to the depths of 30–40 m (Fig. 4.2). This layer was as wide as 65 km. A surface flow of even warmer waters with a maximum temperature equal to

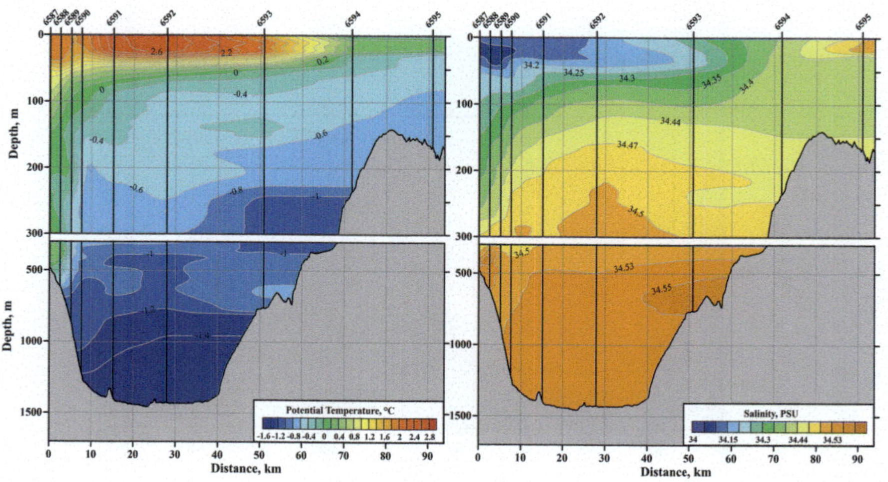

Fig. 4.2 Distribution of potential temperature (°C) (**a, left panel**) and salinity (psu) (**b, right panel**) over the section in the Bransfield Strait

2.77 °C was found in the middle of this layer. This flow was 15-km wide between stations 6591 and 6592. The isotherms, especially the 0 °C isotherm, deepened to the northwest. The surface salinity minimum is displaced closer to the slope of the islands compared to the surface core of warm water. Near the continental slope, the isotherms were located almost vertically, denoting the Bransfield Strait Front. The inclination of isohalines is not as vertical as the slope of isotherms, which are almost vertical. Warm water of the southern ACC branch with a temperature of 0.00–0.26 °C and a salinity of 34.50–34.54 psu is transported here from the Bellingshausen Sea by a jet of the Bransfield Current in the depth range up to 500 m, where the deep water of the southern branch of the ACC occurs. A mass of water with temperatures from −0.76 to −0.90 °C and the maximum salinity observed in this region (34.55–34.56 psu) was found in the central part of the section over the slope of the Antarctic Peninsula. This water can be classified as the water from the eastern part of Bransfield Strait in the depth range of 700–1000 m (Gordon et al. 2000).

The jet of the Bransfield Current is clearly distinguished in the dynamic structure of the strait in Fig. 4.3. The current spreads to a depth of 500 m and even deeper near the slope of the South Shetland Islands. The velocities in its core range from 0.2 to 0.5 m/s. The maximum velocities (0.57 m/s) were recorded by LADCP

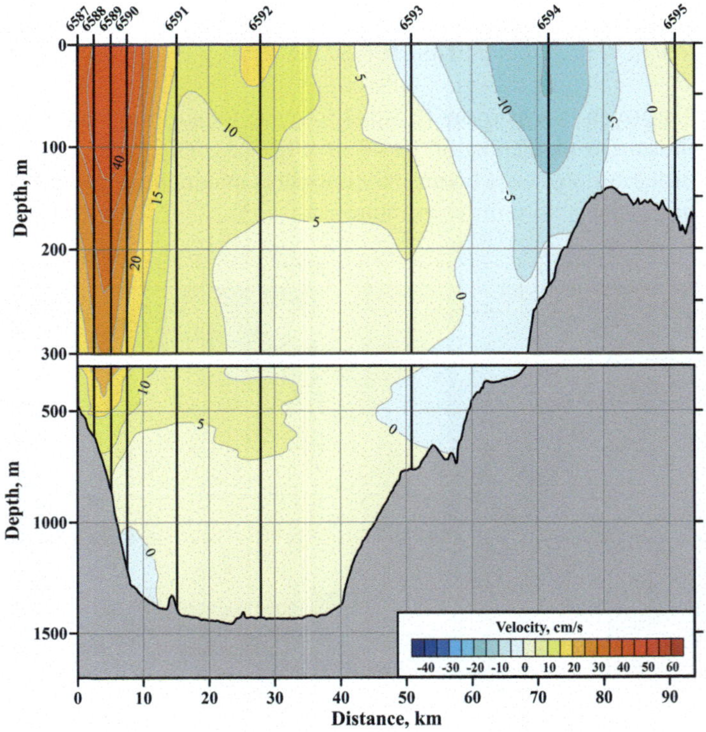

Fig. 4.3 Distribution of the current velocity across the section (in m/s). The northeastern direction is positive

measurements near the surface. A slow current in the southwestern direction with velocities not exceeding 0.12 m/s was revealed over the shelf of the Antarctic Peninsula. Despite the general flow in the southwestern direction, a narrow current was found at station 6594 just over the shelf break of the Antarctic Peninsula. Its velocities were 0.05–0.19 m/s. The entire slope of the peninsula in the depth range of 350–630 m is occupied by the current in the southeastern direction. The velocity of this current is not high, ranging from 0.01 to 0.04 m/s.

Our LADCP measurements in the Bransfield Strait are consistent with the previous reports (Niiler et al. 1991; Zhou et al. 2002, 2006; Sangra et al. 2011) The total transport of the flow directed to the northeast based on the field data of LADCP measurements was 5.03 Sv. The transport by the Bransfield Current was 1.27 Sv. We determine the boundary of the Bransfield Current by the sharp slope of the velocity isolines between stations 6590 and 6591 (near 62°39′ S, 59°25′ W).

The current to the southwest is concentrated in two narrow jets above the shelf break (−0.31 Sv) and at the slope (−0.18 Sv) of the Antarctic Peninsula. The transport by the Bransfield Current exceeds the values reported in Gomis et al. (2002) in the western part of the strait (0.5 Sv), but it agrees well with the estimate of 1 Sv in the central and eastern parts of the strait (Lopez et al. 1999). The differences in measurements in 2005 (Morozov 2007) and 2020 are explained by the strong variability of the flow between measurements taken at different times. In November 2005, the 0.1 m/s current velocity isoline was at a depth of 250 m. The zero velocity isoline was recorded at a depth of 800 m. In January 2020, the 0.1 m/s velocity isoline reached a depth of 800 m. Deeper than 800 m, there were currents with velocities of about 0.1 m/s. The core of the current in 2020 was located deeper, and the current became wider.

Measurements of currents in the central part of the Bransfield Strait revealed that a significant part of the water column was moving to the northeast transported by the current approximately 15-km wide and 500-m thick. The transport was as high as 5 Sv. The transport to the southwest was one order of magnitude smaller (0.49 Sv). The transport by the jet of the Bransfield Current was only a quarter of the total transport. The rest of the transport (about 3.75 Sv) occurred as a wide flow at low velocities between the jet of the Bransfield Current and two jets directed to the southwest.

4.4 Hydrochemical Structure of Waters

The hydrochemical structure of the waters of the Bransfield Strait confirms the presence of different water masses in terms of hydrophysical parameters. The vertical oxygen distribution is a good marker of the difference in water masses in Antarctica. The southern stream of the ACC enters the Bransfield Strait. Its warm waters deepen to a depth of 400 m and spread to the northeast. The core of low oxygen concentrations (less than 6 ml/l) is distinguished by the vertical oxygen distribution (Fig. 4.4a) at a depth of 350–400 m in the northern part of the section.

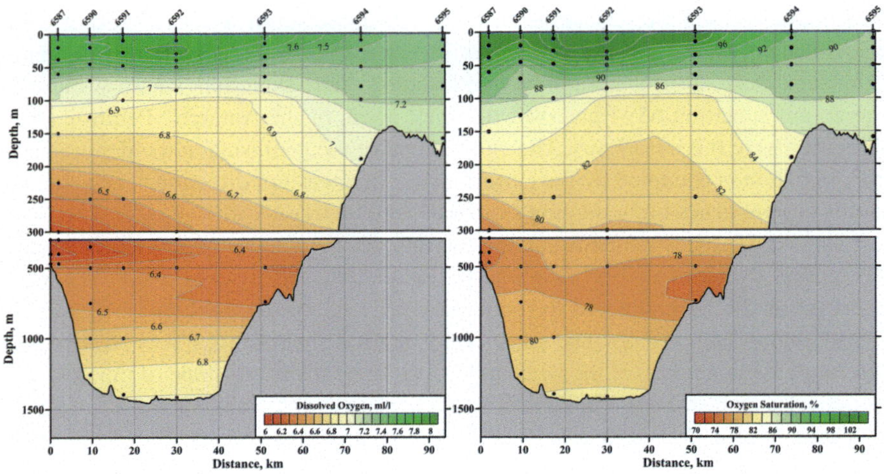

Fig. 4.4 Concentration of dissolved oxygen (ml/l) (**a, left panel**) and saturation with oxygen (**b, right panel**) over the section

In the northeastern part of the Bransfield Strait, surface saline waters from the Weddell Sea flow through the Antarctic Sound and around the Antarctic Peninsula (Garcıa et al. 2002; Gomis et al. 2002). The surface waters of the southern part of the section across the Bransfield Strait are characterized by oxygen concentrations ranging from 7.3 to 7.8 ml/l (Fig. 4.4a), which are also observed in surface waters in the section across the Antarctic Sound Strait. These waters are slightly undersaturated with oxygen due to their increased salinity; the degree of saturation is about 95% (Fig. 4.4b). The shelf waters of the South Shetland Islands in the northwestern part of the Bransfield Strait are subject to local warming and freshwater runoff from the northwest, which is also reflected in the oxygen distribution in the northern part of the section across the Bransfield Strait. Here, the oxygen concentration in the near-surface and surface waters reaches 7.8–8.0 ml/l (Fig. 4.4a), and the degree of oxygen saturation of the waters is 100–104% (Fig. 4.4b).

The range of variations in total alkalinity (TA) was 2.316–2.342 mmol/kg. The normalized total alkalinity (NTA = TA * 35/S) of waters in the northern part of the section was not high (Fig. 4.5). These waters should be classified as shelf waters of the South Shetland Islands. The waters flowing from the Weddell Sea were characterized by high salinity and, as a consequence, by low values of normalized alkalinity. The range of NTA variability was 2.365–2.377 mmol/kg.

Concentration of phosphates was not determined in some of the samples because of high content of suspended matter (probably nutrients) based on the data of the turbidity sensor. In the southern part of the section, on the shelf of the Antarctic Peninsula (station 6595), the phosphate content in the 25–159 m layer ranged from 2.17 to 2.22 μM.

Silicate content in the waters of the study site varied from 69 to 84.5 μM. In the upper 100 m layer, concentration of silicates varied from 74 to 81 μM. In the central and southern parts of the section, silicate concentration ranged from 68 μM in the

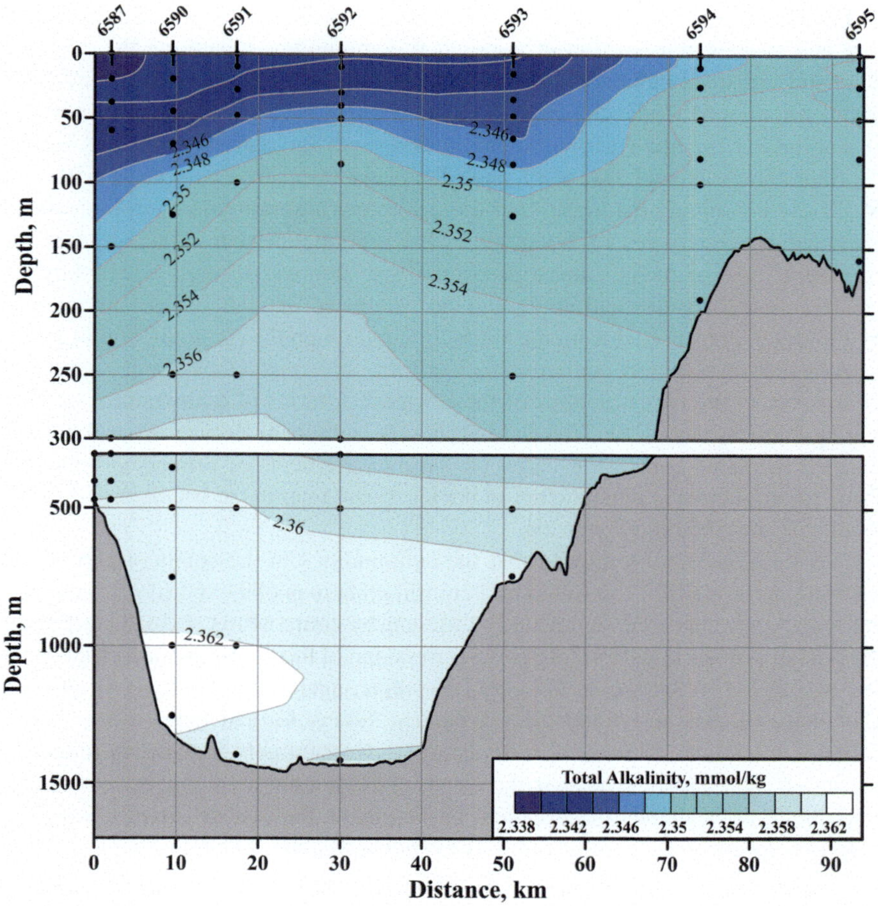

Fig. 4.5 Normalized alkalinity (mmol/kg) over the section

surface layer at station 6594 to 82 μM in the bottom layer at station 6593. Circumpolar Deep Water occupied the depth range 400–500 m based on silicate content (83 μM). Distribution of silicates at station 6594 in the flow from the Antarctic Sound was homogenous deeper than 15 m. The concentration of silicates increased to 76–79 μM near the Antarctic Peninsula, which is associated with the transport from the rocks composing the coast of the peninsula.

4.5 Pelagic Microalgae Communities

Structural features of phytopelagic cenoses at stations of the section across the Bransfield Strait made it possible to distinguish four types of communities, the development of which is macrotopologically related to the heterogeneity of water body in the strait.

Four main types of cenoses were identified according to the quantitative structure and species composition of microalgal communities inhabiting the euphotic zone and statistical indicators of their similarity and difference: "northern coastal" found at station 6587, the closest to the South Shetland Islands; "southern coastal" – at stations on the shelf of the Antarctic Peninsula (at stations 6594 and 6595); "northern" (at stations 6588, 6589, and 6590); and "southern" (at stations 6591 and 6593). By combining the location of these communities with the hydrological and chemical characteristics of water along the section, one can conclude that the "northern" type of cenoses developed in the zone area was strongly affected by the jet streams of the Bransfield Strait, and the "southern" one in the area is influenced by the waters transported from the Weddell Sea through the Antarctic Sound Strait.

Analysis of plankton communities confirms the conclusions about significant differences in the two structures of these waters in terms of temperature, salinity, and hydrochemistry of the Bransfield Strait. The authors of Romanova et al. (2021) reached the same conclusion based on the measurements of primary production. Primary production is much higher in the jet stream than in the southern part of the strait near the Antarctic Peninsula.

Coastal observations showed that the communities in the southern part of the section, on the shelf of the Antarctic Peninsula (stations 6594 and 6595, "southern coastal" type), appeared significantly different by composition and structure from the coastal cenosis in its northern part, on the slope of the South Shetland Islands (at station 6587, "northern coastal" type). However, against the background of high abundance of diatom *Fragilaria* cf. *islandica*, which formed long ribbon-shaped colonies in all three habitats, all coastal cenoses were initially statistically related to one group. The statistical noise in the division of communities along the section was also due to the abundant development of the planktonic diatom *Corethron pennatum*; this species was abundant almost everywhere. After excluding *C. pennatum* cenosis groups from the analysis, the similarity of the "northern coastal" cenosis with the "southern coastal" matched only up to a level of 26.63% by structure (according to the Bray-Curtis index) and 37.63% by composition (according to the Sørensen index). Thus, according to the structure of phytoplankton, four types of cenoses can be eventually distinguished, considering the northern and southern coastal cenoses as independent types.

Numerous short fragments of *Fragilariopsis sublinearis* colonies, representative of the ice flora, as well as chains of *Pseudo-nitzschia turgiduloides* oblong cells, a typical planktonic diatom with a complex of adaptations for floating in water, but richly propagating itself within aggregates as well, and colonizing "clouds" of bacteria populations were found at coastal stations in the southern part of the section within the composition of aggregates (pelagic conglomerates developing on the chaetae of large diatoms and having their own genesis from colonial bacteria and microalgae growing on them). In turn, the benthic diatom *Licmophora gracilis* developed abundantly in aggregates at the northern coastal station. In the cenoses of the "northern" type, aggregates were characterized by the abundance of heterotrophic flagellates *Cafeteria roenbergensis* and *Bodo saltans*, as well as the diatoms

Fragilariopsis nana; flagellates *Bodo designis*, *C. roenbergensis*, and *B. saltans* were found in the cenoses of the "southern" type.

The highest richness of aquatic macrophytes species was recorded at the southernmost point: 29 species grew here, 14 of which were recorded only within the composition of aggregates. There was no pronounced gradient showing any decrease in the number of species from south to north along the section. The smallest number of species was recorded in the cenoses of the "southern" type (4–5 and 3 in aggregates). They were more abundant at the stations affected by the jet stream of the Bransfield Strait (the "northern" type: 7–11 and 6–8 in aggregates). Nine species were found at the northern coastal station (five in aggregates). Cenoses of the "northern" type were more diverse than the "southern" ones, but it was the opposite in the coastal zones: the taxonomic diversity of the community on the Antarctic Peninsula shelf was significantly higher than on the slope of the South Shetland Islands.

Thus, we can conclude that phytopelagic cenoses over the section across the Bransfield Strait also reflected and confirmed the general difference in the water masses and environmental conditions.

4.6 Conclusions

Two regions can be distinguished in the Bransfield Strait based on the dynamics of water, as well as thermohaline and hydrochemical structure: the region of the Bransfield Strait jet stream along the South Shetland Islands and the region along the northeastern coast of the Antarctic Peninsula. The jet stream is about 15-km wide and up to 600-m deep with an average speed of 30–40 cm/s, while the maximum measured speed is 47 cm/s. Deeper than 100 m, the current weakens, and at a depth of 500 m the measured velocity was 20 cm/s. The current carries relatively warm waters from the Bellingshausen Sea. Most of the strait is occupied by modified waters from the Weddell Sea, the temperature of which is less than 0 °C, and the salinity is higher than the salinity in the jet stream. Hydrophysical and hydrochemical characteristics of waters, as well as differences in phytoplankton in different parts of the section through the strait, indicate a significant difference in water masses in the strait. An inflow of water with high concentration of nutrients and transport of phosphates into the photic layer in the northwestern part of the section were observed. The currents recorded in 2020 appeared deeper, and wider than in 2005.

Acknowledgments The study was carried out within the framework of the state assignment of the Shirshov Institute of Oceanology (No. 0128-2021-0017) and RFBR grant 20-08-00246. Analysis of CTD and ADCP data performed by DIF was supported by research project MK-1492.2021.1.5.

References

Egbert G, Erofeeva S (2002) Efficient inverse modeling of barotropic ocean tides. J Atmos Ocean Technol 19(2):183–204

Garcıa M, Castro C, Rıos A et al (2002) Water masses and distribution of physico-chemical properties in the Western Bransfield Strait and Gerlache Strait during austral summer 1995/96. Deep Sea Res Part II 49:585–602

Gomis D, Garcıa M, Lopez O et al (2002) Quasi-geostrophic 3D circulation and mass transport in the western Bransfield Strait during austral summer 1995/96. Deep Sea Res Part II 49:603–621

Gordon A, Mensch M, Dong Z et al (2000) Deep and bottom water of the Bransfield Strait eastern and central basins. J Geophys Res Oceans 105(C5):11337–11346

Huneke WGC, Huhn O, Schröeder M (2016) Water masses in the Bransfield Strait and adjacent seas, austral summer 2013. Polar Biol 39:789–798

Lopez O, Garcıa M, Gomis D et al (1999) Hydrographic and hydrodynamic characteristics of the eastern basin of the Bransfield Strait (Antarctica). Deep Sea Res Part I 46(10):1755–1778

Moffat C, Meredith M (2018) Shelf–ocean exchange and hydrography west of the Antarctic Peninsula: a review. Philos Trans R Soc A 376:20170164. https://doi.org/10.1098/rsta.2017.0164

Morozov EG (2007) Currents in Bransfield Strait. Doklady Earth Sci 415(6):984–986

Niiler P, Amos A, Hu J-H (1991) Water masses and 200 m relative geostrophic circulation in the western Bransfield Strait region. Deep-Sea Res II 38:943–059

Romanova ND, Mosharov SA, Vorobieva OV, Bardukova EB (2021) Quantitative and production characteristics of microplankton in the Powell Basin and Bransfield Strait in summer (Chapter 14). In: Antarctic Peninsula Region of the Southern Ocean. Springer, Cham

Sangrà P, Gordo C, Hernández-Arencibia M et al (2011) The Bransfield Current system. Deep Sea Res Part I 58(4):390–402

Zhou M, Niiler PP, Hu J-H (2002) Surface currents in the Bransfield and Gerlache straits, Antarctica. Deep-Sea Res 49:267–280

Zhou M, Niiler PP, Zhu Y, Dorland RD (2006) The western boundary current in the Bransfield Strait, Antarctica. Deep-Sea Res Part I 53:1244–1252

Chapter 5
Intra-annual Variability of Water Structure in the Atlantic Sector of the Southern Ocean Based on the ECMWF ORA-S3 and OI SST Reanalysis

Yuri V. Artamonov, Elena A. Skripaleva, Alexander V. Fedirko, and Nikolay V. Nikolsky

Abstract Intra-annual variations of the surface temperature field in the western part of the Atlantic sector of the Southern Ocean on seasonal and synoptic scales are considered based on the NOAA OI SST and ECMWF ORA-S3 reanalysis. The relationship between areas of extreme temperature and salinity variability and the areas of their high horizontal gradients is analyzed. It is shown that the highest level of both seasonal and synoptic variability is observed over the coastal shelf of South America. In the open ocean, intense synoptic variability of sea surface temperature can be traced in the areas of maximum horizontal temperature gradients, and its maximum level is observed in the South Subtropical, Subantarctic, and Antarctic Polar frontal zones. The value of the linear correlation coefficient between the mean annual values of the root mean square deviation of the SST synoptic and total gradient of the SST in the water basin generally reaches 0.82. The level of SST seasonal variability is 4–4.5 times higher than the synoptic level in the southern part of the Drake Passage and in the South Subtropical Anticyclonic Gyre zone, it is 4–6 times higher over the coastal shelf of South America, and 2–3 times higher in frontal zones. The minimum values of the ratio of the levels of seasonal and synoptic variability of SST (2–2.5) are observed in an elongated strip covering the northern part of the Drake Passage and the northwestern part of the Scotia Sea, where a low level of SST seasonal variability and an increased level of its synoptic variability are noted. A high level of seasonal variability of temperature and salinity was found in the South Subtropical Front zone, south of the Antarctic Polar Front, and north of

Y. V. Artamonov (✉) · E. A. Skripaleva · A. V. Fedirko · N. V. Nikolsky
Marine Hydrophysical Institute, Russian Academy of Sciences, Sevastopol, Russia

E. G. Morozov et al. (eds.), *Antarctic Peninsula Region of the Southern Ocean*, Advances in Polar Ecology 6, https://doi.org/10.1007/978-3-030-78927-5_5

the Subantarctic Front. The greatest penetration depth of seasonal fluctuations in temperature and salinity fields is observed in the South Subtropical Front, where it reaches 400–600 m and 500–700 m, respectively.

Keywords Atlantic sector of the Southern Ocean · Temperature · Salinity · Seasonal and synoptic variability · Horizontal gradient · Frontal zones

5.1 Introduction

Studies of the variability of water structure in the Southern Ocean on different time scales are among the most urgent problems of modern oceanographic research (Maslennikov 2003; Turner 2004; Simmonds and King 2004; Anisimov et al. 2007; Meredith et al. 2008; Sallée et al. 2008). In recent years, due to the increase of the anthropogenic load on the Antarctic water basin, the need to solve a number of environmental problems associated with assessing the quality of the state of Antarctic ecosystems has increased (Anisimov et al. 2007). The most important abiotic factors that determine the level of biological productivity of waters and affect the state of oceanic ecosystems are thermohaline parameters and hydrological fronts (Sokolov and Rintoul 2007; Venables et al. 2012; Chapman et al. 2020).

Despite a large number of works devoted to the study of the water structure of the Southern Ocean, information on its variability in the deep layers is extremely limited. The data are related to individual areas of the ocean, where a sufficiently large number of hydrological stations have been occupied (such as the Drake Passage, the Scotia and Cooperation seas, and water basins south of Africa and Australia), which makes it possible to draw some conclusions about the peculiarities of the variability of hydrological parameters in the subsurface layers (Valentine et al. 1993; Rintoul et al. 1997; Antipov and Klepikov 2007; Naveira Garabato et al. 2009; Golovin et al. 2011; Antipov et al. 2019; Ionov 2015; Artamonov and Skripaleva 2016). At the same time, the use of various types of data by numerous authors, which are extremely various in space and time, and the use of different methods for their processing and analysis strongly limit the possibility of correct comparison of the results, and, thus, the representations of the variability of thermohaline fields in the South Ocean as a whole and its regional features complicate statistical estimates of the reliability of the revealed regularities.

The appearance of multiple oceanic reanalyses in recent years, which assimilate all available data of contact and satellite measurements and interpolate them into regular grids with identical step, for example, ECMWF ORA, allows us to analyze the large-scale water structure and its seasonal variability – not only at the surface but also in ocean deep layers. In addition, modern reanalyses at the ocean surface with high temporal resolution (daily average) make it possible to estimate the variability of hydrological fields on the mean long-term and synoptic (intra-monthly) scale, which still remains poorly studied.

In this paper based on modern reanalysis ECMWF ORA-S3 and NOAA OI SST, the features of the mean long-term seasonal and synoptic variability of the water structure in the Western Atlantic sector of the Southern Ocean are considered, and the relationship between the intra-annual variability of thermohaline parameters and boundaries between water masses (hydrological fronts) is estimated.

5.2 Data and Methods

The study region is the western part of the Atlantic sector of the Southern Ocean between 30°S and the Antarctica coast and between 30°W and 70°W. We used the daily averaged sea-surface temperature (SST) values at the nodes of the 0.25° regular grid from the NOAA OI SST reanalysis (ftp://eclipse.ncdc.noaa.gov/pub/OI-daily-v2) over the period 1982–2017. When obtaining daily averaged SST values, the method of optimal interpolation of satellite and contact temperature measurements was used, which made it possible to obtain SST data even in the case when the ice-free surface area within the trapezoid of 0.25° × 0.25° was only 5% (Reynolds et al. 2007). According to Monin (1999), a time period of more than 30 years is sufficient to obtain statistically significant climatic norms, and the average daily resolution makes it possible to analyze the synoptic SST variability on an intra-monthly scale. Time series of the SST monthly mean values were obtained at each node of the regular grid based on the daily mean values. To estimate the level of SST seasonal variability by these series, the SST root-mean-square deviations (RMSD) were calculated for each year, which were then averaged over 36 years (RMSDseas SST). Synoptic (intra-monthly) SST variability was estimated from synoptic RMSD, which were calculated by the daily averaged SST values for every month for each year and then averaged over 36 years for January, February, etc. (climatic monthly RMSDsyn values) and over all 432 months (long-term averaged RMSDsyn).

To analyze the seasonal variability of thermohaline parameters in the subsurface layers, we used the monthly mean values of temperature (T) and salinity (S) in the 1-degree grid nodes at depths 5, 15, 25, 35, 45, 60, 75, 90, 105, 120, 135, 150, 165, 180, 195, 211,25, 230, 255, 290, 335, 415, 535, 720, 1020, and 1450 m from the oceanic reanalysis ECMWF ORA-S3 (http://apdrc.soest.hawaii.edu/data/data.php) over the period 1959–2009. The seasonal RMSD of temperature and salinity were calculated at regular grid nodes at each level from the surface to 1450 m. Their vertical distributions were used to analyze the penetration depth of seasonal variations in the temperature and salinity fields, for which the depth was conventionally taken, where the values of the RMSD of parameters decreased by an order of magnitude compared to the surface.

Gradients of temperature and salinity were calculated at each depth from 5 m up to 1450 m, horizontal (meridional MG, zonal ZG and total TG) based on the NOAA OI SST data at the surface and based on the ORA-S3 data to estimate the relationship between the location of areas of the maximum intra-annual variability of temperature, salinity, and the spatial distribution of the penetration depth of seasonal variations in thermohaline fields focusing on the location of water mass boundaries (large-scale fronts). The spatial location of large-scale fronts was determined from the distributions of total horizontal gradients and was specified from the location of the MG extremes on the samples of their values along multiple meridians. A scheme of the location of main temperature fronts on the ocean surface against the background of bottom topography was constructed based on the GEBCO data (http://www.gebco.net) for the optimal choice of meridional profiles crossing the main structural zones of the South Atlantic and interpretation of the vertical distributions of thermohaline characteristics and their intra-annual variability in connection with the bottom topographical features.

5.3 Results

The spatial distributions of long-term mean SST RMSDseas (Fig. 5.1a) and SST RMSDsyn (Fig. 5.1b) show that the general increase in the level of SST intra-annual variability (RMSDseas ~3.2–4.8 °C, RMSDsyn ~0.7–1.1 °C) is observed over the South America coastal shelf. This is due to the fact that the main processes affecting the SST variability (heating and cooling, river runoff effect, intense coastal water dynamics, and surge phenomena) are more pronounced here at all scales of variability. At the same time, the spatial distributions of the maximum values of RMSDseas and RMSDsyn vary somewhat. The region of waters with the maximum SST RMSDseas values is located along the South America coast at depths shallower than 200–300 m (Fig. 5.1a), while waters with high RMSDsyn values are localized deeper in the Falkland and Brazil Currents Convergence zone and further extend to the east in the form of two quasi-zonal "tongues": along the southern periphery of the Southern Subtropical Anticyclonic Gyre (SSAG) and along the northern border of the Falkland Plateau (Fig. 5.1b). Closer to the central part of SSAG the RMSDseas, values increase (up to 2.2 °C), while the RMSDsyn values in this area notably decrease and do not exceed 0.6 °C. On the southern periphery of the SSAG, the level of SST seasonal variability, as well as the synoptic one, increases, and the values of RMSDseas increase to 2.6–2.8 °C (Fig. 5.1a). An increase of the level of SST intra-annual variability is noted in the Drake Passage and northwestern Scotia Sea (RMSDseas ~1.4–1.6 °C, RMSDsyn ~0.6–0.7 °C) (Fig. 5.1a, b). In general, according to the distribution of SST TG values (Fig. 5.1d), an increased level of SST synoptic variability is observed in high-gradient zones (frontal zones), while the position of areas of increased SST seasonal variability does not always coincide with the location of frontal zones.

The lowest level of SST intra-annual variability is observed in the Weddell Sea, where there is a constant presence of pack ice. Here, the values of RMSDseas and RMSDsyn do not exceed 0.6 and 0.15 °C, respectively (Fig. 5.1a, b).

The distribution of the RMSDseas/RMSDsyn ratio values shows that in most of the water area, the level of SST seasonal variability exceeds 3–3.5 times the level of its synoptic variability (Fig. 5.1c). The RMSDseas/RMSDsyn values increase in the

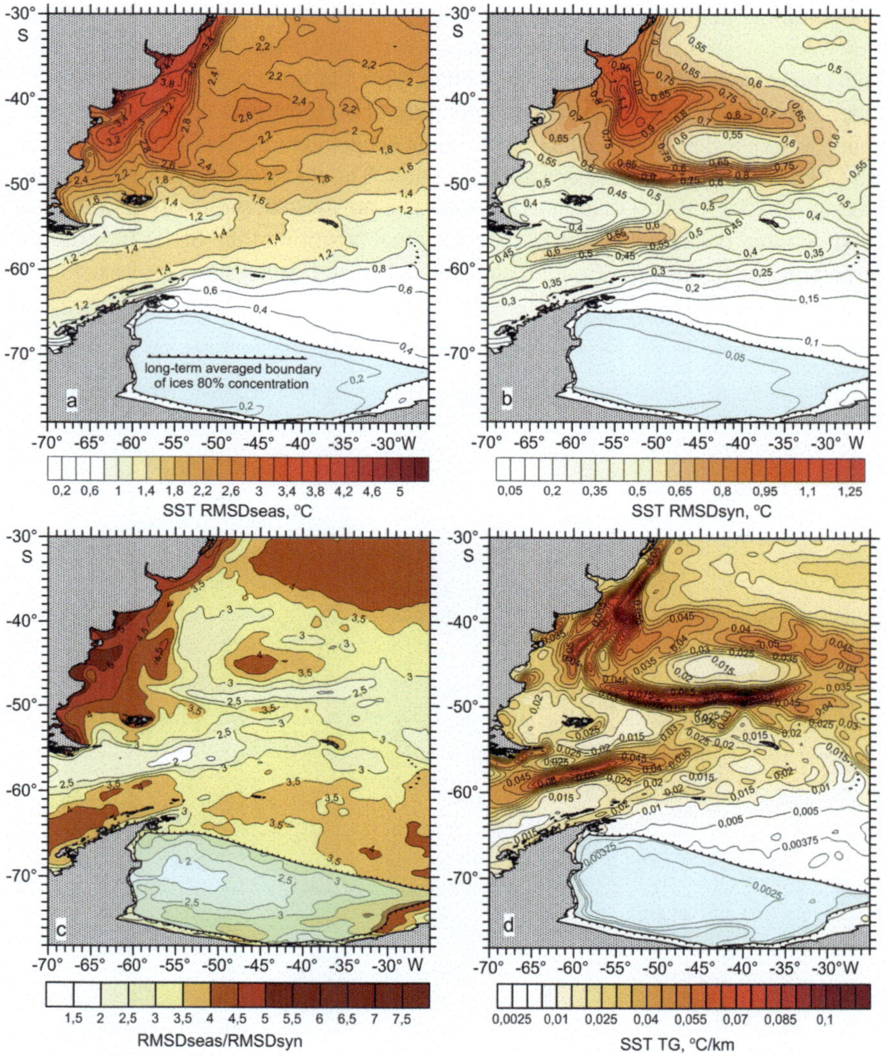

Fig. 5.1 Spatial distributions of the values of SST RMSDseas (**a**), RMSDsyn (**b**), RMSDseas/RMSDsyn (**c**), TG (**d**) according to NOAA OI SST data

southern Drake Passage (up to 4–4.5), over the South America coastal shelf (up to 4–6), and in the SSAG zone (up to 4–4.5). The minimum RMSDseas/RMSDsyn values (2–2.5) are observed in an elongated strip covering the northern Drake Passage and the northwestern Scotia Sea, where a low level of SST seasonal variability and an increased level of its synoptic variability are noted (Fig. 5.1c). In the areas of increased mean annual SST TG corresponding to the frontal zones, there is a slight decrease of the RMSDseas/RMSDsyn values; in these areas the level of SST seasonal variability is 2–3 times higher than the synoptic variability level.

For a more detailed analysis of water structure variability, three meridional profiles were selected, which, in our opinion, on the one hand, intersect all the main fronts presented in Fig. 5.2, and on the other, they characterize the variability of the water structure in various orographic conditions: in the Drake Passage and in the South America shelf zone (59.875°W), in the deeper central (49.875°W), and eastern (39.875°W) parts of the considered region. The distributions of the long-term average SST MG, RMSDseas, and RMSDsyn along these meridians (Fig. 5.3) make possible more detailed analysis of the latitudinal position of the SST MG extremes corresponding to the main fronts with the position of the regions of increased levels of SST seasonal and synoptic variability.

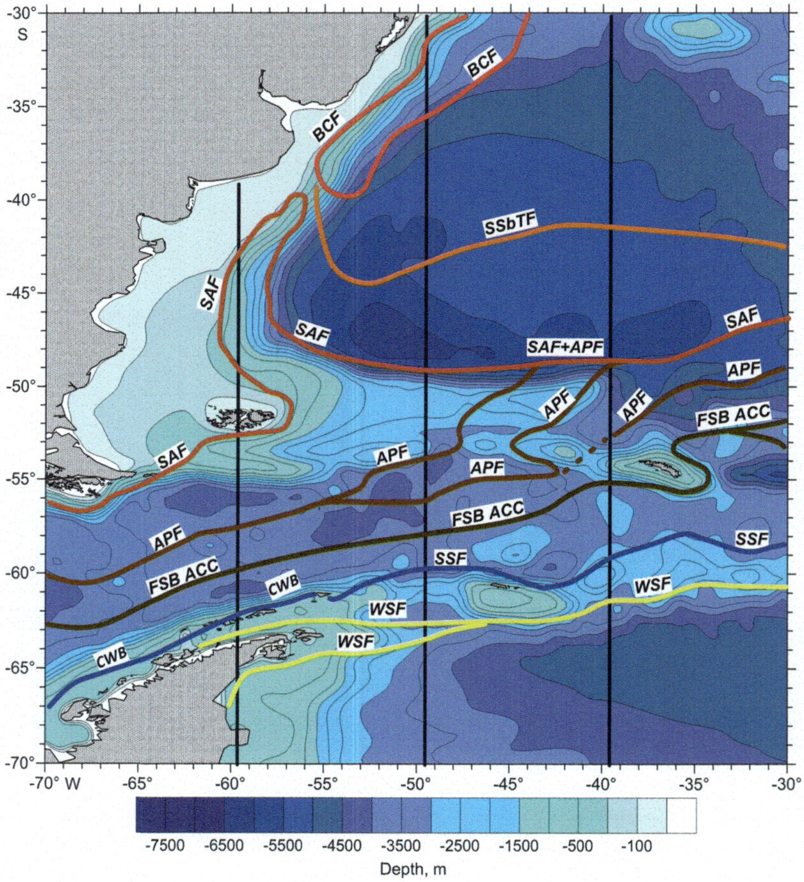

Fig. 5.2 Bottom topography (meters) and schematic location of the main temperature fronts in the Southwest Atlantic. FBT, Brazil Current Front; SSbTF, Southern Subtropical Front; SAF, Subantarctic Front; APF, Antarctic Polar Front; FSB ACC, Front of Southern Branch of the ACC; CWB, Continental Waters Boundary; SSF, Scotia Sea Front; WSF, Weddell Sea Front. The nomenclature is given according to Peterson and Stramma (1991), Maslennikov (2003), and Artamonov and Skripaleva (2005)

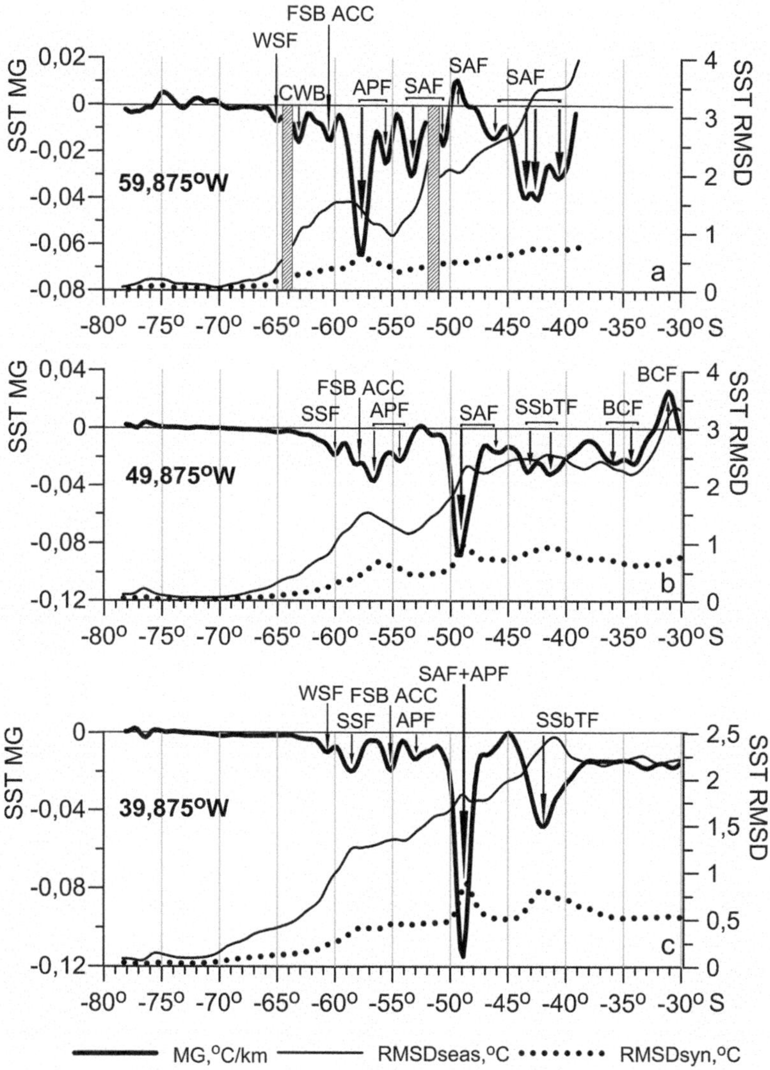

Fig. 5.3 Latitudinal changes of SST mean long-term values: MG (bold line), RMSDseas (thin line), and RMSDsyn (dashed line) along 59,875°W (**a**), 49,875°W (**b**), and 39,875°W (**c**) according to NOAA OI SST data

In the northern part of the profile along 59.875°W (Fig. 5.3a), the Subantarctic Front is well traced above the shelf zone. As a result of the intense meandering of the front along the Patagonian Shelf, several negative extremes of SST MG characterizing the SAF are formed. At 49°S, where the SAF is located around the Falkland Islands to the north, the gradients become positive, since the average annual surface temperature closer to the islands is higher than the temperature of the waters transported from the south by the eastern branch of the Falkland Current (Zyrjanov and

Severov 1979; Peterson and Stramma 1991). South of the Falkland Islands, in the SAF zone, the SST MG values again become negative, reaching a maximum over the island shelf at about 53°S. In the SAF zone, the values of SST RMSDseas and RMSDsyn significantly increase from south to north, while a sharp increase in the level of seasonal and synoptic variability is observed over the shallowest areas of the shelves of the Patagonia and Falkland Islands near 52°, 42–43°, and 38–39°S. In general, the position of the RMSDseas maximums on the shelves of Patagonia and Falklands Islands is slightly displaced to the north relative to the position of the SST MG extremes corresponding to the SAF.

Between 57° and 58°S, an intense SST MG extremum is traced, which we interpret as Antarctic Polar Front manifestation on the surface. The position of the maximum level of SST seasonal variability is shifted relative to the front position to the south to approximately 58–59°S, which is located south of the APF due to an increase in the seasonal variability of the Antarctic Surface water mass. The maximum level of SST synoptic variability (57–58°S) practically coincides with the APF position.

Between the APF and the Antarctic Peninsula, two less intense SST MG extremes are traced and another one to the south of the Antarctic Peninsula in the Weddell Sea, which, according to the established nomenclature, we interpreted as surface manifestations of the Front of Southern Branch of the ACC, Continental Waters Boundary, and Weddell Sea Front, respectively. To the south of the FSB ACC, the levels of seasonal and synoptic variability notably decrease, and they become minimal in the Weddell Sea (Fig. 5.3a).

The distribution of the meridional gradients of SST notably changes along the profile through the central part of the study area (49.875°W) (Fig. 5.3b). In the northern part of the profile (31°S), a maximum of positive MG values is noted, which is replaced by two negative MG extremes in the southern region (34–36°S). This configuration of the SST MG distribution indicates that at these latitudes the Brazil Current and its corresponding front (BCF) are recirculating, turning northward (Peterson and Stramma 1991), and a relatively warmer Subtropical Surface water mass is located within this area. There is a sharp increase of the SST seasonal variability across the front, which reaches a maximum level over the coastal shelf, while the position of the RMSDseas maximum is shifted relative to the position of the SST MG extremum characterizing the BCF.

South of the BCF between 41° and 44°S against the background of a general increase of the SST MG values corresponding to the South Subtropical Front, two MG extremes are traced, which is caused by the meandering of the front. Weak maximums of the seasonal and synoptic variability of SST are observed in the latitudinal range of the SSbTF.

An intense SST MG extremum can be traced around 49°S, where the SAF approaches close to the northern edge of the Falkland Plateau. Here, a notable increase in the levels of the seasonal and synoptic variability of the SST is observed, with the RMSDsyn maximum almost coinciding with the SAF position, while the RMSDseas is slightly shifted to the north. This indicates a more intense seasonal variability of warmer Subtropical Surface waters transported from the north in the turning zone of the Falkland Current to the south, compared with the colder Subantarctic Surface waters located south of the SAF.

A notable increase of SST MG values approximately between 54° and 57°S is an APF surface manifestation. As in the case of the SAF, in the APF zone, an increase of the levels of SST seasonal and synoptic variability is observed, but at the same time, in contrast to the SAF, the RMSDseas maximum is located south of the APF, as well as along the more western profile (59,875°W), which characterizes intense seasonal variability of the Antarctic Surface water mass south of the front.

In the zones of the Front of Southern Branch of the ACC and the Scotia Sea Front at this meridian, the level of SST seasonal and synoptic variability monotonically decreases to the south, and the RMSD extremes are almost absent.

In the eastern part of the study region (39.875°W), the South Subtropical Front is clearly traced at 42°S in the northern part of the profile (Fig. 5.3c). The position of the RMSDseas maximum is notably shifted to the north of the front (approximately to 40°S), while the RMSDsyn maximum coincides with the SSbTF position. The RMSDseas maximum north of the front is explained by the high SST seasonal variability on the SSAG southern periphery.

The highest SST MG value is observed at about 49°S. Meridian 39.875°W crosses the region where the main branch of the APF moves northward and practically confluences with the SAF (Peterson and Stramma 1991), forming a single frontal section at this latitude, which we call the SAF + APF (Artamonov and Skripaleva 2005). Such high SST MG values appear because part of the cold Antarctic Surface waters penetrates through the Falkland Plateau into the Argentine Basin in the eastern Scotia Sea, and temperature contrasts are enhanced. In the SAF + APF zone, the RMSDseas and RMSDsyn values increase.

In zones of less intense fronts located south of the Falklands Plateau (southern branch of the APF, FSB ACC, SSF, and WSF), there are not well-defined extremes of the SST seasonal variability level. A sharp decrease of the SST RMSDseas values is noted south of the SSF. In the FSB ACC and SSF zones, a slight increase of the SST synoptic variability level is observed.

Thus, the analysis of the SST frontal structure and the levels of its seasonal and synoptic variability according to the NOAA OI SST reanalysis data with a high spatial-temporal resolution showed that on the surface the level of the SST synoptic variability is increased in the zones of intense fronts. The position of the SST RMSDseas maximum does not always coincide with the position of fronts.

Analysis of the linear correlations between the mean long-term values of the SST RMSDseas and TG shows that the correlation coefficient R for the entire water area is 0.65 (Fig. 5.4a), and the R value decreases to 0.2 in the APF zone (Fig. 5.4c). The R value between the mean annual values of SST RMSDsyn and TG in the water area increases to 0.82 (Fig. 5.4b), and it is 0.65 in the APF zone (Fig. 5.4d). The correlation between the monthly average values of SST RMSDsyn and TG in the APF zone is significant during the year: the R values vary from 0.60 in winter (Fig. 5.4e) to 0.78 in summer (Fig. 5.4f).

In the subsurface layers, it is currently impossible to estimate the levels of mean annual synoptic variability of temperature and salinity and their relationship with

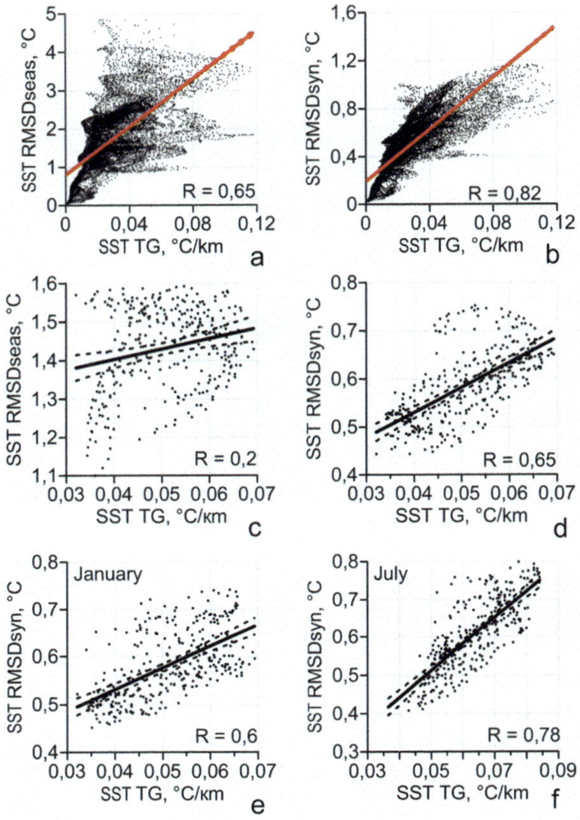

Fig. 5.4 Graphs of the linear correlation between the mean long-term SST RMSDseas and TG values (**a, c**), annual average RMSDsyn and TG values (**b, d**) for the Southwest Atlantic as a whole (**a, b**) and in the APF zone (**c, d**), the monthly average SST RMSDseas and TG values in the APF zone for January (**e**) and July (**f**). Dashed lines show 99% confidence interval boundaries

the frontal structure, since modern oceanic reanalyses have a monthly mean time resolution. The relatively coarse spatial grid step (1-degree) does not allow us to resolve the fine frontal structure in the coastal areas and narrow straits. At the same time, these reanalyses make it possible to estimate the mean long-term seasonal variability of thermohaline fields, while the vertical distributions of thermohaline characteristics in the sections make it possible to interpret the features of this variability in the deep layers in terms of the position of the boundaries of water masses and to estimate the penetration depth of seasonal fluctuations.

Figures 5.5 and 5.6 show the distributions of mean annual meridional gradients of temperature (MGT) and salinity (MGS) at depths of 5, 195, and 290 m, vertical distributions of mean annual temperature (Tann) and salinity (Sann), and their seasonal RMSD (RMSD T and RMSD S) over the sections along 59.5° (left), 49.5° (middle), and 39.5°W (right) based on the ECMWF ORA-S3 reanalysis data.

Two main fronts are well traced in the temperature field: the SAF and APF over the section crossing the Patagonian Shelf and the Drake Passage (59.5°W), (Fig. 5.5). The Subantarctic Front is traced around 42°S. The vertical distribution of temperature in the section clearly shows that this front separates the relatively warm (11–12 °C) waters of the Patagonian Shelf and colder (6–8 °C) Antarctic waters

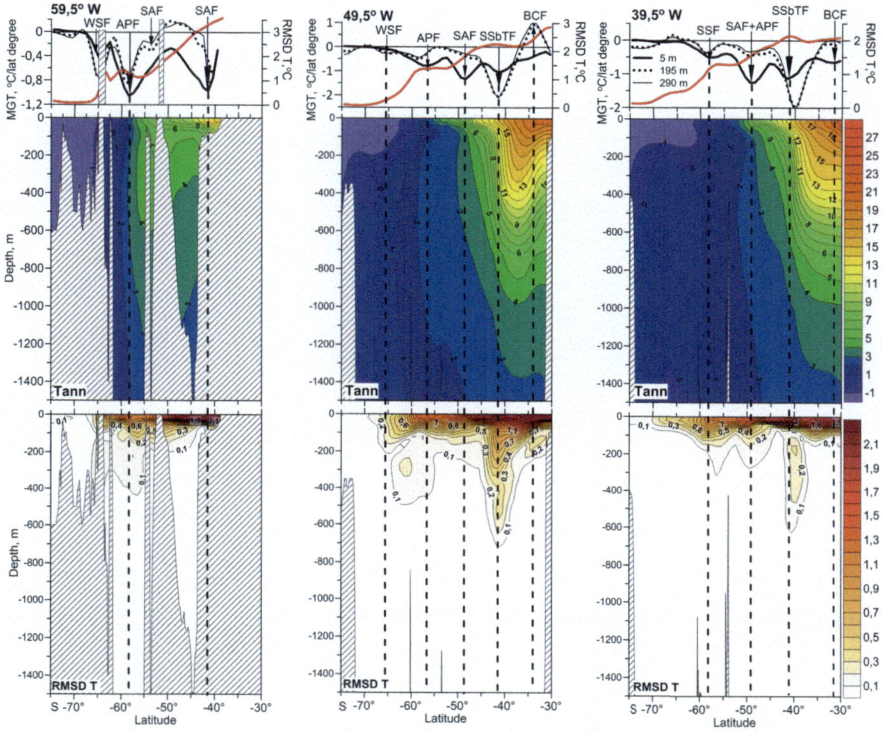

Fig. 5.5 The latitudinal distributions of MGT at 5-m, 195-m, and 290-m levels (black curves) and RMSD T at 5-m levels (red curves) (upper), vertical distributions of annual mean T (Tann) (middle), RMSD T (lower) along 59.5°W (left), 49.5°W (center), and 39.5°W (right). Arrows show the main fronts position

transported northward by the Falkland Current (Fig. 5.5). Another weaker negative MGT extremum is traced at 52–53°S, where the SAF is located around the Falkland Islands from the south. In the climatic salinity field, the SAF is traced at 41–42°S by its weak negative gradient (Fig. 5.6). The maximum RMSD values in the temperature and salinity fields are observed in the upper 100-m layer slightly north of the front (Figs. 5.5 and 5.6).

The Antarctic Polar Front is well traced in the climatic temperature field near 57–58°S as the northern boundary of the Antarctic Winter water mass, which is a winter modification of the Antarctic Surface Water and is characterized by a subsurface temperature minimum that remains in the average annual field (Fig. 5.5). The maximum seasonal temperature fluctuations are observed in the upper 50-m layer slightly south of the front, where the Antarctic Surface Water mass is subject to the greatest seasonal changes. Immediately in the APF zone, the penetration depth of seasonal temperature variations based on the RMSD 0.2 °C isoline is 100–120 m. North of the APF in the Antarctic Convergence zone approximately between 55° and 57°S, the penetration depth of seasonal fluctuations increases and reaches 170–200 m (Fig. 5.5). In the salinity field, in contrast to the temperature field, the

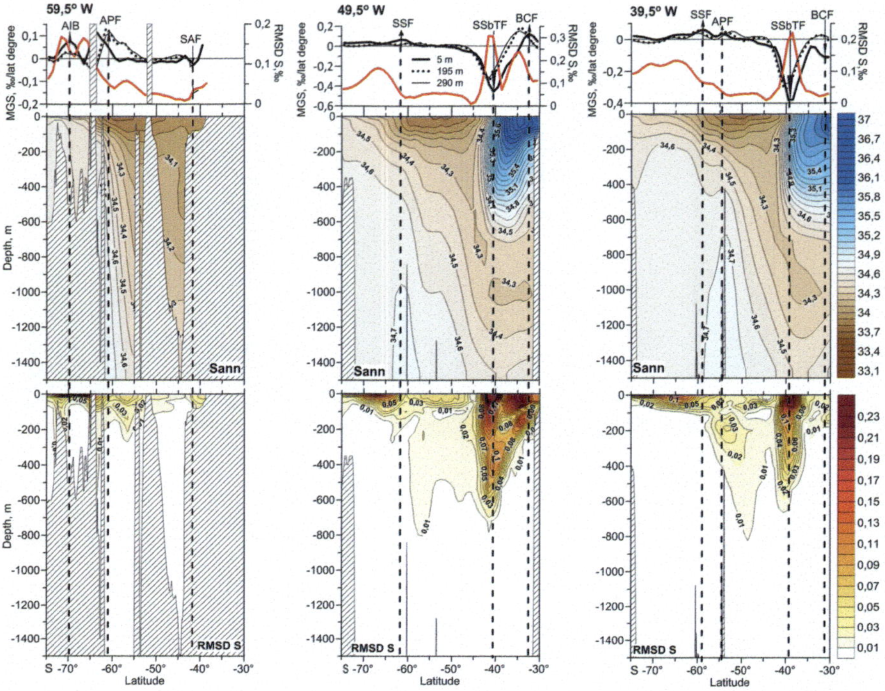

Fig. 5.6 The latitudinal distributions of MGS at 5-m, 195-m, and 290-m levels (black curves) and RMSD S at 5-m level (red curves) (upper), vertical distributions of annual mean S (Sann) (middle), RMSD S (lower) along 59.5°W (left), 49.5°W (center), and 39.5°W (right). Arrows show the main fronts position

APF in the near-surface layer is not pronounced clearly (Fig. 5.6). In the subsurface layer, the APF is traced much farther south than in the temperature field, at about 61°S, in the form of high positive MGS values at 195–290 m. Positive MGS values are caused by the interaction of the Antarctic Intermediate Water characterized by the minimum salinity with the more saline Upper Circumpolar Deep Water located south of the front. The seasonal fluctuations of the maximum salinity are observed in the upper 50-m layer north of the front, approximately between 56° and 58°S. Here, the penetration depth of seasonal salinity fluctuations based on the 0.02-psu isoline reaches 200 m. In the APF zone, the level of seasonal variability of salinity decreases markedly, and the penetration depth of its seasonal fluctuations based on the 0.02-psu isoline decreases to 100 m (Fig. 5.6).

Another notable increase of negative MGT values in the entire 300-m layer is observed right off the coast to the south of the Antarctic Peninsula, which can be interpreted as a manifestation of the Weddell Sea Front (Fig. 5.5). It separates the cold surface waters of the southern Weddell Sea, which are transported by currents at the Weddell Gyre western periphery to the north along the Antarctic Peninsula coast, and the warmer waters of the Antarctic Peninsula eastern shelf. In the WSF zone, the seasonal temperature variations are small and are observed in the upper

50-m layer; south of the front they become minimal. In the salinity field, the WSF is characterized by weak positive MGS near the coast in the near-surface layer and negative MGS at 195–290 m (Fig. 5.6). To the south of the WSF, there is a notable increase of the level of seasonal variability of salinity with seasonal variations covering the upper 50-m layer.

In the salinity field, south of the WSF at about 70°S, there is an intense positive MGS extremum characterizing the Antarctic Ice Boundary (AIB). To the south of AIB (approximately at 71–73°S), the maximum seasonal variability of salinity is observed, due to the processes of intense ice melting and ice formation and the latitudinal displacement of the pack ice boundary during the year. Here, seasonal variations of salinity based on the 0.02-psu isoline propagate to a depth of almost 300 m (Fig. 5.6).

On the section through the central part of the water area (49.5°W), the BCF, SSbTF, SAF, and APF are clearly pronounced in the temperature field. The Brazil Current Front is traced at about 34°S according to the high positive MGT values in the subsurface layer (195–290 m) (Fig. 5.5). In the salinity field, the BCF is characterized by a maximum of positive MGS values located in the near-surface layer at about 32°S and in the subsurface layers, somewhat farther to the south, at about 34–35°S (Fig. 5.6). The positive MGT and MGS values are caused by the fact that the Brazil Current, flowing to the south along the South America coasts, transports on its eastern periphery the waters from the central part of the South Subtropical Gyre, which are warmer and saltier (warmer than 20 °C and more saline than 36.2 psu) than the waters on the Brazilian shelf. The maximum level of seasonal variability of temperature is observed north of the BCF over the shallow coastal shelf. At the latitude of the BCF subsurface position (34–35°S), there is a notable increase in the level of seasonal variability of salinity, while its seasonal variations penetrate to the depths of 500–600 m. Directly in the BCF zone, seasonal temperature variations penetrate to a depth of 200 m, and salinity variations penetrate almost to 300 m.

South of BCF near 42°S, a negative extremum of MGT is observed, and it is the most intense in the subsurface layers (195, 290 m), which characterizes the SSbTF (Fig. 5.5). It separates the warm and salty Subtropical waters from the colder and less saline Subantarctic waters. In the climatic salinity field, in contrast to the temperature field, the SSbTP is most developed in the surface layer (Fig. 5.6). In the subsurface layers, the SSbTF position, like the BCF, is displaced to the south relative to its position on the surface; it is located approximately at 42–43°S. In the SSbTF zone, the maximum seasonal variability of salinity and a slight increase of seasonal variability of temperature are observed. Here, the penetration depths of the seasonal fluctuations in temperature and salinity (based on the RMSD isolines 0.2 °C and 0.02 psu) are maximal and reach 600 m and 700 m, respectively.

The Subantarctic Front along this meridian is well traced only in the temperature field in the surface layer (5 m) in the form of a negative extremum of MGT values at about 48°S (Fig. 5.5). The SAF separates the relatively warm (8–9 °C) Subantarctic Surface waters and colder (below 6 °C) Antarctic Surface waters. The vertical RMSD distribution shows that there is no pronounced maximum of seasonal variability in the frontal zone, and the penetration depth of seasonal temperature fluctuations based on the 0.2 °C isoline does not exceed 150 m.

To the south of the SAF, approximately at 57–58°S, the MGT extremum is traced, which is a manifestation of the Antarctic Polar Front. As in the Drake Passage, it is the northern boundary of the Antarctic Winter water mass (Fig. 5.5). An increase in the level of seasonal fluctuations of temperature occurs notably north of the front, approximately north of 52–53°S, in the formation zone of the Antarctic Intermediate Water, while the penetration depth of seasonal temperature fluctuations based on the RMSD isoline 0.2 °C is about 100 m. In the salinity field, the SAF and APF are almost not manifested on this section (Fig. 5.6).

South of the APF at 65°S, a slight increase in the negative values of MGT in the subsurface layers is observed (Fig. 5.5). At this longitude, this characterizes the southern boundary of the Weddell–Scotia Confluence zone (WSCZ) (Deacon and Moorey 1975, Gordon et al. 1977, Patterson and Sievers, 1980) or the Weddell Sea Front. The northern boundary of WSCZ or the Scotia Sea Front, according to the climatic data on this meridian, is better manifested in the salinity field in the surface layer at about 61°S, where a slight increase in the MGS positive values is observed (Fig. 5.6). The penetration depth of seasonal fluctuations in the temperature and salinity fields in this zone is about 150 m. In the mean annual thermohaline fields, the WSCZ waters in the upper layer represent a transformation between the relatively warm and less saline waters of the ACC southern periphery and the colder and saltier waters of the Weddell Gyre northern periphery.

On the profile crossing the eastern part of the study region (39.5°W), the structure of the fronts is qualitatively preserved. The Brazil Current Front is traced in the surface layer at about 31°S by an increase in the MGT negative values (Fig. 5.5). In the salinity field, in contrast to the section along 49.5°W, the BCF is characterized by negative MGS values. The main seasonal changes are observed in the upper 100–150-m layer. In the salinity field, a RMSD increase is noted somewhat north of the front in the subsurface layer at depths of 100–200 m, which indicates significant seasonal fluctuations of the characteristics of the Subtropical subsurface salinity maximum.

The South Subtropical Front in this section in the temperature field is well traced in the upper 300-m layer at about 40–41°S, and at depths of 195–290 m its position is slightly displaced to the north compared to the surface (Fig. 5.5). In the salinity field, the SSbTF appears to the north at 39°S (Fig. 5.6). It is characteristic that, as on the neighboring section, the maximum MGT values are observed in the subsurface layer, while the maximum MGS values are noted at the surface. This indicates that in the SSbTF zone the interaction of Subantarctic freshened waters with relatively salty Subtropical waters occurs in the surface layer. In the SSbTF zone, an increase in the level of seasonal temperature variability and a maximum level of seasonal salinity variability are noted. At the same time, the maximum penetration depth of seasonal fluctuations in the temperature field is observed north of the front and reaches almost 500 m based on the 0.2 °C isoline; in the salinity field, it is seen in the front zone and reaches 550 m based on the 0.02psu isoline.

Near 49°S the most intense MGT extremum along the given meridian is observed on the surface (Fig. 5.5). It is formed because here, on the northern border of the Falkland Plateau, the SAF and APF approach close to each other, and the interaction of warmer Subantarctic waters with colder Antarctic waters leads to the formation of an intense single front SAF + APF. At depths of 195–290 m, this front is almost not traced. The penetration depth of seasonal temperature fluctuations here does not exceed 50–100 m. Since the main transformation of Subtropical high-saline waters occurs within the BCF and SSbTF zones, and the Subantarctic and Antarctic waters differ slightly in salinity, and the SAF + APF almost does not manifest itself in the salinity field. The level of seasonal variability of salinity in the SAF + APF zone significantly decreases, and its seasonal fluctuations based on the 0.02-psu isoline are not traced deeper than 50 m (Fig. 5.6).

Note that a relatively weak MGS maximum is traced in the salinity field, south of the SAF + APF near 55°S (Fig. 5.6). Probably, this maximum is associated with one branch of the APF, which is formed as a result of the interaction of the ACC with the South Georgia archipelago, and separates various modifications of the Antarctic water mass. Although the level of salinity seasonal variability is low in the area of this MGS extremum, seasonal variations of salinity penetrate to a depth of almost 300 m.

Between 58 and 59°S there are well-defined extremes of negative MGT (Fig. 5.5) and positive MGS (Fig. 5.6). These extremes characterize the boundary between the warmer and less saline waters of the ACC southern periphery and the colder and saltier waters of the Weddell–Scotia Confluence zone, identified as the Scotia Sea Front. At greater depths the SSF is almost not traceable. To the south of the SSF, there is a sharp decrease of the level of seasonal variability of temperature and an increase of the salinity variability. In the SSF zone, the penetration depth of seasonal fluctuations of temperature based on the 0.2 °C isoline is 150 m, and the penetration depth of salinity fluctuations based on the 0.02-psu isoline is 100 m.

5.4 Conclusions

The synoptic and seasonal variability of thermohaline characteristics was investigated in the western part of the Atlantic sector of the Southern Ocean on the basis of the daily averaged SST values at the nodes of the regular 0.25° grid from the NOAA OI SST reanalysis and the monthly averaged temperature and salinity at the nodes of the 1-degree grid from the ECMWF ORA-S3 oceanic reanalysis. The features of the large-scale front manifestation in the temperature and salinity fields are considered, and the penetration depths of the seasonal fluctuations of thermohaline characteristics within various frontal zones are estimated.

A general increase in the level of SST intra-annual variability (RMSDseas ~3.2–4.8 °C, RMSDsyn ~0.7–1.1 °C) over the South America coastal shelf is reported. The area of waters with the maximum SST RMSDseas values is located along the South America coast at sea depths shallower than 200–300 m. High RMSDsyn values are located closer to the greater depths in the convergence zone of the Falkland and Brazil currents and further extend to the east in the form of two quasi-zonal "tongues" along the southern periphery of the Southern Subtropical Anticyclonic Gyre and along the northern border of the Falkland Plateau.

An increase of the level of SST intra-annual variability is observed in the Drake Passage and the northwestern Scotia Sea (RMSDseas ~1.4–1.6 °C, RMSDsyn ~0.6–0.7 °C). The lowest level of SST intra-annual variability is observed in the Weddell Sea, where pack ice is constantly present. Here the RMSDseas and RMSDsyn values do not exceed 0.6 °C and 0.15 °C, respectively.

It is shown that the level of SST seasonal variability exceeds 4–4.5 times the level of synoptic variability in the southern Drake Passage and in the South Subtropical Anticyclonic Gyre zone, on the South America coastal shelf the excess is 5–6 times, and in the zones of fronts it is 2–3 times. The minimum values of the ratio of levels of seasonal and synoptic variability (2–2.5) are observed in an elongated area covering the northern Drake Passage and the northwestern Scotia Sea.

In general, an increased level of SST synoptic variability is observed in high-gradient zones (frontal zones), while the position of areas with an increased level of SST seasonal variability does not always coincide with the position of fronts. The linear correlation coefficient between the mean long-term values of SST RMSDseas and SST total gradients in the entire water area as a whole is 0.65, and in the Antarctic Polar Front zone it decreases to 0.2. The correlation coefficient between the average annual SST RMSDsyn and SST total gradients values in the water area as a whole increases to 0.82 and increases to 0.65 in the APF zone.

A high level of seasonal variability of temperature and salinity was found in the South Subtropical Front zone, south of the Antarctic Polar Front, and north of the Subantarctic Front. The greatest penetration depth of seasonal temperature and salinity fluctuations is noted in the SSbTF zone, where it reaches 400–600 m and 500–700 m, respectively. In the zones of the Subantarctic and Antarctic Polar Fronts, this depth does not exceed 100–150 m.

Acknowledgments The work is performed under State Order No. 0555-2019-0003 "Experimental studies of the variability of hydrophysical, hydrochemical and bio-optical fields at different spatio-temporal scales to identify the features of climatic changes in oceanographic conditions in the Atlantic part of Antarctica."

References

Anisimov OA, Vaughan DG, Callaghan TV, Furgal C, Marchant H, Prowse TD, Vilhjálsson H, Walsh JE (2007) Polar regions (Arctic and Antarctic). In: Climate Change 2007: Impacts, adaptation and vulnerability. Contribution of Working Group II to the Fourth Assessment Report of the Intergovernmental Panel on Climate Change. Cambridge University Press, New York, pp 653–685

Antipov NN, Klepikov AV (2007) Features of the oceanographic regime of Prydz Bay according to AARI expedition data. Arctic Antarctic Res 76:36–48

Antipov NN, Bagriantsev NV, Danilov AI, Klepikov AV (2019) Winter research in the Antarctic zone of the Southern Ocean (in commemoration of the 25th anniversary of Soviet–Russian–

American Weddell 1 drift-ice station). Oceanology 59(2):308–310. https://doi.org/10.31857/S00301574592308-310
Artamonov YuV, Skripaleva EA (2005) The structure and seasonal variability of the large-scale fronts in the Atlantic Ocean on the basis of satellite data. Issledovaniye Zemli iz kosmosa 4:62–75
Artamonov YuV, Skripaleva EA (2016) Oceanographic research of Marine Hydrophysical Institute in the Southern Ocean. Phys Oceanogr 6:56–66. https://doi.org/10.22449/1573-160X-2016-6-56-66
Chapman CC, Lea M-A, Meyer A, Sallée J-B, Hindell M (2020) Defining Southern Ocean fronts and their influence on biological and physical processes in a changing climate. Nature Climate Change, 11 p. doi:https://doi.org/10.1038/s41558-020-0705-4
Deacon GER, Moorey JA (1975) The boundary region between currents from the Weddell Sea and Drake Passage. Deep-Sea Res 22(4):265–268. https://doi.org/10.1016/0011-7471(75)90031-5
Golovin PN, Antipov NN, Klepikov AV (2011) Downflow of the Antarctic shelf water at the shelf and continental slope of the Commonwealth Sea in the summer season and its effect on the bottom water formation in the Southern Ocean. Oceanology 51(3):370–384. https://doi.org/10.1134/S000143701103009X
Gordon AL, Georgi DT, Taylor HW (1977) Antarctic Polar Front Zone in Western Scotia Sea – summer 1975. J Phys Oceanogr 7(3):309–328. https://doi.org/10.1175/1520-0485(1977)007<0309:APFZIT>2.0.CO;2
Ionov VV (2015) Synoptical monitoring of thermic fronts on the sea surface of the Southern Ocean. Trudy VNIRO 157:143–150
Maslennikov VV (2003) Climatic variability and marine ecosystem of the Antarctica. VNIRO, Moscow, 296 p
Meredith MP, Murphy EJ, Hawker EJ, King JC, Wallace MI (2008) On the interannual variability of ocean temperatures around South Georgia, Southern Ocean: forcing by El Niño/southern oscillation and the southern annular mode. Deep-Sea Res II 55(18–19):2007–2022. https://doi.org/10.1016/j.dsr2.2008.05.020
Monin AS (1999) Hydrodynamics of the atmosphere, ocean and earth's interior. Gidrometeoizdat, Saint-Petersburg, 524 p
Naveira Garabato AC, Jullion L, Stevens DP, Heywood KJ, King BA (2009) Variability of Subantarctic mode water and Antarctic intermediate water in the Drake Passage during the late-twentieth and early-twenty-first centuries. J Clim 22(7):3661–3688. https://doi.org/10.1175/2009JCLI2621.1
Patterson SL, Sievers HA (1980) The Weddell-scotia confluence. J Phys Oceanogr 10:1584–1610. https://doi.org/10.1175/1520-0485(1980)010<1584:TWSC>2.0.CO;2
Peterson RG, Stramma L (1991) Upper-level circulation in the South Atlantic Ocean. Prog Oceanogr 26(1):1–73. https://doi.org/10.1016/0079-6611(91)90006-8
Reynolds RW, Smith TM, Liu C, Chelton DB, Casey KS, Schlax MG (2007) Daily high-resolution-blended analyses for sea surface temperature. J Clim 20(22):5473–5496. https://doi.org/10.1175/2007JCLI1824.1
Rintoul SR, Donguy JR, Roemmich DH (1997) Seasonal evolution of upper ocean thermal structure between Tasmania and Antarctica. Deep-Sea Res I 44(7):1185–1202. https://doi.org/10.1016/S0967-0637(96)00125-2
Sallée J-B, Speer K, Morrow R (2008) Response of the Antarctic Circumpolar Current to atmospheric variability. J Clim 21(12):3020–3039. https://doi.org/10.1175/2007JCLI1702.1
Simmonds I, King JC (2004) Global and hemispheric climate variations affecting the Southern Ocean. Antarct Sci 16(4):401–413. https://doi.org/10.1017/S0954102004002226
Sokolov S, Rintoul SR (2007) On the relationship between fronts of the Antarctic Circumpolar Current and surface chlorophyll concentrations in the Southern Ocean. J Geophys Res 112(C07030). https://doi.org/10.1029/2006JC004072
Turner J (2004) Review the El NIÑO – southern oscillation and Antarctica. Int J Climatol 24(1):1–31. https://doi.org/10.1002/joc.965

Valentine HR, Lutjeharms JRE, Brundrit GB (1993) The water masses and volumetry of the southern Agulhas current region. Deep-Sea Res I 40(6):1285–1305. https://doi.org/10.1016/0967-0637(93)90138-S

Venables H, Meredith MP, Atkinson A, Ward P (2012) Fronts and habitat zones in the Scotia Sea. Deep-Sea Res II 59–60:14–24. https://doi.org/10.1016/j.dsr2.2011.08.012

Zyrjanov VN, Severov DN (1979) Water circulation in the Falkland-Patagonian region and its seasonal variability. Oceanology 19(5):768–775

Chapter 6
The Circulation and Mixing Zone in the Antarctic Sound in February 2020

Alexander V. Krek, Elena V. Krek, and Viktor A. Krechik

Abstract Mixing and circulation of water in the Antarctic Sound was studied on the basis of in situ data and satellite observations. Cyclonic circulation was detected, and an intense mixing zone of waters from the Weddell Sea and Bransfield Strait was distinguished. Water inflow from the Weddell Sea to the Antarctic Sound occurs less frequently than the flow from the Bransfield Strait. The main flow of water is directed to the south.

Keywords Antarctic Sound · Weddell Sea · Bransfield Strait · Circulation · Currents · Mixing zone · Satellite data

6.1 Introduction

The Antarctic Sound (AS) is one of the areas of interaction between waters of the Weddell Sea (WS) and waters of the Bransfield Strait (BS). The generally accepted water circulation scheme around the Antarctic Peninsula (AP) is shown in Gordon et al. (2000), Maslennikov et al. (2003), Sangrà et al. (2011), Huneke et al. (2016), Dotto et al. (2016), and Rodrigo et al. (2018). The water from the WS enters the BS along two possible pathways: the main one is around the Joinville Islands and the second through the AS. The periodic absence of ice removal from the AS (Dorschel et al. 2015) indicates a change in the generally accepted seasonal water circulation. Despite the fact that the water exchange pathway through the AS is of the secondary importance, the processes taking place in it may reflect changes in the main circulation of the BS leading to a change in the position of the Peninsula Front (PF) according to Huneke et al. (2016) and productive areas.

The aim of this work is to study circulation and mixing in the AS and their influence on the water exchange between the WS and BS through the sound.

A. V. Krek (✉) · E. V. Krek · V. A. Krechik
Shirshov Institute of Oceanology, Russian Academy of Sciences, Moscow, Russia

E. G. Morozov et al. (eds.), *Antarctic Peninsula Region of the Southern Ocean*, Advances in Polar Ecology 6, https://doi.org/10.1007/978-3-030-78927-5_6

6.1.1 Study Site

The AS is located at the northeastern tip of the AP and separates the Joinville Island group from the AP (Fig. 6.1). The sound is about 55 km long and about 20 km wide in its middle part. According to geomorphological features, the AS is divided into two parts: the shallow northern part with depths of up to 200 m and the deep-water southern part with depths of up to 800–900 m (Dorschel et al. 2015).

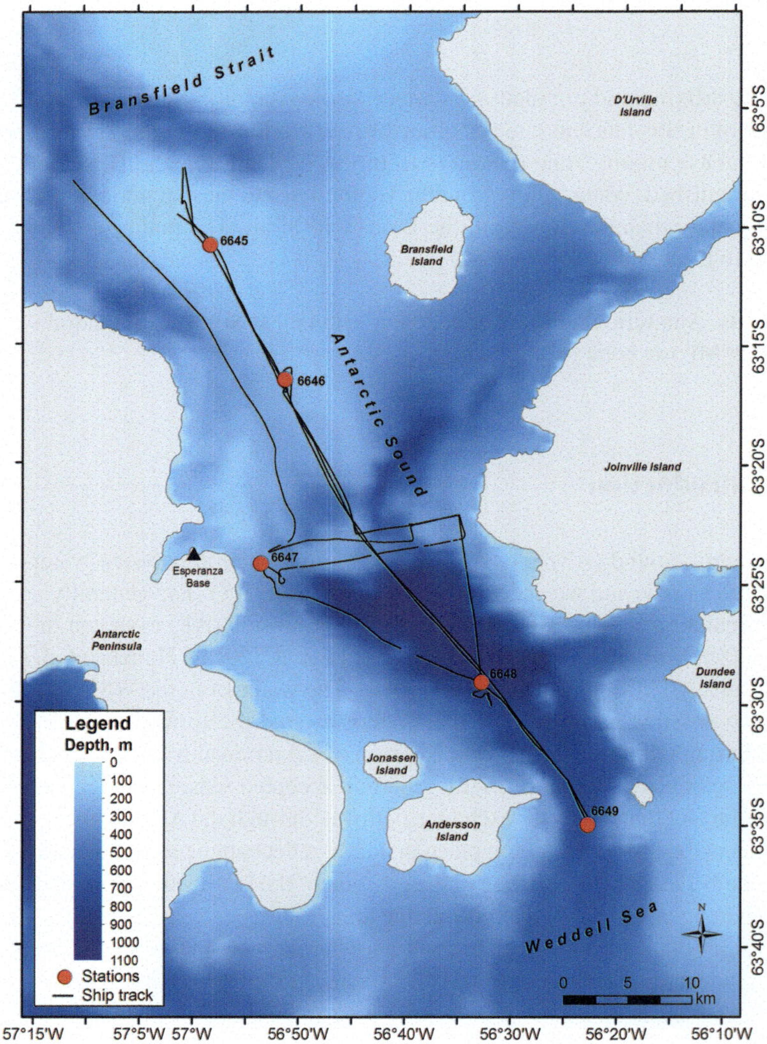

Fig. 6.1 Study site
Red points indicate oceanographic stations occupied on Feb. 13–14, 2020. The black line indicates the ship track. Bathymetry: GEBCO Compilation Group (2020) GEBCO 2020 Grid (doi:https://doi.org/10.5285/a29c5465-b138-234d-e053-6c86abc040b9)

According to Tokarczyk (1987), the AS is located in the zone of the WS surface water spreading. The southern and southeastern parts of the BS are occupied by a subtype of this water that is named Transitional Water with WS influence (TWW). Its potential temperature is usually less than −1 °C, and salinity is 34.3–34.5 psu (Huneke et al. 2016). This water mass is transported to the sound by the Antarctic Coastal Current (ACoC) from the regions located south of the Joinville Island and east of the AP (Collares et al. 2018; Thompson et al. 2009). The mean transport of this current near the northeastern tip of Joinville Island is 1 Sv ± 0.4 Sv (Caspel et al. 2018). In the northwestern WS over the shelf of the AP, the ACoC separates the shelf waters from the open sea waters (Collares et al. 2018; Thompson and Heywood 2008). Thus, the upper 200–400 m of this water constitutes Antarctic Surface Water (AASW) which mixes with the peninsula shelf waters in varying proportions (Thompson and Heywood 2008). Huneke et al. (2016) named this water as Shelf Water (SW). Warm Deep Water (WDW) or its modifications (mWDW) are usually located deeper than AASW (Foster, Carmack 1976). However, shallow depths and significant influence of the shelf isolate waters in the central part of the sea from the study site in the AS. The combination of these factors leads to significant intensification of convective processes in winter. As a result, Winter Water (WW) occupies the bottom and deep layers (Thompson, Heywood 2008; Von Gyldenfeldt et al. 2002).

6.1.2 Materials and Methods

6.1.2.1 Satellite Images

Synthetic Aperture Radar Images

Synthetic aperture radar (SAR) images are efficient for iceberg detection (Viehoff and Li 1995; Willis et al. 1996; Young and Hyland 1998). The success of this technique is mainly due to its high spatial resolution, wide swath, and independence from weather and sunlight. SAR images are successfully used for studying iceberg drift in Antarctica (Collares et al. 2018; Young and Hyland 1998; Williams et al. 1999; Gladstone and Bigg 2002; Silva and Bigg 2005; Wesche and Dierking 2015).

The draft of Antarctic icebergs with lengths of a few hundred meters is in the range 200–300 m. Most of the iceberg is under water; hence the wind contribution to the iceberg drift is insignificant, and its motion is due only to the ocean currents. Therefore, studying iceberg drifts can be useful for studying surface and subsurface currents in the ocean (Williams et al. 1999).

To study the drift of icebergs, SAR images of the Sentinel-1A and Sentinel-1B (European Space Agency, ESA) satellites were used from January 30 to March 8, 2020. A total of 39 SAR images were analyzed. The spatial resolution of SAR images is 10–40 m, and the time resolution is from 15 to 48 h. Several uniquely shaped icebergs in the WS and BS were selected to study currents in the ocean upper layer.

Optical Imaging of the Ocean Surface

Satellite information in the optical range of the spectrum is available in cloudless weather during the daytime.

Two RGB images of the sea surface from the Landsat-8 satellite of Operational Land Imager (OLI) on February 13 and 29, 2020, were used to study the optical characteristics of the sea surface, as well as to track the drift of icebergs. The spatial resolution of the images is 30 m.

The *chlorophyll a* concentration was measured by Moderate Resolution Imaging Spectroradiometer (MODIS) onboard the Terra satellite (NASA) (https://worldview.earthdata.nasa.gov/). The sensor and imagery resolution are 1 km, and the time resolution is 1 day. The MODIS chlorophyll *a* layer provides the near-surface concentration of chlorophyll *a* in milligrams of chlorophyll pigment per cubic meter (mg/m^3) in the ocean.

Total suspended matter (TSM, g/m^3) with a spatial resolution of 60 m was derived from Multi-Spectral Instrument (MSI) aboard the Sentinel-2B satellite (ESA).

6.1.2.2 Ship Measurements

The hydrophysical measurements in the AS were carried out on February 13–14, 2020, during cruise 79 of the R/V "Akademik Mstislav Keldysh". CTD profiles were measured at five stations (Fig. 6.1, Table 6.1). Three stations were located in the shallow part of the sound (stations 6645–6647). Two stations were located in the deep-water part of the sound (6648–6649). The current velocities were measured with shipborne ADCP in underway mode at cross-sections across and along the AS.

Temperature, conductivity, and pressure measurements were made at the stations from the surface to depths of about 5–7 m above the bottom using the SBE 911 CTD system. This system included the SBE 9 CTD Unit and the SBE 11 Deck Unit. The SBE 9 CTD was mounted on a full-size SBE 32 Carousel Water Sampler. The

Table 6.1 Description of the oceanographic stations

Station number	Southern latitude	Western longitude	Date	Time	Depth, m /measurement, m
6645	63°10.850′	56°58.396′	Feb 13, 2020	11:39	130/125
6646	63°16.553′	56°51.403′	Feb 13, 2020	13:43	175/170
6647	63°24.258′	56°53.605′	Feb 13, 2020	18:00	185/179
6648	63°29.161′	56°32.940′	Feb 13, 2020	21:17	900/889
6649	63°35.023′	56°22.832′	Feb 14, 2020	02:05	712/706

SBE 911 system made it possible to obtain the measured parameters of vertical distribution data in real-time mode with a 24 Hz sampling rate.

The zonal and meridional current components in the upper layer of the water column were measured by a Teledyne RDI Ocean Surveyor 75 kHz shipborne acoustic Doppler current profiler (SADCP). The device was used in broadband mode; the vertical cell size and the blanking distance were 8 m. The instrument draft was 5.9 m. Data acquisition was carried out using the VmDas software with a 2-min time averaging. The roll and pitch angles were measured with a Kongsberg Seatex MRU sensor. The collected data were georeferenced in the WGS-84 coordinate system using a Trimble SPS855 GNSS receiver. Data with "percent-good data" values less than 50 were filtered out as unrepresentative. Tidal currents (diurnal and semidiurnal components) were estimated using the global inverse tide model TPXO9.1 (Egbert and Erofeeva 2002). The SADCP data were de-tided with the model results. The current transport was calculated as a product of a 2-min averaging cell area by the velocity in this cell. The length of each cell was calculated as the distance between the start and end coordinates of the 2-min averaging. The velocity in the cell was assumed to be the velocity of the flow normal to the section orientation.

6.2 Results

6.2.1 Satellite Observations

On February 13, 2020, the RGB image from the Landsat-8 (OLI) satellite in the AS waters with different optical properties, as well as their mixing zone, were observed (Fig. 6.2). Waters from the WS were flowing along the Andersson Island, displacing to the AP western coast and penetrating toward the north up to the Esperanza Base in the middle of the AP. This water formed a frontal zone that was located at a distance of 5–7 km northeast of the Andersson Island coast. At the outlet from the AS to the WS, the frontal zone was clearly traced up to 63°40′S.

Water from the Weddell Sea penetrated through the AS along the western coast of the sound, and the countercurrent directed to the south displaced to the coasts of Joinville and Dundee islands under the influence of the Coriolis force. Such circulation in the surface layer led to the formation of a local frontal zone in the southern part of the AS (see Fig. 6.2).

The inflow of waters from the WS into the AS was characterized by the active development of phytoplankton blooms, which was reflected in increased chlorophyll *a* concentration within the distribution of inflow waters (Fig. 6.3b). The area of chlorophyll *a* development in the central part of the sound coincides with the area of the distribution of turbid waters spreading from the coast to the center of the sound to the north from Esperanza Base (to the south from station 6646) (Fig. 6.3a).

In the coastal zone of the AP, an alongshore current was traced, which transported waters with optical properties different to the waters of the sound central

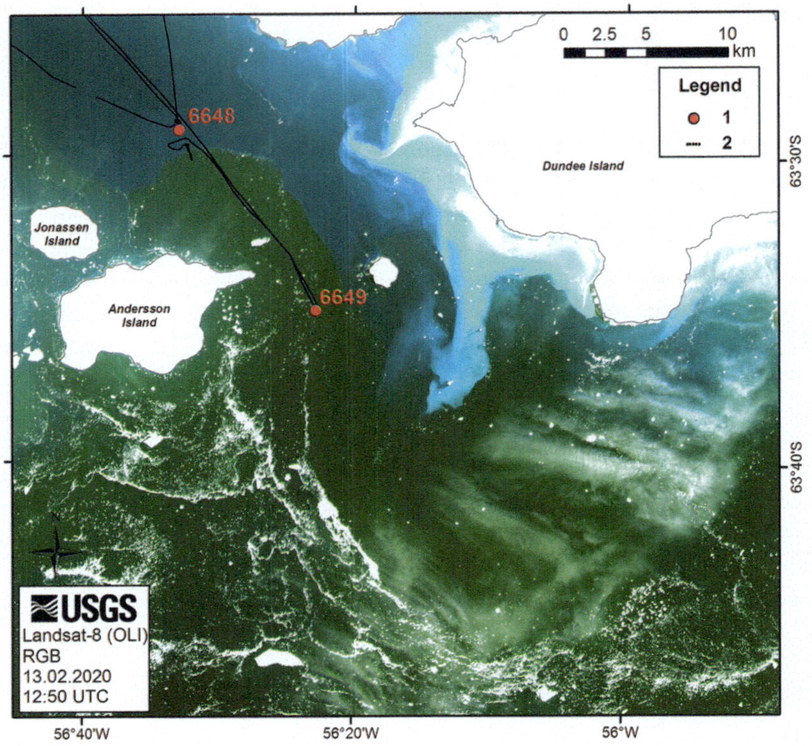

Fig. 6.2 Local frontal zone formed during the interaction of waters from the Weddell Sea with the waters from the Antarctic Sound based on the Landsat-8 (OLI) data
Legend: (1) stations; (2) ship track

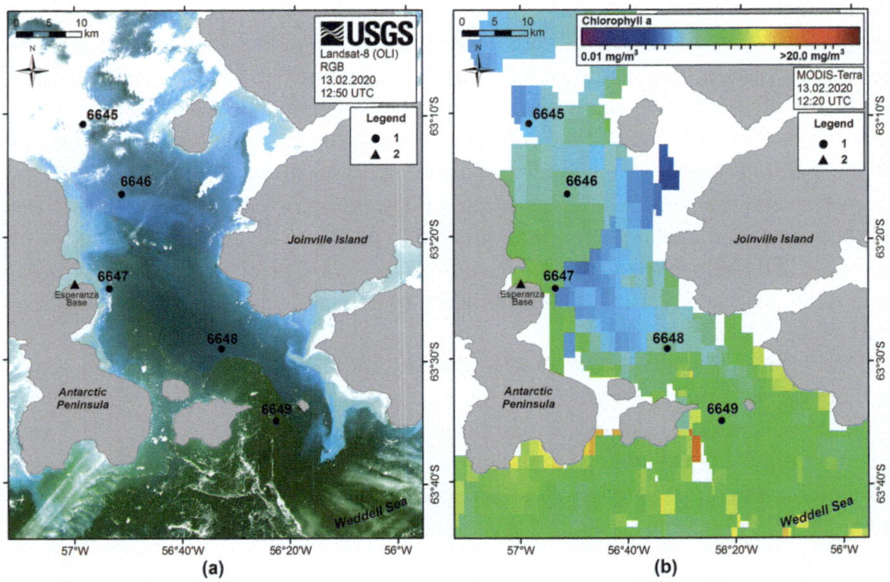

Fig. 6.3 Satellite images of the Antarctic Sound on February 13, 2020: (**a**) RGB, Landsat-8 (OLI), 12:50 UTC, (**b**) Chlorophyll *a* concentration, MODIS-Terra, 12:20 UTC. Legend: 1 – stations; 2 – Esperanza Base

part. The current originates at about 63°30′S and breaks off the coast at about 63°20′S, turning to the sound central part. The current direction was also indicated by the suspended matter propagation (Fig. 6.4).

6.2.2 CTD Results

A significant influence of waters from the Bransfield Strait was noted in the northern part of the Antarctic Sound (station 6645), where the water temperature in the surface layer was maximum over the cross-section and amounted to −0.63 °C (Table 6.1, Fig. 6.5a). Layers subject to abrupt temperature changes were noted at depths of 30–40 and 90–100 m. The water column was characterized by the minimum turbidity values over the entire cross-section of about 10 FTU. The structure of the water column in the vertical distribution of salinity at station 6645 was similar to the distribution at stations 6646–6647 (Fig. 6.5a, b, c).

Stations 6646 and 6647 were located within the zones of interaction of the waters of the central part of the Antarctic Sound with the waters of the alongshore current (Fig. 6.2a). Station 6647 was located close to the shore, which made it possible to

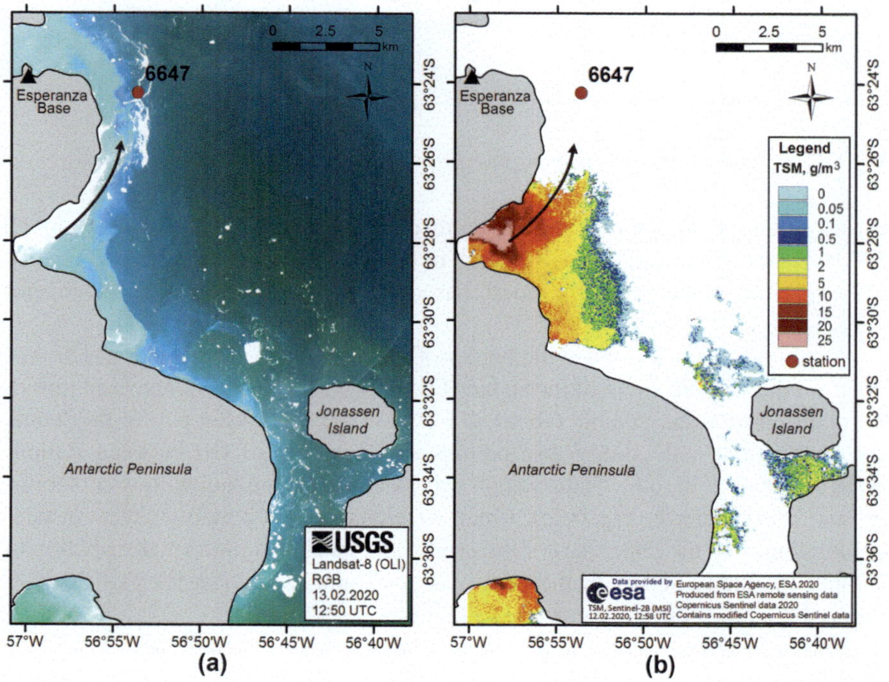

Fig. 6.4 Alongshore current directed north along the Antarctic Peninsula coast according to satellite data: (**a**) RGB-Landsat-8 (OLI), February 13, 2020, 12:50 UTC; (**b**) TSM-Sentinel-2B (MSI), 12.02.2020, 12:58 UTC. Black arrows indicate the direction of current

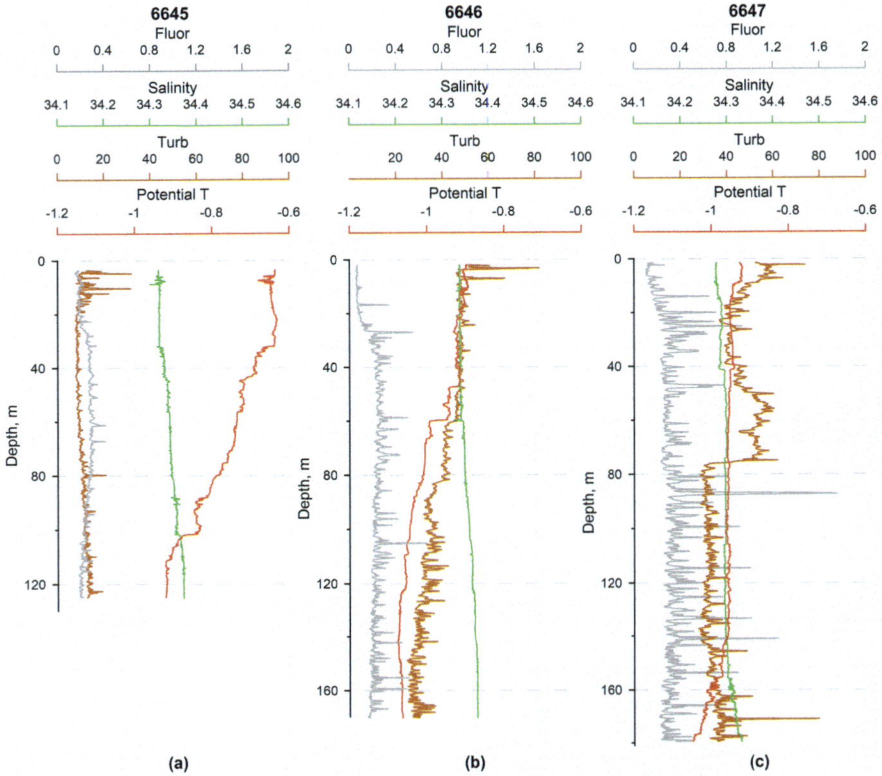

Fig. 6.5 Vertical distribution of hydrological parameters at stations 6645–6647

record a thin (about 10 m) freshened warmer surface layer. A turbidity peak (40–60 FTU) was clearly traced down to a depth of 80 m, confirming the increased content of suspended matter near the coast based on the analysis of satellite images (Fig. 6.4b).

Station 6646 was located at the border of the alongshore current that has broken off from the western shore of the sound (Fig. 6.2a), which caused increased turbidity in the entire water column (about 40 FTU). In the deep-water part of the sound, the temperature and salinity distributions significantly differed between stations 6648 and 6649 (Fig. 6.6). Station 6648 was located in the region with waters transported from the north (Fig. 6.2a), which were characterized by two layers of temperature jump in the upper part of the profile and an almost uniform distribution of salinity. Turbidity values in the entire water column were at a level of 20 FTU increasing in the bottom layer to 60 FTU (Fig. 6.6a).

Station 6649 was located in the waters flowing northward from the Weddell Sea (Fig. 6.2a), which can be clearly seen in hydrological parameters. Thus, an increased variability of thermohaline parameters was noted here throughout the entire water column (Fig. 6.6b). The surface layer with a thickness of about 20 m was significantly freshened due to ice melting. The upper quasi-homogeneous sea layer, that

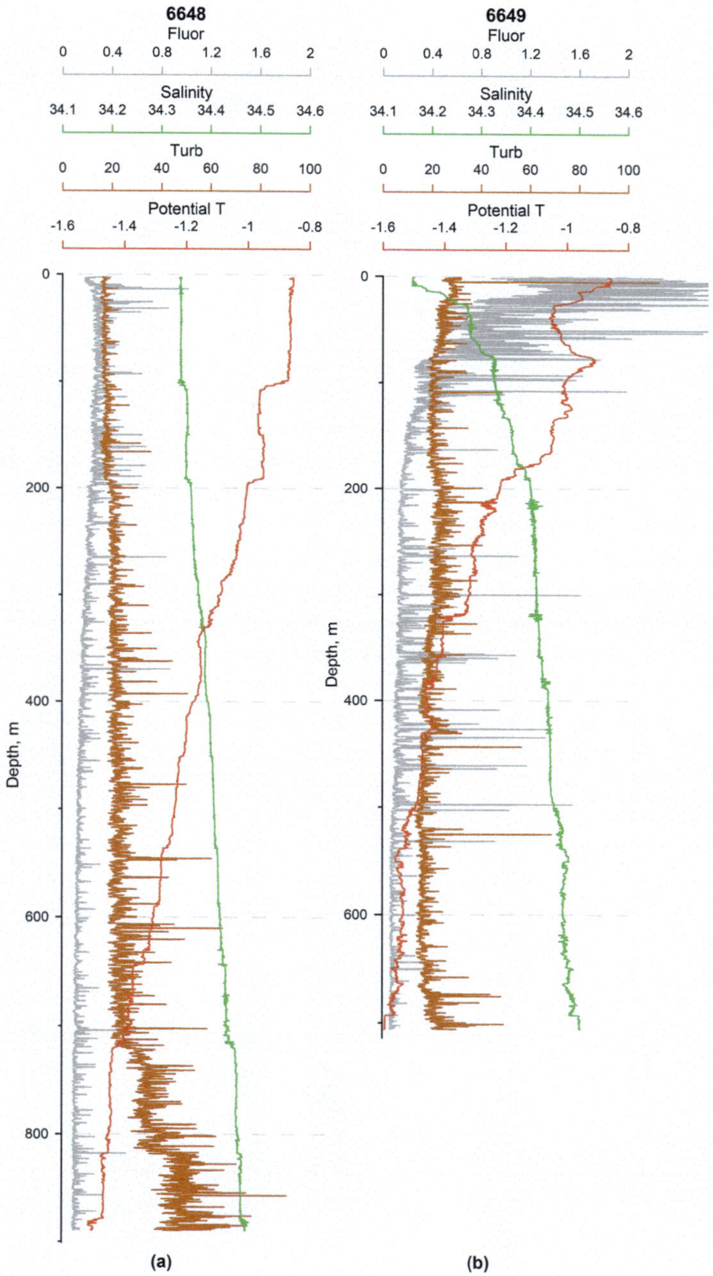

Fig. 6.6 Vertical distribution of hydrological parameters at stations 6648–6649

was recorded at all the previous stations was absent. Turbidity values in the surface layer were higher, than at station 6648 to the depths of 200 m. In the upper 100 m layer, high values of fluorescence were recorded as high as 5–10 times higher than the values at the other stations of the cross-section (Figs. 6.5 and 6.6).

6.2.3 Results of Measurements of Currents Using SADCP

According to the SADCP data, the southeastward current with a velocity of 10–15 cm s^{-1} prevailed over the cross-section across the AS (Fig. 6.7). The maximum current velocities were recorded in the central part of the section at a depth up to 150–180 m (35–42 cm s^{-1} in the core at a depth of about 100 m), as well as at depths ranging between 300 and 450 m (up to 25–31 cm s^{-1}). Near the eastern coast of the sound in the surface layer (up to 120 m), the southeastward current was also detected. Deeper down to a depth of about 230 m, there was a countercurrent jet with a width of about 2 km. The current velocities in the jet core were up to 14 cm s^{-1}. The main stream of the current spread along the western coast to the north-northwest as a jet about 5 km wide and up to 200 m deep.

Cyclonic circulation was also observed over the cross-section along the AS (Fig. 6.8). The northeastward current was located in the north-northwestern part of the cross-section and occupied the water column from the surface to the bottom in the shallow-water part of the sound and up to 600 m in the deep-water part. The jet was about 25 km wide. The velocity of the current directed to the southwest varied from 5 to 20 cm s^{-1} and increased in two cores to 24–29 cm s^{-1}. The jet was also about 25 km wide. The current occupied the entire water column from surface to bottom. The current near station 6645 was also directed to the southwest. The current direction variability in the southeastern part of the cross-section was caused by the water moving along the coast of Andersson Island and the outflow of water from the AS.

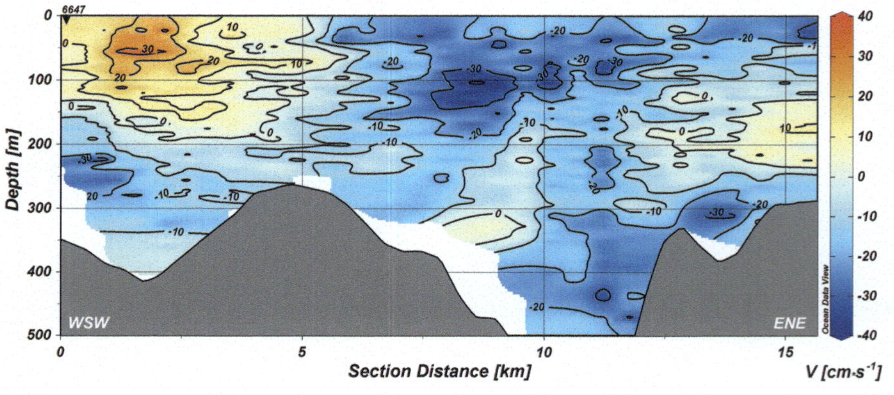

Fig. 6.7 Pattern of currents over the cross-section along the Antarctic Sound

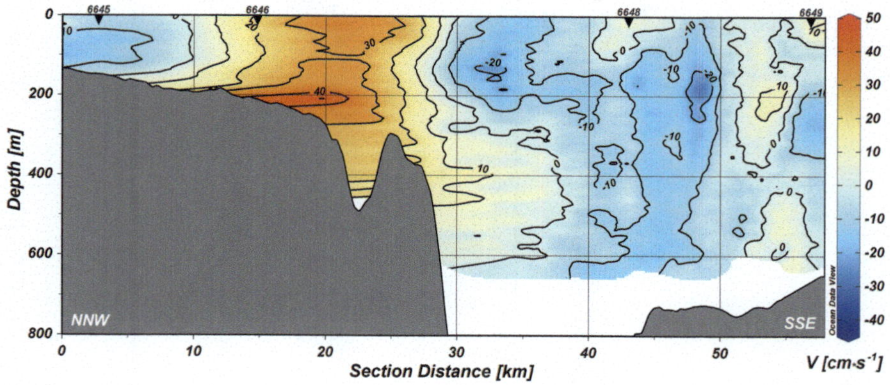

Fig. 6.8 Pattern of currents over the lateral cross-section in the Antarctic Sound

6.3 Discussion

Satellite images on February 13, 2020, clearly showed both the inflow of waters from the WS into the AS along the sound western coast and the outflow of waters from the sound into the WS along the eastern coast. The frontal zone, which was located along the thalweg in the southern part of the sound, separated waters with optically different properties.

The SADCP data confirm the presence of cyclonic circulation centered over the deep-water part of the sound. The inflow current jet along the western coast of the sound and the turn of the current along the frontal zone between stations 6645 and 6646 were also clearly seen. The total transport over the cross-section across the sound was directed to the south-southeast and amounted to −0.43 Sv (0.12 Sv directed to the north-northwest and – 0.55 Sv directed to the south-southeast). The transport over the cross-section along the sound amounted to −0.82 Sv in the south-western direction.

The flow into the AS from the WS was observed at station 6649 (Fig. 6.9). In the surface layer, up to the 27.61 kg^{-3} isopycnal of potential density, water was much colder and fresher due to the melting of icebergs. The water at station 6646 was obviously the main portion of water, that filled the sound in the depth range of 0–200 m. The water from the WS (station 6649) and the water from the sound (station 6646) were identical between isopycnals 27.62 and 27.65 kg m^{-3}. Hence, their origin was the same, and their source was located southwest of the WS shelf areas, which were separated by ACoC from the main sea basin. The water observed at station 6647 was the product of mixing water from the WS and water from the sound. TWW from the southern BS was observed at station 6645. Probably, a small volume of this water was involved in the cyclonic circulation and strongly mixed with the water of the sound during its propagation along the eastern coast of the sound to the southeast. Traces of TWW were noted at station 6648. The water at this station in the depth range of 0–250 m was warmer and saltier than the water of the sound.

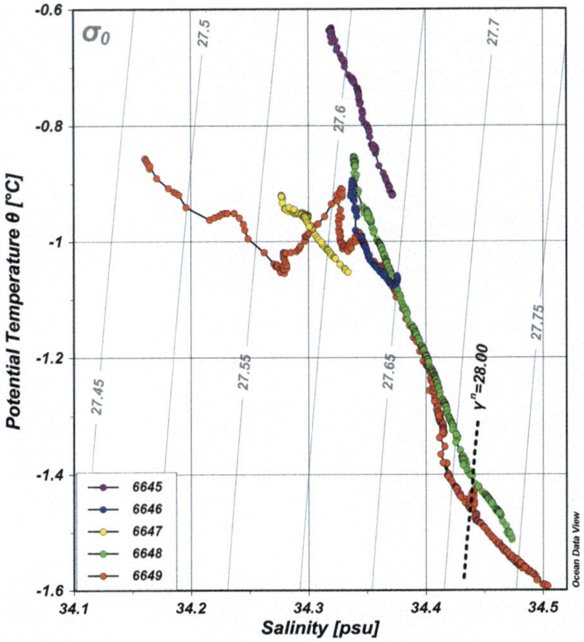

Fig. 6.9 Θ-S diagram for stations in the Antarctic Sound

Deeper than the isopycnal of the neutral density $\gamma^n = 28.00$ kg m^{-3}, which is considered the lower boundary of the AASW propagation (Strass et al. 2020), there was no significant increase in temperature, which would indicate the presence of WDW or mWDW. Thus, the bottom layers of the AS deep-water part were occupied by WW (Fig. 6.10).

Legend: 1 – stations; 2 – general directions of currents; 3 – local front; 4 – mixing zone (MZ).

The main ice source in the study site is the Larsen Ice Shelf. Sea ice is a tracer of surface currents. According to Dorschel et al. (2015), the AS and coastal area east of Joinville Island are often ice-free compared with the neighboring waters at the same latitude. This indicates that "ice-carrying" waters from the WS rarely penetrate to the AS.

Thus, from February 26 to February 29, 2020, no signs of water inflow from the south into the sound were observed (Fig. 6.11). The eroded frontal zone was traced only south of Andersson Island (Fig. 6.11b). This was also evidenced by the low chlorophyll *a* concentration in the sound (Fig. 6.11c).

Analysis of the drift of icebergs near the AP (Fig. 6.12) showed that from February 9 to March 2, 2020, a cyclonic eddy formed southeast of Dundee Island, caused by the weakening of the ACoC. Thereby the preconditions for water penetration into the AS from the south were formed. The frontal zone between the WS waters and the AS waters was located south of Andersson Island from February 8 to 10 (https://worldview.earthdata.nasa.gov/). Apparently, the inflow occurred from February 11 to February 12, 2020, and by February 13, circulation was as follows.

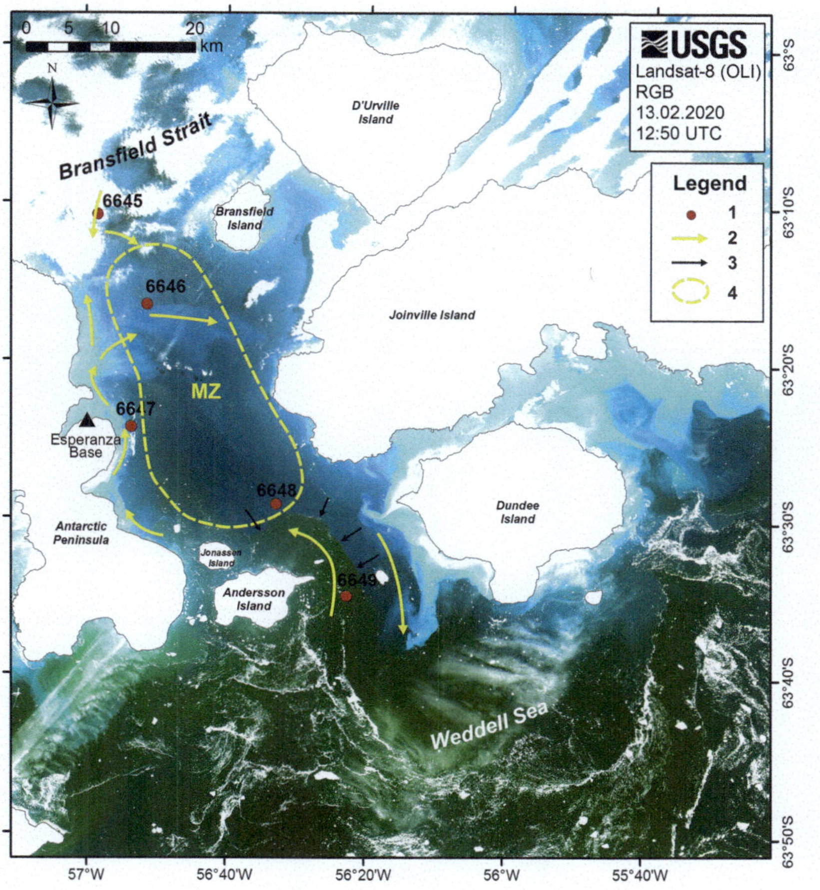

Fig. 6.10 Circulation scheme in the Antarctic Sound on February 13, 2020

There were a general outflow and displacement of the frontal zone to the south, which is confirmed by the SADCP data in the southern part of the sound.

The smaller amount of sea ice in the AS compared to the latitudinal sea ice distribution indicates the inflow of water from the north.

The presence of a polynya east of Joinville Island (according to Dorschel et al. 2015) also indicated the absence of ice transported from the main source in the center of the WS. This was due to the position of the ACoC, which repeated the edge of the shelf. However, the iceberg drift in the period January 30, 2020, to February 12, 2020 (black solid arrow in Fig. 6.12) indicated the existence of at least an episodic current directed to the northeast of Joinville Island. After February 12, 2020, the iceberg stopped and apparently grounded. This iceberg drift direction could be explained by the continuation of the shelf countercurrent along the islands at the northern tip of the AP directed to the east, which was observed in the Bransfield Strait at the end of January (Krechik et al. 2021).

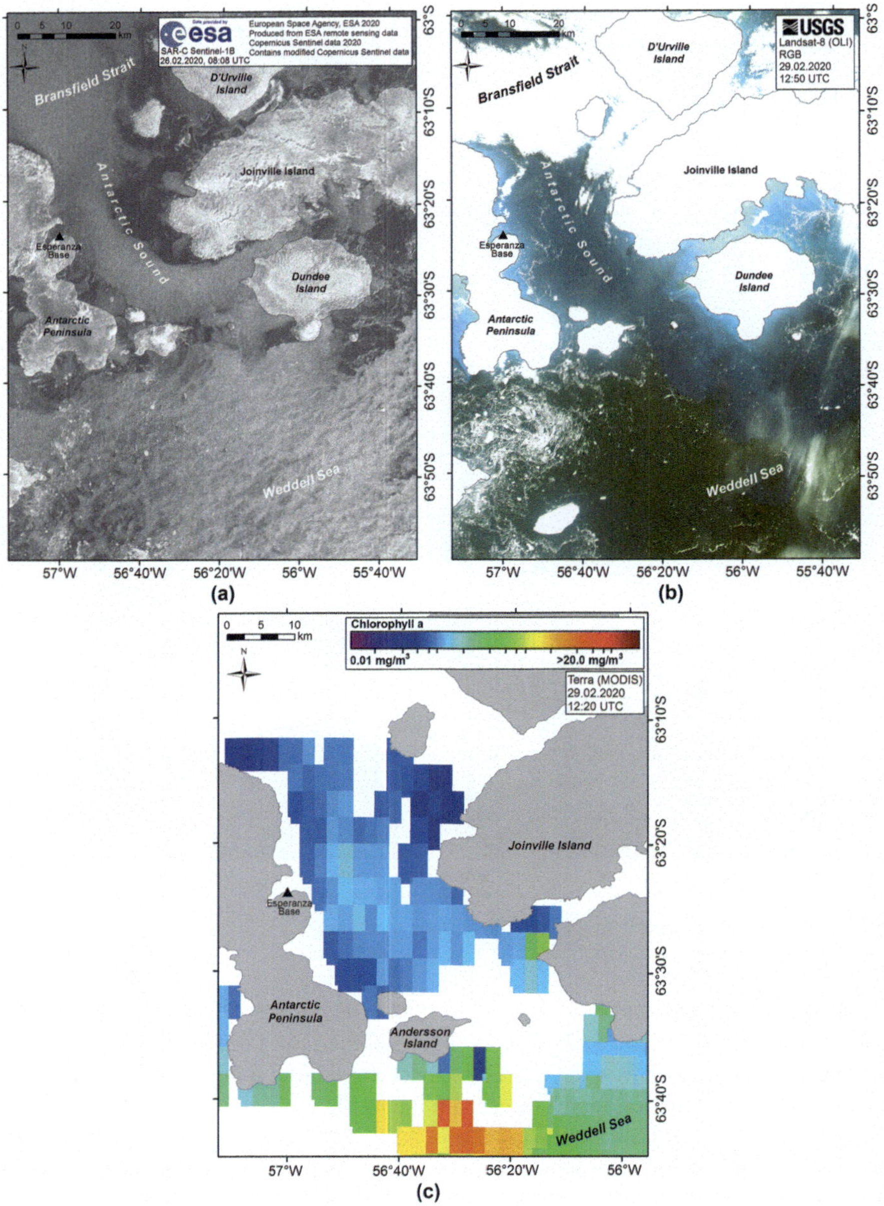

Fig. 6.11 Satellite images of the sea surface on 26.02.2020 and 29.02.2020: (**a**) No ice in the Weddell Sea and in the southern Antarctic Sound according to Sentinel-1B (SAR) 26.02.2020, 08:08 UTC; (**b**) optical image from Landsat-8 (OLI) 29.02.2020, 12:50 UTC; (**c**) low chlorophyll *a* concentration in the Antarctic Sound according to MODIS-Terra, 29.02.2020, 12:20 UTC

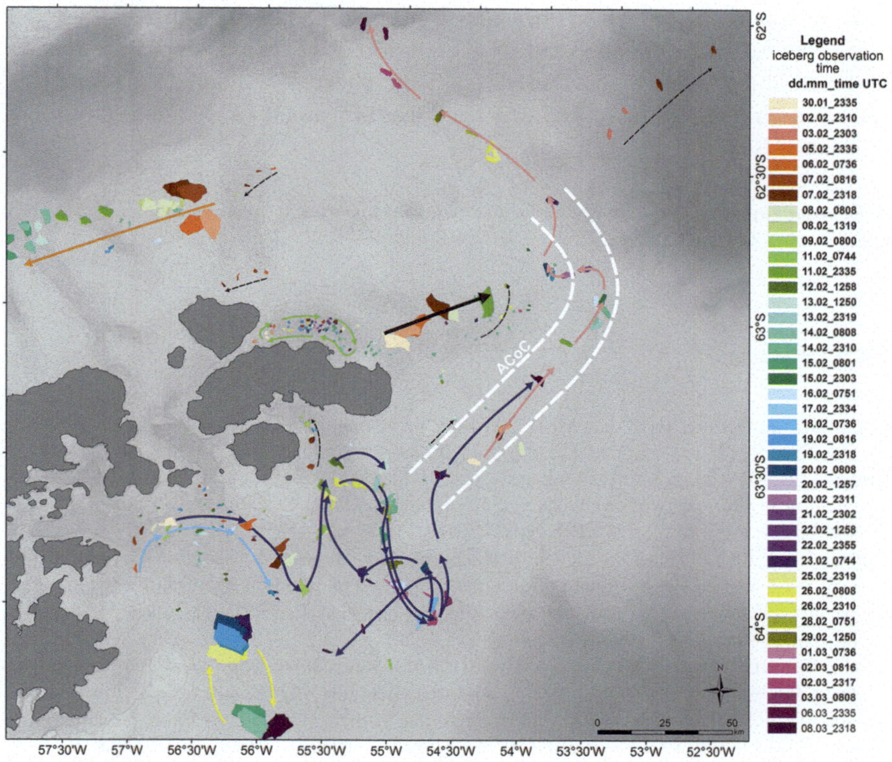

Fig. 6.12 Scheme of the drift of icebergs from January 30, 2020, to March 3, 2020, based on the satellite data

6.4 Conclusions

The data of contact measurements showed the presence of cyclonic circulation in the Antarctic Sound, as well as the presence of water from the shelf areas of the Weddell Sea and Bransfield Strait. The combination of these factors led to the formation of an extensive mixing zone inside the sound and a significant development of phytoplankton blooms, which was reflected in the increased concentration of chlorophyll *a* in the surface layer. Water entered the Antarctic Sound both from the south along the western coast and from the north along the eastern coast, due to the Coriolis force. According to SADCP data, water outflow occurred only in the southern direction. This situation, apparently, became possible due to the inflow of a significant volume of water from the Weddell Sea due to the general circulation in this sea. Later, no signs of water inflow from the south were observed on satellite images, and the concentration of chlorophyll *a* decreased to the values that are typical for the northern shallow part of the Antarctic Sound and shelf areas south of the Bransfield Strait. The southward current in the surface layer of the sound seems to have been dominating, which correlates well with the pure sea ice concentration within the AC as described in Dorschel et al. (2015). Thus, the dynamic situation observed in the sound is, obviously, a periodic phenomenon and at this time was caused by the weakening of the ACoC and the formation of a cyclonic eddy south of Dundee Island.

Acknowledgments The study war performed with the support of the state assignment of IO RAS theme no. 0128-2019-0008.

References

Caspel M, van Hellmer HH, Mata MM (2018) On the ventilation of Bransfield Strait deep basins. Deep-Sea Res II Top Stud Oceanogr 149:25–30. https://doi.org/10.1016/j.dsr2.2017.09.006

Collares LL, Mata MM, Kerr R, Arigony-Neto J, Barbat MM (2018) Iceberg drift and ocean circulation in the northwestern Weddell Sea, Antarctica. Deep-Sea Res II Top Stud Oceanogr 149:10–24. https://doi.org/10.1016/j.dsr2.2018.02.014

Dorschel B, Gutt J, Huhn O, Bracher A, Huntemann M, Huneke W, Gebhardt C, Schröder M (2015) Environmental information for a marine ecosystem research approach for the northern Antarctic Peninsula (RV Polarstern Expedition PS81, ANT XXIX/3). Polar Biol 39(5):765–787. https://doi.org/10.1007/s00300-015-1861-2

Dotto TS, Kerr R, Mata MM, Garcia CAE (2016) Multidecadal freshening and lightening in the deep waters of the Bransfield Strait, Antarctica. J Geophys Res Oceans 121. 16 p. https://doi.org/10.1002/2015JC011228

Egbert GD, Erofeeva SY (2002) Efficient inverse modeling of barotropic ocean tides. J Atmos Ocean Technol 19(2):183–204

Foster TD, Carmack EC (1976) Temperature and salinity structure in the Weddell Sea. J Phys Oceanogr 6(1):36–44. https://doi.org/10.1175/1520-0485(1976)006<0036:TASSIT>2.0.CO;2

Gladstone R, Bigg GR (2002) Satellite tracking of icebergs in the Weddell Sea. Antarct Sci 14:278–287. https://doi.org/10.1017/S0954102002000032

Gordon A, Mensch M, Dong Z et al (2000) Deep and bottom water of the Bransfield Strait eastern and central basins. J Geophys Res Oceans 105(C5):11337–11346

Huneke WGC, Huhn O, Schröeder M (2016) Water masses in the Bransfield Strait and adjacent seas, austral summer 2013. Polar Biol 39(5):789–798. https://doi.org/10.1007/s00300-016-1936-8

Krechik V, Frey D, Morozov E (2021) Water circulation peculiarities in the central part of the Bransfield Strait in January 2020. Dokl Earth Sci 496(1):92–95. https://doi.org/10.1134/S1028334X21010116

Maslennikov VV (2003) Climate fluctuations and the Antarctic marine ecosystem. M.: VNIRO Publications, pp 295

Rodrigo C, Blamey JM, Huhn O, Provost C (2018) Is there an active hydrothermal f lux from the Orca seamount in the Bransfield Strait, Antarctica? Andean Geol 45(3):344–356. https://doi.org/10.5027/andgeoV45n3-3086

Sangrà P, Gordo C, Hernández-Arencibia M et al (2011) The Bransfield current system. Deep Sea Res I Oceanogr Res Pap 58(4):390–402

Silva TAM, Bigg GR (2005) Computer-based identification and tracking of Antarctic icebergs in SAR images. Remote Sens Environ 94:287–297. https://doi.org/10.1016/j.rse.2004.10.002

Strass VH, Rohardt G, Kanzow T, Hoppema M, Boebel O (2020) Multidecadal warming and density loss in the deep Weddell Sea, Antarctica. J Clim 33(22):9863–9881. https://doi.org/10.1175/jcli-d-20-0271.1

Thompson AF, Heywood KJ (2008) Frontal structure and transport in the northwestern Weddell Sea. Deep-Sea Res I 55(10):1229–1251. https://doi.org/10.1016/j.dsr.2008.06.001

Thompson AF, Heywood KJ, Thorpe SE, Renner AHH, Trasviña A (2009) Surface circulation at the tip of the Antarctic peninsula from drifters. J Phys Oceanogr 39(1):3–26. https://doi.org/10.1175/2008JPO3995.1

Tokarczyk R (1987) Classification of water masses in the Bransfield Strait and southern part of the Drake Passage using a method of statistical multidimensional analysis. Polish Polar Res 8(4):333–366

Viehoff T, Li A (1995) Iceberg observations and estimation of submarine ridges in the western Weddell Sea. Int J Remote Sens 16(17):3391. https://doi.org/10.1080/01431169508954636

Von Gyldenfeldt AB, Fahrbach E, García MA, Schröder M (2002) Flow variability at the tip of the Antarctic peninsula. Deep-Sea Res II Top Stud Oceanogr 49(21):4743–4766. https://doi.org/10.1016/S0967-0645(02)00157-1

Wesche C, Dierking W (2015) Near-coastal circum-Antarctic iceberg size distributions determined from synthetic aperture radar images. Remote Sens Environ 156:561–569. https://doi.org/10.1016/j.rse.2014.10.025

Williams RN, Rees WG, Young NW (1999) A technique for the identification and analysis of icebergs in synthetic aperture radar images of Antarctica. Int J Remote Sens 20:3183–3199. https://doi.org/10.1080/014311699211697

Willis CJ, Macklin JT, Partington KC, Teleki KA, Rees WG, Williams RG (1996) Iceberg detection using ERS–1 synthetic aperture radar. Int J Remote Sens 26:1777–1795. https://doi.org/10.1080/01431169608948739

Young NW, Hyland G (1998) Interannual variability of Antarctic snow melt events derived from scatterometer data. In: Proceedings of the 1998 IEEE international geoscience and remote sensing symposium (IGARSS'98), pp. 2261–2263. https://doi.org/10.1109/IGARSS.1998.703806

Chapter 7
Rogue Waves in the Drake Passage: Unpredictable Hazard

Ekaterina G. Didenkulova, Tatiana G. Talipova, and Efim N. Pelinovsky

Abstract It is generally known that the Drake Passage is difficult for navigation; frequent storms occur there, strong winds blow, and huge sea waves are generated. Differentiating storm waves and rogue waves is often difficult. This article provides information about the observations of rogue waves and possible mechanisms for their generation in the Antarctic Circumpolar Current. It has been noted that such anomalous waves can also occur in the depths of the ocean, due to water stratification.

Keywords Storm · Surface and internal rogue waves · Interaction with currents · Drake Passage

---

E. G. Didenkulova (✉)
National Research University Higher School of Economics, Moscow, Russia

Institute of Applied Physics, Russian Academy of Sciences, Nizhny Novgorod, Russia
e-mail: edidenkulova@hse.ru

T. G. Talipova
Institute of Applied Physics, Russian Academy of Sciences, Nizhny Novgorod, Russia

Nizhny Novgorod State Technical University n.a. R.E. Alekseev, Nizhny Novgorod, Russia

E. N. Pelinovsky
National Research University Higher School of Economics, Moscow, Russia

Institute of Applied Physics, Russian Academy of Sciences, Nizhny Novgorod, Russia

Nizhny Novgorod State Technical University n.a. R.E. Alekseev, Nizhny Novgorod, Russia

Special Research Bureau for Automation of Marine Researches, Yuzhno-Sakhalinsk, Russia

University of Southern Queensland, Toowoomba, QLD, Australia

E. G. Morozov et al. (eds.), *Antarctic Peninsula Region of the Southern Ocean*, Advances in Polar Ecology 6, https://doi.org/10.1007/978-3-030-78927-5_7

7.1 Introduction

For a long time, rogue waves were considered as a part of marine folklore, grossly exaggerating the characteristics of the unusually large sea waves rarely seen in the open ocean. In 1840, during the expedition to the Southern Ocean, French navigator Dumont d'Urville observed an unusually huge wave, about 35-m high. On his return, he made a report at a meeting of the French Academy of Sciences, which provoked laughter from those present who did not believe the speaker. The first report on the properties of such waves, called rogue waves, near the east coast of South Africa, which led to ship disasters, appeared in 1972 (Mallory 1974; Lavrenov 2003). The first report counted 11 events in the area, and since then the number of events has increased. By the end of the twentieth century, there was a large number of instrumental observations of rogue waves on stationary oil and gas platforms in various parts of the oceans. The phenomenon of rogue waves was recognized by the scientific and engineering communities. After recording of the so-called New Year's wave, with a height of 26 m on the Draupner platform in the North Sea on January 1, 1995, it became obvious that thorough studies of rogue waves were needed to predict them and possibly mitigate the consequences of encounters with such a wave. Over the past 20 years, reviews and books have appeared on the problem of rogue waves (Dysthe et al. 2008; Kharif et al. 2009; Onorato et al. 2013; Adcock and Taylor 2014; Shamin 2016; Olagnon 2017; Dudley et al. 2019) that demonstrated the wide distribution of rogue waves in the ocean and the possible causes of their occurrence. However, there are still no regulations for mariners and builders of ships and offshore platforms to assess the risks associated with rogue waves. This is due to the insufficiency of long-term instrumental records of sea waves necessary for calculating the probability of the appearance of rogue waves and the variety of the physical mechanisms for the generation of rogue wave under specific hydrometeorological conditions.

The Drake Passage is located between Cape Horn in South America and the South Shetland Islands of Antarctica (Fig. 7.1). It connects the southwestern part of the Atlantic Ocean (Scotia Sea) with the southeastern part of the Pacific Ocean and extends into the Southern Ocean. Its width is about 800 km. There are several straits (the Strait of Magellan and the Beagle Channel) in the southern part of South America, but ships prefer the Drake Passage, which is open water for hundreds of miles. There are no significant land masses at the latitudes of the Drake Passage. This is important for the unimpeded flow of the Antarctic Circumpolar Current, which carries a huge volume of water through the passage and around Antarctica (about 150 million m^3/s). Its speed reaches 4 km/h (more than 1 m/s). The circumpolar current is driven by strong westerly winds (up to 30 m/s) at the latitudes of the Southern Ocean.

In this paper we discuss the significance of the rogue wave phenomenon for the Drake Passage.

Fig. 7.1 Chart of the Drake Passage

7.2 Field Data of Anomalously Large Waves in the Drake Passage

As mentioned, storm conditions very often occur in the Drake Passage (up to Beaufort number 11), caused by winds up to 30 m/s, causing waves above 15 m in height. In addition, icebergs are often observed here. All of these issues lead to difficult navigation conditions and frequent accidents with passenger and cargo ships, tankers, and yachts. It is difficult to distinguish catastrophes associated with rogue waves from those just with storm waves. As is known, the quantitative criterion that determines the rogue wave is its height H_{fr}, which must exceed twice the height of significant wave height H_s (mean of the highest one-third of waves in the wave record):

$$H_{fr} > 2H_s. \tag{7.1}$$

It is very difficult to check from the descriptions whether this criterion is satisfied in specific cases. Therefore, such data on the Drake Passage are very scarce in the existing catalogues of rogue waves (Nikolkina and Didenlulova 2012; Didenkulova 2020, Didenkulova and Pelinovsky 2020, Lui 2007). We managed to supplement this chronicle, and here we give several examples of anomalous wave descriptions that can be considered as rogue waves, although quantitative information is lacking.

- In December 1956, Miles and Beryl departed Melbourne on the sailboat Tzu Hang to visit their daughter Clio at school in England, intending to follow the old clipper route. The journey would take them eastbound around Cape Horn, a voyage that at that time had very rarely been accomplished in small boats. Approaching Cape Horn, the yacht was pitchpoled by a rogue wave. Beryl, who had been at the helm, was tossed from the boat and injured. Tzu Hang was dismasted, partially submerged, and the topsides were severely damaged, but the sailors managed to sail the damaged vessel to Chile, where extensive repairs were undertaken. In 1957, a year later, Miles and Beryl departed again to round Cape Horn. However, approximately at the same position, beset by storms, another dismasting took place. Again, they managed to make their journey to the coast of Chile, and Tzu Hang was shipped to England for repairs. These adventures were published in their book, *Once is Enough* (source: https://en.wikipedia.org/wiki/Miles_and_Beryl_Smeeton).
- There is a description of rogue wave events in 2001: The Bremen and Caledonian Star passenger ships were carrying hundreds of tourists across the South Atlantic. At 5 am, on March 2, the Caledonian Star's first officer saw a 30-m wave bearing down on them. It smashed over the ship, flooding the bridge and destroying much of the navigation and communication equipment. The Caledonian Star limped back to port, her crew and passengers grateful that the engines had kept running. Just days earlier, the cruise liner Bremen had been less fortunate. The ship with 137 German tourists onboard faced an awesome wall of water in the South Atlantic. The impact knocked out all the instrumentation and all power, leaving them helpless in the tumultuous sea. Unable to maintain her course into the waves, there was a real risk the ship could go down, and they knew none of the passengers would survive in lifeboats in such freezing conditions. With emergency power only, the crew battled to restart the engines. When they eventually succeeded, it opened the door to a very lucky escape (source: http://www.bbc.co.uk/science/horizon/2002/freakwave.shtml).
- On February 1, 2007, the Holland America's cruise ship Prinsendam had left Ushuaia and rounded Cape Horn, when it ran into hurricane force winds and then two rogue waves about 12-m high according to the online journal of a passenger. There were around 40 injuries, though none serious, but some people needed hospitalization. This case was independently confirmed by another passenger in a YouTube report (http://www.youtube.com/watch?v=MnoTj7Jx4L4) (source: https://www.cruiseshipsinking.com/2007/02/01/prinsendam-cruise-ship-storm-damage-cape-horn-chile/).
- On December 29, 2006, while operating on a South American cruise itinerary, the vessel Crystal Symphony was crossing through the Drake Passage in rough seas. A rogue wave hit the ship, breaking a passenger cabin window (forward portside on deck 5). As a result, the stateroom was flooded, as well as the adjacent cabins and hallways. No injuries were reported. On December 30, the ship arrived unscheduled and anchored off call port Ushuaia Argentina, where the necessary repairs were made (source: https://www.cruisemapper.com/accidents/Crystal-Symphony-603).

- On December 7, 2010, in 55 mph winds, a 30-ft wave slammed into an Antarctic cruise ship “The Clelia II” with 88 passengers and 77 crew members aboard just northeast of the South Shetland Islands on its way to a port in Argentina. The vessel declared an emergency when it lost power and communications, after a 30-ft wave washed over the deck and took out windows on the bridge causing some electrical malfunctions that temporarily knocked out some communications and affected engine performance. The nearby MS National Geographic Explorer rendered assistance [deep water, Hs = 3.5 m, Hfr = 9 m, WS = 25 m/s, one injury, ship damage, https://www.youtube.com/watch?v=eDTbopUYg20]. In this case, the quantitative criterion (7.1) is satisfied, so this wave is a reliable case of a rogue wave observation.

7.3 Wave-Current Interaction as a Mechanism of Rogue Wave Formation

Bearing in mind the high speed of the Antarctic Circumpolar Current, which can exceed 1 m/s, it is logical to assume that the main mechanism of rogue wave generation is the interaction of wind waves and current. This mechanism has been studied in a number of works (Kharif and Pelinovsky 2003; Toffoli et al. 2013, 2015; Hjelmervik and Trulsen 2009; Lavrenov and Porubov 2006; Onorato et al. 2011; Manolidis et al. 2019) and is here applied to the Antarctic Circumpolar Current conditions. Since the strait is deep enough, the wind waves are short and, in a linear approximation, are described by the dispersion relation.

$$\omega = \sqrt{g\sqrt{k_x^2 + k_y^2}} + k_x U, \tag{7.2}$$

where the flow is assumed to be unidirectional (along the *x* axis), which can depend on both coordinates *U(x, y)*, *g* is acceleration of gravity, k_x and k_y are components of the wave vector, and ω is wave frequency. As usual, we assume that the scales of the current variability exceed the length of the wind wave, which locally can be assumed to be monochromatic with a frequency determined by Eq. (7.2). In the general case, the frequency and wavenumber are also functions of the coordinates. The variability of the flow is described by the kinematic Whitham equations (Whitham 1974; Lavrenov 2003; Kharif et al. 2009):

$$\frac{\partial k'}{\partial t} + \nabla \omega = 0, \tag{7.3}$$

$$rot\left(\vec{k}\right) = \frac{\partial k_y}{\partial x} - \frac{\partial k_y}{\partial y} = 0. \tag{7.4}$$

Wave amplitude of a quasi-monochromatic wave in this approximation is described by the conservation equation:

$$\frac{\partial A}{\partial t} + div\left[\vec{c}_{gr} A\right] = 0, \tag{7.5}$$

where

$$\vec{c}_{gr} = \frac{d\omega}{d\vec{k}} \tag{7.6}$$

is group velocity. The kinematic Whitham equations are partial differential equations having a Hamiltonian form. They can be solved by the method of characteristics. Let us consider two important cases of wave-current interaction.

7.3.1 Unidirectional Wave Propagation

Let us assume that the current varies along the x-axis and does not depend on the transverse y-coordinate. A quasi-monochromatic wave also propagates along the x-axis and can move in the same or the opposite direction. In this case, the wave frequency is constant, and only its wavenumber changes, which follows from Eq. (7.3). Taking into account that the wave vector has only one component, we rewrite the dispersion relation in the form

$$\omega = \sqrt{gk} + skU = const, \tag{7.7}$$

where s is +1 for following and −1 for counter propagation. As the flow velocity changes, the wavenumber (and its wavelength) changes in accordance with the formula (Phillips 1977):

$$\sqrt{\frac{k_0}{k}} = \frac{1}{2} + \frac{1}{2}\sqrt{1 + \frac{4sU}{\sqrt{g/k_0}}} \tag{7.8}$$

as follows from Eq. (7.7). Similarly, the solution of Eq. (7.5) gives the wave amplitude:

$$\frac{A}{A_0} = \frac{\sqrt{g/k_0}}{\sqrt{\sqrt{g/k}\left(\sqrt{g/k} + 2sU\right)}}, \tag{7.9}$$

where k_0 is the wavenumber and A_0 is the wave amplitude on still water. It is convenient to illustrate these formulas qualitatively (Fig. 7.2). At fixed frequency, the wavenumber decreases when the following current increases ($s = +1$), and the wave

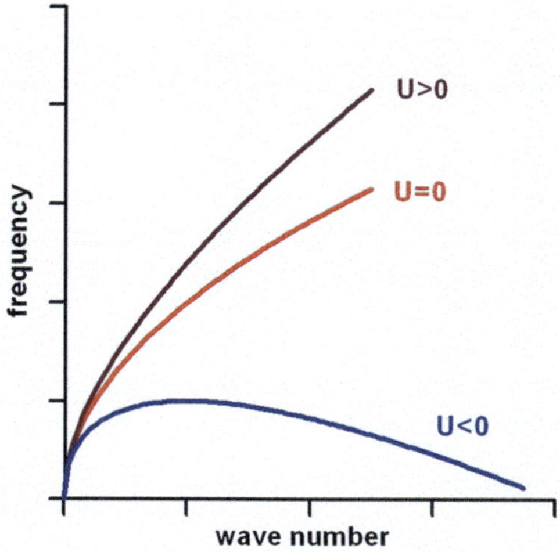

Fig. 7.2 Dispersion relation for unidirectional wave propagation

amplitude also decreases. The following current does not lead to wave amplification. If the speed of the following flow decreases, then the wavenumber increases with the amplitude of the wave. In general, however, these effects are not very significant. Thus, when a wave has a length of 15 m, and an amplitude of 1 m and passes from calm water to a current of 1 m/s, its length increases to 21 m, and its amplitude decreases to 0.7 m.

More dramatic wave changes occur on the counter flow ($s = -1$). If the speed of the counter flow increases, then, as can be seen from Fig. 7.2 and Eqs. (7.8) and (7.9), the wavenumber increases up to the critical value at which the wave is blocked by the flow. The blocking condition has a clear physical description:

$$c_{gr} = \frac{1}{2}\sqrt{\frac{g}{k_0}} = \frac{g}{2\omega} = U. \tag{7.10}$$

At this point, the wave stops, and its amplitude, according to Eq. (7.9), tends to infinity. In the region of the Antarctic Circumpolar Current with a speed of 1 m/s, wind waves with a period of 1.3 s or less are blocked by the current. Such scales are typical for wind waves in the range of collapse, which is often observed in nature. Such waves are dangerous for navigation. But even if we remain within the framework of the linear theory and assume that the wave does not break, it cannot remain at the blocking point and will be reflected from it as a backward wave (the descending branch of the dispersion relation in Fig. 7.2). The wave will become shorter and shorter, and its amplitude will increase. This process is not described in terms of the equations just outlined; for details, see Basovich and Talanov (1977) and Basovich and Bakhanov (1979). It is important to emphasize that there will be steeper waves, often with collapse, in the vicinity of the blockage zone, and their danger is obvious.

The transformation of a wave on a counter current can lead to the formation of large dangerous waves. However, such waves do not belong to the category of rogue waves, which are characterized by their unexpectedness and short duration of appearance. Such waves arise in the field of wind waves, which contain many spectral components, but not in the case of one quasi-monochromatic wave. Since each spectral component propagates with its own group velocity, their interference can lead to anomalously high amplitudes at a certain moment in time and at a certain point in space. This mechanism is called dispersive focusing and is detailed in Kharif et al. (2009) for waves in the ocean without flow. The flow strongly affects the conditions of wave packet focusing, and this case should be considered separately.

The optimal focusing of wave packets is described by a self-similar solution to the one-dimensional version of Eq. (7.3), which can be rewritten in terms of the group velocity:

$$c_{gr}\left(k\right)=\frac{x_0-x}{T-t}, \tag{7.11}$$

where x_0 is the coordinate of the focal point and T is the time of focusing (Fig. 7.3). At the initial moment of time, wave packets should be located according to their speed: slowly moving packets are in front, and faster ones are behind. During the propagation, fast packets catch up with slow ones, and strong interference (focusing) occurs. Then, the faster packets go forward, and the wave field becomes weak again. The appearance of packets with different speeds at different points in space is associated with the spatial inhomogeneity of the wind field above the waves, which generates waves with different lengths (and therefore different speeds) and amplitudes. Leaving the generation zone, the waves move with different speeds and interfere with each other. This mechanism explains the short duration and randomness of rogue wave, since the appearance of an optimal scenario in the field of a random wind is a rare event (Pelinovsky et al. 2011).

The optimal arrangement of wave packets in space at the initial moment of time, as follows from Eq. (7.11), is the following (in terms of the wavelength):

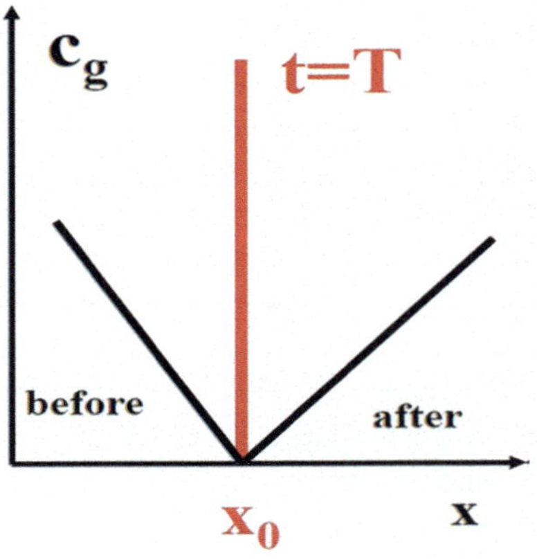

Fig. 7.3 Optimal focusing process of wave trains

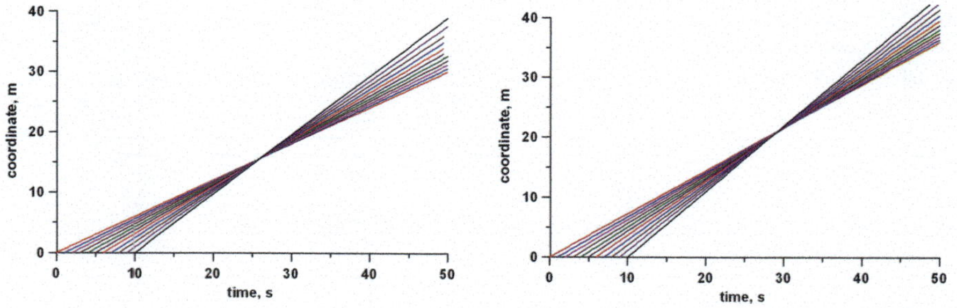

Fig. 7.4 Trajectories of focusing wave packets in the absence (left) and presence (right) of a following current

$$\lambda(x) = \frac{\pi}{2g}\left(\frac{x_0 - x}{T} - sU\right)^2, \tag{7.12}$$

in the domain where the expression in the parenthesis is positive.

$$x < x_0 - sUT. \tag{7.13}$$

In laboratory conditions, the wave field is set on the wave generator, so the wave frequency is variable. A linear dependence of frequency on time is optimal for focusing in a stationary medium:

$$\omega_0(t) = \omega_{max}\left(1 - \frac{\omega_{max} - \omega_{min}}{\omega_{max}}\frac{t}{T}\right). \tag{7.14}$$

In the case of following current, such a change is not optimal, and the focus becomes distributed (Fig. 7.4). This process has been studied experimentally in our research studies (Touboul et al. 2006, 2007; Kharif et al. 2008).

7.3.2 Wave Refraction on a Shear Flow

Now let us consider the jet flow in the horizontal region *U(y)*, when the monochromatic wave approaches the flow zone at a certain angle measured from the *x* axis (Fig. 7.5). In this case, it follows from Eqs. (7.3) and (7.4) that

$$\omega = const, k_x = \frac{\omega^2}{g}\cos\theta_0 = const, \tag{7.15}$$

and, therefore, only the transverse component of the wavevector $k_y(y)$ changes. Its value is found from the dispersion relation (7.2):

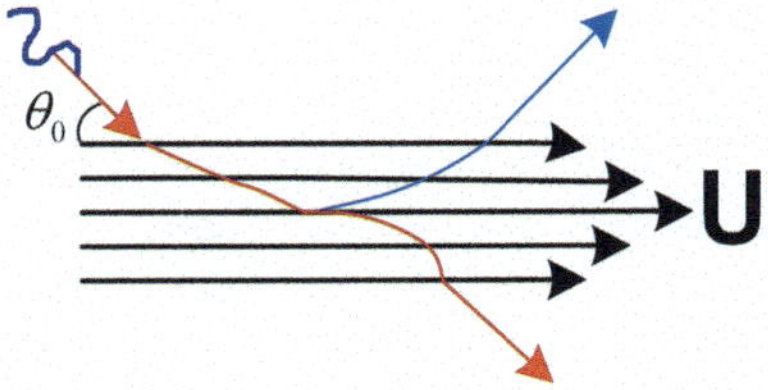

Fig. 7.5 Schematic representation of the wave transformation on the shear flow

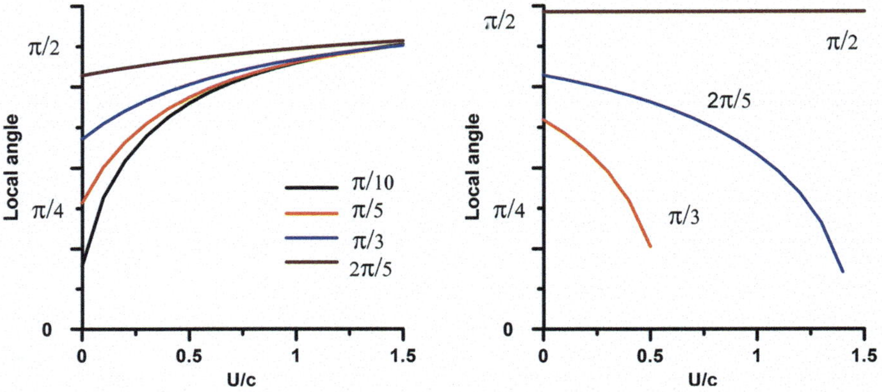

Fig. 7.6 Change of the local angle of refraction depending on the speed of the current (on the left in the opposite current, on the right in the following current)

$$k_y^1 = \frac{(\omega - s k_x U)^4}{g^2} - k_x^2, k_x = \frac{\omega^2}{g}\cos\theta_0. \tag{7.16}$$

In fact, from this we derive an equation for the local angle of refraction in the flow (Fig. 7.6):

$$\tan^2\theta(y) = \frac{1}{\cos^2\theta_0}\left[1 - \frac{sU(y)}{c}\cos\theta_0\right]^4 - 1, \tag{7.17}$$

where $c = g/\omega$ is the wave celerity outside of the current.

In an opposite current ($s = -1$), the wave unfolds across the flow, passes it, and leaves the flow zone at the same angle. In a following current ($s = +1$), the wave turns with the flow, and if the flow is strong, then the wave cannot overcome it. The critical value of the incidence angle is found from the equation following from Eq. (7.17):

$$\frac{U}{c} = \frac{1 - \sqrt{\cos\theta_0}}{\cos\theta_0} \tag{7.18}$$

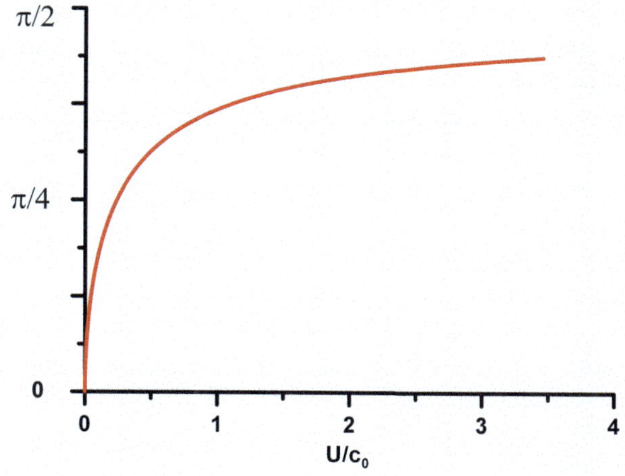

Fig. 7.7 Dependence of the critical angle of waves on the current velocity when the waves approach and reflect from the current

shown in Fig. 7.7. As one can see, only sufficiently long waves, approaching the flow almost perpendicularly, can pass the flow zone. The rest of the waves (shorter ones) are "wrapped" by the strong current; therefore, short waves with a low speed are always reflected from the current. Their interference with approaching waves increases the "crowd" of waves, leading to difficult navigation conditions. If we also take into account that the wind waves are not sinusoidal, then different spectral components of the waves will be reflected from different layers of the current, increasing the irregularity and the surge of waves, leading to the formation of rogue waves in previously unpredictable places. The current also changes the conditions for wave excitation by the wind (Kharif and Abid 2020), which is an additional factor in stimulating the dispersive focusing of irregular wind waves.

The important role of this mechanism in the formation of rogue waves in the ocean has now begun to be recognized (Fedele et al. 2016), but its analysis requires sufficiently complete data on the wind and wave fields. The current leads to a strong irregularity of the wind wave (even if it was sufficiently uniform outside the current) and adds processes that manifest the dispersion mechanism of rogue wave formation.

7.4 Anomalously Large Internal Waves

Data on internal waves of the tidal period in the Drake Passage are presented in Heywood et al. (2007), Morozov (2018), and Khimchenko et al. (2020). The measured amplitudes of internal tides at the frequency of the semidiurnal tide M_2 (12.4 h) vary from 20 m in the southern part of the Drake Passage to 70 m in its northern part. These are waves capable of generating short-scale solitons and high-amplitude breathers, which are observed in the Bransfield Strait south of the Drake Passage (Khimchenko et al. 2020). The theory of the appearance of solitons and breathers in a tidal internal wave has been described in many works (see, e.g.,

Holloway et al. 2001; Pelinovsky et al. 2007; Grimshaw et al. 2010a). Currents play an important role in the variability of internal waves, which can already be seen from the dispersion relation, for example, for internal waves in the ocean with exponential stratification and a constant current at depth, we get

$$\omega = \frac{Nk}{\sqrt{k^2 + \pi^2 / H^2}} + k_x U, \tag{7.19}$$

where N is the Vaisala-Brent frequency and H is the depth. The maximum speed of free internal waves does not exceed 2 m/s and is comparable to the speeds of the currents in the Drake Passage. This means that all the mechanisms described above for the interaction of wind waves with currents are also applicable for internal waves contributing to the formation of internal rogue waves. However, additional mechanisms of the formation of anomalously large internal waves are also possible described in Grimshaw et al. (2010b), Talipova et al. (2011), Talipova (2011), Didenkulova (2019), Shurgalina (2018), Pelinovsky and Shurgalina (2015, 2016), and Shurgalina and Pelinovsky (2016), which we will not discuss here.

7.5 Conclusions

We have compiled data on the unusual sea waves called rogue waves in the Drake Passage. Their origin, in our opinion, is associated with the interaction of sea waves with the Antarctic Circumpolar Current, the speed of which reaches 1 m/s. The theory of such an interaction has been described, and it has been shown that the mechanism of dispersive focusing of wave packets can lead to the formation of rogue waves. It has been noted that internal waves of large amplitude are also observed in the Drake Passage, which can also excite rogue waves, when interacting with the current.

Acknowledgments Analysis of field data of rogue waves was performed with the support from the Russian Science Foundation (grant No. 19-12-00253). The study of wave-current interaction and anomalously large internal waves was supported by the Russian Foundation for Basic Research (grants 19-05-00161 and 19-35-60022).

References

Adcock TA, Taylor PH (2014) The physics of anomalous ('rogue') ocean waves. Rep Prog Phys 77:105901

Basovich AYa, Bakhanov VV (1979) On transformation of surface waves in nonuniform current. Proc Acad Sci USSR, FAO 15(6):660–665. (in Russian)

Basovich AYa, Talanov VI (1977) On transformation of surface waves in nonuniform current. Proc Acad Sci USSR, FAO 13(7):660–665. (in Russian)

Didenkulova E (2020) Catalogue of rogue waves occurred in the World Ocean from 2011 to 2018 reported by mass media sources. Ocean Coast Manag 188:105076
Didenkulova (Shurgalina) EG (2019) Numerical modeling of soliton turbulence within the focusing Gardner equation: rogue wave emergence. Phys D 399:35–41
Didenkulova E, Pelinovsky E (2020) Freak waves in 2011–2018. Dokl Earth Sci 491(1):187–190
Dudley JM, Genty G, Mussot A, Chabchoub A, Dias F (2019) Rogue waves and analogies in optics and oceanography. Nat Rev Phys 1:675–689
Dysthe K, Krogstad HE, Muller P (2008) Oceanic rogue waves. Annu Rev Fluid Mech 40:287–310
Fedele F et al (2016) Real world ocean rogue waves explained without the modulational instability. Sci Rep 6:27715
Grimshaw R, Pelinovsky E, Talipova T, Kurkina O (2010a) Internal solitary waves: propagation, deformation and disintegration. Nonlinear Process Geophys 17:633–649
Grimshaw R, Pelinovsky E, Talipova T, Sergeeva A (2010b) Rogue internal waves in the ocean: long wave model. Eur Phys J Spec Top 185:195–208
Heywood KJ, Collins JL, Hughes CW, Vassie I (2007) On the detectability of internal tides in Drake Passage. Deep-Sea Res 54(11):1972–19849
Hjelmervik KB, Trulsen K (2009) Freak wave statistics on collinear currents. J Fluid Mech 637:267
Holloway P, Pelinovsky E, Talipova T (2001) Internal tide transformation and oceanic internal solitary waves. In: Grimshaw R (ed) Environmental stratified flows. Kluwer Academic Publishers, New York, pp 29–60
Kharif C, Abid M (2020) Miles theory revisited with constant vorticity in water of infinite depth. J Mar Sci Eng 8:623
Kharif C, Pelinovsky E (2003) Physical mechanisms of the rogue wave phenomenon. Eur J Mech B Fluids 22(6):603–634
Kharif C, Giovanangeli J-P, Touboul J, Grare L, Pelinovsky EN (2008) Influence of wind on extreme wave events: experimental and numerical approaches. J Fluid Mech 594:209–247
Kharif C, Pelinovsky E, Slunyaev A (2009) Rogue waves in the ocean. Springer, Berlin, p 216
Khimchenko EE, Frey DI, Morozov EG (2020) Tidal internal waves in the Bransfield Strait, Antarctica. Russ J Earth Sci 20:ES2006
Lavrenov IV (2003) Wind-waves in oceans. Springer, Berlin, p 376
Lavrenov I, Porubov AV (2006) Three reasons for freak wave generation in the non-uniform current. Eur J Mech B Fluids 25:574–585
Liu PC (2007) A chronology of freaque wave encounters. Geofizika 24:57–70
Mallory JK (1974) Abnormal waves in the southeast coast of South Africa. Int Hydrogr Rev 51:99–129
Manolidis M, Orzech M, Simeonov J (2019) Rogue wave formation in adverse ocean current gradients. J Mar Sci Eng 7:26
Morozov EG (2018) Oceanic internal tides. Observations, analysis and modeling. A global view. Springer, Cham. 316 p
Nikolkina I, Didenkulova I (2012) Catalogue of rogue waves reported in media in 2006–2010. Nat Hazards 61:989–1006
Olagnon M (2017) Rogue waves: anatomy of a monster. Adlard Coles Nautical, London, p 160
Onorato M, Proment D, Toffoli A (2011) Triggering rogue waves in opposing currents. Phys Rev Lett 107:184502
Onorato M, Residori S, Bortolozzo U, Montina A, Arecchi F (2013) Rogue waves and their generating mechanisms in different physical contexts. Phys Rep 528:47–89
Pelinovsky EN, Shurgalina EG (2015) Two-soliton interaction in the frameworks of modified Korteweg – de Vries equation. Radiophys Quantum Electron 57(10):737–744
Pelinovsky EN, Shurgalina EG (2016) Formation of freak waves in a soliton gas described by the modified Korteweg–de Vries equation. Dokl Phys 61(9):423–426
Pelinovsky EN, Slunyaev AV, Polukhina OE, Talipova TG (2007) Internal solitary waves. In: Grimshaw R (ed) Solitary waves in fluids. WIT Press, Southampton/Boston, pp 85–110

Pelinovsky E, Shurgalina E, Chaikovskaya N (2011) The scenario of a single freak wave appearance in deep water – dispersive focusing mechanism framework. Nat Hazards Earth Syst Sci 11:127–134
Phillips OM (1977) The dynamics of the upper ocean. Cambridge University Press, New York
Shamin RV (2016) Mathematical problems of killer waves. LENAND/URSS, Moscow, p 168
Shurgalina EG (2018) The mechanism of the formation of freak waves in the result of interaction of internal waves in stratified basin. Fluid Dyn 53(1):59–64
Shurgalina EG, Pelinovsky EN (2016) Nonlinear dynamics of a soliton gas: modified Korteweg-de Vries equation framework. Phys Lett A 380(24):2049–2053
Talipova TG (2011) Mechanisms of the internal rogue wave formation. Fundam Appl Hydrophys 4(4):58–70
Talipova TG, Pelinovsky EN, Kharif C (2011) Modulational instability of long internal waves of moderate amplitudes in stratified and horizontally inhomogeneous ocean. JETP Lett 94(3):182–186
Toffoli A, Waseda T, Houtani H, Kinoshita T, Collins K, Proment D, Onorato M (2013) Excitation of rogue waves in a variable medium: an experimental study on the interaction of water waves and currents. Phys Rev E 87:051201
Toffoli A, Waseda T, Houtani H, Cavaleri L, Greaves D, Onorato M (2015) Rogue waves in opposing currents: an experimental study on deterministic and stochastic wave trains. J Fluid Mech 769:277–297
Touboul J, Giovanangeli JP, Kharif C, Pelinovsky E (2006) Experiments and simulations of freak waves under the action of wind. Eur J Mech B Fluids 25(5):662–676
Touboul J, Pelinovsky E, Kharif C (2007) Nonlinear focusing wave group on current. J Korean Soc Coast Ocean Eng 9(3):222–227
Whitham GB (1974) Nonlinear waves. John Wiley and Sons, New York, p 635

Chapter 8
Water Mass Transformation in the Powell Basin

Alina A. Fedotova and Svetlana V. Stepanova

Abstract We analyze CTD and hydrochemical data from more than 30 oceanographic stations occupied during cruise 79 of the R/V *Akademik Mstislav Keldysh* (AMK-79) in January–February 2020. Sampling stations were organized into a spatial grid in the Powell Basin, Hesperides Trough, Phillip Ridge, Phillip Passage, South Orkney Plateau, and South Scotia Ridge. The paper includes preliminary results of the hydrological and hydrochemical research of water mass properties and their transformation in the region of Antarctic Bottom Water outflow. Three modifications of Warm Deep Water with various thermohaline and hydrochemical characteristics were observed in various parts of the Powell Basin; we suggest that they have dissimilar mechanisms of formation. We propose a concept of Weddell Sea Slope Water that is a result of the mixing of shelf waters formed in the Weddell Sea and Warm Deep Water over the slope south of the Joinville Ridge and research the hydrological role of its inflow into the Powell Basin. The admixture of Weddell Sea Slope Water to Weddell Sea Deep Water transforms water properties at depths of 1500–2500 m over the slope in the Powell Basin. Weddell Sea Deep Water is getting fresher, but colder and denser, than observed in the central deep part of the Powell Basin. We introduce a scheme of pathways of regional water masses based on the analysis of AMK-79 data, high-resolution historical CTD data from World Ocean Database 2018, and literature sources.

Keywords Northwestern Weddell Sea · Powell Basin · AMK-79 · Weddell Sea Deep Water · Warm Deep Water · Weddell Front · Water mass transformation

---

A. A. Fedotova (✉)
Lomonosov Moscow State University, Moscow, Russia

Shirshov Institute of Oceanology, Russian Academy of Sciences, Moscow, Russia

S. V. Stepanova
Shirshov Institute of Oceanology, Russian Academy of Sciences, Moscow, Russia

E. G. Morozov et al. (eds.), *Antarctic Peninsula Region of the Southern Ocean*, Advances in Polar Ecology 6, https://doi.org/10.1007/978-3-030-78927-5_8

8.1 Introduction

The characteristics of deep water outflowing from the Weddell Sea are determined by complex interactions between the atmosphere, which determines sea ice processes, and the heat flows, sub-ice shelf water mass transformations (e.g., Gordon 1971; Grosfeld et al. 1997), and the ocean dynamics at the continental shelf break and slope (Foldvik et al. 1985; Muench and Gordon 1995).

Over the past 50 years, much attention has been paid to studying the processes in the Weddell Sea. Fewer observations have been made in the Powell Basin and in the Hesperides Trough. Production and export of Antarctic Bottom Water as a component of the global climate system is one of the highlights of polar oceanography. Review of recent studies has shown that Antarctic Bottom Water formation in the Weddell Sea is a more complex process than reported in Carmack and Foster, (1975) and Foldvik et al. (1985). Gordon et al. (2001) suggested two types of deep water and two types of bottom water in the northwestern Weddell Sea. These are the more saline, less oxygenated Weddell Sea Deep Water (WSDW) and Weddell Sea Bottom Water (WSBW), which is most likely advected from the southwestern Weddell Sea and is a fresher form of WSDW and WSBW, designated as ventilated WSDW and WSBW (VWSDW/VWSBW). These fresher modifications of WSDW and WSBW are formed due to the inflow of cold and highly oxygenated shelf water descending along the continental slope from the northern tip of the Antarctic Peninsula (NTAP) (Gordon et al. 2001).

The study region off the northern tip of the Antarctic Peninsula (NTAP) includes the Powell Basin, Hesperides Trough, and Phillip Passage, which are important regions of AABW outflow. The Powell Basin hydrography is dominated by the Weddell Sea, the major Antarctic Bottom Water (AABW) source region (Deacon 1937; Gill 1973; Orsi et al. 2002). The WSBW and WSDW flow into the Powell Basin, and WSDW outflow through the Philip Passage is estimated at 0.7 ± 0.4 Sv (Naveira Garabato et al. 2002). The region off NTAP is characterized by strong tidal currents (Robentson 2001; Padman et al. 2008; Flexas et al. 2015), complicated bottom topography, and interaction between water masses from few Southern Ocean regions such as the Weddell Sea, Bransfield Strait, and Drake Passage. Tides and generated internal waves cause upper-layer mixing (Flexas et al. 2015) and deep water mixing (Munk and Wunsch 1998). Robertson (2005) suggested tides as a source of mixing and formation of WSDW in the Weddell Sea. The part of the WSDW exported into the Scotia Sea upwells due to high turbulence, which creates a "blender" effect to mix ocean water masses in this region (Franco et al. 2007).

The freshening signal of decadal trends in deep and bottom water properties over the Southern Ocean remains unexplained, and the changes in this region are not uniform (Rintoul 2007; Purkey and Johnson 2013). According to Fahrbach et al. (2004), WSBW was becoming warmer and losing volume in the 1990s. These changes indicate regional variations or even restructuring in ABBW formation and export processes. Therefore, better knowledge of AABW formation and its pathways based on oceanographic observations is required for a better understanding of changes in the bottom water properties of the Southern Ocean.

The paper includes the hydrological and hydrochemical study in the AABW export region based on the analysis of oceanographic data from cruise 79 of the R/V *Akademik Mstislav Keldysh* (AMK 79) during austral summer 2020 and high-resolution historical conductivity-temperature-depth (CTD) data from World Ocean Database 2018 (WOD18). The objective of this study is to describe the characteristics of the regional water masses and their transformation. We also outline the preliminary results of the study of hydrological role of the Weddell Sea waters inflow into the Powell Basin.

8.2 Regional Oceanography

Antarctic Surface Water (AASW) consisting of relatively warm, fresh, and well-mixed surface layer of Summer Water (SSW) and the subsurface temperature minimum layer or Winter Water (WW) characterizes the upper-layer in this zone. In this paper, we term Shelf Water (SW) all waters that are formed over the Weddell Sea shelf such as Winter Water (WW, the residual of the previous winter mixed layer), Low Salinity Shelf Water (LSSW, 34.5 < S < 34.6 psu), High Salinity Shelf Water (HSSW, S > 34.6 psu, whose potential temperature is close to the surface freezing point $T_f \sim -1.9$ °C), and Ice Shelf Water (ISW, potential temperature below the freezing point). WW, LSSW, and HSSW are produced during sea ice formation over the Antarctic shelf, predominantly in the coastal and ice front polynyas in winter months. The ISW is the coldest water mass of the World Ocean; it is formed from LSSW and HSSW contacting underside of the ice shelf, under Larsen and Ronne-Filchner ice shelves in the Weddell Sea (Gordon et al. 2001). The LSSW, HSSW, and ISW are shown in rectangles marked with dashed gray line in the θ/S diagram in Fig. 8.2b.

According to Thompson and Heywood (2008), the inflow from the Weddell Sea into the Powell Basin is dominated by three barotropic northward flowing currents: the Antarctic Coastal Current (ACoC), Antarctic Slope Front (ASF), and Weddell Front (WF). Less dense newly formed WW flows with ACoC over the shelf between the 400 and 500 m isobaths (Heywood et al. 2004) along the Antarctic Peninsula to Joinville Island (Fig. 8.1). Thereafter, part of this water continues to flow northward over the Philip Ridge, and the rest of water turns west into the Bransfield Strait (Whithworth et al. 1994). Denser SW ($28.0 < \gamma_n < 28.27$) is denser than the ambient water, and it cascades down the slope as an underflow (Amblas and Dowdeswell 2018), but its density is not sufficient to quickly occupy the isopycnal layer of the deep water. This SW flows along the slope and forms the Antarctic Slope Current (ASC). The border between cold fresh shelf water and the surrounding warmer saline Warm Deep Water (WDW) is termed the Antarctic Slope Front (ASF). According to Thompson et al. (2009), the pathway of the ASF is shown as a gray line in Fig. 8.1. The Weddell Front (WF) associated with the dense SW flow was

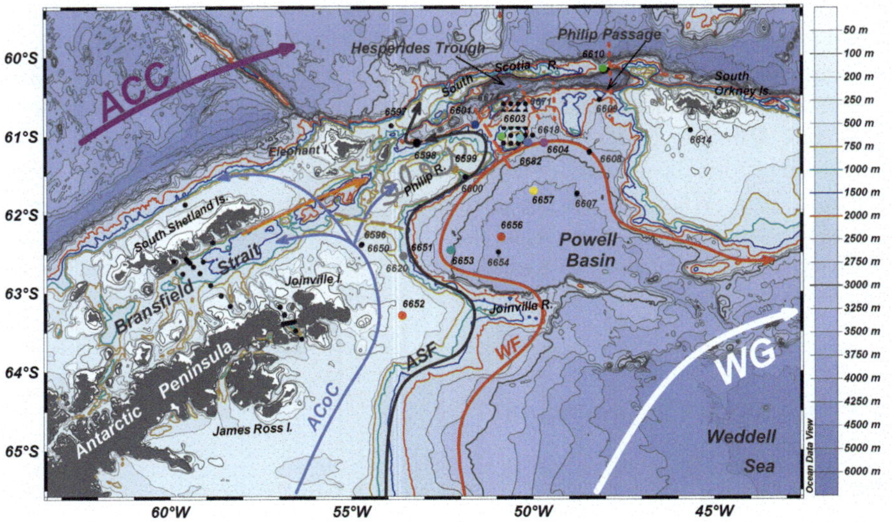

Fig. 8.1 Study site near the Antarctic Peninsula. Oceanographic stations and their numbers are indicated. Color lines of the paths of the Antarctic Coastal Current (ACoC/blue line, Whitworth et al. 1994), Antarctic Slope Front (ASF/dark grey line, Thompson et al. 2009), Weddell Front (WF/dark red line, Thompson and Heywood 2008), Bransfield Waters (orange line), Antarctic Circumpolar Current (ACC/violet line), and Weddell Gyre (WG/white line) are shown. Mixing of water mases over the Philip Ridge according to Von Gyldenfeldt et al. (2002) is indicated by a light gray helix. Channels with depths ~2000 m are shown as red dashed lines; a channel with depth of ~750 m is indicated by a dashed ochre line. Stations shown in Fig. 8.2 indicated by color dots. Bathymetry is based on GEBCO-2014 (Weatherall et al. 2015)

found by Thompson et al. (2009) between 2500 and 3000 m. According to Thompson et al. (2009) and Thompson and Heywood (2008), the ASF and WF are distinct currents.

Foster and Carmack (1976a) suggested a three-stage process of AABW formation in the Weddell Sea. First, relatively warm salty low-oxygen WDW is modified by admixture with overlying colder and fresher WW producing mWDW. Second, mWDW spreads westward in the Weddell Gyre and forms Weddell Sea Bottom Water (WSBW, $\theta < -0.7$ °C) by mixing with the densest near-freezing point shelf water, HSSW. Eventually, WSBW flows northward and mixes again with WDW to form Weddell Sea Deep Water (WSDW, $0 < \theta < -0.7$ °C). Therefore, AABW is commonly considered to consist of two separate water masses in the Weddell Sea.

Warm Deep Water (WDW, with $\theta > 0$ °C) is the modified Circumpolar Deep Water (CDW) entrained from the Antarctic Circumpolar Current (ACC) into the major element of the Weddell Sea circulation, the Weddell Gyre (WG) (Carmack and Foster 1975; Foster and Carmack 1976b; Gordon 1998) in the Southwest Indian Ridge between 25 and 32° W (Orsi et al. 1993). The WDW is transported westward near the Antarctic continent and mixes with WW and SW over the shelf or slope transforming or forming new deep water.

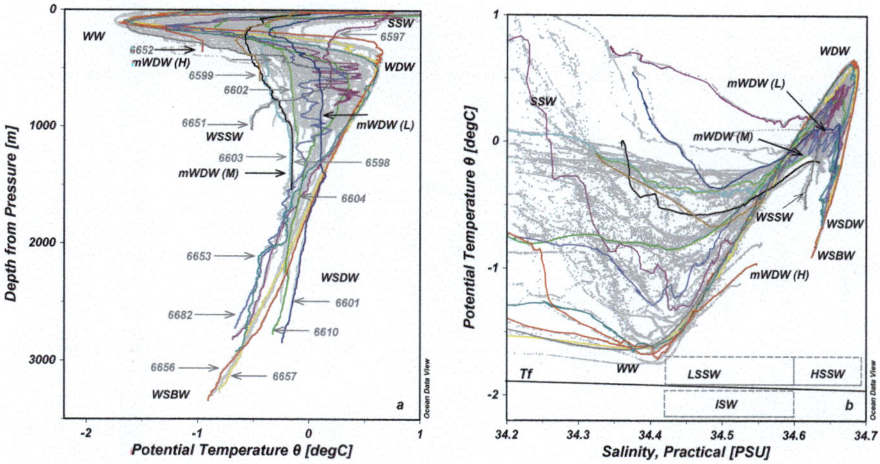

Fig. 8.2 Vertical distribution of potential temperature (**a**) and θ/S diagram (**b**) with marked regional water masses and numbers of stations. Vertical distribution of potential temperature (**a**) and θ/S diagram (**b**) at all CTD stations of the study site occupied during the AMK-79 cruise are shown in the background as light gray points. The location of stations is shown in Fig. 8.1. The lower line Tf represents the surface freezing temperature as a function of salinity (Tf, freezing curve by Miyake, 1939). Weddell Sea shelf waters (HSSW, high salinity shelf water; LSSW, low salinity shelf water; ISW, ice shelf water) are shown in rectangles marked with dashed gray lines

8.3 Data and Layer Definitions

In this paper, we analyze data from more than 30 oceanographic stations occupied during the AMK-79 cruise in January–February 2020. Sampling stations were organized into a spatial grid. Oceanographic stations were occupied in the Powell Basin, Hesperides Trough, Phillip Ridge, Phillip Passage, South Orkney Plateau, and South Scotia Ridge. Locations of stations are shown in Fig. 8.1. In addition, we analyze high-resolution historical CTD data from WOD18 (Boyer et al. 2018) for studying the distribution of WSDW and WSBW in the Powell Basin (dark red and red dots in Fig. 8.3). The data consist of field measurements from R/V *Polarstern* (April, 1998; March, 2005) and R/V *Nathaniel B. Palmer* (August, 1997) during the international oceanography research program Deep Ocean Ventilation Through Antarctic Intermediate Layers (DOVETAIL) (Padman et al. 2006).

Hydrographic data were collected using a Sea-Bird 911plus CTD system, a fast optical DO sensor RINKO III (ARO-CAV), and a SBE 32 Carousel Water Sampler with 12 5-l bottles. The measurements were carried out to the depths of 3–7 m above the bottom. Seawater was sampled at levels of 0, 50, 100, 200, 500, 750, 1000, and 2000 m, and at the bottom and, in addition, at levels selected according to the presence of extremums on the θ/S curves. Bottle samples were analyzed for dissolved oxygen, phosphate, nitrate, nitrite, ammonia nitrogen, silicate, total alkalinity, and pH. However, in this paper, we analyze only dissolved oxygen, phosphate, nitrate, and silicate as additional indicators of the water mass distribution and regional

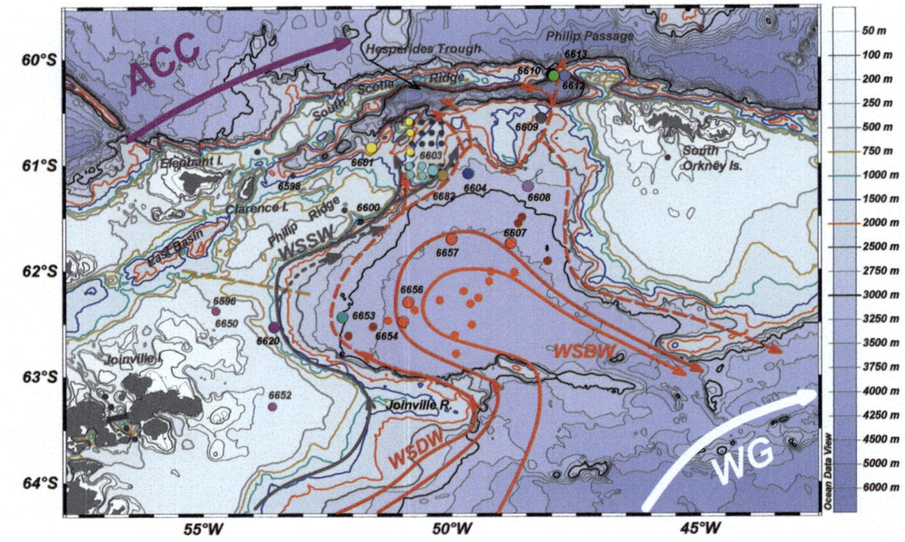

Fig. 8.3 Suggested regional water mass pathways at depths below 700 m and various modifications of WDW in the Powell Basin with additional background information of current pathway based on (Von Gyldenfeldt et al. 2002). The pathways of water masses are schematically marked as a gray line (WSSW, Weddell Sea Slope Water), dark red line (WSDW, Weddell Sea Deep Water), dashed dark red (VWSDW), and as a red line (WSBW, Weddell Sea Bottom Water). The CTD data of WOD18 are shown as small dark red (VWSDW) and red (WSBW) dots. The stations with high modification of WDW (mWDW(H)) are shown as small pink dots (6596, 6650, and 6652); the stations with middle modification of WDW (mWDW(M)) are shown as small cyan dots (6602, 6603, 6667, 6678, 6679); the stations with low modification of WDW (mWDW(L)) are shown as yellow small dots (6674, 6675, and 6677). Selected stations from Fig. 8.4 are shown as color dots. Bathymetry is based on GEBCO-2014 (Weatherall et al. 2015)

modifications. Nutrients have peculiar features of distribution in the Antarctic region and are often used for water mass identification. Concentrations of phosphate, nitrate, and silicate were determined by a Hach Large DR 2800 spectrophotometer. Dissolve oxygen concentrations were determined by the Winkler titration method.

Following Whitworth et al. (1998), Orsi et al. (1999), and Naveira Garabato et al. (2002), we analyzed the boundaries between major water masses based on neutral density (γ^n) (McDougall 1987; Jackett and McDougall 1997). The units of neutral density kg/m^3 are omitted throughout the paper. The boundaries of the regional water masses for the preliminary study were taken from literature sources (Table 8.1).

The method for determining water mass boundaries based on the gradient of dissolved oxygen concentrations is traditionally used for studying water columns in the Antarctic region. The boundary between the surface mixed layer of SSW and WW is marked by the strong seasonal thermocline, halocline, and pycnocline. Following Franco et al. (2007), we defined isopycnal $\gamma^n = 27.9$ as the boundary between summer and winter waters, subdividing AASW into two layers (Table 8.1). In addition to neutral density (see Table 8.1), we located the boundary between WW

Table 8.1 Ranges of neutral density and potential temperature for regional water masses according to Foster and Carmack (1976b[1]), Foldvik et al. (1985[2]), Fahrbach et al. (1995[3]), Whitworth et al. (1998[4]), Orsi et al. (1999[5]), Arhan et al. (1999[6]), Gordon et al. (2001[7]), Naveira Garabato et al. (2002[8]), and Franco et al. (2007[9])

Water mass acronyms	Θ,°C	γ^n, kg/m³	Oxygen, ml/l
WW	−1.5… − 1.9[1]	<28.0[8] 27.9–28.0[9]	Max
WDW	> 0[8,9]	28.0–28.26[8,9] 28.0–28.20[9] 28.1–28.27[4]	Min
WSDW	−0.7[1,3,5]…0[3] −0.7… − 0.2[9]	28.26–28.4[8,9]	
UWSDW LWSDW		28.26–28.31[6,7] 28.31–28.4[6,7]	
WSBW	< −0.7[3,7] < −0.8[2] −1.3[1]	>28.4[8,9]	Local max(B)
AABW	−0.8–0[1]	>28.27[4,5]	Local max(B)

and WDW based on the strong oxygen gradient associated with separating the highly oxygenated WW (>7 ml/l) from the oxygen-depleted WDW (<5 ml/l). The lower boundary of WW is marked also by relatively strong vertical thermohaline gradients, but it is not so evident in the vertical density structure.

8.4 Preliminary Results

8.4.1 Mixing and Transformation of WDW and WW

The cores of WDW and WW are located in the deep part and on the shelf in the Weddell Sea, respectively. The notable characteristics of these waters were traced at the closest to the source stations 6654, 6656, and 6652 (stations in Figs. 8.1 and 8.2).

In this paper, we indicate three modifications of WDW (mWDW) in the Powell Basin. The high/mWDW(H), middle/mWDW(M), and low/mWDW(L) are the waters with various thermohaline and hydrochemical characteristics observed in various parts of the Powell Basin (Table 8.2, Fig. 8.2). We suggest that modifications of WDW have dissimilar mechanisms of formation and various proportions of mixing water mases.

High modification of WDW (mWDW(H)) is the mixture of WW (major portion) and WDW (admixture). The MWDW(H) is observed on the shelf and shelf break near the bottom at stations 6596, 6650, and 6652 (Figs. 8.1 and 8.3). Vertical distribution of θ and θ/S diagram at station 6652 are shown as selected stations in Fig. 8.2, while the locations of stations are shown in Fig. 8.1. The thickness of mWDW(H) column near the bottom at depths of 250–350 m is less than 100 m. This is the coldest and freshest modification of WDW (Θ = −0.8… − 0.5 °C), and most likely

Table 8.2 Thermohaline and hydrochemical characteristics of a water mass core observed in January–February 2020. Blue color indicates data from Sea-Bird 911plus CTD system; red color indicates data from bottles; (B) indicates near-bottom samples

Water Mass Acronyms	S, psu	Θ,°C	γ^n, kg/m³	Oxygen, ml/l	Phosphate, µm	Nitrate, µm	Silicate, µm	Depth of core, m	Station
WW	34.37	-1.67	27.96					91	6656
				7.50	1.73	37.50	70.08	80	
				6.91	1.84	37.87	72.22	100	
WDW	34.68	0.65	28.14					350	6656
				4.87 min	2.06	36.72	95.29	300	
WDW (H)	34.55	-0.96	28.12	6.21	2.02	37.86	80.81	356	6652
WDW (M)	34.63	-0.15	28.22					990-1539	6598
				5.52	2.28	29.59	96.89	1000	
				5.47 min	2.55	34.01	86.72	1537 (B)	
	34.61	-0.13	28.16					770-1740	6603
				5.69	2.16	33.45	94.74	1000	
				5.59 min	2.24	39.97	95.28	1500	
WDW (L)	34.65	0.1	28.18	5.21 min	2.29	41.88	102.78	1000	6601
WSSW	34.61	-0.51	28.23 max	5.77 local max	1.92	35.46	95.28	1028 (B)	6651
WSDW	34.65	-0. 46	28.34					2800	6656
				5.26	2.05	36.23	115.10	2000	
WSBW	34.62	-0.91	28.41	6.15 local max	1.75	42.60	89.93	3342 (B)	6656

mWDW(H) is the first stage of Foster/Carmack Process of AABW formation (Foster and Carmack 1976a). The MWDW(H) is as cold as but more saline than WW, and hydrochemical characteristics (Table 8.2) confirm the admixture of WDW to WW. This modification of WDW is characterized by an oxygen content exceeding 6 ml/l and lower silicate concentrations. The observed neutral density range of mWDW(H) is 28.09–28.12.

Middle modification of WDW (mWDW(M)) is observed near the seamount at a depth of 500 m in the northwestern part of the Powell Basin (Fig. 8.3). A well-mixed layer with θ = −0.14... −0.16 °C was observed at stations 6602, 6603, 6667, 6678, and 6679 (shown as cyan dots in Fig. 8.3), however, at various depths and locations. Stations 6602 (shown as a selected station in Fig. 8.2), 6667, 6678, and 6679, where mWDW(M) was observed, are located on the slope south and east of the seamount at depths of 400–1000 m (Figs. 8.1 and 8.3).

Middle modification is more saline (S = 34.61–34.63 psu) and dense (γ^n = 28.11–28.21) than high modification of WDW. The oxygen concentration here

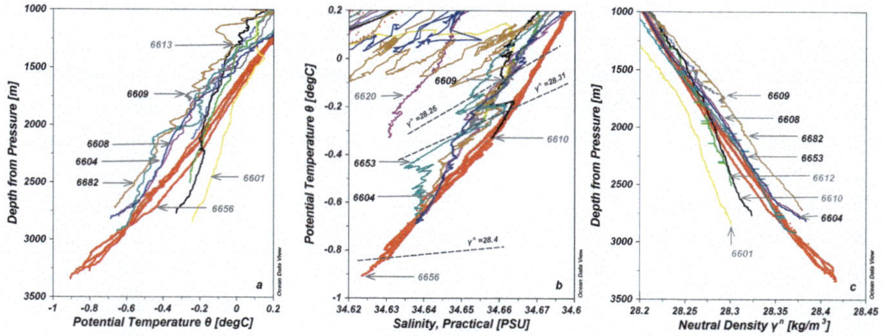

Fig. 8.4 Transformation of deep water in the Powell Basin. Vertical distribution of potential temperature (**a**), θ/S diagram (**b**), and vertical distribution of neutral density (**c**) in the Powell Basin with marked boundaries (γ^n) of water masses (see more details in Table 8.1). The stations located over the deep slope, which we associate with ventilated WSDW, are shown as colored curves (stations 6604, 6608, 6653, and 6682). The stations located in the deep part of the basin (stations 6607, 6654, 6656, and 6657), which we associated with typical WSDW, are shown as red curves. The location of the stations is shown in Fig. 8.3

is about 5.5 ml/l, which is lower than in mWDW(H) and higher than in WDW. We suggest that transformation of WDW and formation of mWDW(M) occur due to the tides and rough bottom topography with seamounts and the channel (~51° W, 61° S, shown as a red dashed line in Fig. 8.1) with depths greater than 1500 m, which connects the Powell Basin with the Hesperides Trough. Seamounts generate submesoscale turbulence during tides (Ruan et al. 2017) that promote diapycnal mixing between WW and WDW. The Philip Ridge is an important isopycnic mixing area where the waters from the Weddell Sea and the Bransfield Strait with different oceanographic properties interact. Similar mixture occurs in the formation of mWDW(M). The region of mixing over the Philip Ridge is indicated by light gray helix in Fig. 8.1 (Von Gyldenfeldt et al. 2002). In addition, the channel is a pathway for the ASF toward the Hesperides Trough (Fig. 8.1), which also provides conditions for mixing.

We assume that all these phenomena contribute to the formation of mWDW(M) in this basin. Strong WDW mixing and intrusions toward mWDW (M) were observed at stations 6682 and 6604 (shown as blue and pink curves in Fig. 8.2).

Station 6598 (Figs. 8.1 and 8.2) is located near the channel, which connects the Hesperides Trough and the Bransfield Strait. However, the thermohaline and hydrochemical characteristics of the water observed at this station are similar to mWDW(M) at station 6603 (Table 8.2). Therefore, this water can be associated with the influence of the Powell Basin rather than with the Bransfield Strait.

Low modification of WDW (mWDW(L)) is the water with θ = 0–0.15 °C and S = 34.62–34.65 psu which was observed at station 6601 (Figs. 8.1 and 8.2) at depths of 500–1500 m. This modification is characterized by a lower oxygen content and higher concentrations of silicate and nitrates in comparison with other

WDW modifications (Table 8.2). All stations with observed mWDW (L) are located near the channel (~51° W, 61° S).

We suppose that mWDW(L) observed at stations 6601, 6674, 6675, and 6677 (shown as yellow small dots in Fig. 8.3) is the mixture of mWDW(M) and WDW entering anticyclonic circulation over the seamount revealed by Thompson et al. 2009.

As WDW is transformed due to diapycnal mixing with WW, the extremum of WW becomes less pronounced. The minimum of θ observed on the shelf off Joinville Island and in the deep-water part in the Powell Basin is less than −1.5 °C with respect to $\theta = -0.5... - 0.2$ °C observed at the stations in the northwestern part of the Powell Basin (e.g., at stations 6598 and 6601 in Fig. 8.2).

As just stated, the northern tip of the Antarctic Peninsula region is characterized by rough bottom topography, strong tidal currents, and interaction of water masses from different regions of the Southern Ocean. As a result, intense isopycnal and diapycnal mixing of waters leads to the transformation of water masses and the formation of several zones with different water properties within a relatively small area. Based on the analysis of the distribution of thermohaline and hydrochemical characteristics, we select nine stations (6597, 6598, 6599, 6601, 6651, 6652, 6653, 6656, and 6682) for representation of the water column in the NTAP region (Figs. 8.1 and 8.2).

8.4.2 Weddell Sea Slope Water and Hydrological Role of Its Inflow into the Powell Basin

Middleton and Humphries (1989) found an anomalous water mass on the continental slope near 50° E (Enderby Land). This water with properties centered at T = 0 °C and S = 34.65 psu appeared to be a mixture of relatively dense shelf water from Prydz Bay and Circumpolar Deep Water. This water was named Prydz Bay Bottom Water (PBBW). Similarly, we found relatively fresh cold waters at three stations 6620 (994 m), 6651 (1026 m), and 6600 (1578 m) over the continental slope with properties near the seafloor $-0.53 < \theta < -0.23$ °C, $34.61 < S < 34.64$ psu, $28.21 < \gamma^n < 28.24$, $5.8 <$ oxygen<6.3 ml/l. This fresh cold water represents the densest water at these depths found in the Powell Basin; it is located below WDW deeper than 800 m at stations 6651 and 6620 and deeper than 950 m at station 6600. The thickness of the water layer is different at each of these stations: 6651, 6620, and 6600, 200 m and 630 m, respectively. A minimum potential temperature θ_{min} was observed on the slope near the bottom at station 6651(Fig. 8.2, Table 8.2). There is a colder and fresher water mass near the bottom at all three stations over the slope.

The fresh cold water at station 6600 is warmer and more saline than at stations 6651 and 6620. We assume that this water (station 6600) is more transformed than the water found at stations 6651 and 6620, and it is a mixture of SW with high proportion of ambient WDW.

Since cascading of SW on this slope is not observed (Amblas and Dowdeswell 2018), we suggest identifying this water as the water advected around the Joinville Ridge from the Weddell Sea and believe that it is a mixture of LSSW and WDW over the slope located south of the Joinville Ridge in the Weddell Sea. Fahrbach et al. (1995) proposed that low-saline deep and bottom waters are formed near the Larsen Ice Shelf.

The water formed by mixing of SW and WDW that flows over the Antarctic slope has various names in different publications. As a rule, this water is not identified as the same water mass in different parts of the Antarctic slope. According to Middleton and Humphries (1989), it is the densest modification of CDW, and it is called Modified Shelf Water in Wong and Riser (2013). We designate this water as Weddell Sea Slope Water (WSSW). The WSSW is formed when plumes of dense SW cascade down continental slopes and mix with ambient WDW. Gradually mixing by cross-slope advection and flowing to the deeper depths WSSW, it continues flowing northward within ASC at depths greater than 800 m along the western slope of the Powell Basin and becomes part of WSDW ventilating and transforming it.

Thermohaline properties of mWDW(M) at stations 6603 (Fig. 8.2) and 6600 (slope) are similar. However, modifications of WDW were formed due to different processes. Modification of WDW at station 6603 as observed mWDW(M) at stations 6602, 6603, 6667, 6678, and 6679 was formed by diapycnal mixing with overlying WW. The MWDW(M) at station 6600 most likely was formed by the admixture of WSSW under WDW (400–1000 m) layer.

The inflow of WSSW was also observed at stations 6604, 6608, 6653, and 6682 located on a deep slope in the Powell Basin. The most pronounced hydrological effect was identified at stations 6653 and 6682 (Fig. 8.4). According to Foster and Carmack (1976b) and Orsi et al. (1993), dense SW can flow with slope currents deeper than 2000 m reaching the northeastern tip of the Antarctic Peninsula and continuing eastward along the southern flank of the South Scotia Ridge. Based on the analysis of thermohaline and oxygen concentration fields from measurements collected during DOVETAIL program (1997), Gordon et al. (2001) suggested two types of deep water and two types of bottom water in the northwestern Weddell Sea. These are typical and ventilated forms of WSDW and WSBW (VWSDW/VWSBW). We also observed two types of WSDW in the Powell Basin (Figs. 8.3 and 8.4). Vertical distribution of potential temperature and neutral density and θ/S diagram of VWSDW (stations 6604, 6608, 6653, and 6682) are shown as colored curves in Fig. 8.4; the same distributions of WSDW (stations 6607, 6654, 6656, and 6657) are shown as red curves in Fig. 8.4. The location of stations where VWSDW and WSDW was observed during the analysis of WOD18 CTD data are marked as dark red and red small dots in Fig. 8.3, respectively. Summing up the results of our analysis of CTD data of AMK 79 and WOD18, we introduce a scheme of regional water mass pathways (Fig. 8.3).

The VWSBW branch continues to be transformed by WSSW influx and diapycnal mixing, and passing through the Powell Basin it splits into two streams.

It should be noted that distribution of hydrochemical parameters at station 6682 (selected station in Figs. 8.1 and 8.2) differs significantly from the deep-water stations (6656, 6657, and 6607) in the layer from a depth of 1500 m to the bottom. This is also seen in the distribution of thermohaline characteristics (Fig. 8.4a, b) and, in particular, neutral density (Fig. 8.4c). Station 6682 is the densest station below the 1000-m depth and at the bottom (2721 m) from stations on the deep slope. The station is characterized by high oxygen concentrations deeper than 1500 m, especially in the layer 1500–2500 m (5.35 ml/l at 1500 m at station 6682, 5.07 ml/l at station 6656). The silicate concentrations are lower (100μM or less) than at stations 6656, 6657, and 6607 (silicates content at station 6656 at depths of 1500–2000 m is 115μM). The concentrations of phosphate are higher (more than 2.2μM) than at deep-water stations (PO_4 at station 6656 at depths of 1500–2000 m is 2.1μM and less). We associate such thermohaline and hydrochemical characteristics with WSSW influx. For sure, one should also take into account the high-amplitude internal tides (Robentson 2004) and topographically generated submesoscale turbulence (Ruan et al. 2017), which contribute to diapycnal mixing in this region.

The spatial distribution of hydrochemical parameters in deep water layers of the Powell Basin also indicates the inflow of WSSW. Increased oxygen and phosphates and decreased silicates were observed near the bottom and water column below 1500 m at the deep slope at stations 6604, 6608, 6653, and 6682 and the most southwestern deep station 6656 (since hydrochemical parameters were not measured at station 6654, we analyzed the nearest station 6656). These observations are relative to the concentrations observed at the northern deep stations 6657, 6607, 6608, and 6604 in the Powell Basin. High oxygen concentrations (6.15 ml/l) and lower silicate concentrations (90μM) in comparison with stations 6657 and 6607 (O_2 6 ml/l, Si 98.5μM) were observed in the bottom layer at station 6656. Therefore, the most intense WSSW influx seems to occur in the southwestern part of the Powell Basin.

8.5 Conclusions

During the multidisciplinary expedition onboard the R/V *Akademik Mstislav Keldysh*, the cold fresh water, designated as Weddell Sea Slope Water (WSSW), with a density exceeding the density of the surrounding waters was observed at depths of 800–1500 m on the western slope of the Powell Basin. We propose a concept of WSSW which is a result of mixing of the shelf water formed in the Weddell Sea and WDW over the slope southward off the Joinville Ridge. Gradually being mixed by cross-slope advection and flowing into the deeper depths, WSSW continues flowing northward transforming surrounding WDW and WSDW at depths greater than 800 m along the western slope of the Powell Basin. We assume that the admixture of WSSW to WSDW and tidally forced cross-slope advection transform WSDW in the depth range of 1500–2500 m over the slope in the Powell Basin. The WSDW is getting fresher but colder and denser than that observed at depths of 1500–2500 m in the central deep part of the Powell Basin. Thereby, WSDW is additionally ventilated in the Powell Basin, far from the source of formation of the shelf water.

Three modifications of WDW with various thermohaline and hydrochemical characteristics were identified in the Powell Basin. Based on the analysis of oceanographic properties of water mass and geophysical conditions of this water formation, we assume that the modifications of WDW were formed due to the different hydrological processes. However, their influence upon the formation of deep water in this region is generally undetermined. Therefore, we consider that studying the processes over the shelf and western slope in the Powell Basin together with the mixing process in the Hesperides Trough is needed. Study of hydrological processes in the channel with depths of more than 1500 m, which connects the Powell Basin and the Hesperides Trough, would provide appropriate data.

We suggest a scheme of regional water mass pathways at depths below 700 m. The WSSW flows deeper than 800 m northward, gradually deepening along the western slope of the Powell Basin. Part of this water at depths of 1000–1500 m turns west and flows into the channel that connects the Powell Basin and the Hesperides Trough. The part of WSSW that flows deeper than 1500 m continues flowing eastward near a seamount and influences the transformation of surrounding waters. The VWSBW branch passes through the Powell Basin and splits into streams. One stream of VWSDW flows northward along the western slope in the Powell Basin and through the channels at depths greater than 2000 m into the Scotia Sea and the Hesperides Trough. The second part of VWSDW flows southward along the western slope of the South Orkney Plato. The WSDW penetrates into the Powell Basin but does not extend northward off the deepest part of this basin. This deep water enters the cyclonic circulation as well as WSBW and returns southward to the Weddell Gyre.

Acknowledgments The authors are grateful to S.A. Dobrolubov and E.G. Morozov for valuable comments to the paper and to the shipboard scientific-technical crew of the R/V *Akademik Mstislav Keldysh* for their help in collecting the data. The study was performed in the framework of the state assignment of Russia (theme No. 0128-2029-0008).

References

Amblas D, Dowdeswell JA (2018) Physiographic influences on dense shelf-water cascading down the Antarctic continental slope. Earth-Sci Rev 185:887–900

Arhan M, Heywood KJ, King BA (1999) The deep waters from the Southern Ocean at the entry to the Argentine basin. Deep-Sea Res II 46(1–2):475–499

Boyer TP, Baranova OK, Coleman C, Garcia HE, Grodsky A, Locarnini RA, Mishonov AV, Paver CR, Reagan JR, Seidov D, Smolyar IV, Weathers K, Zweng MM (2018) World Ocean Database 2018. A.V. Mishonov, Technical Ed., NOAA Atlas NESDIS 87

Carmack EC, Foster TD (1975) On the flow of water out of the Weddell Sea. Deep-Sea Res 22:711–724

Deacon GER (1937) The hydrology of the Southern Ocean. Discov Rep 15:3–122

Fahrbach E, Rohardt G, Scheele N, Schroder M, Strass V, Wisotzki A (1995) Formation and discharge of deep and bottom water in the northwestern Weddell Sea. J Mar Res 53:5–538

Fahrbach E, Hoppema M, Rohardt G, Schroder M, Wisotzki A (2004) Decadal-scale variations of water mass properties in the deep Weddell Sea. Ocean Dyn 54:77–91

Flexas MM, Schodlok MP, Padman L, Menemenlis D, Orsi AH (2015) Role of tides on the formation of the Antarctic slope front at the Weddell-Scotia confluence. J Geophys Res Oceans 120:3658–3680

Foldvik A, Gammelsrod T, Torresen T (1985) Circulation and water masses on the southern Weddell Sea shelf. Antarctica Res Ser 43:5–20

Foster TD, Carmack EC (1976a) Frontal zone mixing and Antarctic bottom water formation in the southern Weddell Sea. Deep-Sea Res I 23:301–317

Foster TD, Carmack EC (1976b) Temperature and salinity structure in the Weddell Sea. J Phys Oceanogr 6:36–44

Franco BC, Mata MM, Piola AR, Garcia CAE (2007) Northwestern Weddell Sea deep outflow into the Scotia Sea during the austral summers of 2000 and 2001 estimated by inverse methods. Deep Sea Res I Oceanogr Res Pap 54(10):1815–1840

Gill AE (1973) Circulation and bottom water production in the Weddell Sea. Deep-Sea Res 20(2):111–140

Gordon AL (1971) Oceanography of Antarctic waters. In: Reid JL (ed) Antarctic oceanology I, Antarctic research series, 15. American Geophysical Union, Washington, DC, pp 205–221

Gordon AL (1998) Western Weddell Sea thermohaline stratification. In: Ocean ice and atmosphere: interactions at the antarctic continental margin, Antarctic research series 75. American Geophysical Union, Washington, DC, pp 215–240

Gordon AL, Visbeck M, Huber BA (2001) Export of Weddell Sea deep and bottom water. J Geophys Res 106(C5):9005–9018

Grosfeld K, Gerdes R, Determann J (1997) Thermohaline circulation and interaction between ice shelf cavities and the adjacent open ocean. J Geophys Res Oceans 102(C7):15595–15610

Heywood KJ, Naveira Garabato AC, Stevens DP, Muench RD (2004) On the fate of the Antarctic slope front and the origin of the Weddell front. J Geophys Res 109:C06021

Jackett DR, McDougall TJ (1997) A neutral density variable for the World's oceans. J Geophys Res 27:237–263

McDougall TJ (1987) Neutral surfaces. J Phys Oceanogr 17:1950–1964

Middleton JH, Humphries SE (1989) Thermohaline structure and mixing in the region of Prydz Bay Antarctica. Deep-Sea Res 36:1255–1266

Miyake (1939) Chemical studies of the western Pacific Ocean I and II. The chemical composition of the oceanic salt parts 1 and 2. Bull Chem Soc Jpn 14(29–35):55–58

Muench RD, Gordon AL (1995) Circulation and transport of water along the western Weddell Sea margin. J Geophys Res 100(C9):18,503–18,515

Munk W, Wunsch C (1998) Abyssal recipes II: energetics of tidal and wind mixing. Deep-Sea Res 45:1977–2010

Naveira Garabato AC, Heywood KJ, Stevens DP (2002) Modification and pathways of Southern Ocean deep waters in the Scotia Sea. Deep Sea Res I 49:681–705

Orsi AH, Nowlin WD Jr, Whitworth T III (1993) On the circulation and stratification of the Weddell gyre. Deep-Sea Res I Oceanogr Res Pap 40:169–203

Orsi AH, Johnson GC, Bullister JL (1999) Circulation mixing and production of Antarctic bottom water. Prog Oceanogr 43:55–109

Orsi AH, William MSJ, Bullister JL (2002) On the total input of Antarctic waters to the deep ocean: a preliminary estimate from chlorofluorocarbon measurements. J Geophys Res 107(C8):31–31,22

Padman L, Howard SL, Muench RD (2006) Internal tide generation along the south Scotia ridge. Deep-Sea Res II Top Stud Oceanogr 53:157–171

Padman L, Muench RD, Howard S, Mueller R (2008) Tidal impacts on oceanographic and sea-ice processes in the Southern Ocean. American Geophysical Union, Fall Meeting, abstract id. OS13C-1208

Purkey SG, Johnson GS (2013) Antarctic bottom water warming and freshening: contributions to sea level rise, ocean freshwater budgets, and global heat gain. J Clim 26:6105–6122
Rintoul SR (2007) Rapid freshening of Antarctic bottom water formed in the Indian and Pacific oceans. Geophys Res Lett 34:L06606
Robertson R (2001) Internal tides and baroclinicity in the southern Weddell Sea: 1 model description. J Geophys Res Oceans 106(C11):27001–27016
Robertson R (2005) Baroclinic and barotropic tides in the Weddell Sea. Antarct Sci 17(03):461–474
Ruan X, Thompson AF, Flexas MM, Sprintall J (2017) Contribution of topographically generated submesoscale turbulence to Southern Ocean overturning. Nat Geosci 10(11):840–845
Thompson AF, Heywood KJ (2008) Frontal structure and transport in the northwestern Weddell Sea. Deep Sea Res I Oceanogr Res Pap 55(10):1229–1251
Thompson AF, Heywood KJ, Thorpe SE, Renner AHH, Trasviña (2009) A surface circulation at the tip of the Antarctic peninsula from drifters. J Phys Oceanogr 39:3–26
Von Gyldenfeldt AB, Fahrbach E, García MA, Schröder M (2002) Flow variability at the tip of the Antarctic peninsula. Deep Sea Res II Top Stud Oceanogr 49(21):4743–4766
Weatherall P, Marks KM, Jakobsson M, Schmitt T, Tani S, Arndt JE, Rovere M, Chayes D, Ferrini V, Wigley R (2015) A new digital bathymetric model of the world's oceans. Earth Space Sci 2:331–345. https://wwwgebconet/data_and_products/gridded_bathymetry_data/gebco_30_second_grid. The latest release of the GEBCO 30 arc-second global grid of elevations, GEBCO_2014 Grid, was published in 2014
Whitworth T III, Orsi AH, Kim SJ, Nowlin WD, Locarnini RA (1998) Water masses and mixing near the Antarctic slope front. In: Ocean ice and atmosphere: interactions at the Antarctic continental margin, Antarctic research series 75. American Geophysical Union, Washington, DC, pp 1–27
Whitworth T, Nowlin WD, Orsi AH, Locarnini RA, Smith SG (1994) Weddell Sea shelf water in the Bransfield Strait and Weddell-Scotia confluence. Deep-Sea Res I Oceanogr Res Pap 41:629–641
Wong APS, Riser SC (2013) Modified shelf water on the continental slope north of Mac Robertson land East Antarctica. Geophys Res Lett 40(23):6186–6190

Chapter 9
Interannual Variations of Water Mass Properties in the Central Basin of the Bransfield Strait

Alina A. Fedotova and Sergey V. Kashin

Abstract We analyze CTD data from the annual transect across the central basin in the Bransfield Strait. The measurements were sampled during the 61st (2016), 62nd (2017), 63rd (2018), and 65th (2020) Russian Antarctic Expeditions (RAE). Monitoring of the interannual variations of Bransfield Deep Water (BDW) properties in 2016, 2017, 2018, and 2020 confirmed the existence of an unidirectional trend of warming and density loss described in previous research studies. In the paper, it is shown that density loss of BDW in 2016, 2017, 2018, and 2020 is due to various changes of BDW properties. A neutral density loss of 0.03 kg/m^3 was caused by an increase in potential temperature, which was observed at depths of 500–1300 m, while potential temperature in the bottom layer remained the same in 2016 and 2017. In 2018, a strong density loss of 0.1 kg/m^3 caused mainly by a decrease in salinity of 0.03 psu and less by warming of 0.12 °C was observed in BDW. The density loss due to a decrease in salinity and an increase in warming continued in 2020. We found that potential temperature in the near-bottom layer increased by 0.18 °C and salinity decreased by 0.04 psu in the last 5 years. Interannual variations of the water mass properties caused a decrease in neutral density in the bottom layer from 28.5 in 2016 to 28.42 in 2020 and a decrease of layer thickness of BDW with $\gamma^n > 28.4$. Thus, by 2020 the deep layer in the central basin of the Bransfield Strait was almost completely occupied by water with $28.27 < \gamma^n < 28.4$.

Keywords Bransfield Strait · Russian Antarctic Expedition · Bransfield Deep Water · Annual transect · Water mass transformation · Density loss of deep water

A. A. Fedotova (✉)
Lomonosov Moscow State University, Moscow, Russia

Shirshov Institute of Oceanology, Russian Academy of Sciences, Moscow, Russia

S. V. Kashin
Arctic and Antarctic Research Institute, St Petersburg, Russia

E. G. Morozov et al. (eds.), *Antarctic Peninsula Region of the Southern Ocean*, Advances in Polar Ecology 6, https://doi.org/10.1007/978-3-030-78927-5_9

9.1 Introduction

The central basin of the Bransfield Strait is located in the region of the Western Antarctic Peninsula; it is characterized by the interaction of water masses with different oceanographic properties from the Weddell Sea and the Pacific Water column (Gordon et al. 2001) including waters from the Antarctic Circumpolar Current and the Bellingshausen Sea. According to Garcia et al. (2002), the Bransfield Strait is best defined as a transition zone. They suggested naming the main local water as Transitional Zonal Waters with Bellingshausen Sea influence (TBW) and Transitional Zonal Waters with Weddell Sea influence (TWW) depending on the dominant source waters. These water masses are separated in the Bransfield Strait by two fronts, the so-called Bransfield and Peninsula fronts (Garcia et al. 2002; Sangra et al., 2011; Huneke et al. 2016). The Bransfield Front (BF) extends along the South Shetland Islands, and the Peninsula Front (PF) occupies the eastern part of the strait along the western coast of the Antarctic Peninsula (Huneke et al. 2016). Bransfield Deep Water (BDW) with salinity of more than 34.5 psu and potential temperature (θ) below -1 °C occupies the deep layer of the central basin under TWW and TBW (Garcia et al. 2002). According to Garcia et al. (2002), this water is an isopycnal mixture of Weddell Sea Shelf Water and Warm Deep Water, which is advected into the eastern part of the Bransfield Strait and may be further modified in situ by winter convection.

The hydrography of the Bransfield Strait, including the pathways of the circulation, depends on uneven bottom topography. The currents transporting TWW to the southwest along the western Antarctic Peninsula shelf and the Bransfield Current flowing to the northeast form cyclonic circulation in the Bransfield Strait (Huneke et al. 2016). An approximate scheme of currents in the Bransfield Strait according to Morozov (2007) is shown with blue arrows in Fig. 9.1. Thereby, the properties of the Bransfield Strait waters are a result of complex interactions between the atmosphere, which determines sea ice processes and the heat flows, dynamic processes, and water masses exported from the Weddell Sea, Antarctic Circumpolar Current, and Bellingshausen Sea.

The international program Biological Investigations of Marine Antarctic Systems and Stocks (BIOMASS), US, and Spanish National Antarctic programs were formed to comprehensively identify the oceanographic characteristics of the Bransfield Strait waters. The main features of regional water masses and circulation patterns studied on the basis of the analysis of various measurements have been reported in many publications (Gordon and Nowlin 1978; Grelowski et al. 1986; Tokarczyk 1987; Niiler et al. 1991; Garcıa et al. 1994; Hofmann et al. 1996; Lopez et al. 1999; Gordon et al. 2000; Garcıa et al. 2002; Gomis et al. 2002; Huneke et al. 2016; Avelina et al., 2020).

However, the Bransfield Strait is sensitive to climate change (e.g., Huneke et al. 2016; Kerr et al. 2018) and, according to Dotto et al. (2016), is an important proxy region for monitoring changes in the dense Weddell Sea Shelf waters, which are an important precursor of Antarctic Bottom Water (AABW). According to Dotto et al.

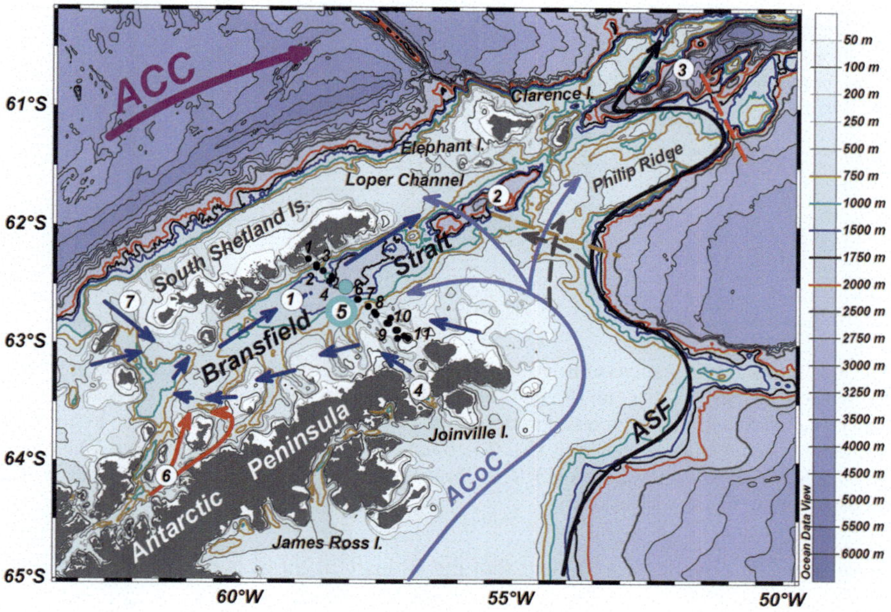

Fig. 9.1 Study site of the Bransfield Strait. Locations of oceanographic stations during 61st (2016), 62nd (2017), 63rd (2018), and 65th (2020) Russian Antarctic Expeditions (RAE) from R/V (icebreakers) "*Akademik Fedorov*" and "*Akademik Tryoshnikov*" are shown as black dots. Color lines schematically show the pathways of water from the Gerlache Strait (red lines, Antarctic Coastal Current (ACoC/blue line, Whitworth et al. 1994), Antarctic Slope Front (ASF/dark gray line, Thompson et al. (2009)), approximate scheme of currents (blue arrows, Morozov, 2007), and Antarctic Circumpolar Current (ACC/violet line). The dashed gray arrows indicate the alternating dominance of shelf-derived water or modified WDW in the channel between the Powell Basin and the Bransfield Strait (Von Gyldenfeldt et al. 2002). The channel with depth ~ 750 m is shown as an ochre dashed line. The channel with depths ~2000 m is shown as a red dashed line. (1) Central Basin; (2) East Basin; (3) Hesperides Trough; (4) Antarctic Sound; (5) the selected station in Figs. 9.2 and 9.3 is shown as a light blue dot; (6) Gerlache Strait; (7) Boyd Strait. Bathymetry is based on GEBCO-2014 30 sec grid (Weatherall et al. 2015).

(2016) and Henley et al. (2019), long-term series of hydrographic data showed warming and freshening of the deep-water masses of the Bransfield Strait in the 1960s–2010s. Dotto et al. (2016) suggested that warming and density loss in the deep layer of the Bransfield Strait are caused by large freshwater inputs originating from the western shelf of the Weddell Sea. It should be noted that the warming during the last 30 years below 700-m depths in the interior Weddell Sea and the decrease of salinity in the outflow branch of the Weddell Gyre and in the newly formed bottom water were also detected in a recent study by Strass et al. (2020). A brief report on the warming of waters in the Bransfield Strait over the past 5 years was published by Antipov et al. (2020). They proposed that the warming could be caused both by changes of transport of the waters from the Bellingshausen and Weddell seas, and by changes in the water properties, which may reflect the effect associated with global

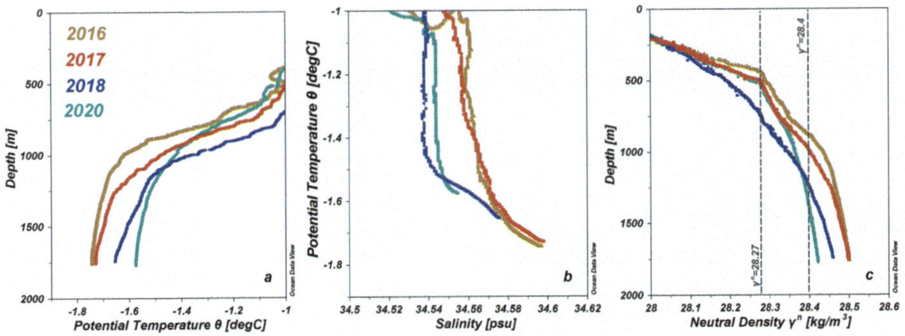

Fig. 9.2 Interannual variations of Bransfield Deep Water properties. Vertical distribution of potential temperature (**a**), θ/S diagram (**b**), and vertical distribution of neutral density (**c**) based on the data at station 5 in the annual transect in the central basin of the Bransfield Strait. The location of the station on the chart is shown in Fig. 9.1. Boundaries (γ^n) between Warm Deep Water and Weddell Sea Deep Water (28.27), Weddell Sea Deep Water, and Weddell Sea Bottom Water (28.40) are shown in (c) as gray dashed lines

warming. Thereby, despite the fact that the Bransfield Strait is one of the Antarctic Peninsula regions being intensely studied, further research studies are required.

In this paper we analyze the CTD (conductivity-temperature-depth) data from the annual transect across the central basin in the Bransfield Strait. The location of the transect is shown in Fig. 9.1. The measurements were sampled during 61st (2016), 62nd (2017), 63rd (2018), and 65th (2020) Russian Antarctic Expeditions (RAE) from R/V (icebreakers) "*Akademik Fedorov*" and "*Akademik Tryoshnikov*" under the leadership of the Arctic and Antarctic Research Institute (AARI, St. Petersburg).

9.2 Data and Methods

Oceanographic data were collected using a Sea-Bird 911plus CTD system and SBE 32 Carousel Water Sampler with 12 5-l bottles. The accuracy of temperature and electrical conductivity measurements declared by the manufacturer is ±0.001 °C and ± 0.0003 S/m, respectively. Salinity samples were taken to monitor electrical conductivity sensor at a few stations of the transect and were determined by using the high-precision salinity meter Autosal 8400. The annual transect was sampled during austral autumn on April 25, 2016, April 9–10, 2017, April 23–24, 2018, and March 9–10, 2020. The location and numbers of stations are shown in Fig. 9.1.

We analyzed the vertical distribution of potential temperature (a) and θ/S diagram (b) below isotherm the −1 °C, defined by Garcia et al. (2002) as the shallowest extent limit of BDW, and vertical distribution of neutral density (c) below isopycnal

Table 9.1 Thermohaline properties of BDW over the annually repeated transect. The average value of thermohaline characteristics of BDW at four deep stations (3, 4, 5, 6) is marked in red. Thermohaline characteristics of BDW at station 5 are marked in blue. (B) is the core of BDW (minimal **θ** and S) near the bottom. ID is the depth of the isotherm (**θ**) at station 5 indicated by thick black lines in Fig. 9.2

	2016	2017	2018	2020	2016	2017	2018	2020
- 1.05°C ID	633	656	812	709	564	622	805	518
-1.50°C ID	951	1003	1130	1199	882	1029	1129	1184
-1.60°C ID	1026	1125	1437	-	973	1154	1437	-
θ (B), °C	-1.725	-1.687	-1.563	-1.537	-1.744	-1.727	-1.657	-1.578
S (B), psu	34.592	34.598	34.560	34.550	34.597	34.597	34.575	34.556
γ^n, kg/m^3	28.497	28.496	28.402	28.465	28.5	28.498	28.46	28.422

(γ^n = 28.0) at the deepest station 5 for interannual variations of BDW properties research (Fig. 9.2).

The average values of potential temperature, salinity, and neutral density near the bottom were calculated at four deep stations (3, 4, 5, 6) to obtain a spatial assessment of interannual changes in the bottom layer (Fig. 9.1). The depths of isotherms (−1.00, −1.05, −1.50, −1.60) are also determined to estimate the layer thickness of BDW and for identification of changes of vertical distribution of potential temperature. It should also be noted that the −1 °C isotherm, which is a boundary of BDW, has been meandering, especially in 2016 and 2020 (Fig. 9.3a and b), which is related to the mixing processes. A uniform decrease of θ was observed below the −1.05 °C isotherm. Therefore, we used the depths of the −1.05 °C isotherm to estimate the layer thickness of the BDW (Table 9.1).

As mentioned before, the waters in the central basin of the Bransfield Strait are a result of isopycnal spreading of water masses (Whitworth et al. 1994; Lopez et al. 1999) from the Weddell Sea (major part, Gordon et al. (2000)) and from the Bellingshausen Sea. Therefore, we suggest using the isopycnal analysis based on neutral density (γ^n, Jackett and McDougall, 1997). We propose to analyze isopycnals 27.90, 28.00, 28.27, and 28.40, which are the boundaries used for identification of water masses in the Weddell Sea. They are the boundaries of water masses between Summer Surface Water/Winter Water (Franco et al. 2007), Winter Water/Warm Deep Water (Naveira Garabato et al. 2002), Warm Deep Water/Weddell Sea Deep Water (Whitworth et al. 1998), and Weddell Sea Deep Water/Weddell Sea Bottom Water (Gordon et al. 2001), respectively. The isopycnals were marked as red dashed lines in the annual transect in Fig. 9.3.

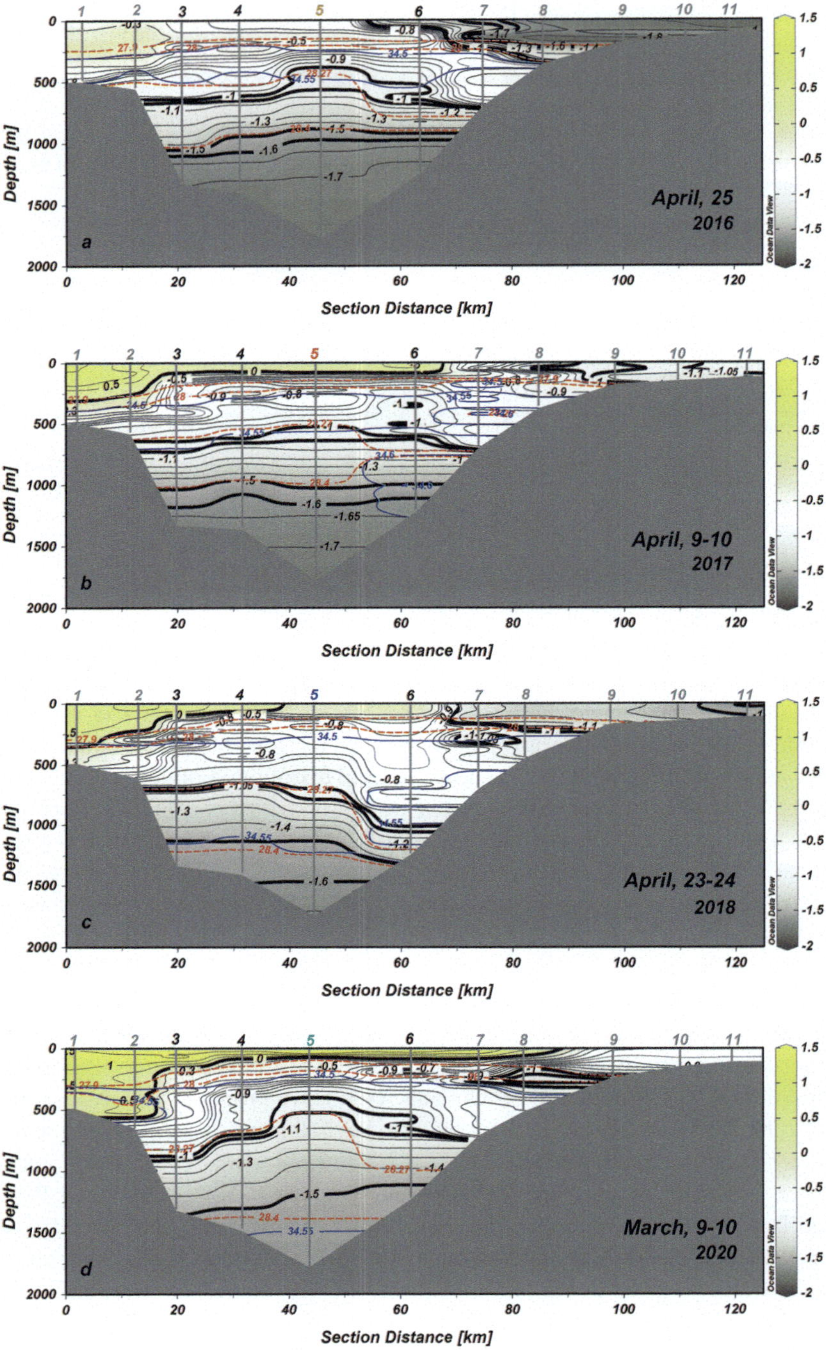

Fig. 9.3 The annual transect sampled during 61st (2016), 62nd (2017), 63rd (2018), and 65th (2020) Russian Antarctic Expeditions (RAE) from R/V (icebreakers) "*Akademik Fedorov*" and "*Akademik Tryoshnikov*" with selected deepest station #5. The location of the annual transect and numbers of stations are shown in Fig. 9.1. Isopycnals (γ^n) are marked as red dashed lines. Isohalines are marked as blue lines. Isotherms (−1.05, −1.50, −1.60, −1.70 °C) presented in Table 9.1 are indicated by thick black lines

9.3 Density Loss of Bransfield Deep Water in the Central Basin of the Bransfield Strait

During a period of 4 years, the approximate scheme of the water masses distribution in the central basin of the Bransfield Strait described in previous articles (Smith et al. 1999; Gordon et al. 2000, Garcia et al. 2002, Huneke et al. 2016) was also observed and confirmed. The position of isopycnal 28.0 and isohaline 34.5 within surface layer up to 300 m has remained approximately stable (Fig. 9.3).

The highest temperatures ($\theta > 1$ °C) in Summer Surface Water (SSW) were observed in 2020 (Fig. 9.3d). The warmer temperatures were likely to have been caused because the transect was performed in March, while in previous years observations were carried out in April. The subsurface layer of TBW was characterized by $\theta = 0.0$–1.5 °C in 2017, 2018, and 2020 (Fig. 9.3b, c, d). The minimum temperatures were observed in subsurface layers of TBW ($\theta < 0$ °C) and TWW ($\theta < -1.5$ °C) in 2016 (Fig. 9.3a). The water with a near-freezing potential temperature of −1.88 °C and salinity of 34.3 psu which is associated with the Weddell Sea Shelf waters inflow was observed at a depth of 150 m at stations 7–11. Locations of the stations are shown on the chart in Fig. 9.1 and along the transect in Fig. 9.3a. Winter Water (WW) is formed from SSW at the time of sea-ice formation during winter months and advected from the Weddell Sea shelf with the Antarctic Coastal Current (shown as blue line in Fig. 9.1). The potential temperature of this water did not decrease below −1.50 °C in 2018 and − 1.15 °C in 2017 and 2020. We assume that the water observed over the slope and shelf in the eastern part of the central basin in 2017, 2018, and 2020 is WW with an admixture of Warm Deep Water (WDW). This water was formed over the shelf north off Joinville Island (a presumptive mixing zone is shown as gray dashed arrows in Fig. 9.1). The coldest water observed in 2016 is WW without an admixture of WDW. Thus, we associate variations in potential temperatures (θ) in WW in 2016, 2017, 2018, and 2020 with the presence or absence of an admixture of WDW rather than with the interannual variability of the Weddell Sea shelf waters.

The depths of the −1.05 °C isotherm remained stable (Table 9.1, Fig. 9.3) with the exception of 2018 when the −1.05 °C isotherm deepened and the layer thickness of BDW reduced. However, thermohaline properties of BDW below this isotherm significantly changed. This is clearly seen from the depths of the −1.5, −1.6, −1.7 °C isotherms, 34.50, 34.55 psu isohalines, and 28.27 and 28.40 isopycnals shown in transects in Fig. 9.3.

The thermohaline characteristics of the bottom layer of BDW in 2016 and 2017 remained unchanged (Table 9.1, Fig. 9.2). However, relative to 2016, an increase of potential temperature by 0.1 °C and as a result a decrease in the neutral density loss by 0.03 kg/m^3 at the same salinity were observed in 2017 only at depths of 500–1300 m (Fig. 9.3a and c). As seen in Fig. 9.3b, the water with $\theta < -1.7$ °C was observed only at the deepest station 5 near the bottom (deeper than 1500 m) in 2017. The −1.7 °C isotherm was located at a depth of 1200 m in 2016 (Fig. 9.3a).

Warming and salinity decreases were observed in BDW in 2018 simultaneously with the reduction of the BDW layer with its upper limit at –1.05°C, (Fig. 2, Fig. 3c). The –1.5°C isotherm and 34.55 psu isohaline descended to a depth of 1,100 m (Table 1, Fig. 3), and the water with $\theta \leq -1.7$°C was not found in the transect in 2018. The layer of BDW continued to lose its density due to a decrease in salinity and an increase in potential temperature in 2020 relative to 2016, 2017, and 2018 (Table 1). However, layer thickness with its upper limit at –1.05°C was restored to that observed in 2016.

9.4 Discussion

We revealed a correlation with ice conditions, unlike that of 2016–2017 when excessive ice cover was observed. In 2018, the strait was completely ice-free (Kashin et al. 2018). In the beginning of March 2020, the operations were carried out in the ice-free sea without floating ice. However, there was a large number of icebergs in the strait produced by the giant A68 iceberg, which was drifting at a distance of 100 miles from the section near Elephant Island (Antipov et al. 2020). Large pieces of the broken iceberg transported by the coastal current formed an impassable cluster of 9–10 icebergs in the southeastern part of the section (near the Antarctic Peninsula), and we had to relocate our stations. These icebergs were relatively young without traces of melting and wave erosion, which indicated that they were parts of the broken A68.

The, potential temperature in the bottom layer (stations 3, 4, 5, 6) increased by 0.188 °C in the central basin and by 0.166 °C at the deepest station 5 during the past 5 years. Salinity in the near-bottom layer decreased by 0.04 psu (Table 9.1).

The water layer of BDW with $\gamma^n > 28.4$ has been shrinking for 5 years. The 28.40 isopycnal was observed at a depth of 900 m in 2016 (Fig. 9.3a). This densest water was observed only at the two deepest stations (4, 5) deeper than 1400 m in 2020. Thus, by 2020, the deep layer in the central basin of the Bransfield Strait was almost completely occupied by water with $28.27 < \gamma^n < 28.4$. The neutral density in near-bottom layer at the deepest station 5 decreased from 28.50 in 2016 to 28.42 in 2020.

Based on the analysis of temperature, salinity, oxygen, and chlorofluorocarbon (CFC) data, Gordon et al. (2000) defined the water type of the deep layer in the central basin of the Bransfield Strait. The components (and their percentages) are Weddell Deep Water (11%), low-salinity shelf water (24%), high-salinity shelf water (60%), to Pacific pycnocline (5%). Thus, according to Dotto et al. (2016), Antipov et al. (2020), and Strass et al. (2020), we suggest that the observed density loss of BDW in the central basin of the Bransfield Strait is due to a decrease of salinity and an increase of potential temperature during the past 5 years; this is caused by increased proportion of low-salinity shelf water and the changing properties of the water masses exported from the Weddell Sea. We also propose that an increase in temperature of WW flowing within ACoC from the Weddell Sea shelf observed during these 5 years in the surface layer above the shelf and the slope of the eastern part of the Bransfield Strait along the Antarctic Peninsula, which we associate with admixture of WDW to WW over the shelf north off Joinville Island, and a shortening of the ice-formation season, reduces the depth of convection and interrupts the increase of density of the underlying waters and BDW.

9.5 Conclusions

Monitoring of the interannual variations of Bransfield Deep Water (BDW) properties in 2016, 2017, 2018, and 2020 confirmed the existence of a unidirectional trend of warming and density loss described in the previous research. It is shown that the density loss of BDW in 2016, 2017, 2018, and 2020 is due to various changes of BDW properties. The neutral density loss of 0.03 kg/m^3 was caused by an increase in potential temperature which was observed at depths of 500–1300 m with the same characteristics of the bottom layer in 2016 and 2017. In 2018, a strong density loss of 0.1 kg/m^3 was mainly caused by a decrease of salinity by 0.03 psu and to a lesser extent by a warming by 0.12 °C observed in BDW. The density loss due to a decrease of salinity and warming continued in 2020.

We revealed that the potential temperature in the bottom layer has increased by 0.18 °C and salinity had decreased by 0.04 psu during the past 5 years. Interannual variations of the water mass properties caused a decrease of neutral density in the bottom layer from 28.50 in 2016 to 28.42 in 2020 and a reduction of layer thickness of BDW with $\gamma^n > 28.40$. Thus, by 2020 the deep layer in the central basin of the Bransfield Strait was almost completely occupied by water with $28.27 < \gamma^n < 28.40$.

In correspondence with previous research studies, we suggest that the observed unidirectional trend of warning and density loss in the deep layer in the central basin of the Bransfield Strait caused by an increased proportion of low-salinity shelf water and changing properties of the water masses exported from the Weddell Sea. We also propose that an increase of temperature of WW flowing within the ACoC from the Weddell Sea shelf (as observed during these 5 years in the surface layer above the shelf and the slope of the eastern part of the Bransfield Strait along the Antarctic Peninsula) reduces the depth of convection and interrupts the increase of density of the underlying waters and BDW. We associate this with an admixture of WDW to WW over the shelf north off Joinville Island and shortening of the ice-formation season.

Acknowledgments This work is a contribution to project 5.2 "Complex Integrated oceanographic, climatic, glaciological and geophysical studies of Antarctica and the Southern Ocean" funded by Roshydromet of the Russian Federation. The authors are grateful to N.N. Antipov for valuable comments to the article and to shipboard scientific-technical crew of the R/V "Akademik Fedorov" and "Akademik Tryoshnikov" for their help in obtaining data. This research was supported by the Russian state Task 0128-2919-0008.

References

Antipov NN, Kashin SV, Molchanov MS (2020) Oceanographic research performed during the cruise 14th R/V Akademik Tryoshnikov. Russ Polar Res 2:32–35. (RUS)

Avelina R, da Cunha LC, Farias CDO, Hamacher C, Kerr R, Mata MM (2020) Contrasting dissolved organic carbon concentrations in the Bransfield Strait, Northern Antarctic Peninsula: Insights into ENSO and SAM effects. J Mar Syst (J Pre-proof)

Dotto TS, Kerr R, Mata MM, Garcia CAE (2016) Multidecadal freshening and lightening in the deep waters of the Bransfield Strait, Antarctica. J Geophys Res Oceans 121:3741–3756

Franco BC, Mata MM, Piola AR, Garcia CAE (2007) Northwestern Weddell Sea deep outflow into the Scotia Sea during the austral summers of 2000 and 2001 estimated by inverse methods. Deep Sea Res I Oceanogr Res Pap 54(10):1815–1840

Garcıa MA, Lopez O, Sospedra J, Espino M, Gracia V, Morrison G, Rojas P, Figa J, Puigdefabregas J, S-Arcilla A (1994) Mesoscale variability in the Bransfield Strait region (Antarctica) during austral summer. Ann Geophys 12(9):856–867

García MA, Castro CG, Rıos AF, Doval MD, Roson G, Gomis D, Lopez O (2002) Water masses and distribution of physico-chemical properties in the Western Bransfield Strait and Gerlache Strait during austral summer 1995/96. Deep Sea res. Part II top. Stud Oceanogr 49:585–602

Gomis D, Garcıa MA, Lopez O, Pascual A (2002) Quasi-geostrophic 3D circulation and mass transport in the western sector ofthe south Sheltland Islands. Deep Sea Res II 49:603–621

Gordon AL, Nowlin WD (1978) The basin waters of the Bransfield Strait. J Phys Oceanogr 8:258–264

Gordon AL, Mensch M, Zhaoqian D, Smethie W, de Bettencourt J (2000) Deep and bottom water of the Bransfield Strait eastern and central basins. J Geophys Res Atmos 105(C5):11337–11346

Gordon AL, Visbeck M, Huber BA (2001) Export of Weddell Sea deep and bottom water. J Geophys Res 106(C5):9005–9018

Grelowski A, Majewicz A, Pastuszak M (1986) Mesoscale hydrodynamic processes in the region of the Bransfield Strait and the southern part of Drake Passage during BIOMASS-SIBEX 1983/84. Polish Polar Research 7:353–369

Henley SF, Schofield OM, Hendry KR, Schloss IR, Steinberg DK, Moffat C, Peck LS, Costa DP, Bakker DCE, Hughes C, Rozema PD, Ducklow HW, Abele D, Stefels J, van Leeuwe MA, Brussaard CPD, Buma AGJ, Kohut J, Sahade R, Friedlaender AS, Stammerjohn SE, Venables HJ, Meredith MP (2019) Variability and change in the West Antarctic peninsula marine system: research priorities and opportunities. Prog Oceanogr 173:208–237

Hofmann E, Klinck J, Lascara MC, Smith AD (1996) Water mass distribution and circulation west of the Antarctic peninsula and including Bransfield Strait. Antarctica Res Ser 70(1):61–80

Huneke WGC, Huhn O, Schröeder M (2016) Water masses in the Bransfield Strait and adjacent seas, austral summer 2013. Polar Biol 39(5):789–798

Jackett DR, McDougall TJ (1997) A neutral density variable for the World's oceans. J Geophys Res 27:237–263

Kashin SV, Molchanov MS, Chistyakov IA (2018) Oceanography operations carried out during the expedition. In: 63rd RAE VII international Conf "Marine res and education": conference proceedings, vol III (IV), pp 113–116

Kerr R et al (2018) Carbonate system properties in the Gerlache Strait, northern Antarctic peninsula (February 2015): II. Anthropogenic CO2 and seawater acidification. Deep Sea Res II 149:171–181

Lopez O, Garcia MA, Gomis D, Rojas P, Sospedra J, S-Arcilla A (1999) Hydrographic and hydrodynamic characteristics of the eastern basin of the Bransfield Strait (Antarctica). Deep Sea Res I 46:1755–1778

Morozov EG (2007) Currents in Bransfield Strait. Dokl Earth Sci 415A(6):984–986

Naveira Garabato AC, Heywood KJ, Stevens DP (2002) Modification and pathways of Southern Ocean deep waters in the Scotia Sea. Deep Sea Res I 49:681–705

Niiler P, Amos A, Hu J-H (1991) Water masses and 200 m relative geostrophic circulation in the western Bransfield Strait region. Deep Sea Res II 38:943–959

Sangra P, Gordo C, Hernandez-Arencibia M, Marrero-Diaz A, Rodriguez-Santana A, Stegner T, Martínez-Marrero A, Pelegrí JL, Pichon A (2011) The Bransfield current system. Deep Sea Res I Oceanogr Res Pap 58(4):390–402

Smith AD, Hofmann E, Klinck MJ, Lascara MC (1999) Hydrography and circulation of the West Antarctic peninsula continental shelf. Deep Sea Res I Oceanogr Res Pap 46(6):925–949

Strass VH, Rohardt G, Kanzow T, Hoppema M, Boebel O (2020) Multidecadal warming and density loss in the deep Weddell Sea, Antarctica. J Clim 33(22):9863–9881

Thompson AF, Heywood KJ, Thorpe SE, Renner AHH, Trasviña (2009) A surface circulation at the tip of the Antarctic peninsula from drifters. J Phys Oceanogr 39:3–26

Tokarczyk R (1987) Classification of water masses in the Bransfield Strait and southern part of the Drake Passage using a method of statistical multidimensional analysis. Polish Polar Res 41:629–641

Von Gyldenfeldt AB, Fahrbach E, García MA, Schröder M (2002) Flow variability at the tip of the Antarctic peninsula. Deep Sea Res II Top Stud Oceanogr 49(21):4743–4766

Weatherall P, Marks KM, Jakobsson M, Schmitt T, Tani S, Arndt JE, Rovere M, Chayes D, Ferrini V, Wigley R (2015) A new digital bathymetric model of the world's oceans. Earth Space Sci 2:331–345. https://wwwgebconet/data_and_products/gridded_bathymetry_data/gebco_30_second_grid. The latest release of the GEBCO 30 arc-second global grid of elevations, GEBCO_2014 Grid, was published in 2014

Whitworth T, Nowlin WD, Orsi AH, Locarnini RA, Smith SG (1994) Weddell Sea shelf water in the Bransfield Strait and Weddell-scotia confluence. Deep-Sea Res I Oceanogr Res Pap 41:629–641

Whitworth T III, Orsi AH, Kim SJ, Nowlin WD, Locarnini RA (1998) Water masses and mixing near the Antarctic slope front. In: Ocean ice and atmosphere: interactions at the Antarctic continental margin, Antarctic research series 75. American Geophysical Union, Washington, DC, pp 1–27

Chapter 10
Sea Surface Temperature and Ice Concentration Analysis Based on the NOAA Long-Term Satellite and Sea-Truth Data in the Atlantic Antarctic

Viktor V. Zamshin and Vladislav A. Shliupikov

Abstract In our research, we used the NOAA long-term satellite and sea-truth data to study sea ice concentration (SIC) and sea surface temperature (SST) in the Atlantic Antarctic. The research was performed in a region limited by 53° and 67° south latitude and 75° and 37° west longitude. The research time frame was from September 1, 1981, to December 31, 2019. A total of ~14,000 thousand measurements of SST and the same number of SIC measurements were considered in each of ~1000 spatial cells in the zone of interest. Different zones of various annual SIC concentrations were delineated in 40-year dynamics, and the area of interest was segmented into parts that have different, mostly negative SST trends. The results of the analysis made it possible to observe in detail an unusual process: cooling in the Drake Passage in the twenty-first century against the background of global warming.

Keywords Sea surface temperature · Sea ice concentration · Remote sensing · Satellite oceanography · Atlantic Antarctic · Global warming · Climate · Longterm data

10.1 Introduction

Remote sensing is a good tool to study vast, hard-to-reach water basins (Bondur 2004; Robinson 2010). The methods of remote sensing are very promising when combined with mathematical modeling (Bondur 2005, 2011) and in situ measurements (Keeler et al. 2004; Bondur and Tsidilina 2005). Combined use of

V. V. Zamshin (✉) · V. A. Shliupikov
AEROCOSMOS Research Institute for Aerospace Monitoring, Moscow, Russia

E. G. Morozov et al. (eds.), *Antarctic Peninsula Region of the Southern Ocean*, Advances in Polar Ecology 6, https://doi.org/10.1007/978-3-030-78927-5_10

remote sensing and mathematical modeling with experimental results allows us to carry out mutual validation of heterogeneous data (Bondur and Tsidilina 2005; Laxon et al. 2013) and to implement comprehensive research programs, as well as to create, operate, and develop multicomponent systems for monitoring the sea basins (Bondur and Savin 1993; Bondur and Zamshin 2018).

One approach for comprehensively processing the results of remote sensing data and sea-truth measurements, when combined with mathematical modeling, is long time series of spatial distributions of significant marine environment parameters formed on a regular spatiotemporal grid. An example of such a dataset is the "NOAA OISST" (Reynolds et al. 2008).

The NOAA 1/4 degree daily Optimum Interpolation Sea Surface Temperature (OISST) provides complete ocean temperature fields constructed by combining bias-adjusted observations from various platforms (satellite, ships, buoys) on a regular global grid with gaps filled in by interpolation. Satellite data from the Advanced Very-High-Resolution Radiometer (AVHRR) provides the main input which makes possible the high temporal-spatial coverage beginning in late 1981 up to the present (Reynolds et al. 2008). The "NOAA OISST" product also includes the 7-day median values of daily sea ice concentrations, which serve as the basis for proxy SSTs in the marginal ice zone where observations are lacking.

In our research, we used this dataset to study sea ice concentration (SIC) and sea surface temperature (SST) in the Atlantic Antarctic. Natural resources of the Southern Ocean are in equal measure vulnerable, highly in demand, and understudied. In this regard, research programs focused on the study of the Antarctic seas are currently being implemented. The suggested study is part of the efforts described in Morozov et al. (2020) and Zamshin et al. (2020).

10.2 Initial Data

The research was performed in a region limited by 53° and 67° south latitude and 75° and 37° west longitude. The area of the site of interest is ~2.8 mln. km^2.

The NOAA OISST (daily SST distribution, °C and SIC, %) dataset used in our research was preprocessed using Google Earth Engine (https://earthengine.google.com/).

During the preprocessing, we downloaded sets of daily raster fields sorted by years and fragmented by the site of interest (see Fig. 10.1). The research time frame was from September 1, 1981, until December 31, 2019.

Figure 10.2 shows example graphs of sea surface temperature and ice concentration in 1982 and 2019 in the ~61° S, 50° W cell that has been visited by the R/V *Akademik Mstislav Keldysh* during its 79th expedition (Morozov et al. 2020).

Analysis of graphs in Fig. 10.2 reveals quite similar ice and temperature variations in 1982 and 2019 in a single spatial cell. The beginning of a calendar year is characterized with gradual decrease of temperature from 1.5–2.5 °C to negative values. Also, beginning from April–May, nonzero values of sea ice concentration

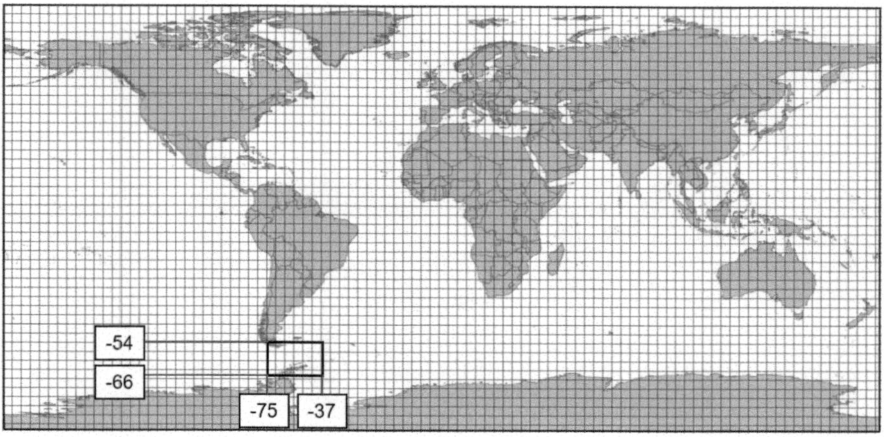

Fig. 10.1 Schematic map of the location of the study region with a land mask

appear. The greatest ice concentrations (exceeding 70%) occur in June–September. Then, ice concentration decays to November–December, and the temperature increases up to ~2 °C.

The cited example illustrates initial data fragments out of the overall spatial and temporal context. During almost 40 years of research, a total of 14,001 measurements of SST and the same number of SIC measurements were recorded (excluding a small number of gaps in the observations that were tracked and filtered out) in each of ~10^3 spatial cells in the zone of interest. We will make an attempt to analyze the dynamics of these indicators taking into account geospatial aspect of the data (results as quick maps).

10.3 Data Analysis of Sea Ice Concentration

With the goal in mind to develop information products suitable for analyzing both the spatial and temporal components of sea ice concentration dynamics within each calendar year, each cell was checked for agreement with the following conditions:

$$\sum_{n=1}^{D} k_n / D > 70 \tag{10.1}$$

$$70 \geq \sum_{n=1}^{D} k_n / D > 1 \tag{10.2}$$

where k_n is the ice concentration in the studied cell on the current day, n, and D is the number of measurements in the current year.

Condition (10.1) corresponds to the cells, in which annual mean sea ice concentration exceeds 70% (almost complete ice cover). Condition (10.2) describes

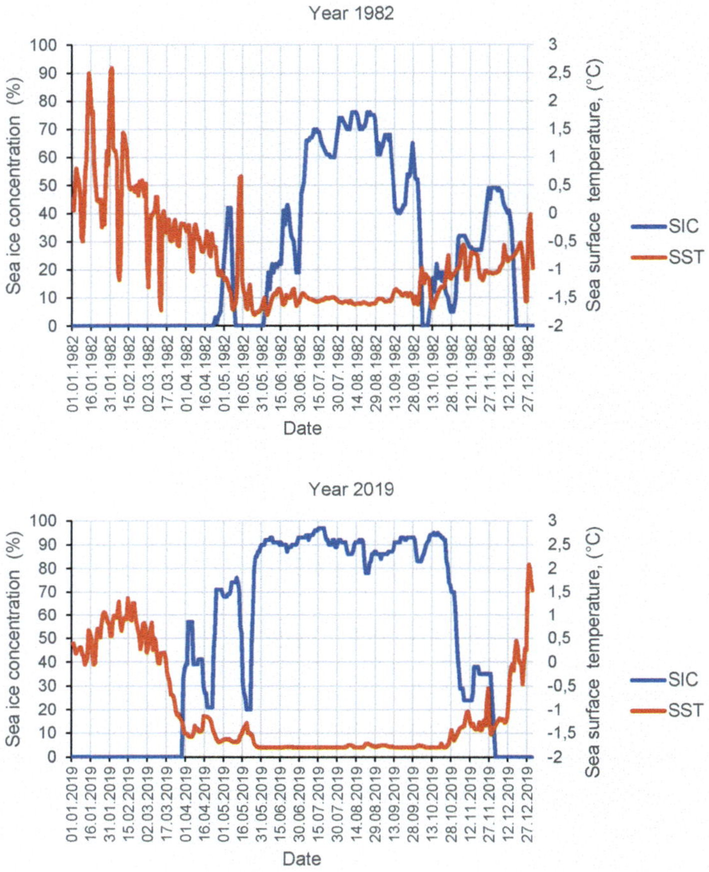

Fig. 10.2 Sea ice concentration (SIC) (blue line) and sea surface temperature (SST) (red line). Dynamic examples for one of the spatial cells (61° S, 50° W) of the studied dataset in 1982 (top) and 2019 (bottom)

the regions with middle and low annual mean ice concentration. Regions with annual mean ice concentration smaller than 1% are considered completely ice-free surface.

The maps in Figs. 10.3a and 10.3b were plotted on the basis of the stated conditions (10.1 and 10.2) for each year of the period (1981–2019).

The dynamics of areas with high and low ice concentration (expressed through the number of spatial cells) in the selected zones is given in Fig. 10.4.

Analysis of Figs. 10.3a, 10.3b and 10.4 provides evidence about the increase of ice cover area due to the growth of the "green zone" (SIC higher than 70%) from ~400 cells up to ~1000 cells and higher, which is especially evident since 2000 (green graphs in Fig. 10.4). The abnormally higher ice concentration in the "green zone" in 1981 was apparently related to the fact that the observations started only on September 1. The zone of moderate and low ice concentration (blue graphs in

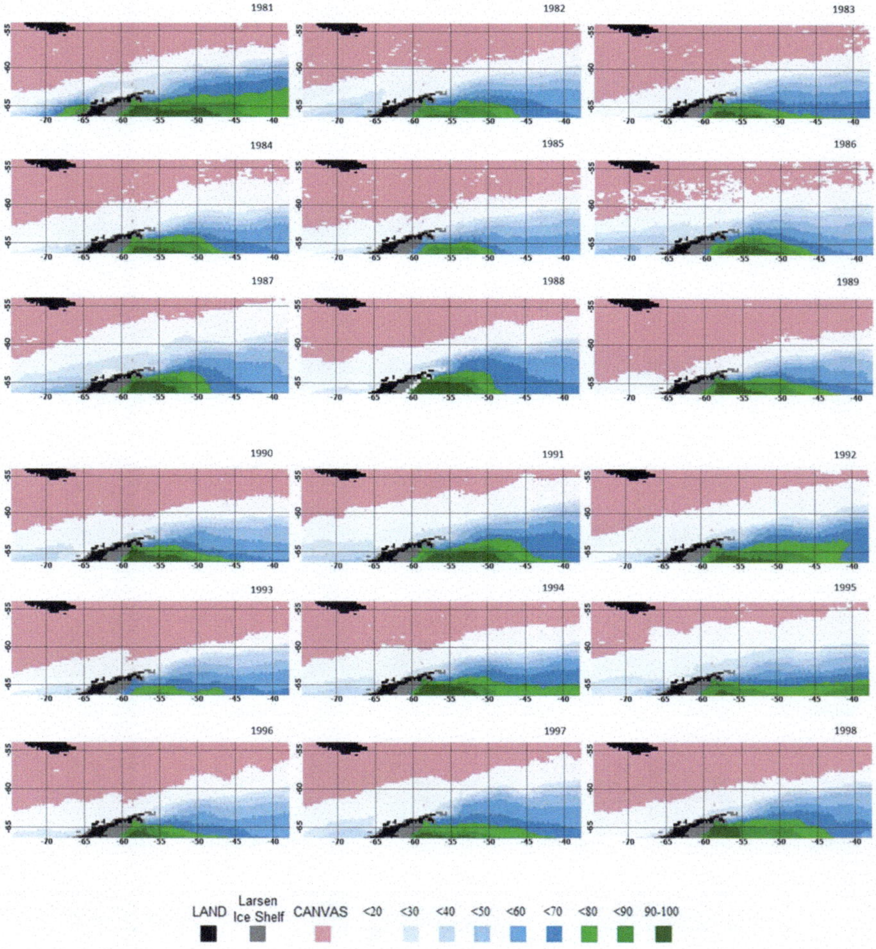

Fig. 10.3a Time series of maps illustrating the location of regions with average annual SIC exceeding 70% (green colors) over a year in the period 1981–1998, as well as with moderate and low annual SIC (blue colors turning to white). Pink regions correspond to ice-free water surface

Figs. 10.2 and 10.4) is characterized by high variability and an almost-zero area trend. Consequently, the total ice cover area (combining the "blue" and "green zones") increases.

For further analysis of multiannual ice dynamics, maps were generated that show how often the regions of the studied basin were attributed to one or another class corresponding to conditions (10.1) or (10.2) (see Fig. 10.5).

Taking into account the duration of the period covered by the initial data, the distributions from Fig. 10.5 illustrate the probability of an event when a spatial cell belonging to one or another water area meets condition (10.1) (average annual sea ice concentration exceeding 70%) or condition (10.2) (moderate and low average annual sea ice concentrations).

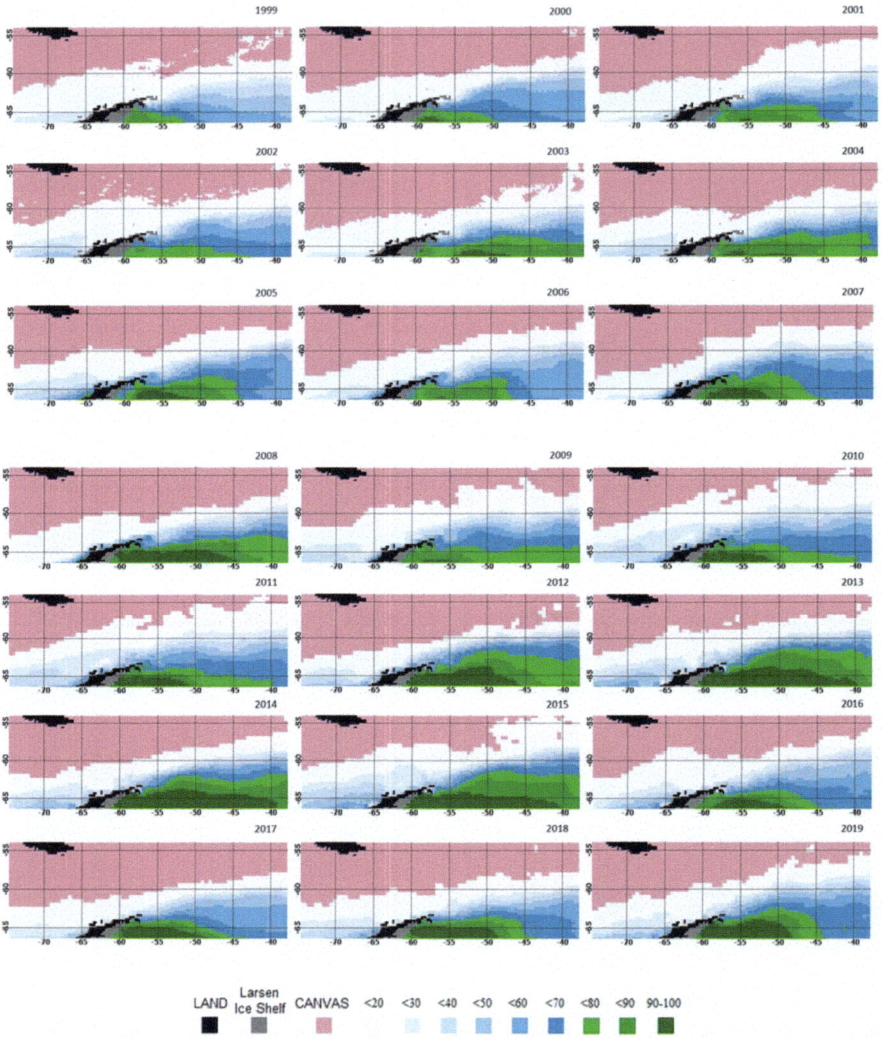

Fig. 10.3b Time series of maps illustrating the location of regions with average annual SIC exceeding 70% (green colors) over a year in the period 1999–2019, as well as with moderate and low annual SIC (blue colors turning to white). Pink regions correspond to ice-free water surface

Analysis of Figs. 10.3a, 10.3b and 10.5 shows that the zone of high SIC is located only east (not west) of the Antarctic Peninsula, localizing mainly at the latitudes south of 63° S in a range of longitudes between 60° and 45° W. The longitude of the northern boundary of moderate and low SIC zone monotonously increases from 62° W in the west to 55° W in the east.

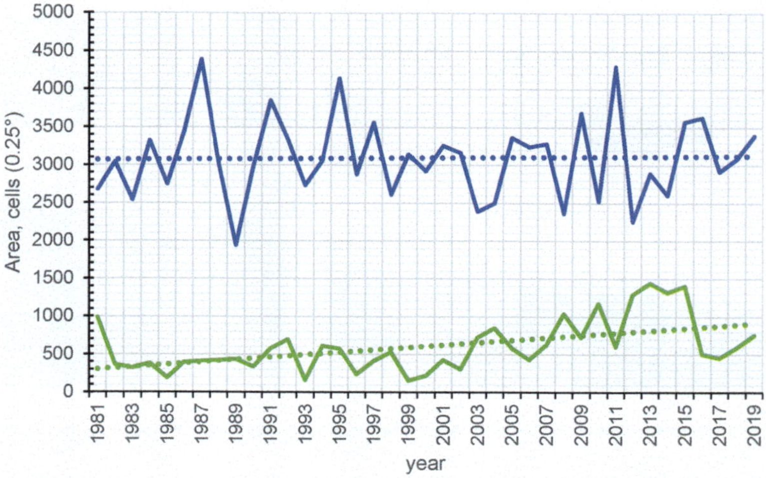

Fig. 10.4 Areas of zones with average annual SIC higher than 70% (green) and with moderate and low one (blue). The dotted lines are trends

10.4 Data Analysis of Sea Surface Temperatures

We estimated standard deviations (stddev) and linear regression coefficients for sea surface temperature (SST) series that characterize the variability and long-term trends of this significant water environment parameter in the studied region for each spatial cell.

The data processing yielded maps (spatial distributions) of linear regression coefficients and stddev (Fig. 10.6a, b, respectively) of SST time series.

Vast areas of SST increase trends (red regions) are seen in the map of linear regression coefficients (Fig. 10.6a). At the same time, there are also regions where SST decreases quickly (blue).

In Fig. 10.6b, two large regions (blue and red) draw our attention. These regions correspond to relatively low (0.0–0.5 °C) and relatively high (1.8–2.3 °C) values of SST stddev, respectively.

A two-dimensional (2D) scatter plot of stddev and linear regression coefficients of the SST series was considered in the comprehensive analysis of sea surface temperature (Fig. 10.7).

As seen from Fig. 10.7a, the point cloud represented in the coordinate system SST linear regression coefficient (y-axis) vs stddev (x-axis) can be segmented into zones that correspond to certain areas on the map (Fig. 10.7b), zones shown in appropriate colors in the 2D scatter plot (Fig. 10.7a) and map (Fig. 10.7b).

The "red" zone corresponds to those spatial cells within the dataset under processing where the temperature increases in the long term (positive linear regression coefficient). The "green" zone is characterized with low stddev values, and the "blue" one is characterized with quite high values of SST rate of decrease.

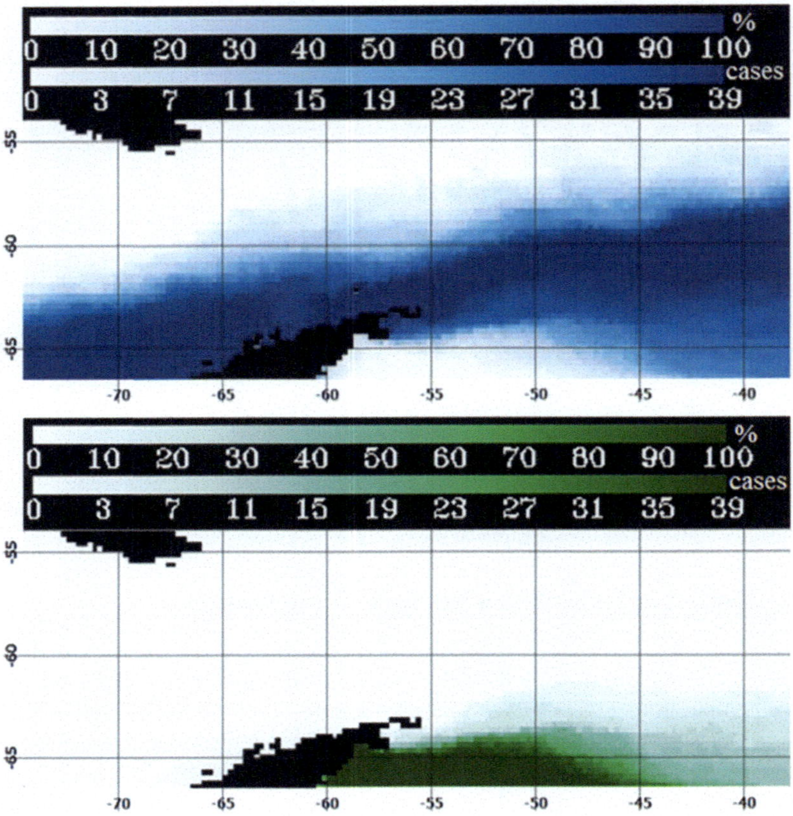

Fig. 10.5 Maps illustrating how often the regions of the study basin were attributed to the "average annual SIC exceeding 70% (green)" and "moderate and low average annual SIC (blue turning to white)" classes. The scales show the relative (%) and absolute numbers of cases when a region was attributed to a selected class

In general, a decrease in SST was found in all zones, except for the "red" in the long-term retrospective.

It should be noted that the "red class" (warming) divides the remaining unclassified vast water area into two parts. Yellow and cyan colors were assigned to these parts.

Analysis of Fig. 10.7 shows that the approach that was applied to study SST dynamics enabled us to delineate five zones within the studied ocean region. The mutual locations and configurations of these zones generally agree with the ice conditions (as previously noted) and other oceanographic properties of the region.

Figure 10.8 shows the average annual SST dynamics in five selected zones, and Table 10.1 summarizes the main SST characteristics of the same zones. The most pronounced trend in the region of SST decrease draws our attention (the blue polyline in Fig. 10.8). It is seen from Fig. 10.8, Table 10.1, and Fig. 10.7a that the

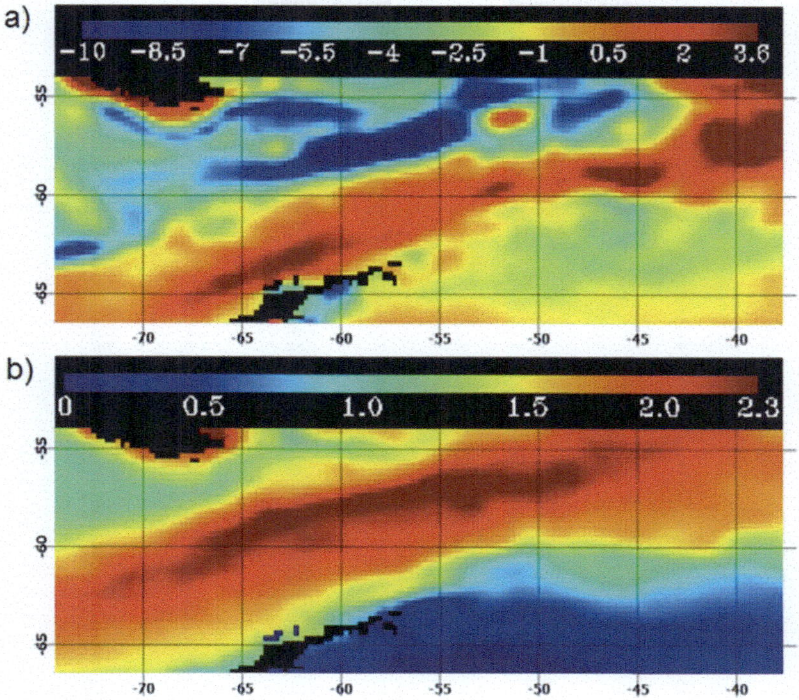

Fig. 10.6 Maps of linear regression coefficients (**a**) (10^{-5} °C) and stddev (**b**) in (°C) for the SST time series

SST change rate in the "red" zone (warming) is almost an order of magnitude smaller than cooling in the "blue" zone.

Zones with the most stable SST (low stddev values, i.e., green and cyan in Figs. 10.7 and 10.8, and in Table 10.1) are attributed to the regions characterized by the presence of ice with average annual concentration of about 50% and higher (Figs. 10.3a, 10.3b and 10.5).

The largest area of the study region corresponds to the "yellow" class (see Fig. 10.7b). Like most other classes, the "yellow" one is characterized by an SST decrease at a moderate rate of reduction. Also, we see here high standard deviation values, i.e., strong variability (see Table 10.1 and Fig. 10.6). The water basin corresponding to the "yellow" class is remarkable because inside it (in terms of the relative position in space) there is a relatively small area of the already mentioned "blue" class water region segmented into two parts. In the graph in Fig. 10.8a, the similarity (high-correlated) of SST dynamics in these water basins is well traced; however, we once again emphasize the peculiar anomaly of the "blue" class water basin: the SST decrease at a rate of −0.0000736 °C per day. Thus, over the observation period (since 1981), the SST of this water area decreased by 1.03 °C with a mean value of 2.95 °C at present.

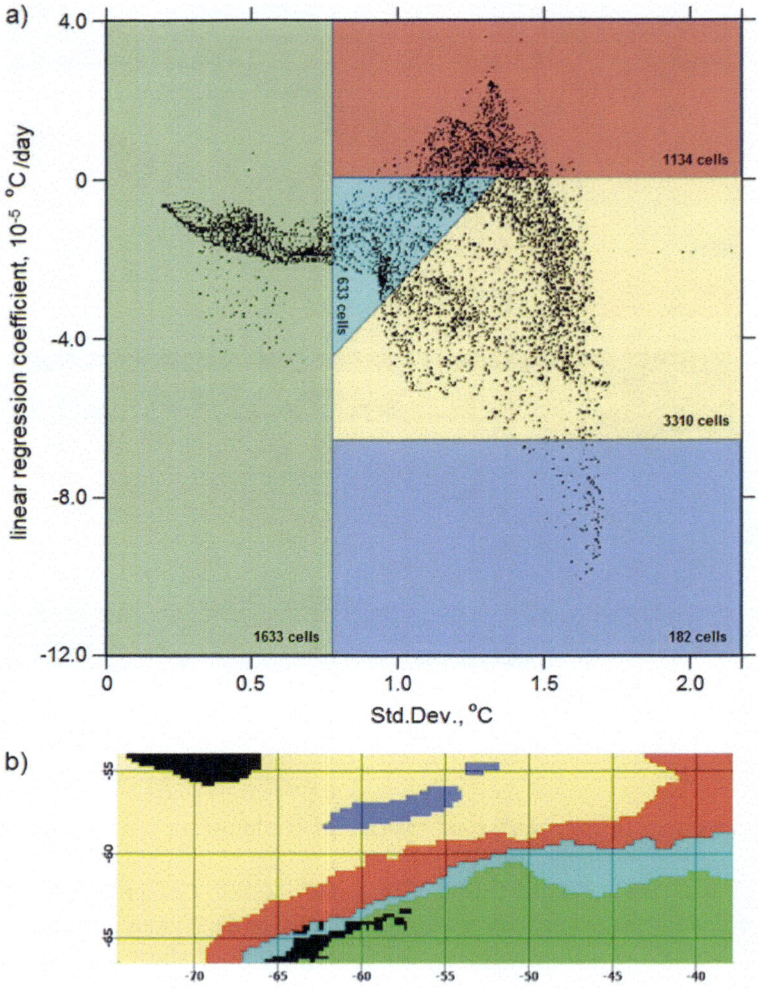

Fig. 10.7 A 2D scatter plot of the SST linear regression coefficient vs stddev and its division into classes (**a**). A map of classes selected in the 2D scatter plot (**b**)

As seen from these data, in general, we observe a decrease of the SST in the area of interest based on the results of our investigation. Thus, we have the opportunity to identify one of the few areas of the globe in which, despite global warming, a negative temperature trend is observed. The obtained result is in agreement with the works (Donev et al. 2019; NOAA 2020) and is of interest for further research with a view to a deeper interpretation.

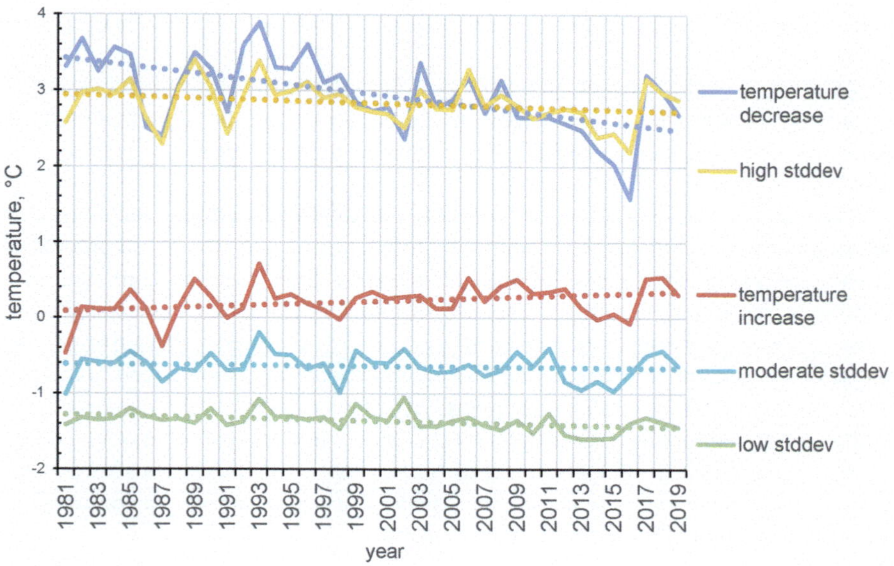

Fig. 10.8 Average annual temperature dynamics in five zones of the study area selected on the basis of the comprehensive analysis of multiannual trends and standard deviations of SST

Table 10.1 Quantitative indicators, mean (°C), trend (10^{-5} °C per day), and stddev (°C) in five zones of the study region selected on the basis of the comprehensive analysis of multiannual trends and standard deviations of SST

zones	mean (°C)	trend (10^{-5} °C per day)	stddev (°C)
temperature increase	0.22	0.93	1.30
temperature decrease	2.95	-7.36	1.61
high stddev	2.84	-2.35	1.36
moderate stddev	-0.64	-1.07	1.01
low stddev	-1.36	-1.46	0.52

10.5 Conclusions

Satellite remote sensing combined with sea-truth observations and mathematical modeling methods are able to provide unique data on the ocean environment over wide spatial and temporal ranges. In this research, we have tried to explore and analyze a NOAA OISST dataset that is a uniquely long-term set of regular data on the sea surface temperature (SST) and sea ice concentration (SIC) with 0.25 deg. cell size. This archive comprises almost 40 years of data.

Since 2000, we observe an increase in the area covered with sea ice in the Atlantic Antarctic. This is especially true for the zone with annual SIC exceeding 70% (the area of this zone has almost doubled during the observation period). The growing zone of high SIC is a factor in the gradual invasion of the ice-covered area deep into the ocean.

As a whole, a decrease of SST has been observed in the research ocean zone, against the background of an increase in ice concentration. In some fast-cooling areas, the trend indicates that SST has decreased by more than 1 °C during the observation period.

The trends we have observed are probably due to global warming, including the warming of the overall Antarctic continent. It is fair to assume that these processes are the drivers of increasing volumes of ice separating from the continent and drifting in the Atlantic Antarctic. One of the significant phenomena associated with these processes is the separation and drift of the extremely large iceberg A68. Thus, the sea ice zone, which is expanding due to global warming, should lead to a local decrease in SST, which we had the opportunity to investigate in this framework.

It is worth mentioning that the observed complex map of spatial inhomogeneity of the decrease rate in SST is subject to further study. The most probable factors causing such inhomogeneity are the local pattern of ocean currents in conjunction with sea-level anomalies in the Atlantic Antarctic.

Acknowledgments The study was supported by the State Assignment of Russia no. 0588-2019-0030.

References

Bondur VG (2005) Complex satellite monitoring of coastal water areas. In: Proceedings, 31st International Symposium on Remote Sensing of Environment, ISRSE. p 7

Bondur VG, Tsidilina MN (2005) Features of formation of remote sensing and sea truth databases for the monitoring of anthropogenic impact on ecosystems of coastal water areas. In: Proceedings, 31st International Symposium on Remote Sensing of Environment, ISRSE. pp 192–195

Bondur VG (2004) Aerokosmicheskie metody v sovremennoy okeanologii (Aerospace methods in modern oceanology). In: New Ideas in Oceanology V.1. Physics. Chemistry. Biology. Nauka, Moscow, pp 55–117

Bondur VG (2011) Satellite monitoring and mathematical modelling of deep runoff turbulent jets in coastal water areas. In: Waste water – evaluation and management. InTech, Croatia, pp 155–180

Bondur VG, Savin AI (1993) Design a system to monitor the environment for purposes relating to ecology and natural-resources. Sov J Remote Sens 10:1078–1093

Bondur VG, Zamshin VV (2018) Comprehensive ground-space monitoring of anthropogenic impact on Russian Black Sea coastal water areas. In: Anisimov KV, Dub AV, Kolpakov SK et al (eds) Proceedings of the Scientific-Practical Conference "Research and Development – 2016". Springer, Cham, pp 625–637

Donev J, Stenhouse K, Sheardown A, Hanania J (2019) Global surface temperature anomaly – Energy Education. https://energyeducation.ca/encyclopedia/Global_surface_temperature_anomaly#cite_note-4. Accessed 6 Jan 2021

Keeler RN, Bondur VG, Vithanage D (2004) Sea truth measurements for remote sensing of littoral water. Sea Technol 45:53–58

Laxon SW, Giles KA, Ridout AL et al (2013) CryoSat-2 estimates of Arctic sea ice thickness and volume. Geophys Res Lett 40:732–737. https://doi.org/10.1002/grl.50193

Morozov EG, Spiridonov VA, Molodtsova TN et al (2020) Investigations of the Ecosystem in the Atlantic Sector of Antarctica (Cruise 79 of the R/V Akademik Mstislav Keldysh). Oceanology 60:721–723

NOAA (2020) Global climate report for annual 2019. https://www.ncdc.noaa.gov/sotc/global/201913. Accessed 6 Jan 2021

Reynolds RW, Banzon VF, Program NC (2008) NOAA Optimum Interpolation 1/4 Degree Daily Sea Surface Temperature (OISST) analysis, version 2

Robinson IS (2010) Discovering the Ocean from Space: the unique applications of satellite oceanography. Springer-Verlag, Berlin Heidelberg

Zamshin VV, Matrosova ER, Morozov EG et al (2020) Satellite remote sensing information support for the R/V "Akademik Mstislav Keldysh" South Ocean mission. Issledovanie Zemli iz kosmosa 4:74–85. https://doi.org/10.31857/S0205961420040077

Part II
Chemical Oceanography, Seawater Optical Properties, Productivity and Microbial Processes

Chapter 11
Hydrochemical Structure of Waters in the Northern Weddell Sea in Austral Summer 2020

Svetlana V. Stepanova, Alexander A. Polukhin, Gennadii V. Borisenko, Anna L. Chultsova, Evgeniia N. Marina, Oleg S. Popov, Anna M. Seliverstova, Anna V. Vidnichuk, and Petr P. Tishchenko

Abstract Studies of the hydrochemical parameters of water in the northern part of the Weddell Sea were carried out in January and February 2020. The integral section from west to east of the study site shows spatial variability of the abiotic characteristics of the ecosystem: oxygen content, nutrients (dissolved phosphates, silicates, mineral forms of nitrogen), and parameters of the carbonate system (pH and total alkalinity). The average content of phosphates varied from 1.7 to 2.1μM, silicates from 66.5 to 119.5μM, and nitrates from 29 to 38μM. Water properties in the study were typical for the waters of the Weddell Sea, and five regions differing in the distribution of hydrochemical parameters were identified. In the west, we found the influence of waters from the Bransfield Strait; in the east, on the shelf of the South Orkney Islands, the upper layer of the water column was influenced by the waters of the southern front of the Antarctic Circumpolar Current. The exchange of waters through the Philip Passage was recorded mainly in the upper 500-m layer. In addition, the basin was influenced by the "ambient" high-latitude waters of the Weddell Sea. Significant variability in the concentrations of nutrients, especially phosphates, which is associated with the genesis of waters and the peculiarities of circulation is a peculiar feature of the basin.

S. V. Stepanova (✉) · A. A. Polukhin (✉) · G. V. Borisenko · A. L. Chultsova · A. M. Seliverstova
Shirshov Institute of Oceanology, Russian Academy of Sciences, Moscow, Russia
e-mail: polukhin@ocean.ru

E. N. Marina · O. S. Popov · P. P. Tishchenko
V.I. Ilyichev Pacific Oceanological Institute, Russian Academy of Sciences, Moscow, Russia

A. V. Vidnichuk
Marine Hydrophysical Institute, Russian Academy of Sciences, Moscow, Russia

E. G. Morozov et al. (eds.), *Antarctic Peninsula Region of the Southern Ocean*, Advances in Polar Ecology 6, https://doi.org/10.1007/978-3-030-78927-5_11

Keywords Weddell Sea · Hydrochemical structure · Water masses · Variability of nutrients

11.1 Introduction

The coastal zone of the Antarctic Peninsula is characterized by unique productivity richness and biodiversity. The waters surrounding the peninsula are rich in krill, *Euphausia superba* (Siegel and Watkins 2016). Krill serves as food for cetaceans, seals, and seabirds and is an important object of fishery. Primary production in the waters of the Antarctic Peninsula is greater than in the adjacent waters. The flow of nutrients to the bottom is also large, which indicates a large amount of suspended matter in the water (Nedwell and Walker 1995).

It is known that the region is characterized by a complex scheme of water circulation, i.e., the flow of various water masses from the Bransfield Strait, the Weddell Sea, and the Drake Passage (Orsi et al. 1999).

The Antarctic Peninsula is the northernmost tip of Antarctica, separated from South America by the Drake Passage. The region adjacent to the Antarctic Peninsula supports the productivity of the marine ecosystems of the Scotia and Weddell seas. The waters surrounding the Antarctic Peninsula are rich in dissolved iron and nutrients, which in turn supports high primary production (Helbling et al. 1991).

The Powell Basin is part of the Weddell Sea, bounded in the west by the Antarctic Peninsula shelf and the eastern part of the Bransfield Strait, in the north by the South Scotia Ridge, and in the east by the South Orkney underwater plateau whose mark points are the South Orkney Islands. In the southern abyssal part (over 3000 m deep) the Powell Basin is partially isolated from the vast basin of the Weddell Sea by underwater elevations with depths shallower than 2000 m. By its geological origin, this is an independent micro-oceanic basin, which began to form during the Eocene, when the rift zone between the Antarctic Plate and the South Orkney microcontinent began to deepen (Coren et al. 1997).

The northeastern shelf of the Antarctic Peninsula is characterized by high values of summer primary production. However, due to the fact that the area is covered by drifting ice during a significant part of the year, the production of phytoplankton in the water column on the annual scale is low (Ardyna et al. 2017).

The primary production is directly affected by the content of nutrients (N, P, Si) in the water, in euphotic zone of the water column. In turn, phytoplankton is the main food source for crustaceans. This is how the Antarctic food chain is formed: nutrients-phytoplankton-krill-large mammals and birds. Currently, there are few literature data on the hydrochemical conditions in the region. Relatively isolated due to its geographical location, the Weddell Sea has an independent cyclonic circulation system. Cold and oxygen-rich waters are formed here ("pure" waters of the high-latitude modification (Arzhanova and Artamonova 2014)), which, due to the peculiarities of circulation, are transported far to the north of the mainland and have a significant impact on the basin of the South Atlantic. The Weddell Sea is also a key area for the formation of Antarctic Bottom Water (AABW), which plays an important role in the ventilation of the deep layers of the World Ocean (Foster and Carmack 1976; Frey et al. 2019; Morozov et al. 2020b). The study site in the Weddell Sea is important for understanding and monitoring the productivity of Antarctica as the basin important for fishery (Arzhanova 1990).

The goal of this article is to describe the details of horizontal and vertical distribution of hydrochemical parameters in the Powell Basin and to explain their formation, the influence of ice cover, and the flow of waters from the Bransfield Strait.

11.2 Materials and Methods

The work in the Powell Basin was carried out from the R/V "Akademik Mstislav Keldysh" on January 22–February 2 and February 14–20, 2020 (Morozov et al. 2020a). Samples for hydrochemical analysis were taken at 26 hydrological stations using 5-l plastic Niskin bottles of the Rossette complex from various depths, determined from the previously measured temperature and fluorescence profiles. In the samples, the contents of the main forms of nutrients were determined: inorganic forms of silicate, mineral phosphorus, nitrate, nitrite and ammonium nitrogen, as well as dissolved oxygen, total alkalinity (*Ta*), and pH. Further processing of samples for the study of all these parameters was carried out in the onboard laboratory immediately after sampling. The station layout is shown in Fig. 11.1. A total of 20 stations with measurements at 216 depths were performed during the stage of work (January 2–February 2, 2020) and 6 stations with 77 samplings during the stage (February 14–20, 2020).

Silicate, mineral nitrogen, and phosphates were determined by standard colorimetric methods. Total alkalinity *Ta* was measured by direct titration in the Bruevich modification with visual determination of the equivalence point (Pavlova et al. 2008). The hydrogen index was determined using a cell of a liquid-free compound (Tishchenko et al. 2001). The oxygen content in the water was determined by titration using the Winkler method modified by Chernyakova (Bordovsky and Chernyakova 1992).

Temperature and salinity measurements were carried out in situ using a CTD probe SBE 19plus with a cassette of bottles and additional sensors (Morozov et al. 2020a).

11.3 Results

In order to consider and compare water properties in different parts of the basin, we constructed an integral section passing through the northern part of the study site (Fig. 11.1). In the eastern part of the study site in the Philip Passage region, a relatively warm surface layer was observed (the temperature reached 2 °C) with a thickness of about 10–20 m, less often down to 50 m (Fig. 11.2).

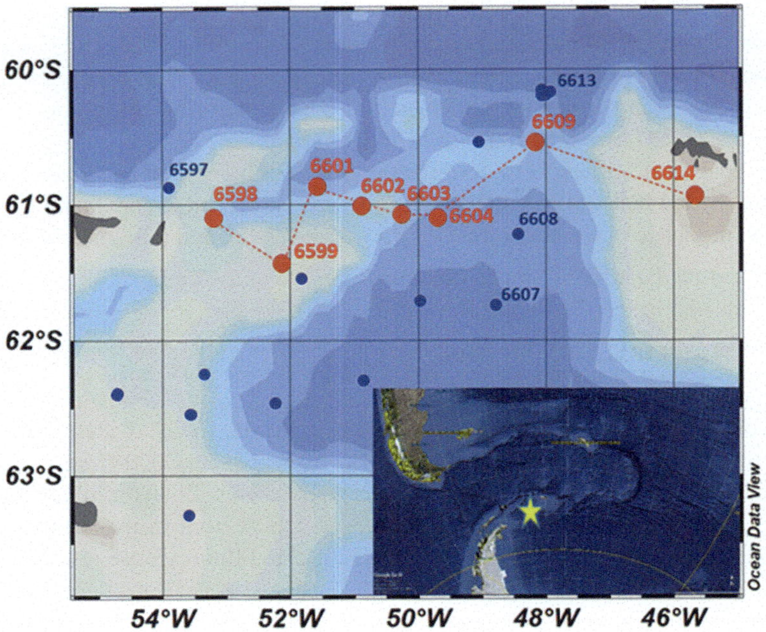

Fig. 11.1 Research area (northern Weddell Sea) with locations and numbers of oceanographic stations. Red dots indicate stations included in the integral cross section. The yellow star in the inset indicates the location on a broader map

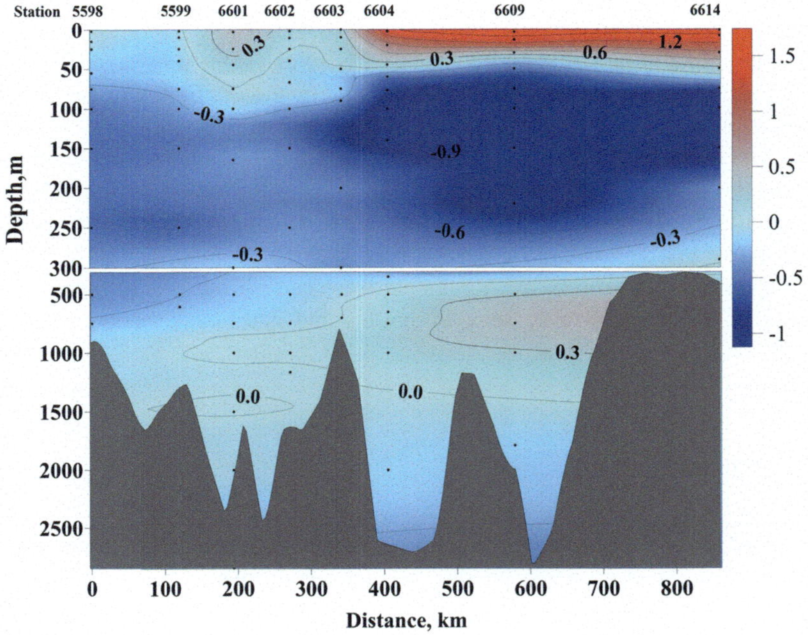

Fig. 11.2 Vertical distribution of temperature (°C) over the integral section

The salinity in the region was mostly below 34 psu, which indicates some desalination of surface waters in comparison with the northwestern part of the basin (Fig. 11.3).

The surface layer was characterized by high oxygen saturation, which exceeds 100% and indicates the activity of the photosynthesis process (Fig. 11.4). The highest concentrations of nitrite nitrogen were recorded (up to 0.3μM, Fig. 11.5), while concentrations of ammonium nitrogen (Fig. 11.6) were also high (up to 3.5μM). The concentration of phosphates (Fig. 11.7) was less than in the western part of the study site (less than 1.5μM). The lowest values of phosphates (less than 1μM) were recorded at station 6609. At the same station, lower values of nitrates (Fig. 11.8) were also observed (about 22μM), the pH value (Fig. 11.9) reached the maximum values of 8.15 (total scale), and the relative oxygen content was about 107%. At other stations in the considered area, the nitrate values were higher (up to 30μM). The silicate concentration varied in the range from 54 to 80μM. The highest concentrations were observed at station 6607.

A sharp increase in the gradients of hydrochemical parameters was found under the upper productive layer, which generally exists in Winter Water (WW) located below. The decrease in oxygen concentration in the WW was 1.5 ml/l (Fig. 11.10), and the concentration varied from approximately 7 ml/l at the upper boundary to 5.5 ml/l at the lower one. There was a gradual increase in silicates, and extremes were notable in the distribution of nitrates and phosphates in the WW. The gradient

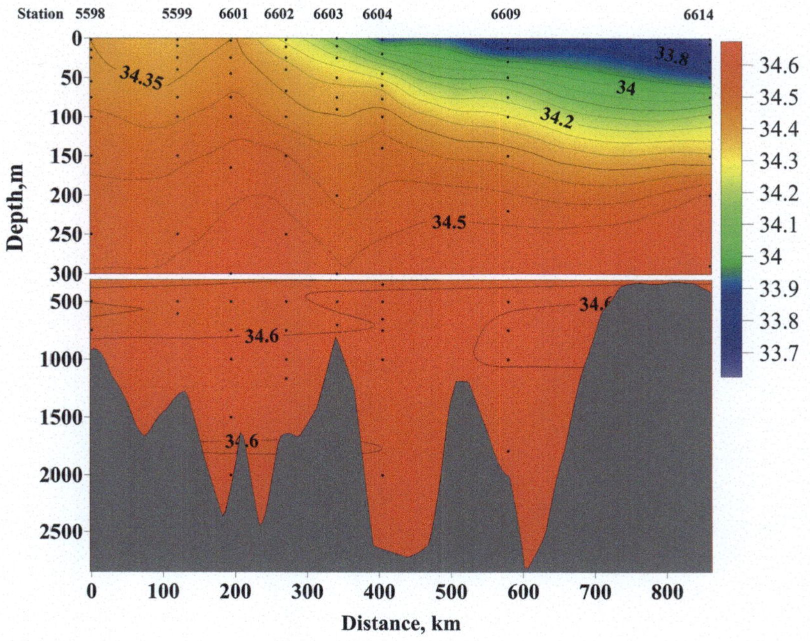

Fig. 11.3 Vertical distribution of salinity (PSU) over the integral section

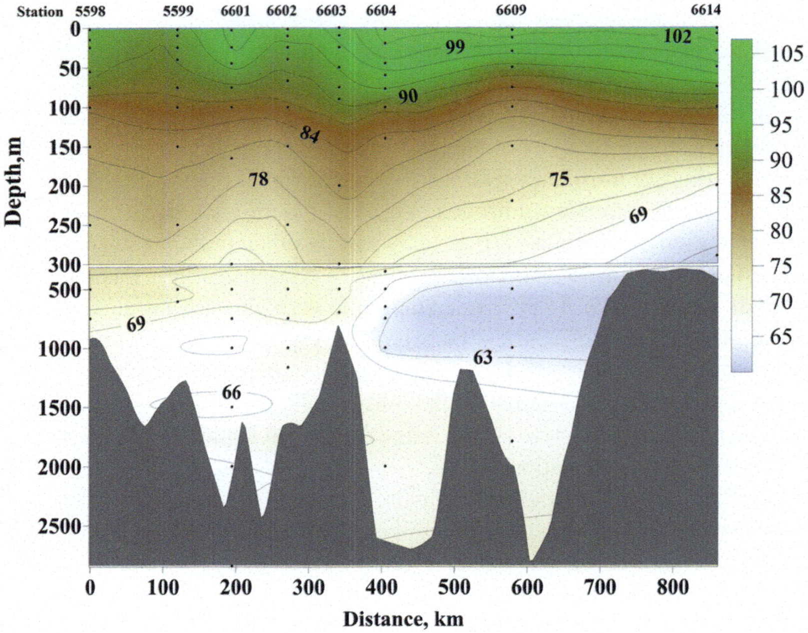

Fig. 11.4 Vertical distribution of oxygen saturation (%) over the integral section

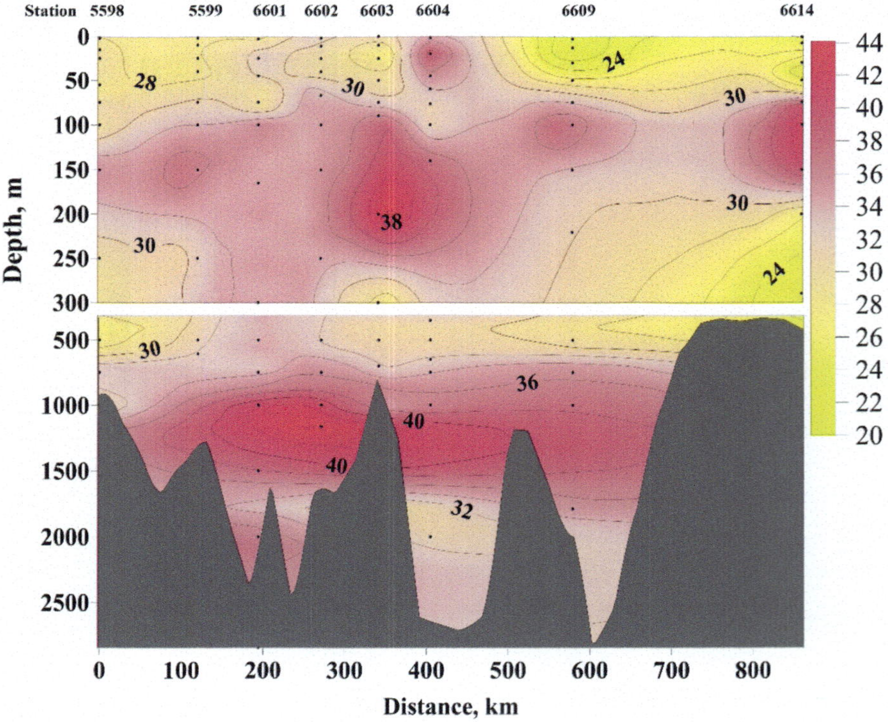

Fig. 11.5 Vertical distribution of nitrite nitrogen (μM) over the integral section

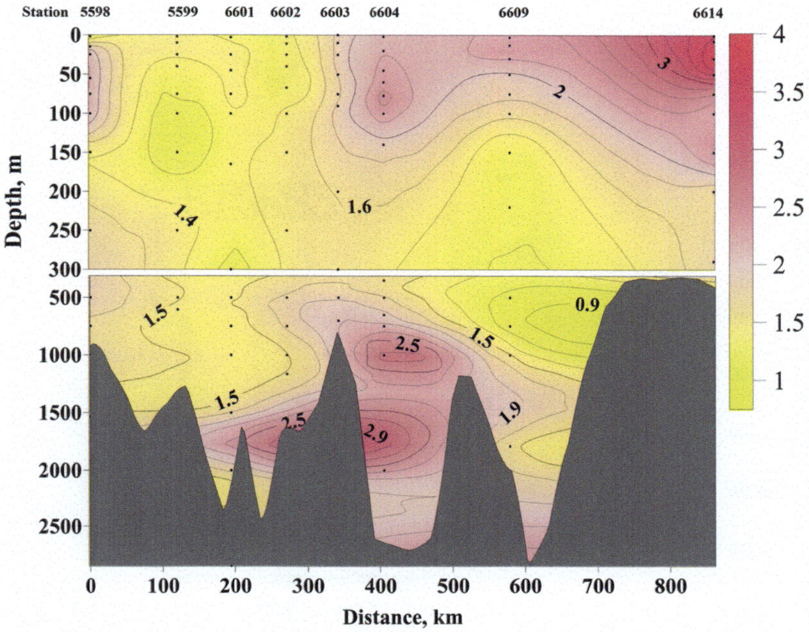

Fig. 11.6 Vertical distribution of ammonium nitrogen (μM) over the integral section

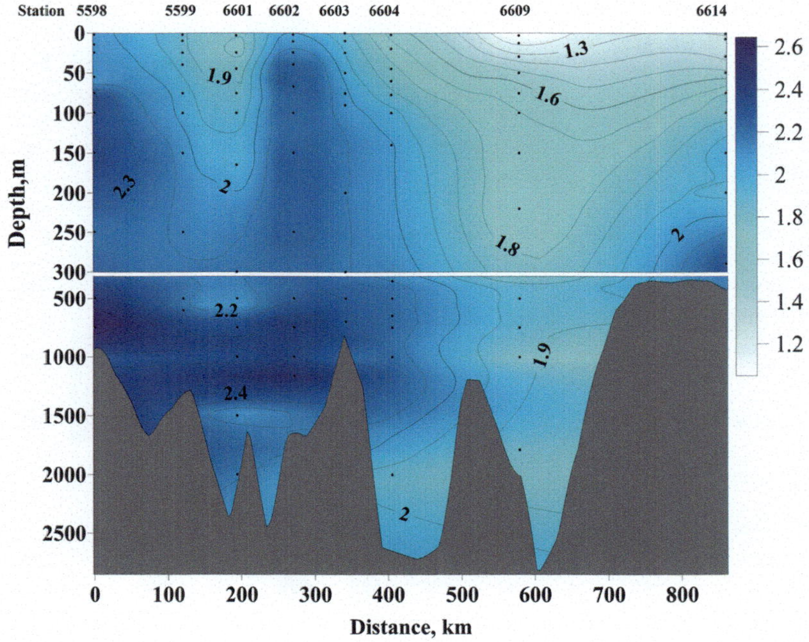

Fig. 11.7 Vertical distribution of phosphates (μM) over the integral section

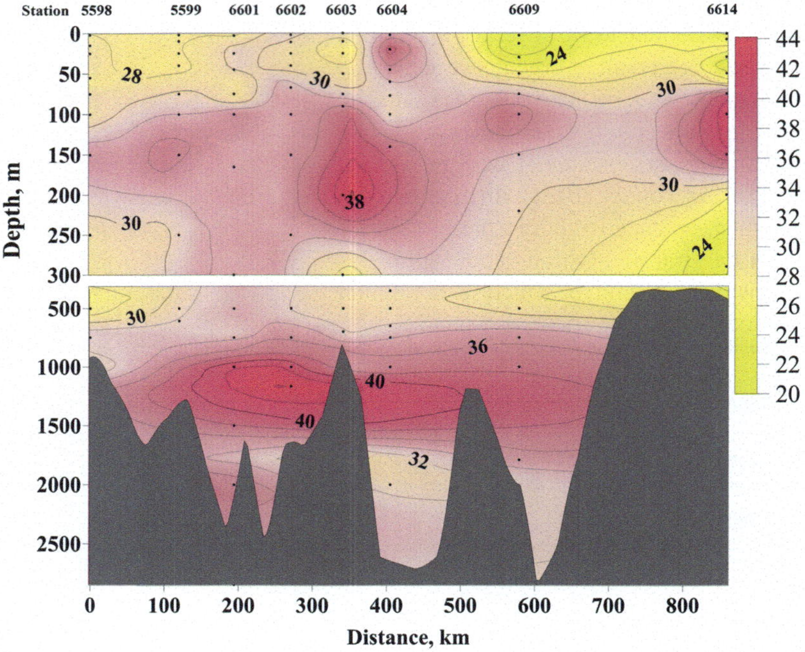

Fig. 11.8 Vertical distribution of nitrate nitrogen (μM) over the integral section

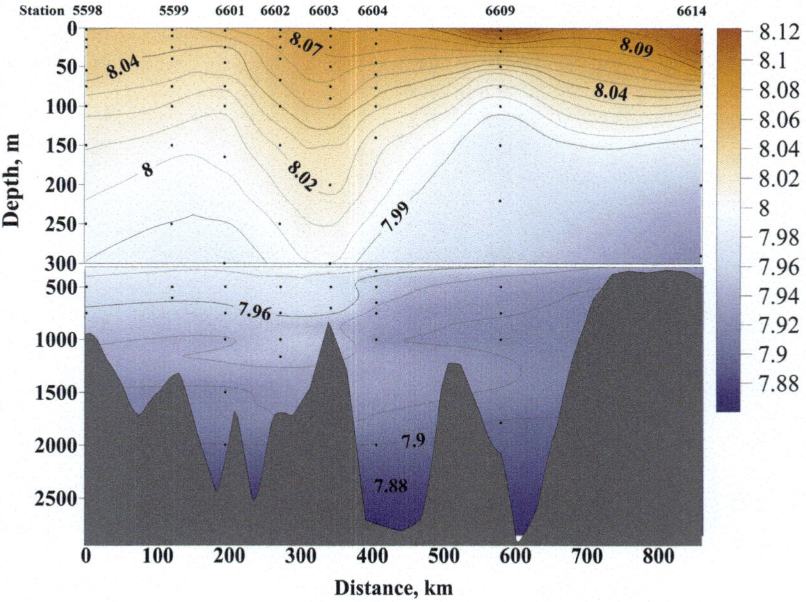

Fig. 11.9 Vertical distribution of pH (total scale) over the integral section

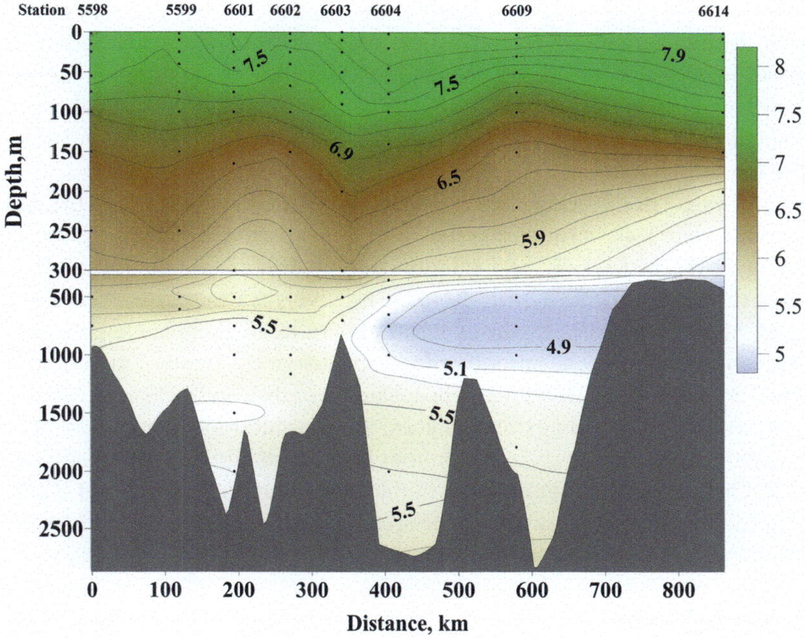

Fig. 11.10 Vertical distribution of dissolved oxygen (ml/l) over the integral section

of parameters at the lower boundary of the upper quasi-homogeneous layer and the WW, as well as in WW layer, was generally higher than in other regions.

At depths of 100–300 m, there was some differentiation in the vertical distribution of nutrients. Thus, station 6614, located on the shelf of the South Orkney Islands, significantly differs in the content of nutrients. This station at depths of 200–300 m was characterized by higher concentrations of silicates and phosphates and a lower content of nitrates in comparison with deep-water stations in the area of the Philip Passage. At station 6608, lower values of silicates were revealed, which were also found even at greater depths. Concentration of phosphates in Winter Water at station 6613 was high and reached 2μM, and the phosphate content gradually increased with the depth. This may indicate differences in the conditions of WW formation.

A layer of Warm Deep Water (WDW) was located below WW, which was formed from North Atlantic Deep Water (Arzhanova and Artamonova 2014). A specific feature of these waters is a low oxygen content so that the oxygen concentration in the core located at a depth of 500–600 m decreases to 4.7 ml/l (Fig. 11.10).

These waters were characterized by temperatures exceeding 0.3 °C and salinity exceeding 34.6 psu. As for the distribution of other hydrochemical parameters, an increase in the total alkalinity and silicate concentration was observed in WDW as the depth increases. The distributions of phosphates, nitrates, nitrites, and ammonium nitrogen were quite complex; they were characterized by alternating extremes, with the minima of phosphates usually corresponding to the maxima of nitrate nitrogen. It can be assumed that there is a contact zone of waters that differ in the content of nutrients, as well as an uneven "jet" flow of water in the deep layers. We also note some isolation of station 6613. Deeper, in the bottom waters, a decrease in the concentrations of nutrients was observed. The absolute oxygen content increases slightly toward the bottom.

Surface waters of the deepest investigated part of the Powell Basin to the west of station 6607 were less warm during the observation period. The temperature in the surface layer reached −1 °C, and salinity decreased to 33 psu and even less. The waters of this part of the area differ from the relatively warm transformed waters in the eastern part of the basin by the concentration of hydrochemical parameters. We note an increase in the concentration of phosphates in the western direction along the entire profile. The distribution of phosphates over the depth range becomes more uniform, without significant extreme values; the concentration increases with depth (up to 2000 m) and decreases in the bottom layer. It should be also noted that this area was characterized by an increase in the concentration of nitrates at depths of more than 200 m, in comparison with the area of the South Orkney Islands shelf and the Philip Passage, especially in its northern part. The distribution of nitrates in deep waters remains uneven, and the presence of a maximum in WW and two maximums in WDW was observed.

On the Antarctic Peninsula shelf, salinity varied in the range from 33.8 to 34.2 psu in the upper quasi-homogeneous layer, whose thickness at some stations reaches 50 m and increases toward the west and north. The surface temperature dropped to −1 °C. This area is in an intermediate position in terms of the content of hydrochemical parameters such as oxygen (95–98%) and nitrites (less than 0.2μM), as well as phosphates (less than 2μM) in the surface layer. The content of silicates and ammonium nitrogen is generally the same as in the warmer "productive" area, and the concentration of nitrates, on the contrary, is closer to that in the northwestern area. An increase in the concentration of nutrients and a decrease in the oxygen content with depth were recorded.

The hydrochemical and hydrophysical structure in the northwestern part of the basin is very different from the other areas. In fact, a significant part of this area was occupied by the contact zone of waters of various origin, which extends between stations 6658 and 6603. The mixing zone was clearly seen in the vertical distributions of most hydrochemical parameters over the integral section, such as silicates and pH (Figs. 11.9 and 11.11).

The surface layer in this part of the study site was characterized by increased salinity and intermediate temperature values relative to the areas previously described. The influence of supercooled water on the surface at stations 6598 and 6599 was found. This area was characterized by the lowest oxygen content (the concentration decreased below 90%) and high concentrations of phosphates and silicates. The concentration of nitrates is generally higher than in the previously described "warm" area (up to 30μM or greater). The concentrations of nitrites and ammonium were lower here than in other regions, which were below 0.15 and 2μM, respectively.

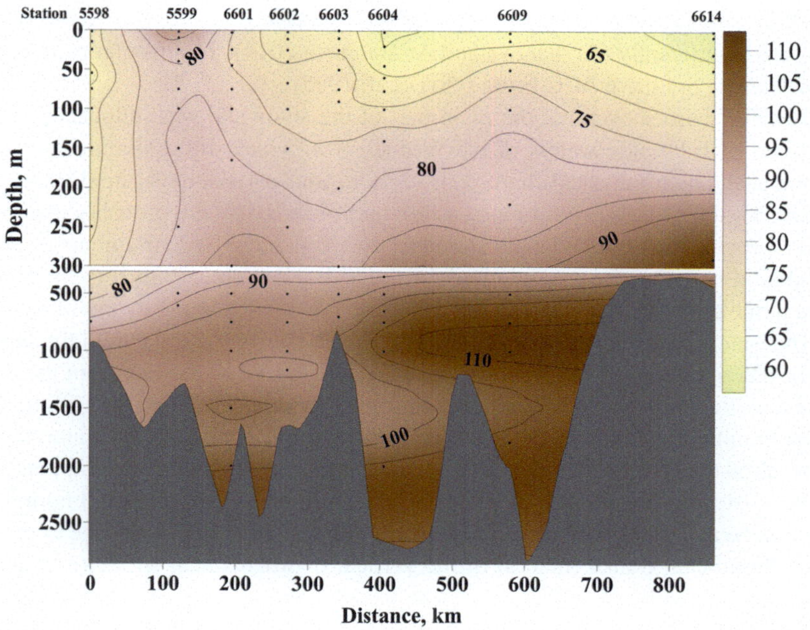

Fig. 11.11 Vertical distribution of silicates (μM) over the integral section

The layer of WW was poorly observed, especially at stations 6598–6599, in comparison with the eastern part of the Powell Basin (Fig. 11.7). We did not find high gradients typical for this layer and the boundary between WW and surface well-mixed Summer Waters (SSW). This also applies to station 6597, where the chemical properties differed from those in the other parts of the Powell Basin.

The WDW was poorly distinguished based on the hydrochemical parameters in the western part of the Powell Basin. Here we observed a gradual decrease in dissolved oxygen concentrations, pH, and an increase in silicate, nitrate, and phosphate concentrations to the bottom. The bottom water in the western part of the investigated region was also somewhat different from the rest of the area; in particular, high values of phosphates were found (Fig. 11.7).

11.4 Discussion

The Si/P ratio (index) of the concentrations of silicates and phosphates is considered a reliable indicator of the influence of high-latitude sea waters, especially in the surface layer (Arzhanova and Artamonova 2014). There is an increase in the index values from north to south from less than 20 in the waters of the Antarctic Circumpolar Current (ACC) to 40 and more in the adjacent high-latitude waters. The range of the Si/P ratio from 20 to 35 is inherent in the waters of the Weddell Scotia Confluence Zone (WSCZ) or Secondary Frontal Zone of Antarctica, although depending on the intensity of photosynthesis the extreme limits of Si/P in the WSCZ can vary significantly (Arzhanova and Artamonova 2014). This frontal zone is also known as the Southern ACC Front (Orsi et al. 1995).

The Si/P ratio shows that the main part of the study site was influenced by the modified high-latitude waters of the Weddell Sea, which differ slightly in hydrochemical and hydrophysical characteristics in various parts of the basin (Fig. 11.12). The high values of this ratio in the eastern part of the basin are probably explained by the processes of organic matter production (Arzhanova and Artamonova 2014). We should note that the lowest values of the Si/P index are confined to the frontal zone in the northwestern part of the basin, which is related to the dynamic factors.

The waters of the Bransfield Strait spread into the northwestern part of the Powell Basin. These extremely cold and freshened waters extend to the east from the tip of the Antarctic Peninsula and, according to some estimates, can be traced all the way to the South Sandwich Islands, and thereafter they lose their specific features eastward (Tarakanov and Gritsenko 2018). As it was previously noted, the main features of the western study site are the absence of a pronounced oxygen minimum layer and a decrease in the gradients of hydrochemical parameters in the upper 500-m layer. Significant transformation of the vertical hydrochemical structure is due to the contact of the Bransfield and northern Weddell Sea waters in the Phillip Passage area, as well as the influence of bottom topography and tidal activity, which increase diapycnal mixing (Chukharev et al. 2017). For more detailed information on the transformation of Warm Deep Water and Winter Water in this area, as well as

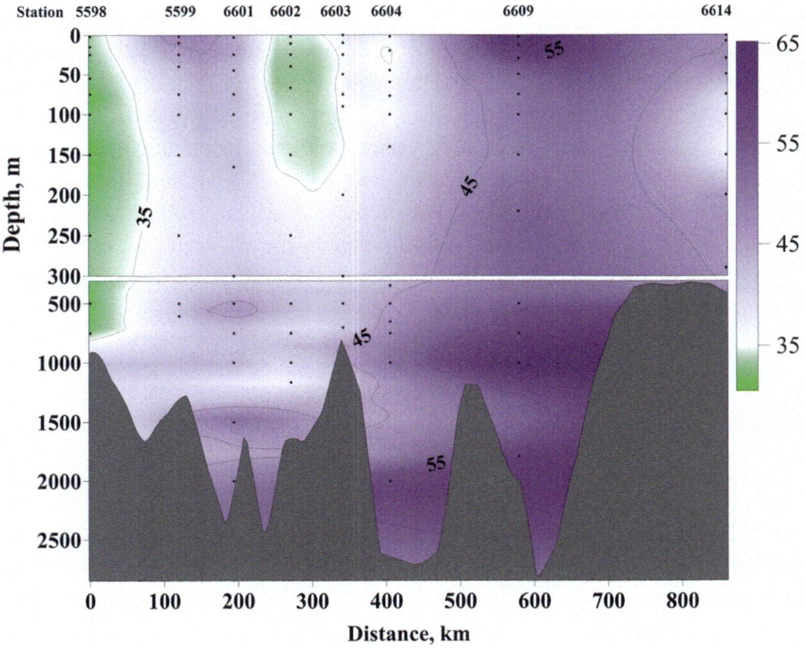

Fig. 11.12 Vertical distribution of Weddell Sea Index (Si/P) over the integral section

hydrochemical features corresponding to various modifications of the Circumpolar Deep Water (CDW), see Fedotova and Stepanova, this volume.

According to the available data, water exchange was observed in the area of the Philip Passage in the upper 500-m layer, which was confirmed by the peculiarities of the circulation in the regions. In general, the influence of the Powell Basin waters can be traced down to the depths of about 2000 m. Sharp changes in the hydrochemical structure are observed at the outlet of the Philip Passage (station 6613) at depths exceeding 2000 m. Concentrations of nitrate decrease to the bottom to values less than 30μM. A significant increase in silicate concentrations is observed. Thus, the silicate content in this area reached 150μM at depths exceeding 2500 m, while in the basin waters the silicate content at these depths was about 100μM. The content of dissolved oxygen and phosphate seems to decrease toward the bottom. We assume that the deep waters here had a different origin and the influence of the Powell Basin waters was limited at these depths.

The present type of distributions of the concentration of nutrients in the Powell Basin shows the large-scale structure of Antarctic waters, which is represented by several water masses. It is accepted to delineate three main water masses by their vertical distribution in Antarctica: Antarctic Surface Water (AASW), Circumpolar Deep Water (CDW), and Antarctic Bottom Water (AABW) (Orsi et al. 1995). The AASW occupies a layer from 0 to 150–300 m. In AASW, a relatively warm and freshened, oxygen-rich surface layer of Summer Water (SSW) is located, extending to the depths of about 50 m. According to the long-term data, the oxygen concentration in this layer ranges within 7.5–8.0 ml/l, which corresponds to 95–105% oxygen saturation (Arzhanova and Artamonova 2014; Artamonova et al. 2015). Below is a layer of Winter Water (WW) with a core of minimum temperature at depths of about 100–150 m. In this layer, the oxygen concentration gradually decreases, and the concentrations of nutrients increase with depth. The CDW located under the AASW is represented by the Upper CDW with a core at a depth of 500–600 m, where the layer of maximum temperatures and minimum oxygen concentrations (on average 4.7–5 ml/l) is located, and Lower CDW with a core of maximum salinity at a depth of 800 m, which is characterized by a gradual increase in oxygen concentration (Maslennikov 2004). A significant transformation occurs in Weddell Sea CDW, and the differences between Upper CDW and Lower CDW are smoothed out. In the literature, one can find a term that reflects these local features: Warm Deep Water (WDW).

Usually, there are two extremes of the vertical distribution of nutrients in CDW: the maximum of nitrates and phosphates in Upper CDW and the minimum in Lower CDW, while the silicates increase with depth. However, in the areas affected by the waters of high-latitude seas, such as the Weddell Gyre (WG), the extremes of nutrients (phosphates and nitrates) are smoothed out, remaining as traces. In the vertical distribution of silicates, there is a significant increase in its content in the upper layer of deep waters close to AASW, which is associated with upwelling of deep waters of the Weddell Gyre (Arzhanova and Artamonova 2014). Such features were generally typical for WDW in the Powell Basin at the time of the research. However, the recorded maxima of nutrients did not always coincide with the core of WDW.

The complex uneven vertical distribution of hydrochemical parameters, primarily phosphates and nitrates, in the deep layers of the Powell Basin is due to the influence of Weddell waters and the specifics of their inflow and transformation. Thus, Weddell Sea Bottom Water (WSBW) and Weddell Sea Deep Water (WSDW) enter the Powell Basin through deep-water passages. In addition, WSDW passes over the Joinville Ridge at depths of about 2000 m, transforming during its motion. The thermohaline characteristics of the basin and the distribution of hydrochemical parameters in the water column above the western slope of the Powell Basin at depths of about 800 m and deeper are influenced by shelf waters of various origins, primarily the so-called Weddell Sea Slope Water (WSSW) (Fedotova and Stepanova this volume) and the associated Antarctic Slope Current (ASC) and Antarctic Slope Front (ASF). These waters, which have different origins, are transformed as they mix with the surrounding waters before entering the basin and further within the basin; they transport different amounts of nutrients. We assume that these processes lead to the formation of extremes of nutrients in the deep layers of the basin during the interaction of various water masses. In addition, such a discontinuous distribution of hydrochemical parameters, especially over the slope, may be due to the dynamic instability of the Antarctic Slope Current and Antarctic Slope Front.

The average values of the concentrations of the most significant nutrients based on our data are shown in Table 11.1. Comparison of our data with the real range of variability of hydrochemical characteristics, based on the analysis of the entire available array of hydrochemical data presented in Batrak (2008), shows that the values are not outliers from the general climatic standards in the Antarctic region. In addition, the values of nutrients such as O_2, PO_4, NO_3, and Si in the main structural zones are close to the long-term mean values given in Arzhanova and Artamonova (2014).

It is reported in the publications on hydrochemistry of the Antarctic region that nutrients such as dissolved oxygen, silicates, and the Si/P ratio play an important role in differentiating the existing modifications of Antarctic waters and detecting the frontal zone of the Southern Front of the ACC (Arzhanova and Artamonova 2014). However, the analysis of our data revealed that in the Powell Basin, which is mainly under the influence of high-latitude seas, the distribution of oxygen and silicates is characterized by similarity, and the range of changes in these parameters is

Table 11.1 Average and standard deviation of concentrations of nutrients at various depths calculated from the data in January–February, 2020 in the northern Weddell Sea

Depths	O_2, ml/l	Si, μM	PO_4, μM	NO_3, μM
Surface water (0–100 m)	7.8 (±0.3)	66.5 (±5)	1.7 (±0.3)	29 (±3)
Winter water (100–300 m)	6.4 (±0.6)	77.0 (±9)	2.0 (±0.3)	35 (±5)
Warm deep water (300–1500 m)	5.2 (±0.4)	100 (±13)	2.1 (±0.3)	37 (±5)
>2000 m	5.5 (±0.3)	119.5 (±18)	1.95 (±0.3)	38 (±5)

small, while the content of nitrates and phosphates can play an important role as the tracers in studying the transformation and classification of water masses in the study site.

11.5 Conclusions

Thus, modified high-latitude waters of the Weddell Sea were found in the study site, which determines some common features in the vertical distribution of hydrochemical parameters. Indeed, the vertical hydrochemical structure of the Powell Basin corresponds to the existing concepts, and the range of nutrients did not exceed the average range typical for the Antarctic region. Nevertheless, the distribution of hydrochemical and hydrophysical characteristics of the basin revealed patterns that enabled us to identify five distinguished areas. This differentiation in the distribution of hydrochemical characteristics is due primarily to the origin of waters and the peculiarities of their transformation and contact during the open water period. It can be argued that the hydrochemical structure of the basin is influenced by the waters transported from the Bransfield Strait, while the influence of waters of the ACC is limited and probably possible only when this flow weakens. In addition, the basin is affected by the "pure" high-latitude waters of the Weddell Sea. There is also an exchange of water through the Philip Passage mainly in the upper 500-m layer. An interesting feature of the basin is the significant variability of nutrients, especially phosphates, which is associated with the genesis of water masses and circulation peculiarities. Thus, the waters of the deep-water basin have lower values of phosphates, in comparison with the "ambient" waters of the Weddell Sea entering the basin and with the area at the outlet of the Philip Passage.

Acknowledgments The work was carried out within the framework of the state assignment of the Shirshov Institute of Oceanology, RAS (No. 0128-2019-0008), state assignment of the Pacific Oceanological Institute, RAS (No. FWMM-2019-0007, state assignment, add. Agreement No. 075-03-2019-562 / 2 of August 30, 2019), and state assignment of the Marine Hydrophysical Institute, RAS (No. 0555-2019-0003). The authors express their gratitude to the captain and crew of the R/V "Akademik Mstislav Keldysh" (79th cruise), the head of the expedition Dr. E.G. Morozov for organizing the cruise, and D.I. Frey for providing hydrophysical data and sampling.

References

Ardyna M, Claustre H, Sallée JB, d'Ovidio F, Gentili B, Van Dijken G, D'Ortenzio F, Arrigo KR (2017) Delineating environmental control of phytoplankton biomass and phenology in the Southern Ocean. Geophys Res Lett 44(10):5016–5024

Artamonova KV, Gangnus IA, Maslennikov VV, Torgunova NI (2015) Hydrochemical research in Antarctic waters in the 59th Russian Antarctic expedition. Oceanology 55(5):866–866

Arzhanova NV (1990) Hydrochemical indicators of frontal zones in South Atlantic. In: Biological resources: conditions, perspectives and problems of their rational use, Electrona carlsbergi in southern polar frontal zone, vol 2, pp 12–20

Arzhanova NV, Artamonova KV (2014) Hydrochemical structure of waters in the Antarctic krill Euphausia superba Dana fishing areas. Proc VNIRO 152:118–132. (In Russian)

Batrak KV (2008) Hydrochemical characteristic of different modifications of Antarctic waters. Oceanology 48(3):349–356

Bordovsky OK, Chernyakova AM (1992) Modern methods of hydrochemical studies of the ocean. Shirshov Institute of Oceanology, Moscow, p 200. (in Russian)

Chukharev AM, Runovsky KV, Kulsha OE (2017) Modeling of turbulent patches statistical distribution in the stratified ocean layers. Phys Oceanogr 5:31–41

Coren F, Ceccone G, Lodolo E, Zanolla C, Zitellini N, Bonazzi C, Centonze J (1997) Morphology, seismic structure and tectonic development of the Powell Basin, Antarctica. J Geol Soc 154(5):849–862

Fedotova AA, Stepanova SV Water mass transformation in the Powell Basin. In: Antarctic peninsula region of the Southern Ocean. Oceanography and ecology. Springer, Cham. (This book, Chapter 8)

Foster TD, Carmack EC (1976) Frontal zone mixing and Antarctic bottom water formation in the southern Weddell Sea. Deep-Sea Res Oceanogr Abstr 23(4):301–317

Frey DI, Morozov EG, Fomin VV, Diansky NA, Tarakanov RY (2019) Regional modeling of Antarctic bottom water flows in the key passages of the Atlantic. J Geophys Res Oceans 124:8414–8428. https://doi.org/10.1029/2019JC015315

Helbling EW, Villafañe V, Holm-Hansen O (1991) Effect of iron on productivity and size distribution of Antarctic phytoplankton. Limnol Oceanogr 36(8):1879–1885

Maslennikov VV (2004) Climate fluctuations and the Antarctic marine ecosystem. PhD thesis, Moscow. (In Russian)

Morozov EG, Spiridonov VA, Molodtsova TN, Frey DI, Demidova TA, Flint MV (2020a) Investigations of the ecosystem in the Atlantic sector of Antarctica (Cruise 79 of the R/V Akademik Mstislav Keldysh). Oceanology 60(5):721–723. https://doi.org/10.1134/S0001437020050161

Morozov EG, Frey DI, Tarakanov RY (2020b) Antarctic bottom water flow through the eastern part of the Philip passage in the Weddell Sea. Oceanology 60(5):680–684. https://doi.org/10.1134/S000143702005015X

Nedwell DB, Walker TR (1995) Sediment-water fluxes of nutrients in an Antarctic coastal environment: influence of bioturbation. Polar Biol 15(1):57–64

Orsi AH, Whitworth T III, Nowlin WD Jr (1995) On the meridional extent and fronts of the Antarctic circumpolar current. Deep Sea Res I 42(5):641–673

Orsi AH, Johnson GC, Bullister JL (1999) Circulation, mixing, and production of Antarctic bottom water. Prog Oceanogr 43(1):55–109. https://doi.org/10.1016/S0079-6611(99)00004-X

Pavlova GY, Tishchenko PY, Volkova TI, Dickson A, Wallmann K (2008) Intercalibration of Bruevich's method to determine the total alkalinity in seawater. Oceanology 48(3):438–443. https://doi.org/10.1134/S0001437008030168

Siegel V, Watkins JL (2016) Distribution, biomass and demography of Antarctic krill, Euphausia superba. In: Biology and ecology of Antarctic krill. Springer, Cham, pp 21–100

Tarakanov RY, Gritsenko AM (2018) Jets of the Antarctic circumpolar current in the Drake Passage based on hydrographic section data. Oceanology 58(4):503–516

Tishchenko PYA, Wong CS, Pavlova GYU, Johnson VK, Kang DD, Kim KR (2001) Measuring the pH of seawater using a liquid-free cell. Oceanology 41(6):849–859

Chapter 12
Features and Processes of the Oxygen and pCO_2 Dynamics in the Surface Waters in the Western Parts of the Weddell and Scotia Seas (Southern Ocean)

Natalia A. Orekhova, Anna V. Vidnichuk, and Sergey K. Konovalov

Abstract The Southern Ocean, being a major sink of CO_2, has a crucial role in mediating the global climate. Moreover, this ocean plays an equally important role in the budget of oxygen in the World Ocean. It is one of two major areas where downwelling of cold and oxygen-saturated waters transports oxygen to the deepest layers of the ocean. Thus, the dynamics of oxygen and carbon dioxide in the surface waters of the Southern Ocean is one of the most important issues for the global and regional marine biogeochemistry and climatology. In this work, we study pCO_2 and oxygen distributions and their dynamics as well as key processes controlling their spatial variation in the basin of the Weddell-Scotia Confluence Zone including the Antarctic Sound, Powell Basin, and Bransfield Strait, which is the convergence zone of the Weddell-Scotia seas. Data was collected during cruise 79 of the R/V "*Akademik Mstislav Keldysh*" in the austral summer of 2020.

It was found that the Powell Basin and the Bransfield Strait reveal mainly pCO_2 undersaturation and O_2 oversaturation. This is an evidence for active photosynthesis as a controlling process, and high nutrient concentrations support it. However, the Antarctic Sound waters are CO_2 oversaturated and O_2 depleted. They are the source of CO_2 for the atmosphere. These features provide evidence for recent and ongoing prevailing processes of destruction of organic carbon accompanied by oxygen consumption and CO_2 production, while primary production processes remain inactive in this area.

Based on oxygen and pCO_2 dynamics, we have identified three end-members, which are very likely determined by the origin of the water masses and biogeochemical processes.

Keywords Oxygen · pCO_2 · Weddell Sea · Scotia Sea · Southern Ocean

N. A. Orekhova (✉) · A. V. Vidnichuk · S. K. Konovalov
Marine Hydrophysical Institute, Russian Academy of Sciences, Moscow, Russia
e-mail: natalia.orekhova@mhi-ras.ru

E. G. Morozov et al. (eds.), *Antarctic Peninsula Region of the Southern Ocean*, Advances in Polar Ecology 6, https://doi.org/10.1007/978-3-030-78927-5_12

12.1 Introduction

The Southern Ocean plays a significant role in mediating the global climate (Keller et al. 2016; Roden et al. 2013; Deppeler and Davidson 2017), though it occupies only 10% of the World Ocean (Takahashi et al. 2002; Arrigo et al. 2008; Wang and Moore 2012). This region is known as a major sink of CO_2 (Deppeler and Davidson 2017; Mongwe et al. 2018; Landschützer et al. 2015) reaching up to 30–40% of the total consumption by oceans (Deppeler and Davidson 2017; Mongwe et al. 2018; Takahashia et al. 2002; Wang and Moore 2012). It is suggested that without this sink, the atmospheric CO_2 concentration would be ~50% higher than it is today (Deppeler and Davidson 2017). The Southern Ocean plays an equally important role in the budget of oxygen in the World Ocean. It is one of the two major areas where cold and oxygen-saturated waters descend and supply oxygen to the deepest layers of the ocean (Schmidtko et al. 2017; Frölicher et al. 2015). Thus, the dynamics of oxygen and carbon dioxide in the surface waters of the Southern Ocean is one of the most important issues for global and regional marine biogeochemistry and climatology (Gruber et al. 2001; Schuster et al. 2013; Carrillo et al. 2004).

Several abiotic and biotic processes govern the oxygen and carbon dioxide dynamics in the sea (Makkaveev 2013). The solubilities and partial pressures of these gases in the atmosphere and their equivalents in waters together with the meteorological conditions and the state of the sea surface govern fluxes between the atmosphere and surface waters (Gruber et al. 2001; Kihm and Körtzinger 2010; Wang and Moore 2012; Verdy et al. 2007; Takahashi et al. 2002). Production, melting, and the presence or absence of ice at the sea surface make these fluxes physically possible or impossible (Roden et al. 2013). The biogeochemical cycles of oxygen and carbon are coupled in the marine environment through the processes of the production of organic carbon and photosynthetic oxygen in the euphotic zone and organic carbon oxidation (Eq. (12.1)) while it sinks to the deeper layers, which comprises the so-called biological pump (Carrillo et al. 2004).

$$6CO_2 + 6H_2O \leftrightarrow 6H^+ + 6HCO_3^- \leftrightarrow C_6H_{12}O_6 + 6O_2 \qquad (12.1)$$

Seasonal extremes in productivity are observed due to the seasonal ice variation (as a result of its production or melting), air–sea CO_2 flux (Roden et al. 2013), and solar radiation intensity in the Southern Ocean region (Deppeler and Davidson 2017; Mongwe et al. 2018). The system becomes more complicated as various waters undergo transport and mixing due to strong currents and vertical motions in the Southern Ocean.

In summer, phytoplankton blooms support intense production of organic carbon and oxygen (Deppeler and Davidson 2017; Roden et al. 2013), while warming of water and melting of ice make possible the exchange of oxygen and carbon dioxide between the sea and atmosphere. Metabolic activities also affect biogeochemical cycles involving CO_2 transformations (Deppeler and Davidson 2017; Roden et al. 2013; Takahashi et al. 2002) in line with (Millero 1996, 2007):

$$CO_2(gas) \leftrightarrow CO_2(aq) \leftrightarrow CO_2(aq) + H_2O \leftrightarrow H^+ + HCO_3^- \leftrightarrow 2H^+ + CO_3^{2-} \quad (12.2)$$

$$Ca^{2+} + CO_3^{2-} \leftrightarrow CaCO_3(s) \quad (12.3)$$

The marine environment of the convergence zone of the Weddell and Scotia seas is climatically sensitive and biologically extremely productive on the seasonal timescale making this area important and interesting to investigate the interactions between biological, chemical, and physical processes (Sanchez et al. 2019; Carrillo et al. 2004). Moreover, the area includes intra- and interannual variable ice zone, open ocean areas, and shelf and coastal waters (Sanchez et al. 2019; Carrillo et al. 2004; Deppeler and Davidson 2017).

In this work, we study pCO_2 and oxygen distributions and their dynamics as well as key processes controlling their spatial variation in the Weddell-Scotia seas confluence including the Antarctic Sound, Powell Basin, and Bransfield Strait, which is the Weddell-Scotia Confluence Zone. The ratio of O_2 to pCO_2 is used to elucidate processes controlling pCO_2 dynamics. The difference of pCO_2 in the surface waters and near water atmosphere makes it possible to estimate the direction of the CO_2 flux and the ability of this area to absorb CO_2 from the atmosphere. Multidisciplinary investigations of physical (temperature, salinity, wind speed, etc.) and biochemical (oxygen and pCO_2) characteristics help to highlight the main processes in the region.

12.2 Materials and Methods

The area of the Bransfield Strait, Antarctic Sound, and Powell Basin (Fig. 12.1) was studied during cruise 79 of the R/V "*Akademik Mstislav Keldysh*" in the austral summer of 2020. The study site includes zones covered with ice or undergoing deicing, open ocean areas, shelf waters, and coastal regions.

Solid lines in Fig. 12.1a show transects: blue line is transect A, black line is transect B, and red line is transect C; rhombs denote stations of transect A, triangles denote stations of transect B, and rings denote stations of transect C. Dashed light-blue regions in Fig. 12.1a denote regions covered with ice in winter.

Water samples were taken with a CTD 911 Rossette of 24 5-l Niskin bottles. Sea surface temperature (SST) and salinity were measured using SeaBird 911 profiler. Water samples were used for several biogeochemical analyses (e.g., oxygen). Additionally, pCO_2 in the atmosphere near the surface of seawater (10 m above the water surface) and in the surface waters (3 m below the water surface) was measured using the Li-7000 DP instrument.

The occupied oceanographic stations were aggregated in three transects for the analysis: transect A: stations 6595, 6594, 6593, 6592, 6591, 6590, and 6587 (Bransfield Strait); transect B: stations 6622, 6625, and 6627 (Antarctic Sound) and stations 6596, 6600, 6603, 6606, 6609, and 6619 (Weddell-Scotia Confluence

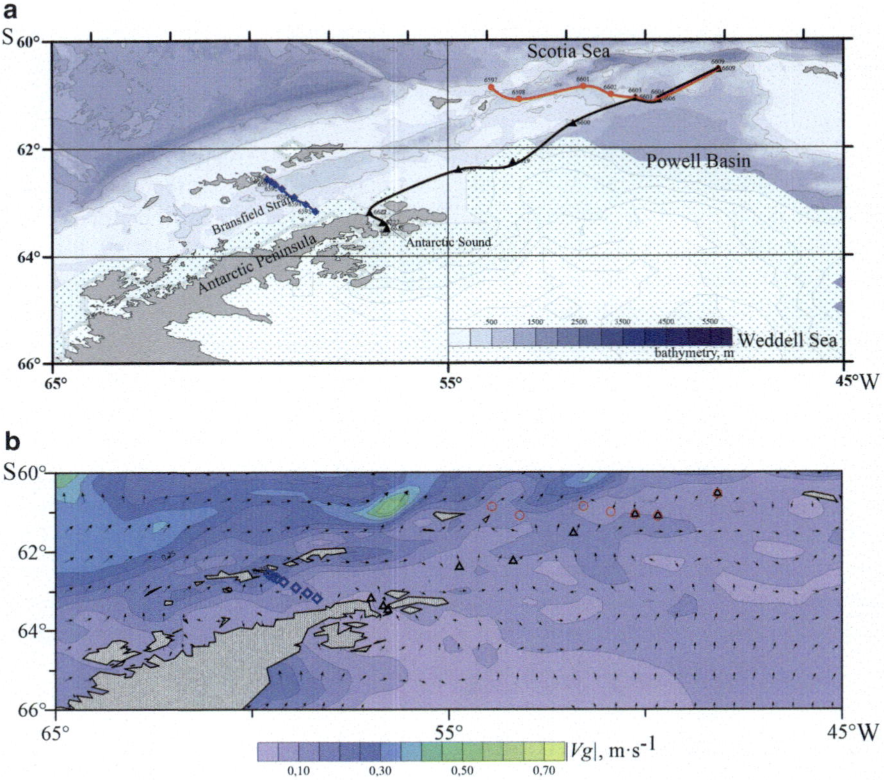

Fig. 12.1 Oceanographic stations (**a**) and reanalyzed geostrophic currents – *Vg* (**b**) in cruise 79 of the R/V "*A. M. Keldysh*" (January–March 2020), based on the data of Artamonov and Skripaleva (private communication) (**b**)

Zone); and transect C: stations 6597, 6598, 6601, 6602, 6603, 6604, and 6609 (Fig. 12.1).

Dissolved oxygen was titrated in line with Winkler's method modified for low oxygen concentration (Carpenter 1965). Oxygen saturation was calculated using equations proposed by Wess (Weiss 1970) and data on temperature and salinity.

The surface water pCO_2 was determined applying a flow-through system with a continuous flow and a cylinder-type equilibrator (10 cm in diameter and 250 cm in height) filled with plastic balls and enclosed with a headspace. The water flow was 3.7–4.0 l min^{-1}. The renewal time was 3–5 min. After equilibration and dehydration, the CO_2 mole fraction was measured with a Li-Cor spectrometer (Li-7000 DP). The Li-Cor was calibrated daily, using pure argon (0 ppm) and a gas mixture with a CO_2 molar fraction of 440 ppm. The error of determination is 1% (Khoruziy 2010).

12.3 Results

12.3.1 The Oxygen Distribution

Oxygen dynamics is governed by abiotic (changes in solubility, water dynamics, evolution of water masses, air–sea exchange, etc.) and biotic (biogeochemical processes of organic carbon production and degradation; Eq. (12.1)) processes. The solubility of oxygen is higher, but the intensity of chemical and biological processes is lower in cold waters. The presence of the ice cover dramatically restricts the air–sea exchange and results in gradual decrease in the oxygen content.

The vertical (0–100 m) and spatial oxygen distributions in the study site (Fig. 12.2) reveal the role of the water masses origin and transportation.

The chemical structure of the Bransfield Strait waters (Fig. 12.2a) reveals dependence on the basic scheme of water dynamics (Skripaleva 2004; Morozov 2007, Thompson and Youngs 2013). Relatively warm and fresh waters from the Bellingshausen Sea invade the western part of the Bransfield Strait (Fig. 12.2a, right panel) and proceed to the northeast (Skripaleva 2004; Morozov 2007), forming the western coastal current (Maslennikov 2003; Skripaleva 2004; Morozov 2007). The current along the coast of the South Shetland Islands transports relatively warm waters (Maslennikov 2003); the water temperature at the surface reaches 2.6–2.8 °C (Fig. 12.2a, right panel). The oxygen concentration is high here and reaches 7.95 $ml \cdot l^{-1}$, while saturation reaches 104% (Fig. 12.2a, left panel). In the southeastern part of the Bransfield Strait, the surface cold (down to −0.1 °C) and saline waters from the Weddell Sea penetrate through the Antarctic Sound and around the Antarctic Peninsula (Maslennikov 2003; Skripaleva 2004; Morozov 2007; Thompson and Youngs 2013), proceed along the southern part of the strait to the west along the northwest coast of the Antarctic Peninsula, and then turn to the northeastward with the boundary current near the South Shetland Islands (Skripaleva 2004; Morozov 2007). The surface waters of the southern part of the transect in the Bransfield Strait are characterized by a decline of the oxygen concentration from 7.44 to 7.32 $ml \cdot l^{-1}$, while saturation drops to 90–92% (Fig. 12.2a, left panel). Similar oxygen concentrations are also observed in the surface waters along the transect in the Antarctic Sound (Fig. 12.2b, left panel).

The oxygen distribution along transect B is a result of the inflow of cold waters from the Weddell Sea through the Antarctic Sound and their spreading in the northwestern direction (Maslennikov 2003; Morozov 2007). At station 6609, waters are warmer (Fig. 12.2b, right panel). The oxygen concentration varies from 7.52 $ml \cdot l^{-1}$ in the surface waters of the Antarctic Sound to 8.31 $ml \cdot l^{-1}$ in the Powell Basin (at station 6609), while oxygen saturation increases from 93% to 107% (Fig. 12.2b, left panel). Vertical oxygen and temperature profiles suggest the presence of end point waters interacting in this region.

Stations of transect C are under the influence of waters from the Southern branch of the Antarctic Circumpolar Current (ACC). The oxygen concentration varies from

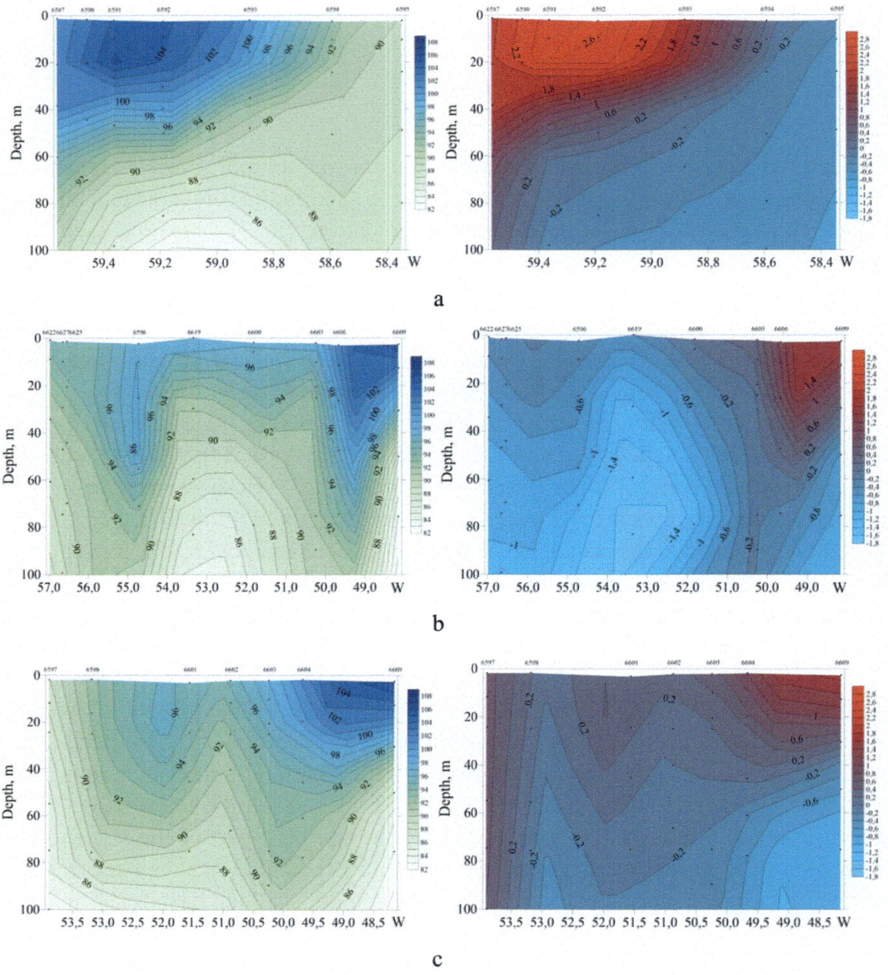

Fig. 12.2 Oxygen saturation (left panel) and temperature values (right panel) along transects A (a), B (b), C (c)

7.03 to 8.31 ml·l^{-1}, while saturation increase from 89% to 107% (Fig. 12.2c, left panel), following the water temperature changes (Fig. 12.2c, right panel).

12.3.2 pCO_2 Dynamics

Carbon dioxide (CO_2) is a counter partner of oxygen. CO_2 is consumed in photosynthetic processes resulting in production of organic carbon and oxygen, while the opposite processes of oxidation of organic carbon result in consumption of oxygen and production of CO_2 (Eq. 12.1). The partial pressure of carbon dioxide (pCO_2) in marine waters is a result of biogeochemical redox processes, chemical balances in the carbonate system (Ciais et al. 2013); and similar to the processes of air–sea exchange and water dynamics and transformation (Gruber et al. 2001; Carrillo et al. 2004; Keller et al. 2016).

Partial pressure of CO_2 is an important component determining the ability of an ecosystem to absorb CO_2 from the atmosphere and affecting climate change (Wang and Moore 2012; Carrillo et al. 2004). The pCO_2 distribution in the surface waters of the investigated area depends on biogeochemical processes, which depends on water temperature and nutrients enrichment supporting primary production (Makkaveev 2013; Wang and Moore 2012). Our data reveal that the pCO_2 varies from 308 to 453μatm in the surface water, while the mean atmospheric pCO_2 is 415 ± 4μatm. The lowest pCO_2 value (308μatm, deep blue in Fig. 12.3) is detected at the last station of transects B and C in the zone of the Scotia and Weddell Sea confluence. The highest pCO_2 value (452μatm, deep red in the figure) is in the waters of the Antarctic Sound.

The surface waters oversaturated with pCO_2 are traced along the Antarctic Sound and at the first station of transect C. The first one is explained by restricted air–sea exchange of CO_2 due to ice coverage in the Antarctic Sound. The ice cover supports CO_2 storage, as a result of organic carbon degradation, according to Eq. (12.1).

Oversaturated sea surface pCO_2 (421μatm) at 6597 station (transect C, Figs. 12.1, 12.3) is likely a result of the organic carbon decomposition supported by mixing of warm water masses, saturated with nutrients, from the southern branch of ACC (Thompson and Youngs 2013) with the warm coastal current formed in the Bransfield Strait, which spreads to the northeast (Maslennikov 2003). pCO_2 in the surface waters of the Bransfield Strait varies within 394–396μatm in the north of the strait at stations 6594 and 6495, which is the influence of waters from the Weddell Sea transported through the Antarctic Sound.

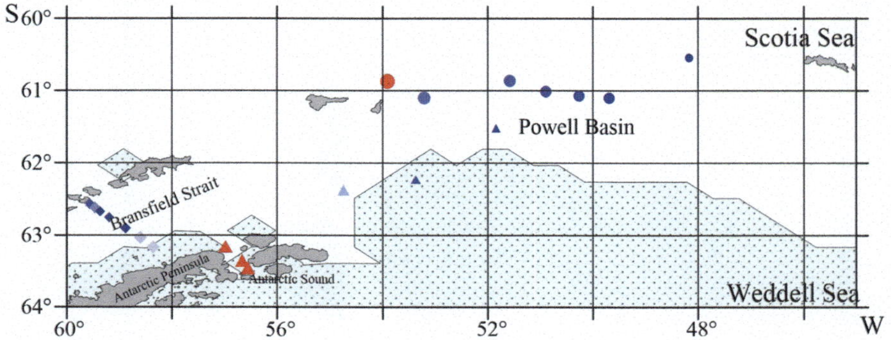

Fig. 12.3 The sea surface pCO_2 variation (symbols denote stations of transects A, B, C; blue color is related to the stations where the seawater pCO_2 values are lower than the pCO_2 in the atmosphere; red color is related to the stations where the seawater pCO_2 is higher than the pCO_2 in the atmosphere; size and color intensity of symbols is proportional to the value; dotted light-blue region denotes region covered with ice in winter)

The intensity of air–sea CO_2 exchange is proportional to the difference between the pCO_2 value in the surface waters and in the atmosphere (Landschützer et al. 2015). Our data reveal that the investigated area mostly serves as a sink for atmosphere CO_2, except for the Antarctic Sound (Fig. 12.3).

12.4 Discussion

Spatial variations in the oxygen and pCO_2 distribution make it possible to identify areas with different biogeochemical processes and thermohaline characteristics of waters, which are end-members for the waters in the study site. An increase in the intensity of biogeochemical processes of oxygen consumption with increasing temperature contradicts the revealed increase in apparent oxygen utilization (AOU) with decreasing temperature (Fig. 12.4). One end-member is warmer waters, where oxygen production prevails due to active photosynthesis. This should result in the lower AOU values. In cold waters covered with ice, the rate of photosynthesis is low, if any, and oxygen utilization in organic carbon oxidation (Eq. (12.1), reverse reaction) results in higher AOU values. This is another end-member. A transformation of one end-member to another end-member results in the observed temperature–AOU relations (Fig. 12.4). Different slopes of inclination at the graph (Fig. 12.4) may indicate different rates of processes in which oxygen is involved, due to the waters of different origin and characteristics (temperature, salinity, nutrient concentration), which interact and undergo transformations.

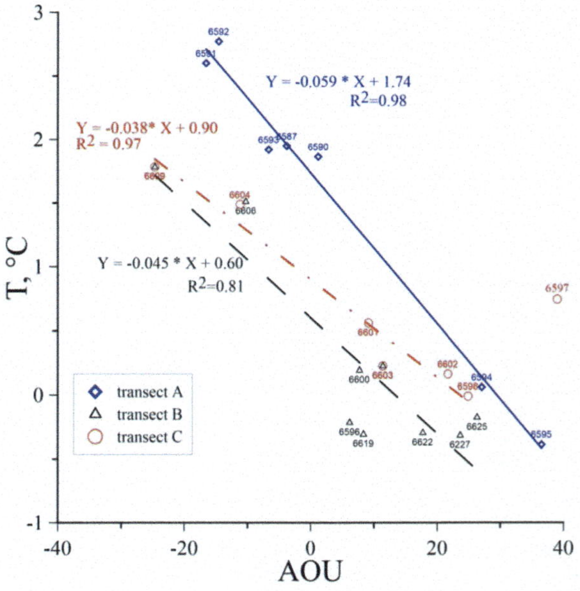

Fig. 12.4 Apparent oxygen utilization versus temperature at transects **a** (blue rhombs), **b** (black triangles), **c** (red rings)

Carbon dioxide (CO_2) is a counter partner of oxygen. The relationship between values of AOU and (pCO_2 seawater–pCO_2 air) is a good marker of the main biogeochemical processes and fluxes of CO_2 in the study site. If CO_2 is a complete twining counter partner of oxygen, this relationship should be similar to those for the already discussed temperature–AOU relations (Fig. 12.4). Yet, we can identify three, rather than two end-members (Fig. 12.5b). The first one is for warmer waters, where oxygen production and CO_2 consumption are observed. The second end-member is for cold waters, where oxygen consumption and CO_2 production occur (Fig. 12.5). The third end-member is water, where the sea surface pCO_2 is close to pCO_2 in the air, but oxygen is consumed according to Eq. (12.1).

The difference between end-members 2 and 3 is explained by the intensity and possibility of CO_2 exchange between the sea surface waters and the atmosphere. End-member 2 is cold water covered with ice or that has recently become ice-free. This water is enriched with CO_2 and oxygen depleted. End-member 3 is also cold water with unrestricted CO_2 exchange between the surface water and the atmosphere.

An intrusion of warmer waters from the Bellingshausen Sea supports photosynthesis intensification and oxygen production (Eq. 12.1) in the coastal waters of the South Shetland Islands (Figs. 12.2 and 12.4) in the Bransfield Strait. On the contrary, cold water masses enriched with nutrients from the Weddell Sea remain oxygen depleted. In this region, environmental conditions do not support intense photosynthesis.

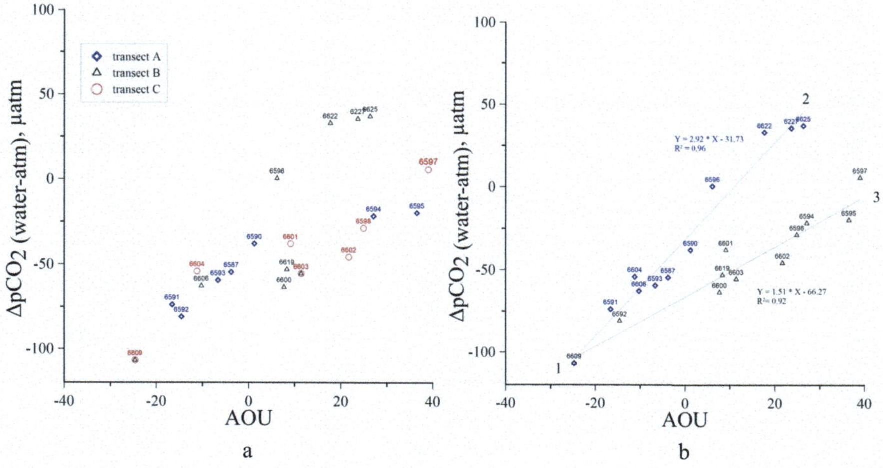

Fig. 12.5 Apparent oxygen utilization via pCO_2

12.5 Conclusions

The Powell Basin and the Bransfield Strait reveal mainly pCO_2 undersaturation and O_2 oversaturation. This is an evidence of active photosynthesis as a controlling process. High nutrient concentrations support it. However, the Antarctic Sound waters are CO_2 oversaturated and O_2 depleted. They are the source of CO_2 for the atmosphere. These features provide evidence for recent and ongoing prevailing processes of destruction of organic carbon accompanied by oxygen consumption and CO_2 production, while the primary production processes remain inactive in this area.

Based on oxygen and pCO_2 dynamics, we have identified three end-members that are very likely determined by the origin of the water masses. The first one is warmer waters where oxygen production and CO_2 consumption are observed. The second end-member is cold waters where oxygen consumption and CO_2 production occur. The third end-member is waters where the sea surface pCO_2 is close to pCO_2 in the air, but oxygen is consumed.

Acknowledgments The work was supported by the state assignments of FRC MHI # 0555-2019-0003 “Experimental studies of the variability of hydrophysical, hydrochemical and bio-optical fields at different spatial and temporal scales to identify the features of climatic changes in oceanographic conditions in the Atlantic Antarctic” and # 0827-2019-0003 “Fundamental studies of oceanological processes that determine the state and evolution of the marine environment under the influence of natural and anthropogenic factors, based on observation and modeling methods.”

References

Arrigo KR, Gert van Dijken, Long M (2008) Coastal Southern Ocean: a strong anthropogenic CO2 sink. Geophysical Res Lett 35:L21602. https://doi.org/10.1029/2008GL035624

Carpenter JH (1965) The Chesapeake Bay institute technique for the Winkler dissolved oxygen method. Limnol Oceanogr 10:141–143

Carrillo CJ, Smith RC, Carl DM (2004) Processes regulating oxygen and carbon dioxide in surface waters west of the Antarctic peninsula. Mar Chem 84:161–179

Ciais P, Sabine C, Bala G, Bopp L, Brovkin V, Canadell J, Chhabra A, DeFries R, Galloway J, Heimann M, Jones C, Le Quéré C, Myneni RB, Piao S, Thornton P (2013) Carbon and other biogeochemical cycles. In: Stocker TF, Qin D, Plattner G-K, Tignor M, Allen SK, Boschung J, Nauels A, Xia Y, Bex V, Midgley PM (eds) Climate change 2013: the physical science basis. Contribution of Working Group I to the fifth assessment report of the Intergovernmental Panel on Climate Change. Cambridge University Press, Cambridge/New York, pp 465–570

Deppeler SL, Davidson AT (2017) Southern Ocean phytoplankton in a changing climate. Front Mar Sci 4:40. https://doi.org/10.3389/fmars.2017.00040

Frölicher TL, Sarmiento JL, Paynter DJ, Dunne JP, Krasting JP, Winton MJ (2015) Dominance of the Southern Ocean in anthropogenic carbon and heat uptake in CMIP5 models. Climate 28(2):862–886. https://doi.org/10.1175/JCLI-D-14-00117.1

Gruber N, Gloor M, Fan S-M, Sarmiento JL (2001) Air-sea flux of oxygen estimated from bulk data: implications for the marine and atmospheric oxygen cycles. Glob Biogeochem Cycles 15(4):783–803

Keller DP, Kriest I, Koeve W, Oschlies A (2016) Southern Ocean biological impacts on global ocean oxygen. Geophys Res Lett 43:6469–6477. https://doi.org/10.1002/2016GL069630
Khoruziy DS (2010) Use of the AS-C3 instrument complex to determine partial pressure of carbon dioxide and concentration of inorganic carbon in seawater. In: Environmental safety of coastal and shelf zones and integrated use of shelf resources, vol 23, pp 260–272. (in Russian)
Kihm C, Körtzinger A (2010) Air-sea gas transfer velocity for oxygen derived from float data. J Geophys Res 115:C12003. https://doi.org/10.1029/2009JC006077
Landschützer P, Gruber N, Haumann FA, Rödenbeck C, Bakker DCE, Heuven S, Hoppema M, Metzl N, Sweeney C, Takahashi T, Tilbrook B, Wanninkhof R (2015) The reinvigoration of the Southern Ocean carbon sink. Science 349(6253):1221–1224. https://doi.org/10.1126/science.aab2620
Makkaveev PN (2013) Dissolved inorganic carbon in the ocean and climate. Water Res 40(7):197–204
Maslennikov VV (2003) Climatic variability and marine ecosystem of the Antarctic. M.:VNIRO Publishing, pp 295
Millero FJ (1996) Chemical oceanography. CRC Press, Boca Raton, p 469
Millero FJ (2007) The marine inorganic carbon cycle. Chem Rev 107:308–341. https://doi.org/10.1021/cr0503557
Mongwe NP, Vichi M, Monteiro PMS (2018) The seasonal cycle of pCO_2 and CO_2 fluxes in the Southern Ocean: diagnosing anomalies in CMIP5 earth system models. Biogeosciences 15:2851–2872. https://doi.org/10.5194/bg-15-2851-2018
Morozov E (2007) Currents in Bransfield Strait. Dokl Earth Sci 415(2):984–986. https://doi.org/10.1134/S1028334X07060347
Roden NP, Shadwick EH, Tilbrook B, Trull TW (2013) Annual cycle of carbonate chemistry and decadal change in coastal Prydz Bay, East Antarctica. Mar Chem 155:135–147. https://doi.org/10.1016/j.marchem.2013.06.006
Sanchez N, Reiss CS, Holm-Hansen O, Hewes CD, Bizsel KC, Ardelan MV (2019) Weddell-Scotia confluence effect on the iron distribution in waters surrounding the South Shetland (Antarctic peninsula) and South Orkney (Scotia Sea) islands during the austral summer in 2007 and 2008. Front Mar Sci 6:771. https://doi.org/10.3389/fmars.2019.0077
Schmidtko S, Stramma L, Visbeck M (2017) Decline in global oceanic oxygen content during the past five decades. Nature 542:335. https://doi.org/10.1038/nature21399
Schuster U, McKinley GA, Bates N, Chevallier F, Doney SC, Fay AR, Gonzalez-Davila M, Gruber N, Jones S, Krijnen J, Landschutzer P, Lefevre N, Manizza M, Mathis J, Metzl N, Olsen A, Rios AF, Rodenbeck C, Santana-Casiano JM, Takahashi T, Wanninkhof R, Watson AJ (2013) An assessment of the Atlantic and Arctic Sea–air CO_2 fluxes, 1990–2009. Biogeosciences 10:607–627. https://doi.org/10.5194/bg-10-607-2013
Skripaleva EA (2004) Features of the water structure and geostrophic circulation in the shelf zone of the Weddell Sea and Bransfield Strait. In: Ecological safety of coastal and shelf zones and integrated use of shelf resources, vol 10, pp 70–77. (in Russian)
Takahashi T, Sutherland SC, Sweeney C, Poisson A, Metzl N, Tilbrook B, Bates N, Wanninkhof R, Feely RA, Sabine C, Olafsson J, Nojiri Y (2002) Global Sea-air CO_2 flux based on climatological surface ocean pCO_2, and seasonal biological and temperature effects. Deep-Sea Res II 49:1601–1622
Thompson AF, Youngs MK (2013) Surface exchange between the Weddell and Scotia seas. Geophys Res Lett 40:1–6. https://doi.org/10.1002/2013GL058114
Verdy A, Dutkiewicz S, Follows MJ, Marshall J, Czaja A (2007) Carbon dioxide and oxygen fluxes in the Southern Ocean: mechanisms of interannual variability. Glob Biogeochem Cycles 21:GB2020. https://doi.org/10.1029/2006GB002916
Wang S, Moore JK (2012) Variability of primary production and air-sea CO_2 flux in the Southern Ocean. Glob Biogeochem Cycles 26:GB1008. https://doi.org/10.1029/2010GB003981
Weiss RF (1970) The solubility of nitrogen, oxygen and argon in water and seawater. Deep-Sea Res 17:721–735

Chapter 13
Earth's Insolation and Spatiotemporal Variability of Albedo in the Antarctic

Anton A. Bukatov and Margarita V. Babiy

Abstract The spatiotemporal dependence of albedo fluctuations on the variability of the floating sea ice concentration and total cloud cover in the Antarctic seas from 1982 to 2014 is studied. Estimates and analysis of temporary trends in albedo and total cloud cover are carried out. It is found that the largest area with a maximum correlation coefficient between albedo and sea ice concentration (~0.85) is observed in the Southern Ocean in December. A high feedback between cloudiness and albedo is observed mainly in the areas where the high positive trend of sea ice and albedo is determined. Also, the insolation variability at the upper boundary of the Earth's atmosphere during the year due to astronomical parameters over the period 1950–2150 is considered. Incoming solar radiation to the Earth's surface is estimated taking into account the transparency coefficient.

Keywords Antarctica · Southern Ocean · sea ice concentration · Earth's insolation · total cloud cover · albedo

13.1 Introduction

Variability of the hydrometeorological conditions of the environment, especially in recent decades, requires comprehensive study and determination of causes of their global and regional variations. These investigations are necessary to assess possible climate transformations in the future and to study the contribution of various physiographic characteristics. In this regard, the processes occurring in the polar regions have an impact on the climatic variability of the entire planet; they require constant

A. A. Bukatov (✉) · M. V. Babiy
Marine Hydrophysical Institute, Russian Academy of Sciences, Sevastopol, Russia

E. G. Morozov et al. (eds.), *Antarctic Peninsula Region of the Southern Ocean*, Advances in Polar Ecology 6, https://doi.org/10.1007/978-3-030-78927-5_13

monitoring and analysis (Baidin and Meleshko 2014; Frolov et al. 2010; Lagun et al. 2010; Ionov and Lukin 2017; Bukatov et al. 2016a, b, 2017).

Solar energy transported to the Earth is an important global factor affecting its climate. The influence is due to numerous direct and feedback processes transforming solar radiation. Such transformations include dispersion, reflection, and absorption, when interacting with atmospheric aerosol, atmospheric gases that absorb energy in the microwave range, cloud cover, etc., as well as interaction with the heterogenous underlying Earth's surface (Loginov 2015; Nazarov et al. 2008; Bukatov and Babiy 2019; Gudkovich et al. 2009; Ermakova et al. 2007). Earth's insolation is also affected by astronomical factors (the orbit eccentricity, the axis tilt to the equatorial plane, vernal equinox anomaly, etc.), which have variability periods of tens of thousands of years (Milankovitch 1939; Monin 1982; Fedorov 2018; Melnikov and Smulsky 2004; Lorenz et al. 2006).

Global climate models are used for a better understanding of the Earth's climatic system variability in the past (Hunt 2006). The data from the results of drilling sedimentary layers and ice masses containing information to construct paleoclimatic series are analyzed (Vasil'ev et al. 2016). Such studies make it possible to reconstruct the paleoclimate in order to try to study the cause-effect relationships and find the mechanisms of modern climatic trends.

At present, the amount of observational data reflecting changes in the natural components of the climate system is increasing. Albedo is one of the factors determining the interaction of downward solar radiation with the surface. Its value measures the ratio of the amount of reflected energy to the amount of energy falling at this surface and depends on its physical properties (Ivanov et al. 2009). In the South Polar region, the ice-snow- albedo relationship is an important indicator of the manifestation of climate change because Antarctica is almost completely covered with ice and snow and reflects a significant part of solar energy (Seo et al. 2016). At the same time, not only the continent but also the Southern Ocean waters that surround it are covered with ice for many months. Hence, it is important to analyze the albedo variability in the Antarctic seas and assess its relationship with floating sea ice of various concentrations, as well as cloudiness, which affects the effective radiation and albedo of the atmosphere-ocean-sea ice system (Frolov et al. 2010).

In this paper, the variability of solar radiation arriving at the upper atmosphere and the Earth's surface during the year, taking into account the influence of astronomical factors, is investigated. The spatiotemporal dependency of the albedo fluctuations on the floating sea ice concentration and variability of the total cloud cover in the Antarctic region are analyzed.

13.2 Method and Data

Investigation is performed on the basis of the NCEP Reanalysis data on sea ice concentration and total cloud cover and the NOAA Reanalysis albedo data from 1982 to 2014 (http://nomad1.ncep.noaa.gov; https://www.esrl.noaa.gov). To assess the relationship between the albedo and sea ice concentration, and the albedo and total cloud cover, correlation analysis of the detrended data series was performed and correlation coefficients significant at a level of 95% were determined for each one-degree grid node. Their significance is determined by the *F* value of the probability of the correlation absence within the confidence interval. Here, $F \leq 0.05$ for all values of correlation coefficients.

The estimates of solar radiation incoming to the upper atmosphere of the Earth and its localization in latitude and over time from 1950 to 2150 are performed on the basis of Milankovitch theory, the study of the astronomical causes of climate fluctuations (Milankovitch 1939): variability over time of the vernal equinox point, determined by the precession of the Earth's axis, the orbit eccentricity and the Earth's axis tilt, and the solar constant. To calculate the downward solar radiation to the Earth's surface, an atmospheric transparency coefficient from 0.6 to 0.9 is introduced. The coefficient value ≈ 0.6 is close to the average atmospheric transparency at midlatitudes.

13.3 Analysis of the Results

The Earth revolution around the Sun and the axis tilt of its rotation are among the most important factors determining seasonal change, especially over short duration time intervals (hundreds of years) (Fedorov 2018). Seasonal changes are manifested in the variability of the Earth's insolation in latitude, which affects all climatic processes. However, this influence is insignificant. Figure 13.1a shows the distribution of average daily insolation at the upper boundary of the Earth's atmosphere in 2020, calculated from the equation (Milankovitch 1939):

$$Q = \frac{S}{\pi \rho^2}\left(\sin\ \varphi\ \sin\delta\ \tau_0 + \cos\ \varphi\ \cos\ \delta\ \sin\ \tau_0\right),$$

where, *S* is solar constant, φ is latitude, δ is the Sun's declination, τ_0 is the solar-hour angle, and ρ is the distance from the Earth to the Sun.

It can be seen that in the warm season (summer months) the influx of solar radiation increases, in the Northern and Southern hemispheres, and two insolation maxima are observed – one at the pole, the second at ~43°. During the cold season in the polar regions from 90° latitude to the southern and northern polar circles, solar radiation does not arrive during a maximum period of up to 6 months.

Calculation of the Earth's insolation changes over 200 years from 1950 to 2150 shows that the influx of solar radiation to the upper boundary of the atmosphere during the warm seasons in the Southern and Northern hemispheres slightly decreased and in the cold seasons it increased (Fig. 13.1b).

Solar radiation downward to the Earth's surface is subject to variability due to the spatiotemporal inhomogeneity of the atmosphere. This arises from the

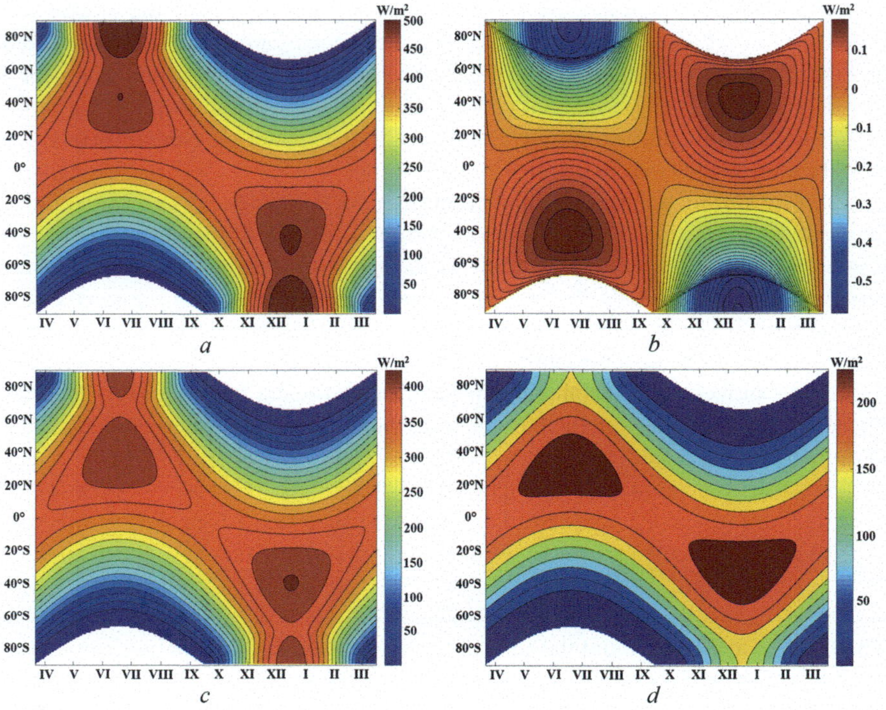

Fig. 13.1 Earth's insolation: at the upper boundary of the Earth's atmosphere in 2020 (**a**); variation in insolation over 200 years (**b**); insolation on the Earth's surface in 2020 for $p = 0.9$ (**c**) and $p = 0.6$ (**d**)

atmospheric influence on solar radiation and the insolation of the Earth's surface using the equation (Milankovitch 1939):

$$Q_p = \frac{S}{\pi \rho^2} \int_{-\omega_0}^{\omega_0} p^{\sec z} \cos z \, dz. \tag{13.1}$$

Here, p is the atmospheric transparency coefficient, ω_0 is sunset hour angle, $-\omega_0$ is sunrise hour angle, and z is the zenith distance to the Sun.

The insolation distribution on the Earth's surface in 2020 is shown in Fig. 13.1c for $p = 0.9$ and in Fig. 13.1d for $p = 0.6$ as calculated from Eq. (13.1). The analysis of the spatial distributions shows that a decrease in the transparency coefficient leads to a decrease in the surface illumination. The greatest insolation occurs in the equatorial belt throughout the year and at midlatitudes in the summer season. At high latitudes, the insolation notably weakens with a decrease of coefficient transparency.

The interaction of the solar radiation with the structure of the Earth's surface determines the spatiotemporal variability of the underlying surface albedo (Vasilyeva and Svyashchennikov 2003). The albedo trend in the Southern Ocean in 1982–2014 is shown in Fig. 13.2. The color scale characterizes the value of the albedo trend by percent over 10 years. The numbers indicate the months. The largest albedo changes of 10% have been observed in February in the Weddell Sea and in December in the Somov Sea. In the annual distribution, the area with a significant positive trend over 9 months (November–July) is located in the Somov Sea. Areas with a positive trend are also recorded in other months. As illumination decreases, they displace to the north to the Southern Ocean border in the Risen-Larsen, Cosmonauts, and Ross seas. Areas with a negative trend over the period under review also exist. There are more such areas in the Western Antarctic than in the Eastern. The largest areas are located in the Amundsen, Bellingshausen, Scotia, and Mawson seas. The maximum decrease in albedo (10%) in February is recorded in the Bellingshausen Sea.

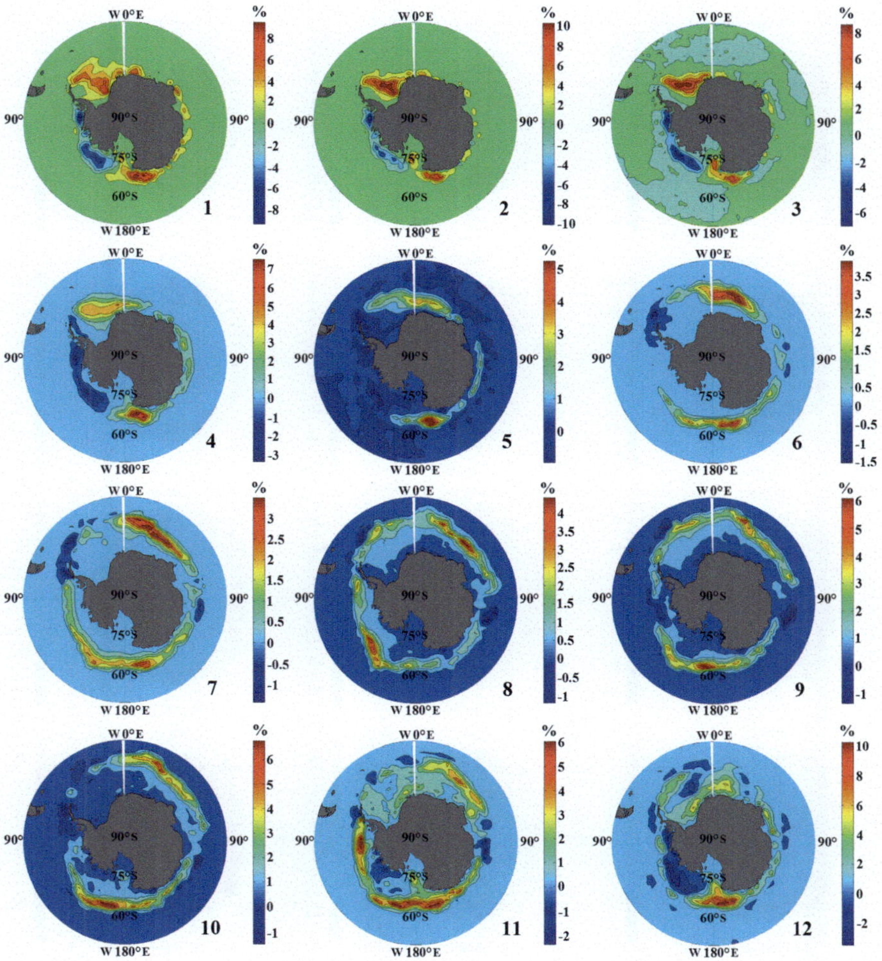

Fig. 13.2 Albedo trend in the Southern Ocean in the period 1982–2014. The color scale is the trend in percent over 10 years. Numbers indicate months of the year

The intra-annual regional distribution of the correlation coefficients (R) between sea ice concentration and albedo in the period from 1982 to 2014 is shown in Fig. 13.3. The color scale characterizes the value of R. It can be seen that in February, the month with a minimum area of floating ice around Antarctica, a high correlation with $R \sim 0.85$ is observed in the areas with ice concentration from 10 to 40%. A weak correlation is recorded along the edge of the ice sheet, where the concentration is less than 10%.

In March, the regional distribution of $R > 0.8$ remains close to that of February, with a slight increase in the area of large R in Western Antarctic.

Since April, the influence of the polar night in the Southern Ocean begins to manifest itself in the Ross and Weddell seas. A decrease or absence of surface illumination results in a significant decrease in the correlation coefficients between the

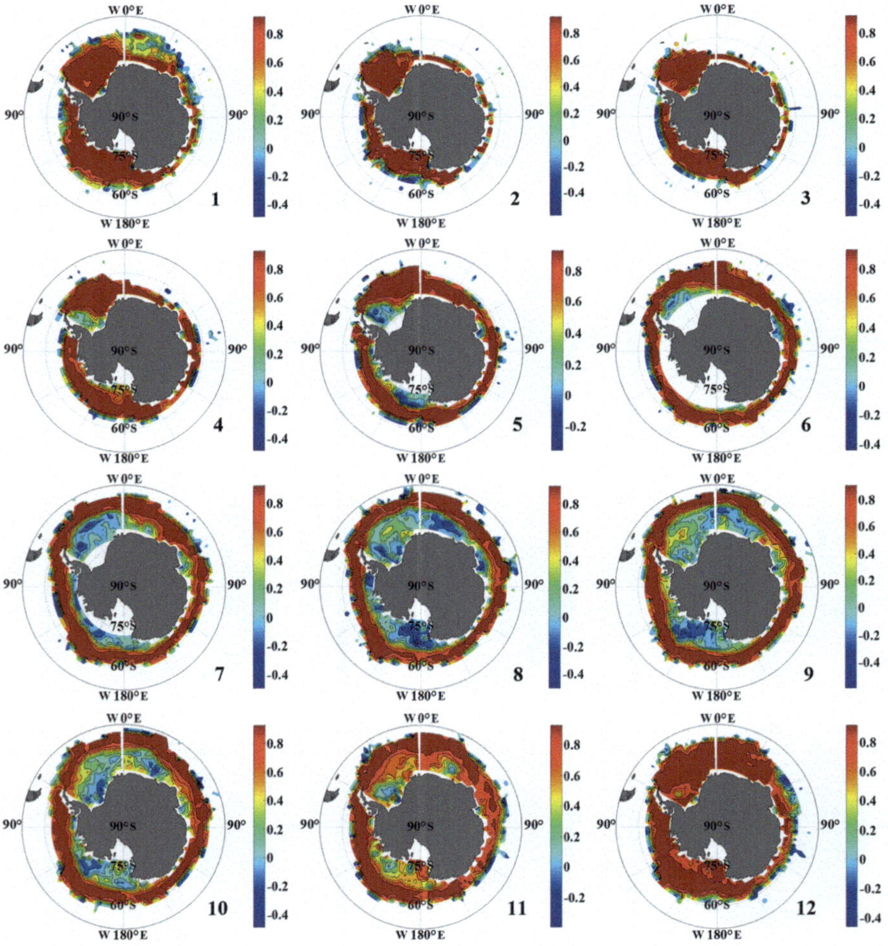

Fig. 13.3 Distribution of the correlation coefficients (R) between the sea ice concentration and albedo. Numbers indicate months of the year

sea ice concentration and albedo. Gradually, the unilluminated surface expands. The regions of significant coefficients displace to the north. They are located along the perimeter of the ice sheet (Fig. 13.4). This continues until September, the month when the sea ice edge position is at the maximum distance from Antarctica.

In October, the regional distribution of R between albedo and ice concentration remains qualitatively similar to September. A high correlation region is situated closer to the outer boundary of the sea ice. In other areas where sea ice is present,

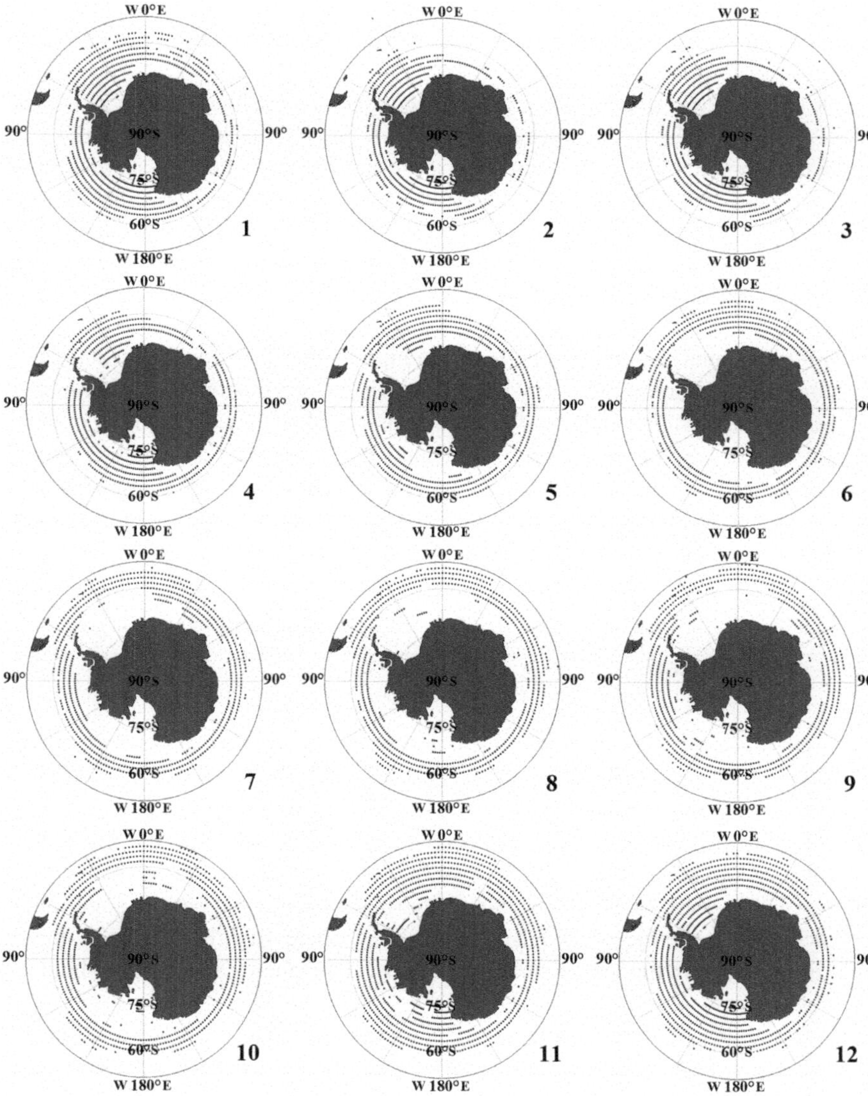

Fig. 13.4 Distribution of probability values of the correlation absence F between the sea ice concentration and albedo for $F \leq 0.05$. Numbers indicate months

the correlation both direct and reverse is low. This is because the main changes in the sea ice concentration in October occur along the outer boundary.

In November, when the ice begins to melt intensely, the sea ice area with a concentration from 60 to 80% increases. As a result, the areas increase where $R > 0.6$.

In December, in most areas covered with sea ice, the concentration is strongly reduced. At the same time, the total area with $R \sim 0.85$ between albedo and sea ice is the largest in the considered intra-annual distribution.

In January, ice melting reaches its maximum. High significant correlation coefficients are determined in almost the entire territory of Western Antarctic. In Eastern Antarctic, they are located in the coastal zone. In other areas, the values of coefficient decrease but remain significant.

One of the factors that influence the albedo variability is cloudiness (Ivanov et al. 2009). Research of the cloudiness variability from 1982 to 2014 considered as a single atmospheric layer in the Southern Ocean showed that a large positive trend in the total cloud cover is recorded in the Ross Sea in January and November and in the Lazarev Sea in January, February, July, August, and October. A significant trend in the absolute value in the annual distribution is observed in the Mawson and Somov seas, in January–February in the Bellingshausen Sea and near the Antarctic Peninsula.

Correlation analysis shows that there is a high feedback between the total cloud cover and albedo in most areas of the Southern Ocean, where high positive trends of the sea ice concentration and albedo are determined. Moreover, the number of areas with high feedback, where the coefficients are significant at a 95% level, is greater in Western Antarctic than in Eastern Antarctic. For most of the year, they are located near the ice edge, where the concentration does not exceed 50%. The maximum correlation coefficient by absolute value ~0.8 is observed in June and August.

13.4 Conclusion

The distribution of incoming solar radiation at the upper boundary of the atmosphere is considered on the basis of the Milankovitch theory, taking into account astronomical factors: variability over time of the vernal equinox point, as determined by the precession of the Earth's axis, the orbit eccentricity and the Earth's axis tilt, and the solar constant. The distribution of the change in insolation over 200 years from 1950 to 2150 is obtained. During the indicated period, solar radiation incoming at the upper boundary of the atmosphere slightly decreased during the warm seasons in the Southern and Northern Hemispheres and increased during the cold seasons.

The variability of albedo in the Antarctic seas is investigated, and its relationship with floating sea ice of various concentrations and total cloud cover over the period 1982–2014 is analyzed. Calculation and analysis of temporary trends in albedo and total cloud cover are carried out. The correlation coefficients between albedo, total cloud cover, and sea ice concentration are obtained. Maps of the spatial distribution of significant correlation coefficients at a 95% level are constructed. Analysis of the results shows that in the intra-annual distribution, a significant positive trend in albedo exists in the Somov Sea during a period of 9 months (November–July). The number of areas with a negative trend is greater in Western Antarctica than in Eastern Antarctica. Between albedo and sea ice concentration, the largest area with a maximum correlation coefficient ~ 0.85 is observed in the Southern Ocean in December. A high feedback between total cloud cover and albedo is observed mainly in the areas where the high positive trends of the sea ice concentration and albedo are determined.

Acknowledgments This study is carried out within the framework of the State Order No. 0555-2019-0003.

References

Baidin AV, Meleshko VP (2014) Response of the atmosphere at high and middle latitudes to the reduction of sea ice area and the rise of sea surface temperature. Russ Meteorol Hydrol 39(6):361–370. https://doi.org/10.3103/S1068373914060016

Bukatov AA, Babiy MV (2019) Influence of downward solar radiation flux on the processes in the Antarctic. Process GeoMedia 2:159–166. [In Russian]

Bukatov AE, Bukatov AA, Babiy MV (2017) Spatial and temporal variability of the Arctic sea ice distribution. Earth's Cryosphere 1:85–92. https://doi.org/10.21782/KZ1560-7496-2017-1(85-92)

Bukatov AE, Bukatov AA, Babiy MV (2016a) Regional variability of Antarctic sea ice extent. Russ Meteorol Hydrol 41(6):404–409. https://doi.org/10.3103/S1068373916060042

Bukatov AE, Bukatov AA, Babiy MV (2016b) Response of influence of internal and external factors in the latitudinal sea ice edge displacement in the Arctic basin. Phys. Oceanogr 6:24–31. https://doi.org/10.22449/1573-160X-2016-6-24-31

Ermakova VI, Okhlopkov VP, Stozhkov YI (2007) Influence of cosmic dust on the Earth's cloudiness, albedo, and climate. Mosc Univ Phys Bull 5(62):308–312. https://doi.org/10.3103/S0027134907050086

Fedorov VM (2018) Earth insolation and modern climate change. PHIZMATLIT, Moscow, p 232. [In Russian]

Frolov IE, Gudkovich ZM, Karklin VP, Smolyanitsky VM (2010) Climate change in the Arctic and Antarctic – result of natural causes. Arct Antarct Res 85(2):52–61. [In Russian]

Gudkovich ZM, Karklin VP, Smolyanitsky VM, Frolov IE (2009) On the character and causes of the Earth's climate changes. Arct Antarct Res 1(81):15–23. [In Russian]

http://nomad1.ncep.noaa.gov

https://www.esrl.noaa.gov

Hunt BG (2006) The medieval warm period, the little ice age and simulated climatic variability. Clim Dyn 27:677–694. https://doi.org/10.1007/s00382-006-0153-5

Ionov VV, Lukin VV (2017) The 10 years long monitoring of climatically significant thermic characteristics of the surface of the Southern Ocean by means of in situ measurements and satellite remote sensing. Arct Antarct Res 3(113):5–15. [In Russian]

Ivanov BV, Andreev OM, Bezgreshnov AM, Polyakov SP (2009) New data about albedo of nature and artificial snow-ice covers around of Antarctic station Novolazarevskaya. Arct Antarct Res 3(83):28–36. [In Russian]

Lagun VY, Klepikov AV, Danilov AI, Korotkov AI (2010) On the warming over the Antarctic peninsula. Arct Antarct Res 85(2):90–101. [In Russian]

Loginov VF (2015) The influence of solar activity and other external factors on the Earth's climate. Fundam Appl Climatol 1:163–182. [In Russian]
Lorenz SJ, Kim J-H, Rimbu N, Schneider RR, Lohmann G (2006) Orbitally driven insolation forcing on Holocene climate trends: Evidence from alkenone data and climate modeling. Paleoceanography 21:PA1002. https://doi.org/10.1029/2005PA001152
Melnikov VP, Smulsky JJ (2004) The astronomical factors of influence on the Earth's cryosphere and the problem of their research. Earth's Cryosphere 1(VIII):3–14. [In Russian]
Milankovitch M (1939) Mathematical climatology and astronomical theory of climate fluctuations, vol 208. GONTI, Moscow–Leningrad. [In Russian]
Monin AS (1982) Introduction to the theory of climate. Gidrometeoizdat, Leningrad, p 246. [In Russian]
Nazarov BI, Abdullaev SF, Maslov VA (2008) Assessment of influence of the direct of solar radiation and albedo surface to global climate change. Rep Acad Sci Repub Tajikistan 9(51):665–671. [In Russian]
Seo M, Kim H-C, Huh M et al (2016) Long-term variability of surface albedo and its correlation with Climatic Variables over Antarctica. Remote Sens 8:981. https://doi.org/10.3390/rs8120981
Vasil'ev NI, Dmitriev AN, Lipenkov VYA (2016) Results of the 5G borehole drilling at Russian Antarctic station «Vostok» and researches of ice cores. J Min Inst 218:161–171. [In Russian]
Vasilyeva DA, Svyashchennikov PN (2003) Albedo dependence on low clouds amount. Vestnik of St. Petersburg State University, Seriya 7 3(23):129–131. [In Russian]

Chapter 14
Quantitative and Productional Characteristics of Microplankton in the Powell Basin and Bransfield Strait in Summer

Nadezda D. Romanova, Sergey A. Mosharov, Olga V. Vorobieva, and Elena V. Bardyukova

Abstract Spatial distribution and abundance of prokaryotes, chlorophyll *a*, and rates of primary production as well as the main factors determining the development of microplankton were analyzed in the northwestern Weddell Sea and in the Bransfield Strait. The average number of prokaryotic cells in the euphotic layer at various stations varied from 85 to 354 × 10^3 cells/ml in the Weddell Sea and from 171 to 379 × 10^3 cells/ml in the Bransfield Strait. The values of bacterioplankton abundance that we observed in January–February, 2020, were generally lower than those normally observed in the austral summer. These differences may be associated with a more extended peak of phytoplankton development in December 2019. The integral concentration of chlorophyll *a* in the euphotic layer at various stations varied from 17.4 to 78.6 mg/m^2. The integral primary production in the northern part of the Powell Basin varied from 26.7 to 654.6 mgC m^{-2} day^{-1} and from 116 to 205.7 mgC m^{-2} day^{-1} in the Bransfield Strait. The relatively low share of chlorophyll *a* in the euphotic layer compared to the water column at most of the stations may indicate the underutilization of the photosynthetic potential of primary producers. A positive correlation was observed between the bacterioplankton concentration and chlorophyll *a* content in the euphotic layer. No consistent pattern was observed in phytoplankton characteristics with regard to the analyzed environmental factors.

Keywords Primary production · Chlorophyll · Bacterioplankton · Weddell Sea · Bransfield Strait

N. D. Romanova (✉) · S. A. Mosharov · E. V. Bardyukova
Shirshov Institute of Oceanology, Russian Academy of Sciences, Moscow, Russia

O. V. Vorobieva
Russian Federal Research Institute of Fisheries and Oceanography, Moscow, Russia

E. G. Morozov et al. (eds.), *Antarctic Peninsula Region of the Southern Ocean*, Advances in Polar Ecology 6, https://doi.org/10.1007/978-3-030-78927-5_14

14.1 Introduction

The Weddell Sea is in the western part of the Weddell Gyre of the Southern Ocean. The Antarctic Circumpolar Current north of the Weddell Sea has a high potential for primary production (Whitehouse et al. 2012). Currently the most conspicuous characteristic of the Weddell Sea is the abundant and persistent sea ice cover and its extreme seasonal variability, shrinking to a third of its September maximum area by the following April (Vernet et al. 2019).

High summer phytoplankton biomass, measured as chlorophyll *a* (chl *a*) concentration, is usually found in the open ocean (1.5–4μg/l) and is generally associated with the marginal ice zone (Vernet et al. 2019; Bathmann et al. 1997). However, the highest reported concentrations (more than 10μg/l) were confined to the Weddell Sea polynyas in the northwestern part of the region (Cape et al. 2014).

An essential part of the Southern Ocean including the Weddell Gyre is related to high-nutrient low-chlorophyll (HNLC) regions. It is assumed that the relatively low primary production in the presence of high-nutrient concentrations is determined mainly by the availability of light and iron (Boyd et al. 2010). Moreover, the influence of these two factors on the photosynthetic activity of phytoplankton is interrelated at the physiological level. Iron limitation causes dysfunction of pigment synthesis and inefficiency of the electron transport system, reducing the photosynthetic rate per unit chlorophyll (Behrenfeld et al. 1996). These disturbances in photosynthetic processes greatly reduce the capacity of phytoplankton to adjust to low-light conditions (Alderkamp et al. 2012; Galbraith et al. 2010).

The light conditions for the development of phytoplankton are determined not only by the surface radiation and the daylight duration but also by the stratification of the water column. Depending on the depth of the upper mixed layer (UML) and its position within the boundaries of the euphotic layer, the conditions for the development of phytoplankton also change. In the areas of the marginal ice zone, where the melt water is the main factor of water column stratification, the primary production and concentration of chlorophyll can be significantly higher compared to the drifting ice zone and open sea areas far from the ice edge (Mathot et al. 1992). The regions distant from the melt-water sources are characterized by a relatively deep UML extending beyond the euphotic layer. Therefore, a large part of the Antarctic pelagic region is characterized by low values of primary production and phytoplankton growth rates (Sakshaug and Holm-Hansen 1986).

One more crucial component of the microbial food web is prokaryotes. This group contributes to the primary production assimilation and transformation of organic matter. It was shown that bacterial production rate is closely related to the development of phytoplankton in the Weddell Sea region (Billen and Becquevort 1991; Kuparinen and Bjorsen 1992; Vaque et al. 2002).

In this work we review the spatial distribution and the abundance of prokaryotes, chlorophyll *a*, and rates of primary production in the northwestern Weddell Sea and in the Bransfield Strait and reveal the main factors determining the development of microplankton in this area.

14.2 Materials and Methods

Study Site and Sample Collection Studies of the quantitative and productional characteristics of microplankton were carried out on cruise 79 of the R/V *Akademik Mstislav Keldysh* in the northern part of the Weddell Sea, in the Bransfield Strait, and in the Antarctic Sound during the austral summer from January 16 to February 6, 2020. The following areas of the northern Weddell Sea were investigated during the cruise: the eastern outer shelf of the Antarctic Peninsula, the continental slope, the deep-sea basin of the Powell Basin, and the elevations of the South Scotia Ridge separating the basin to the north of the Hesperides Deep and the Scotia Sea (Fig. 14.1). The further route to the central part of the Powell Basin was limited by the northern boundary of the drifting ice. Drifting ice was encountered also over the continental slope of the Antarctic Peninsula in the western part of the study site.

Abundance of prokaryotes, chlorophyll *a* concentration, and primary productivity were determined at 16 stations in the northern Weddell Sea, at 9 stations in the Bransfield Strait, and at 3 stations in the Antarctic Sound. Water samples for all measured parameters were collected at discrete depths using SBE-32 Carousel with 5-l Niskin bottles. The sampling depths (5–7 per station) were selected according to the following scheme: surface, layer of maximum fluorescence, the lower limit of photosynthesis (1% of the surface PAR), and intermediate depths.

The surface solar irradiance (PAR, photosynthetically active radiation) was measured at 5-minute intervals using a quantum sensor LI-190SA (LI-COR, USA) and integrated daily (mole photon m^{-2} day^{-1}) in the LI-1400 data logger. The PAR

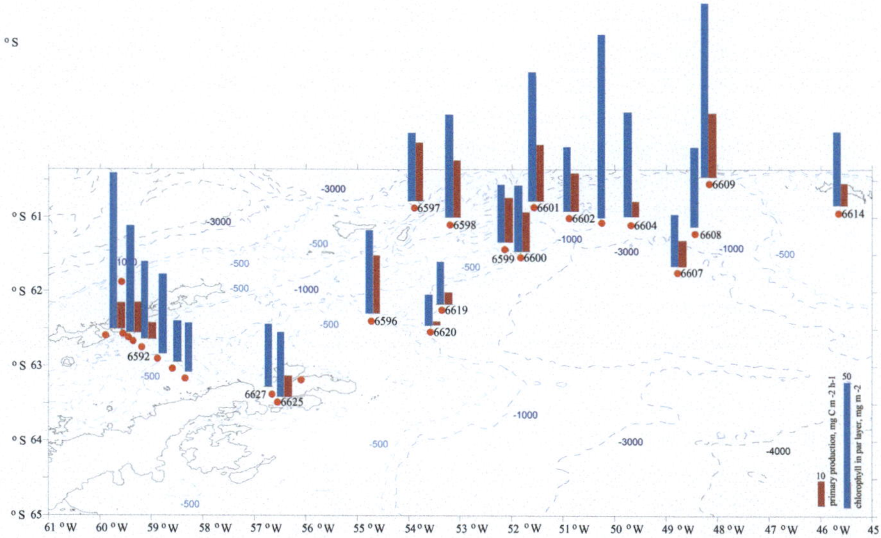

Fig. 14.1 Location of the stations and the values of chl *a*, mg/m^2 and integral primary production, mgC m^{-2} day^{-1} in the euphotic layer

profile in the water column was measured using an underwater quantum sensor LI-193 (LI-COR, USA).

The potential water density was calculated from the temperature, salinity, and pressure data based on the results of CTD profiling. The depth of the upper mixed layer (UML) was defined as the depth at which the density was by 0.02 kg/m^3 greater than at the surface (Alderkamp et al. 2011).

Chl *a* and pheophytin concentrations were measured fluorometrically (Knap et al. 1996). Samples (0.5 l) were filtered through GF/F filters (Whatman) at vacuum $\leq$0.3 mmHg. Then the filters were placed in a 90% acetone solution and exposed to darkness at +4 °C for 24 h. The fluorescence of the extracts was determined with a SMART fluorometer (Moscow State University, Russia) before and after acidification with 1 N HCl. The fluorometer was calibrated using a standard solution of Chl *a* (Sigma).

To determine the prokaryotic abundance, water samples were fixed immediately after collection by adding neutral formaldehyde solution to a final concentration of 1% (Porter and Feig 1980). The 7 ml samples were stained with DAPI concentrated on a 0.2μm Nuclepore black filter and frozen at −18 °C until further analysis. Prokaryotic abundance was determined using a Leica DM 2500 fluorescent microscope at a magnification of 1000. The prokaryotic biomass was calculated after Norland (1993): C cell^{-1} = 0.12 pg x (μm^3 cell^{-1})$^{0.7}$, based on the average prokaryotic cell size for the described area 0.047μm^3/cell (Vaque et al. 2017).

Primary production (PP) rates were determined experimentally by the radiocarbon method (Nielsen 1952). The $NaH^{14}CO_3$ solution was placed in 50-ml vials containing water samples. The vials were exposed by simulated light and temperature conditions in a laboratory incubator with adjustable LED lighting (Gontarev and Mosharov 2015; Mosharov et al. 2015, 2018). The incubation temperature corresponding to the temperature at the sampling point was maintained using a HAILEA-100 chiller and an EHEIM water pump. The irradiance level was assigned for each vial by the adjusted value of direct current flowing through an LED (Mosharov et al. 2015). Samples were exposed in the incubator for 3 h. After incubation, the samples were filtered onto 0.45-μm "Vladipor" membrane filters (Russia). Radioactivity of the samples was determined using a Triathler (Hidex, Finland) liquid scintillation counter.

Biomass-specific PP (P^B) (mg C (mg chl *a*)$^{-1}$ h^{-1}) was calculated by normalizing PP at various depths to the corresponding chl *a* concentration. The proportion of pheophytin was determined as the ratio of the concentration of pheophytin to the sum of the concentrations of chl *a* and pheophytin. This parameter was used as indicator of the phytoplankton functional activity. As was shown in Mosharov and Sergeeva 2018, the proportion of pheophytin less than 40% corresponds to an active state with high productive capability.

The data on the concentration of dissolved inorganic carbon in the water for the calculation of PP were provided by A.A. Polukhin (IO RAS). The integral values of primary production and the content of chl *a* in the water column were calculated using the trapezoidal integration. The lower boundary of the euphotic layer was defined as a depth of 1% surface PAR. Statistical analysis was performed using PAST 3.14 and Primer 6 software.

14.3 Results and Discussion

14.3.1 Bacterioplankton

The prokaryotic abundance in the northern part of the Weddell Sea at multiple depths varied from 40 to 399 × 10^3 cells/ml (0.56–5.63 mgC/m^3). The average abundance in the euphotic layer at various stations ranged from 85 to 354 × 10^3 cells/ml (1.19–4.99 mgC/m^3). At most stations, the number of prokaryotic cells reached the highest values at the depth of the maximum chlorophyll or deeper. Only at the westernmost station of the study site located on the shelf of the Antarctic Peninsula, the maximum value of the prokaryotic abundance was observed in the surface layer. In the northern part of the Weddell Sea, a positive correlation was observed between the bacterioplankton concentration and chl *a* content in the euphotic layer, as well as the integral primary production: $r = 0.68$ and 0.79, respectively, $p < 0.05$.

In the Bransfield Strait, the abundance of bacterioplankton varied by depth from 116 to 438 × 10^3 cells/ml (1.63–6.18 mgC/m^3). On average, in the euphotic layer, prokaryotic abundance at various stations varied from 171 to 379 × 10^3 cells/ml (2.58–5.35 mgC/m^3). Only at two southern stations, the maximal values of the abundance were confined to the surface water. No apparent pattern in the distribution of quantitative characteristics of bacterioplankton was observed either vertically or depending on the abiotic (geographic location, depth, temperature, and salinity) or biological (chlorophyll *a* concentration and primary production) factors considered in this work.

In the Antarctic Sound, the average abundance of prokaryotes in the upper 70-m layer was 94 × 10^3 cells/ml (1.33 mgC/m^3) in the northwestern shallow part of the strait and 240 × 10^3 cells/ml (37.3 mgC/m^3) in the relatively deep southern part of the strait. The pattern of the vertical distribution of quantitative characteristics of prokaryotes was also different. At the southern station, the prokaryotic abundance was uniform over the entire depth of the euphotic layer, while in the north of the strait, the maximum concentration was observed at a depth of 60 m; in the upper 10-meter layer, it was two times smaller, and at a depth of 35 m, the values decreased by one order of magnitude, amounting to only 20 × 10^3 cells/ml.

14.3.2 Chlorophyll

The chl *a* concentration in the northern part of the Powell Basin at various depths of the euphotic layer varied from 0.03 to 3.38μg/l. At most of the stations, the maximum values of chl *a* concentration were observed in the upper 15-m layer. Nevertheless, at some stations, the chl *a* concentration remained at the same level to depths exceeding 50 m, or two concentration peaks were observed (stations 6601 and 6602): in the surface layer and at a depth of about 40 m. In two cases (stations 6607 and 6608), the peaks of chl *a* concentration were at 50 and 75 m depths, respectively. The maximum values of chl *a* at different stations varied from 0.31 to 3.38μg/l. Among them, the lowest values were noted on the shelf of the Antarctic Peninsula: 0.31 and 0.35μg/l at stations 6619 and 6620, respectively. The maximum values, several times higher than that at neighboring stations, were noted at station 6609 (3.38µg/l). Also, high concentrations of chl *a* were observed above the deepwater channel of the Philip Ridge (at stations 6601 and 6613, they reached 1.62 and 1.66µg/l, respectively).

The integral concentration of chl *a* in the euphotic layer at multiple stations varied by more than four times: from 17.4 to 78.6 mg/m^2 (Fig. 14.1). The minimum values were also recorded at the stations in the southwestern part of the study area above the outer shelf of the Antarctic Peninsula: at stations 6619 and 6620, they were 23 and 17.4 mg/m^2. The highest values of the integral concentration of chl *a* were observed in the area of the elevations of the South Scotia Ridge at stations 6603 and 6609, measuring 78.6 and 67.8 mg/m^2, respectively.

The chl *a* concentration in the fluorescence maximum layer exceeded the average value more than two times only at station 6600. The most pronounced maxima of chl *a* concentration (the ratio of the maximum and average concentration for the euphotic layer) were recorded in the eastern deepwater part of the Powell Basin (1.3–1.5µg/l), on the shelf of the South Orkney Islands (1.6µg/l), and in the area of the Philip Ridge elevation in the northwestern part of the study site (1.4–2.1µg/l). Despite the significant variability of the profiles of the vertical distribution of chl *a* concentration, we were unable to identify their relationship with the hydrological factors taken into consideration (including the depth of the euphotic layer and the UML).

The share of pheophytin (inactive form of chl *a*) in the total amount of chl *a* and pheophytin varied from 17 to 62% in the northern part of the Powell Basin at various depths of the euphotic layer. The average value in the euphotic layer varied between stations from 30 to 52%. Values of the pheophytin share exceeding 40% were confined to the slope of the Powell Basin (above depths of more than 1500 m), while in the Scotia Ridge area, on the shelf of the Antarctic Peninsula, and near the South Orkney Islands, this parameter was 29–38%. A high proportion of pheophytin (more than 40%) at most stations in the area of the Powell Basin slope may indicate lower phytoplankton production potential (Mosharov and Sergeeva 2018).

The depth of the UML in the northern part of the Weddell Sea varied from 8 to 102 m. At most stations, this value was less than the depth of the euphotic layer, and only in the northwestern part of the study site (stations 6601, 6599, and 6597), the UML depth was 16–35% greater. In turn, the depth of the euphotic layer at different stations of the study site varied from 30 to 88 m. The euphotic layer depth reached the maximal values in the western part of the region (72–88 m), while in the central and eastern parts of the region, it was significantly smaller (58–69 and 30–54 m, respectively).

It is important to note that the proportion of chl *a* in the euphotic layer considered in this work, relative to the entire water column, varied from 38% to 95%, averaging 64 ± 8%. More than 74% of the total chl *a* content in the euphotic layer of that in the water column was observed in the northeastern part of the study site (stations 6609, 6613, and 6614), as well as in the area of the outer shelf of the Antarctic Peninsula (stations 6619 and 6600). At other stations, this value was below 70%. The relatively low share of chl *a* in the euphotic layer compared to the water column at most stations may indicate the underutilization of the photosynthetic potential of primary producers.

In the Bransfield Strait, the concentration of chl *a* at various depths of the euphotic layer varied from 0.12 to 1.29μg/l. In the majority of cases, the chlorophyll maximum was in the surface layer. The most pronounced chlorophyll maxima (the ratio of the maximum to average concentration in the euphotic layer) were observed in the southern part of the strait (stations 6593 and 6594), amounting to 1.9 and 2.0, respectively. The average content of chl *a* in the euphotic layer varied from 16.7 mg/m^2 in the southern part of the strait to 62.4 mg/m^2 at its northernmost station. The values of chl *a* concentration in the northern part of the Bransfield Strait were close to those observed in the area above the ridges in the northern part of the Powell Basin: 44.3 ± 9.7 and 44.6 ± 8.1 mg/m^2, respectively. Average content of chl *a*, in the euphotic layer in the southern part of the strait, was close to the values at stations above the outer shelf of the Antarctic Peninsula (16.7–19.6 and 12.7–23.1 mg/m^2).

The average share of pheophytin in the euphotic layer in the Bransfield Strait varied from 31 to 50%. The depth of the UML varied from 18 to 46 m, and only at the two southernmost stations it reached 140 and 159 m (stations 6594 and 6595, respectively).

In the Antarctic Sound, the concentration of chl *a* varied from 0.16 to 0.76μg/l. The maximum concentrations in the water column were confined to the surface. The average concentration of chl *a* in the euphotic layer at different stations varied from 23.5 to 33.3 mg/m^2 (24.8 ± 1.4 mg/m^2). The average share of pheophytin in the euphotic layer was 43–47%, which indicates a reduced physiological activity and a low production potential of phytoplankton (Mosharov and Sergeeva 2018). The depth of the UML was 23–55 m.

14.3.3 Primary Production

The values of integral primary production in the northern part of the Powell Basin varied from 26.7 to 654.6 mgC m^{-2} day^{-1} (Fig. 14.1). The minimum values were observed above the outer shelf of the Antarctic Peninsula at stations 6619 and 6620, amounting to 81.3 and 26.7 mgC m^{-2} day^{-1}, respectively. The maximum value of the integral primary production was recorded at station 6609; high values of phytoplankton production (385.1–398 mgC m^{-2} day^{-1}) were also observed on the shelf of the Antarctic Peninsula (station 6596) and over the slope of the deepwater channel east of Clarence Island (stations 6597, 5598, and 6601). There is a notable trend of decreasing primary production eastward along the South Scotia Ridge from 397.8 mgC m^{-2} day^{-1} at station 6597 to 152.5 mgC m^{-2} day^{-1} at station 6614 (Fig. 14.2). Station 6609 was an exception, where, as already mentioned, an anomalously high value of primary production was observed.

The vertical profiles of primary production were mainly characterized by the maximum values at the surface and a decrease with depth at most of the stations. Two stations (6596 and 6597) in the northwestern part of the study site were the exception where the maximum value of primary production was noted at a depth of 10 m, but the difference with the value of surface primary production was less than

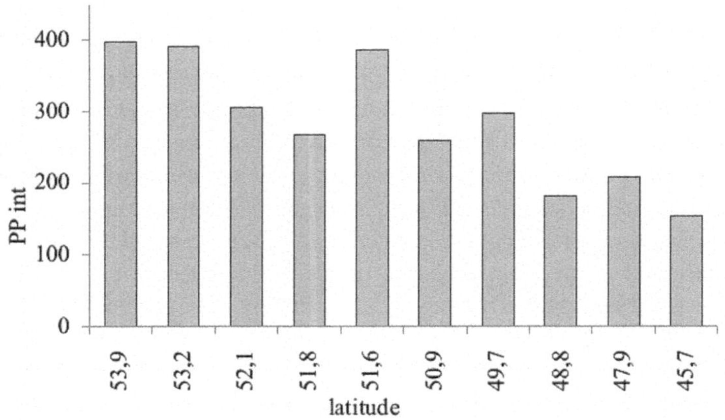

Fig. 14.2 Integral primary production (PP int, mgC m^{-2} day^{-1}) along the South Scotia Ridge

5%. The primary production in the surface layer varied from 2.7 to 63.5 mgC m^{-3} day^{-1}. The lowest values of primary production in the surface layer (2.7–9.2 mgC m^{-3} day^{-1}) were observed at deepwater stations in the Powell Basin. The maximum value (63.5 mgC m^{-3} day^{-1}) was observed at station 6609; relatively high values were also recorded at stations 6600 and 6601 (15.9 and 18.1 mgC m^{-3} day^{-1}, respectively). At other stations, the production values in the surface layer varied from 11.3 to 14.4 mgC m^{-3} day^{-1}.

The maximum values of the biomass specific PP (P^B) in the water column were in most cases confined to the surface, with the exception of station 6597, where this maximum was at 10 m, and station 6602, where the maximal value of P^B was observed at 25 m. The highest values of P^B (1.38 mg C (mg chl *a*)$^{-1}$ h^{-1}) were recorded at stations 6596 and 6597 located in the northwestern part of the region. Also, a high value of P^B was observed at station 6609 (1.36 mg C (mg chl *a*)$^{-1}$ h^{-1}), which was characterized by the maximum production of phytoplankton. The lowest values of P^B were found in the northeastern part of the study site (stations 6613 and 6619; 0.51 mg C (mg chl *a*)$^{-1}$ h^{-1}), as well as above the outer shelf of the Antarctic Peninsula (stations 6619 and 6620; 0.51 and 0.64 mg C (mg chl *a*)$^{-1}$ h^{-1}, respectively).

In the northern part of the Weddell Sea, with an average depth of the euphotic layer of 36 ± 9 m, the main pool of primary production was concentrated in the upper 10-m layer, and already at a depth of 15 m, the production did not exceed 35% of the maximum value in the water column. At the same time, at the depths of the lower boundary of the euphotic layer (averaging 70 ± 6 m), the relative values of primary production higher than 35% remained up to a depth of 25 m. The integral primary production did not depend on the depth of the euphotic layer, varying from 27 to 655 mgC m^{-2} day^{-1} at a depth of euphotic layer from 30 to 41 m and from 81 to 398 mgC m^{-2} day^{-1} at euphotic layer depth from 58 to 88 m.

The values of integral primary production along the transect across the Bransfield Strait varied from 116 to 205.7 mgC m^{-2} day^{-1}. In the southern part, the values of primary production were approximately one and a half times lower than in the northern part. The maximum values were observed in the surface water layer. A similar pattern of the vertical distribution of primary production was also observed in the Antarctic Sound, where the integral value of primary production was 143.8 mgC m^{-2} day^{-1}. The maximum values of P^B were also confined to the upper 10-m water layer. The average value of the P^B of the surface water layer in the Bransfield Strait (0.39 ± 0.07 mg C (mg chl *a*)$^{-1}$ h^{-1}) was minimal compared to the neighboring areas: in the area of uplifts in the northern boundary of the Powell Basin, this indicator averaged 0.99 ± 0.14 mg C (mg chl *a*)$^{-1}$ h^{-1}; on the shelf of the South Orkney Islands and over the continental slope of the Antarctic Peninsula, its average value was 0.53 ± 0.07 mg C (mg chl *a*)$^{-1}$ h^{-1}; and in the Antarctic Sound, it was 0.62 mg C (mg chl *a*)$^{-1}$ h^{-1}.

Statistical analysis did not reveal any relationship in the distribution of quantitative and production characteristics of phytoplankton and the analyzed environmental factors. No consistent pattern was observed in phytoplankton characteristics with regard to the distribution of the dominant zooplankton groups: *Salpa thompsoni* and *Euphausia superba* (Yakovenko et al. this volume).

14.4 Conclusions

Compared to observations in other years, we indicate that the values of bacterioplankton abundance that we observed in the Bransfield Strait and Weddell Sea are somewhat lower than those described in January–February (Karl et al. 1991; Vaque et al. 2002, 2017). According to Vaque et al. (2002, 2017) in the end of January–February 2009, the average bacterioplankton abundance in the upper 100-meter water column was 308 (133–634) × 10^3 cells/ml in the Weddell Sea and 367 (167–649) × 10^3 cells/ml in the Bransfield Strait. The biomass of bacterioplankton in the surface layer and at the depth of the deep fluorescence maximum also slightly differed in the Bransfield Strait, amounting to 5.99 and 6.04 mgC/m^3, respectively. In the Weddell Sea, bacterioplankton biomass at the surface was 2.82–8.1 mgC/m^3 in the surface layer and 3.73–6.3 mgC/m^3 at the depth of the fluorescence maximum. These values are slightly higher than our results. During our work in the Bransfield Strait, the abundance of bacterioplankton in the surface water layer varied in a fairly wide range: from 2.16 to 4.96 mgC/m^3; in the layer of the deep fluorescence maximum, this parameter varied 1.63–6.11 mgC/m^3. In January 2020 in the Weddell Sea, bacterioplankton biomass also varied by an order of magnitude from 0.56 to 5.63 mgC/m^3 in the surface layer and from 1.33 to 4.57 mgC/m^3 in the layer of the deep fluorescence maximum. It is possible that these differences were associated with a more extended peak of phytoplankton development in December 2019 since the active growth of bacteria follows the development of phytoplankton with a time lag of about 15–30 days (Billen and Becquevort 1991).

Our values of the primary production and concentration of chl *a* are comparable with the data already available for the Antarctic regions under consideration. The values of primary production in the northern part of the Weddell Sea in the summer period were 104–410 mgC m^{-2} day^{-1} (El-Sayed and Taguchi 1981), reaching the maximum values in the marginal ice zone. Seasonal changes in the average values of primary production in the Weddell Sea in the area of ice influence are pronounced: in spring 1130 mgC m^{-2} day^{-1} (Mathot et al. 1992), in summer 410 mgC m^{-2} day^{-1} (E1-Sayed and Taguchi 1981), and in autumn, 130 mgC m^{-2} day^{-1} (Smith and Nelson 1990). The concentration of chlorophyll *a* in the ice proximity zone in summer was 28 mg/m^2 (Morán et al. 2001).

The lack of pronounced statistical relationship between our data and environmental factors considered is possibly due to the complicated hydrodynamic structure of the region. Detailed field surveys, as well as appreciation of temporal dynamics of separate components of the food web, are required.

References

Alderkamp A-C, Garcon V, de Baar HJW, Arrigo KR (2011) Short-term photoacclimation effects on photoinhibition of phytoplankton in the Drake Passage (Southern Ocean). Deep-Sea Res I Oceanogr Res Pap 58(9):943–955. https://doi.org/10.1016/j.dsr.2011.07.001

Alderkamp A-C, Kulk G, Buma AGJ, Visser RJW, Van Dijken GL, Mills MM, Arrigo KR (2012) The effect of iron limitation on the photophysiology of Phaeocystis antarctica (Prymnesiophyceae) and Fragilariopsis cylindrus (Bacillariophyceae) under dynamic irradiance: photophysiology under iron limitation. J Phycol 48(1):45–59. https://doi.org/10.1111/j.1529-8817.2011.01098.x

Bathmann UV, Scharek R, Klaas C, Dubischar CD, Smetacek V (1997) Spring development of phytoplankton biomass and composition in major water masses of the Atlantic sector of the Southern Ocean. Deep-Sea Res II Top Stud Oceanogr 44(1–2):51–67. https://doi.org/10.1016/S0967-0645(96)00063-X

Behrenfeld MJ, Bale AJ, Kolber ZS, Aiken J, Falkowski PG (1996) Confirmation of iron limitation of phytoplankton photosynthesis in the equatorial Pacific Ocean. Nature 383(6600):508–511. https://doi.org/10.1038/383508a0

Billen G, Becquevort S (1991) Phytoplankton-bacteria relationship in the Antarctic marine ecosystem. Polar Res 10(1):245–254. https://doi.org/10.1111/j.1751-8369.1991.tb00650.x

Boyd PW, Strzepek R, Fu F, Hutchins DA (2010) Environmental control of open-ocean phytoplankton groups: now and in the future. Limnol Oceanogr 55(3):1353–1376. https://doi.org/10.4319/lo.2010.55.3.1353

Cape MR, Vernet M, Kahru M, Spreen G (2014) Polynya dynamics drive primary production in the Larsen a and B embayments following ice shelf collapse: primary production in the Larsen area. J Geophys Res Oceans 119(1):572–594. https://doi.org/10.1002/2013JC009441

El-Sayed SZ, Taguchi S (1981) Primary production and standing crop of phytoplankton along the ice-edge in the Weddell Sea. Deep Sea Res Part A Oceanogr Res Pap 28(9):1017–1032. https://doi.org/10.1016/0198-0149(81)90015-7

Galbraith ED, Gnanadesikan A, Dunne JP, Hiscock MR (2010) Regional impacts of iron-light colimitation in a global. Biogeochem Model 22

Gontarev SV, Mosharov SA (2015) Incubator and method of incubation of water samples, Patent Ru 2547685 C2 [publication in Russian]

Karl DM, Holm-Hansen O, Taylor GT, Tien G, Bird DF (1991) Microbial biomass and productivity in the western Bransfield Strait, Antarctica during the 1986–87 austral summer. Deep Sea Res Part A Oceanogr Res Pap 38(8–9):1029–1055. https://doi.org/10.1016/0198-0149(91)90095-W

Knap A, Michaels A, Close A, Ducklow H, Dickson A (1996) Protocols for the joint global ocean flux study (JGOFS) core measurements. JGOFS, Reprint of the IOC Manuals and Guides No. 29, UNESCO 1994, 19

Kuparinen J, Bjørnsen PK (1992) Spatial distribution of bacterioplankton production across the Weddell-scotia confluence during early austral summer 1988–1989. In: Hempel G (ed) Weddell Sea ecology. Springer, Berlin Heidelberg, Berlin, Heidelberg, pp 197–204

Mathot S, Dandois J-M, Lancelot C (1992) Gross and net primary production in the scotia-Weddell Sea sector of the Southern Ocean during spring 1988. In: Hempel G (ed) Weddell sea ecology. Springer, Berlin Heidelberg, Berlin, Heidelberg, pp 321–332

Morán X, Gasol J, Pedrós-Alió C, Estrada M (2001) Dissolved and particulate primary production and bacterial production in offshore Antarctic waters during austral summer: coupled or uncoupled? Mar Ecol Prog Ser 222:25–39. https://doi.org/10.3354/meps222025

Mosharov S, Gontarev S, Korsak M (2015) New phyto-incubator with temperature and illumination adjustment for water ecosystems' ecological parameters monitoring. Saf Technosphere 4(4):3–9. https://doi.org/10.12737/14428

Mosharov SA, Sazhin AF, Druzhkova EI, Khlebopashev PV (2018) Structure and productivity of the phytocenosis in the Southwestern Kara Sea in early spring. Oceanology 58(3):396–404. https://doi.org/10.1134/S0001437018030141

Mosharov SA, Sergeeva VM (2018) Estimate of the marine phytoplankton state by fluorescence parameters and ratio of chlorophyll a and pheophytin concentrations. Issues Mod Algol 1(16):10–17

Nielsen ES (1952) The use of radio-active carbon (C14) for measuring organic production in the sea. ICES J Mar Sci 18(2):117–140. https://doi.org/10.1093/icesjms/18.2.117

Norland S (1993) The relationship between biomass and volume of bacteria. Handb Method Aquat Microbial Ecol 1:303–307

Porter KG, Feig YS (1980) The use of DAPI for identifying and counting aquatic microflora. Limnol Oceanogr 25(5):943–948. https://doi.org/10.4319/lo.1980.25.5.0943

Sakshaug E, Holm-Hansen O (1986) Photoadaptation in Antarctic phytoplankton: variations in growth rate, chemical composition and *P* versus *I* curves. J Plankton Res 8(3):459–473. https://doi.org/10.1093/plankt/8.3.459

Smith W, Nelson DM (1990) Phytoplankton growth and new production in the Weddell Sea marginal ice zone in the austral spring and autumn. Limnol Oceanogr 35(4):809–821. https://doi.org/10.4319/lo.1990.35.4.0809

Vaqué D, Guixa-Boixereu N, Gasol JM, Pedrós-Alió C (2002) Distribution of microbial biomass and importance of protists in regulating prokaryotic assemblages in three areas close to the Antarctic Peninsula in spring and summer 1995/96. Deep-Sea Res II Top Stud Oceanogr 49(4–5):847–867. https://doi.org/10.1016/S0967-0645(01)00127-8

Vaqué D, Boras JA, Torrent-Llagostera F, Agustí S, Arrieta JM, Lara E, Castillo YM, Duarte CM, Sala MM (2017) Viruses and protists induced-mortality of prokaryotes around the Antarctic Peninsula during the austral summer. Front Microbiol 8:241. https://doi.org/10.3389/fmicb.2017.00241

Vernet M, Geibert W, Hoppema M, Brown PJ, Haas C, Hellmer HH, Jokat W, Jullion L, Mazloff M, Bakker DCE, Brearley JA, Croot P, Hattermann T, Hauck J, Hillenbrand C-D, Hoppe CJM, Huhn O, Koch BP, Lechtenfeld OJ, Meredith MP, Naveira Garabato AC, Nöthig E-M, Peeken I, Rutgers van der Loeff MM, Schmidtko S, Schröder M, Strass VH, Torres-Valdés S, Verdy A (2019) The Weddell Gyre, Southern Ocean: present knowledge and future challenges. Rev Geophys 57(3):623–708. https://doi.org/10.1029/2018RG000604

Whitehouse MJ, Atkinson A, Korb RE, Venables HJ, Pond DW, Gordon M (2012) Substantial primary production in the land-remote region of the central and northern Scotia Sea. Deep-Sea Res II Top Stud Oceanogr 59–60:47–56. https://doi.org/10.1016/j.dsr2.2011.05.010

Yakovenko V.A., Spiridonov V.A., Gorbatenko K.M., Shadrin N.V., Samyshev E.Z., Minkina N.I. (this volume) Macro- and mesozooplankton in the Powell Basin (Antarctica): species composition and distribution of abundance and biomass in February 2020. This book, Chapter 27

Chapter 15
Detection of Thermophilic Methanotrophic Microbial Communities in the Water Column of the Bransfield Strait (Antarctica)

Anna L. Ponomareva, Nikita S. Polonik, Aleksandra V. Kim, and Renat B. Shakirov

Abstract The paper presents new findings of the thermophilic and mesophilic methanotrophic microorganisms in the study site in the Bransfield Strait (deeper than 1000 m). The gas-geochemical survey revealed single methane vent at a depth of 1397 m bss with a high methane concentration (43 nmol/l) in the bottom water layer. A methane saturated (10.29–14.46 nmol/l) water layer extended to a distance of 30–35 km around the central part of the methane anomaly. The studies of methane oxidation rate parameters revealed a significant microbiological filter within and around the methane venting site. The thermogenic origin of methane is addressed and confirmed by the associated cultivated forms of the thermophilic methanotrophic bacteria in the microbiological filter. The paper is based on original data obtained during the multidisciplinary expedition of the Russian Academy of Sciences (RV "Akademik Mstislav Keldysh," cruise 79, December 2019 – May 2020, head of cruise Dr. Sci. E.G. Morozov).

Keywords Antarctica · Bransfield Strait · Methane · Thermophilic microorganisms

---

A. L. Ponomareva (✉) · N. S. Polonik · R. B. Shakirov
V.I. Il'ichev Pacific Oceanological Institute, Far Eastern Branch of the Russian Academy of Sciences, Vladivostok, Russia
e-mail: ponomareva.al@poi.dvo.ru

A. V. Kim
Far Eastern Federal University, Vladivostok, Russia

Federal Scientific Center of the East Asia Terrestrial Biodiversity, Far Eastern Branch of the Russian Academy of Sciences, Vladivostok, Russia

E. G. Morozov et al. (eds.), *Antarctic Peninsula Region of the Southern Ocean*, Advances in Polar Ecology 6, https://doi.org/10.1007/978-3-030-78927-5_15

15.1 Introduction

The environment of the Antarctic continent has produced a unique biodiversity of microorganisms adjusted to the extreme environmental conditions. Even though the Antarctic climate is mostly cold, geothermal areas provide an ideal environment for the development of thermophilic and hyperthermophilic microorganisms (Flores et al. 2018).

Most of the currently active Antarctic volcanoes are in the West Antarctic Rift System (WARS). Three of them are in Victoria Land (Mount Erebus, Mount Melbourne, and Mount Rittmann), and the fourth is on Deception Island in the South Shetland Islands group, located in the Bransfield Strait. Continental geothermal systems in Antarctica such as Tramway Ridge of Mount Erebus and Deception fumaroles have few taxa in common, especially Chloroflexi, Planctomycetes, and Nitrosopumilales. The bacterial groups of Aquificae, Thermotogae, and Epsilonbacteraeota types were not found on Deception Island. However, thermophilic microorganisms belonging to the Bacillaceae families were recorded there (Mironov et al. 2013; Bendia et al. 2018; Rodrigo et al. 2018).

Several kinds of research have been conducted near Deception Island in the Bransfield Strait to find deep-sea hydrothermal vents and associated ecosystems based on chemotrophs (Bohrmann et al. 1999; Lawver et al. 1995). Most of these research studies were associated with seabed uplifts of volcanic origin (The Hook, The Ax, The Three Sisters) in various parts of the Bransfield Strait and made it possible to find the most promising areas for the hydrothermal venting. A single (focused) hydrothermal vent was found in this area at a significant distance from the previously studied vents. We analyze the entire rift zone along the Bransfield Strait, especially in its deep and poorly investigated area (Polonik et al. 2020).

The formation of methane-oxidizing bacteria in activity-based "oases of life" was observed within the submarine hydrothermal fields. Research on the activity of thermophilic methanotrophic bacteria in Antarctica is limited. In the meantime, this group is a specific bioindicator of the certain methane-rich manifestations.

The article presents data on the relationship between the methane anomalies and thermophilic methanotrophic communities in the bottom water layer of the Bransfield Strait and describes their ability to oxidize methane under mesophilic and thermophilic conditions.

15.2 Study Site, Materials, and Methods

The research was carried out onboard the R/V "Akademik Mstislav Keldysh," cruise 79, 2020. The paper presents the preliminary results on the detected thermophilic and mesophilic methanotrophic microorganisms in the Bransfield Strait (deeper than 1000 m). The study site was located from 62° 33′988 S to 62° 36′725 S and

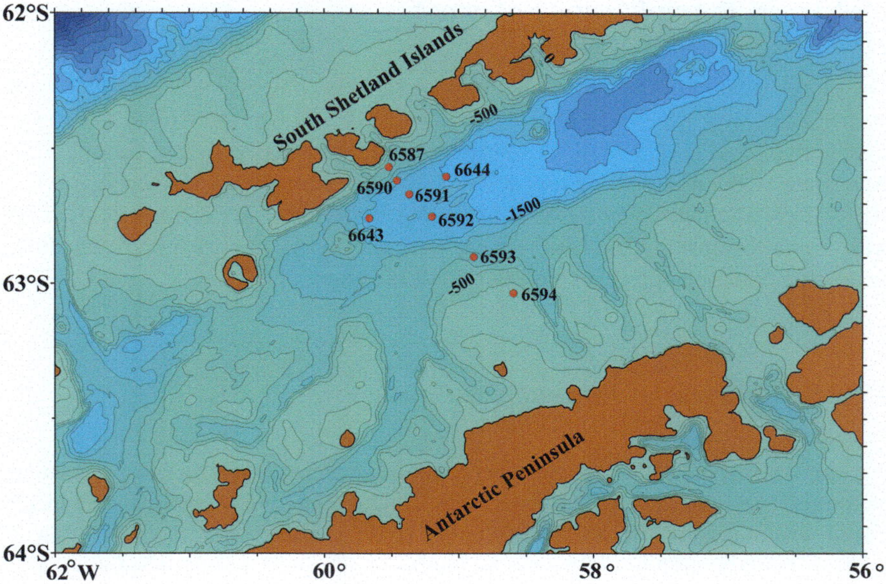

Fig. 15.1 Study site, sections, and stations

from 59° 38′806 W to 59° 05′224 W in the Bransfield Strait, between the South Shetland Islands and the Antarctic Peninsula.

Two sampling profiles were performed: longitudinal and transverse. The cross-profile located in the central part of the Bransfield Strait in the southeast direction from Greenwich Island to the Antarctic Peninsula included six CTD stations; the longitudinal profile consisted just of three stations located along the Livingston, Greenwich, and Robert Islands (Fig. 15.1).

A 24-position Rosette system (USA) equipped with Niskin bottles and combined with a CTD probe was applied to collect water samples. Temperature and salinity profiles were measured. To determine the methane concentration, we sampled at 50, 100, 150, 200, 300, 500, 700, 1000, and 1500 m below sea surface and in the bottom water layer. Samples of 500 ml each for microbiological analysis were taken only at the bottom level of the water column. Seawater samples were filtered through a filter with a pore size of 0.22μm and treated with alcohol.

15.2.1 DNA Extraction

DNA was extracted with the Lintex DNA extraction kit (Russia). The filter membrane was cut out along the edge with a scalpel and transferred into a tube, and DNA was subsequently extracted according to the corresponding protocol. The DNA was eluted with 60 pl buffer TE, after incubating the buffer for 1 min on the membrane.

15.2.2 PCR Amplification of pmoA

The pmoA subunit of the particulate monooxygenase (pMMO) was amplified with primer pair 189f (5/–xx-GGNGACTGGGACTTCTGG-3/) and mb661r (5′-yy-CCGGMGCAACGTCYTTACC-3′) (Holmes et al. 1999; Lyew and Guiot 2003). After initial denaturation for 2 min at 95 °C, DNA was amplified in 30 cycles of 30 s at 95 °C denaturation, 30 s at 55 °C annealing, and 30 s at 72 °C for the extension with a final extension. The results were recorded by the electrophoretic method. Fixing the presence of a gene was qualitative in nature.

15.2.3 Creation of Enrichment Culture

For the cultivation of thermophilic methanotrophic bacteria, we used a nutrient medium NMS ((№ 632 of DSMZ-German Collection of Microorganisms and Cell Cultures) modified by pH of 6.8 (Galchenko 2001) with the addition of a solution of trace elements according to Pfening, bismuth-selenite and 10-vitamin solution. The infecting dose was 1% of the total volume of the medium. The cultivation was carried out at a temperature of +20 and + 65 °C. The methane concentration was at least 10% of the volume of the medium.

15.2.4 Determination of the Rate of Methane Oxidation in the Stored Culture

The change in methane concentrations was determined after 2 days of incubation at a temperature of 20 and 65 °C. The rate was expressed in nM/l per hour.

15.2.5 Gas Chromatography

The methane content in water samples was analyzed by the "headspace" method of equilibrium concentrations (Vereshchagina et al. 2013). For gas chromatographic analysis of the gas composition, we used an EHO-EW mod.2 gas chromatograph (Russia) equipped with a flame ionization detector (FID) and a metal polycapillary column (2 m in length, thickness 0.2–0.3 mm, Hayesep SD phase). Pure helium of "6.0" grade was used as a carrier gas. The methane concentrations were computed according to Yamamoto et al. (1976) modified by Wiesenburg and Guinasso (1979), using the solubility constants.

15.3 Results and Discussion

Gas-geochemical surveys in the Bransfield Strait revealed a single vent at 1397 m depth characterized by the high methane concentration (43 nmol/l) in the bottom water layer. The bottom seawater layer, enriched in methane (10.29–14.46 nmol/l), extends over 30–35 km from the center of the methane anomaly field. The oceanographic profile across the Bransfield Strait was conducted within the area of 191 m to 1417 m below the sea surface (bss). The longitudinal profile included stations 6643 and 6644 located at approximately 15–17 km from each other (1376 and 1507 m bss, respectively). The maximum methane concentrations were found in the bottom water layer and amounted to 14.46 and 12.42 nmol/l, respectively. The water temperature at all stations did not exceed +2 °C (Fig. 15.2).

At all stations in the Bransfield Strait, thermophilic and mesophilic methanotrophic microorganisms were detected in the bottom water layer. The presence of methanotrophic organisms was confirmed by the pmoA monooxygenase gene detection in all water samples.

A large number of thermophilic methanotrophs were found in the eastern side of the Bransfield Strait within the minimum methane concentration zone. This phenomenon can be explained by the pattern of currents in the area. Direct microscopy allowed us to reveal the morphological uniformity of the thermophilic communities of microorganisms. Cultures are represented by oval cells lying freely or united in short chains.

The maximum rate of methane oxidation over this profile at a temperature of +20 °C was observed at stations 6587, 6592, and 6593 (0,26, 0,31, 0,24 nM/l per hour, respectively). The rate of methane oxidation at a temperature of +65 °C was generally lower than at +20 °C. Only at two stations (6591, 6592), the rate of methane oxidation under thermophilic conditions was higher than at mesophilic conditions. At the same time, only at station 6591, thermophilic oxidation of methane

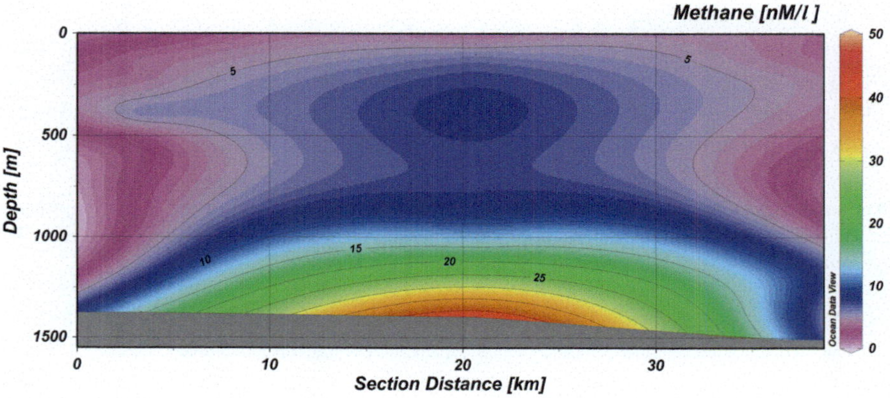

Fig. 15.2 Methane distribution in the water column of the Bransfield Strait (sublongitudinal section, stations 6643, 6591, 6644). Color scale of methane concentration is on the right

Table 15.1 Methane concentration, rate of methanotrophy under thermophilic and mesophilic conditions in the bottom water layer of the Bransfield Strait

				Rate of methane oxidation nmol/l per hour	
Station no.	Depth, m	Temp, °C	CH_4 concentration, nmol/l	+ 20 °C	+ 65 °C
6587	470	0,39	5,84	0,26	0,19
6590	1243	−1,13	5,2	0,11	0,05
6591	1397	−1,50	42,99	0,13	0,84
6592	1417	−1,50	14,36	0,31	0,34
6593	743	−0,83	4,39	0,24	0,11
6594	191	−0,79	3,98	0,16	0,08

significantly prevailed over mesophilic by more than 6 times and at station 6592 by only 1.4 times (Table 15.1).

15.4 Conclusions

The collected data allow us to conclude that thermophilic methanotrophic microorganisms were observed in an active physiological state at two stations. The source of the thermophilic bacteria ingress into the waters of the Bransfield Strait is located in the area close to station 6591. The discussed bacteria are indicators of a newly discovered venting site. Hydrothermal-active vent is a source of high methane concentration and the studied bacteria. The source of methane is most likely a thermogenic one and probably related to the hydrocarbon accumulation.

The comparative analysis of methane oxidation rate in the seawater at different temperatures revealed a strong microbiological filter in the methane anomalous zone in the Bransfield Strait. The thermogenic origin of methane was confirmed by the discovery of cultivated forms of thermophilic methanotrophic bacteria found in the microbiological filter.

Our results make it possible to suggest that a large hydrothermally active zone exists that is associated with the methane emission along the Bransfield Strait. Thermophilic methanotrophic microorganisms related to the methane plume in the water column indicate a unique ecosystem, which should be studied in detail.

Acknowledgments This paper is based on the original data obtained in the multidisciplinary expedition of the Russian Academy of Sciences (R/V “Akademik Mstislav Keldysh,” cruise 79, December 2019 – May 2020, head of cruise Dr. Sci. E.G. Morozov).

References

Bendia AG, Signori CN, Franco DC, Duarte RTD, Bohannan BJM, Pellizari VHA (2018) Mosaic of geothermal and marine features shapes microbial community structure on deception Island Volcano, Antarctica. Front Microbiol 9:899. https://doi.org/10.3389/fmicb.2018.00899

Bohrmann G, Chin C, Petersen S, Sahling H, Schwarz-Schampera U, Greinert J, Lammers S, Rehder G, Daehlmann A, Wallmann K, Dijkstra S, Schenke H-W (1999) Hydrothermal activity at Hook Ridge in the Central Bransfield Basin. Antarctica Geo-Marine Lett 18:277–284

Flores PAM, Correa-Llantén DN, Blamey JM (2018) A thermophilic microorganism from Deception Island, Antarctica with a thermostable glutamate dehydrogenase activity. Biol Res 51:55. Open access. https://doi.org/10.1186/s40659-018-0206-3

Galchenko VF (2001) Methanotrophic bacteria. Geos 496p. ISBN 5-89-118-200-9

Holmes AJ, RoslevP MDIR, Iversen N, Henriksen K, Murrell JC (1999) Characterization of methanotrophic bacterial populations in soils showing atmospheric methane uptake. Appl Environ Microbiol 65:3312–3318

Lawver LA, Keller RA, Fisk MR, Strelin JAS (1995) Bransfield Strait, Antarctic Peninsula active extension behind a dead arc. In: Taylor B (ed) Backarc basins: tectonics and magmatism, pp 315–342. https://doi.org/10.1007/978-1-4615-1843-3_8

Lyew D, Guiot S (2003) Effects of aeration and organic loading rates on degradation of trichloroethylene in a methanogenic-methanotrophic coupled reactor. Appl Microbiol Biotechnol 61:206–213. Open access: https://doi.org/10.1007/s00253-003-1224-8

Mironov VA, Rivkina EM, Gilichinsky DA, Shcherbakova VA (2013) Thermophilic bacteria of the genus geobacillus from permafrost volcanic sedimentary rocks. Microbiology 82(3):389–392

Polonik NS, Ponomareva AL, Shakirov RB (2020) Deep water methane anomaly in the Bransfield Strait (Antarctic) Bull Irkutsk State Univ Ser Earth Sci 32:61–76. Open access: https://doi.org/10.26516/2073-3402.2020.32.61

Rodrigo C, Blamey JM, Huhn O, Provost C (2018) Is there an active hydrothermal flux from the orca seamount in the Bransfield Strait, Antarctica? Andean Geol 45(3):344–356. Open access: https://doi.org/10.5027/andgeoV45n3-3086

Vereshchagina OF, Korovitskaya EV, Mishukova GI (2013) Methane in water columns and sediments of the north western Sea of Japan. Deep-Sea Res II 86-87:25–33. https://doi.org/10.1016/j.dsr2.2012.08.017

Wiesenburg DA, Guinasso NL (1979) Equilibrium solubility of methane, carbon monooxide, and hydrogen in water and sea water. J Chem Eng Data 24(4):356–360

Yamamoto S, Alcauskas JB, Crozier TE (1976) Solubility of methane in distilled water and sea water. J Chem Eng Data 21(1):78–80

Chapter 16
Spectral Bio-optical Properties of Waters in the Bransfield Strait and Powell Basin

Tanya Ya Churilova, Nataliia A. Moiseeva, Tatiana V. Efimova, Vladimir A. Artemiev, Elena Y. Skorokhod, and Anatoly S. Buchelnikov

Abstract In this research, we examined variability of the spectral absorption coefficients of optically significant components in Antarctic waters (Bransfield Strait and Powell Basin) based on the data collected during cruise 79 of the R/V "Akademik Mstislav Keldysh" in the austral summer of 2020 (January 11–February 4). Chlorophyll *a* concentration, spectral light absorption coefficients of phytoplankton, non-algal particles (NAP), and colored dissolved organic matter (CDOM) varied by more than one order of magnitude. Vertical distribution of chlorophyll *a* concentration depends on hydrophysical characteristics. In the case of the depth-dependent distribution of chlorophyll *a* concentration specific light absorption coefficient of phytoplankton ($a^*_{ph}(\lambda)$ (λ)) decreased with depth and especially in the blue spectrum domain, which resulted in the decreasing of blue-to-red peak ratio (R). Values of $a^*_{ph}(678)$ varied from 0.017 to 0.025 $m^2\ mg^{-1}$. The value of R was in the range 1.5–2.8. At one station (the most trophic) in the upper layer (0–15 m), a local maximum was observed at a wavelength of ~545 nm, corresponding to the absorption band of phycoerythrin. The relative NAP absorption at 440 nm was equal to 25 ± 12% of particulate absorption. The spectral slope coefficient (S_{NAP}) varied from 0.006 to 0.016 nm^{-1} with the mean value equal to 0.010 ± 0.002 nm^{-1}. The $a_{CDOM}(440)$ varied from 0.016 to 0.19 m^{-1} without any correlation with chlorophyll *a* concentration. The spectral slope coefficient (S_{CDOM}) varied from 0.009 to 0.022 m^{-1} with an inverse trend of $a_{CDOM}(440)$. The mean value of the S_{CDOM} was 0.013 ± 0.003 nm^{-1}. The absorption budget at 438 nm showed that phytoplankton and CDOM were the main optically active components.

T. Ya. Churilova (✉) · N. A. Moiseeva · T. V. Efimova · E. Y. Skorokhod · A. S. Buchelnikov
A.O. Kovalevsky Institute of Biology of the Southern Seas, Russian Academy of Sciences, Sevastopol, Russia
e-mail: tanya.churilova@ibss-ras.ru

V. A. Artemiev
Shirshov Institute of Oceanology, Russian Academy of Sciences, Moscow, Russia

E. G. Morozov et al. (eds.), *Antarctic Peninsula Region of the Southern Ocean*, Advances in Polar Ecology 6, https://doi.org/10.1007/978-3-030-78927-5_16

Keywords Chlorophyll *a* · Sectral light absorption coefficient · Phytoplankton · Non-algal particles · Colored dissolved organic matter · Bransfield Strait · Powell Basin

16.1 Introduction

The Southern Ocean is a very productive region of the global ocean with high stocks of commercial species and, in particular, stocks of Antarctic krill (Kerr et al. 2018). Due to the influence of climatic and anthropogenic factors, changes are observed in the ecosystem of the Southern Ocean, which negatively affects the stocks of commercial species (Atkinson et al. 2004; Clarke et al. 2007; Ducklow et al. 2007). Long-term monitoring is therefore critical for understanding and predicting possible changes in the ecosystem of the Southern Ocean (Kerr et al. 2018). Ocean color remote sensing is an important tool for operating monitoring of the state of this ecosystem in comparison with in situ observations in terms of spatial and temporal resolution. The food web in the ecosystem is based on the primary producer of organic matter, namely, phytoplankton. Therefore, the assessment of the ecosystem state requires the analysis of abundance of phytoplankton and its photosynthetic potential. Concentration of the photosynthetic-active pigment – chlorophyll *a* – is widely used as a marker of phytoplankton biomass, which can be assessed using satellite data. Previous studies showed that chlorophyll *a* was overestimated by up to two times calculated by the standard algorithm in the Southern Ocean (Dierssen and Smith 2000). Correct assessment of the chlorophyll *a* content from space-based scanners require a thorough understanding of in-water optical properties: absorbance and scattering by optically active components. Spectral light absorption coefficients of phytoplankton, non-algal particles (NAP), and colored dissolved organic matter (CDOM) are major parameters that determine the variability in the light field in the ocean. Hence, understanding their variability with respect to environmental factors is required for the development of the regional bio-optical models for assessment of water quality and productivity indicators based on ocean color remote sensing.

Total absorption $a(\lambda)$ includes contribution of all optically active components (Kirk, 2011):

$$a(\lambda) = a_{ph}(\lambda) + a_{NAP}(\lambda) + a_{CDOM}(\lambda) + a_w(\lambda), \tag{16.1}$$

$$a_p(\lambda) = a_{ph}(\lambda) + a_{NAP}(\lambda), \tag{16.2}$$

where $a_{ph}(\lambda)$ is the light absorption coefficient of phytoplankton at wavelength (λ), $a_{NAP}(\lambda)$ is the NAP light absorption coefficient, $a_{CDOM}(\lambda)$ is the CDOM light absorption coefficient, $a_w(\lambda)$ is the absorption coefficient of pure water, and $a_p(\lambda)$ is the particulate light absorption coefficient.

The overestimation of chlorophyll *a* concentration by standard algorithm is likely to be caused by regional peculiarities in chlorophyll *a*-specific phytoplankton absorption (Ferreira et al. 2017), and high and variable content of CDOM due to melting ice appeared under climate change (Clarke et al. 2007; Ducklow et al. 2007; Kerr et al. 2018). The spectral absorption by all optically active components has recently been studied only in the northwestern and northern parts of the Antarctic Peninsula (Ferreira et al. 2017, 2018). The lack of bio-optical data in the Powell Basin and the limited amount of spectral light absorption data in the Bransfield Strait require further detailed investigation of spectral bio-optical characteristics of waters in these regions of the Southern Ocean. The goal of this study is to assess variability in spectral light absorption coefficients by phytoplankton, NAP, and CDOM in the Bransfield Strait and Powell Basin in austral summer 2020 (January 11–February 4, 2020).

16.2 Methods

Bio-optical data were collected during the scientific cruise of R/V "Akademik Mstislav Keldysh" in the austral summer of 2020 (January 11–February 4). The study site covered the Bransfield Strait (BS), between the Southern Shetland Islands and the Antarctic Peninsula, and the Powell Basin (PB) (Fig. 16.1). Station locations that were sampled are shown in Fig. 16.1. The ocean bottom at stations in the Bransfield Strait varied as follows: 37 m (st 6585), 472 m (st 6587), 1260 m (st 6590), 1417 m (st 6591), 1436 m (st 6592), 748 m (st 6593), 196 m (st 6594), and 165 m (st 6595). In the Powell Basin, the stations were in the deepwater region with a bottom deeper than 1000 m, except for shelf stations 6596 (depth 360 m), 6597 (depth 828 m), 6599 (depth 579 m), and 6614 (depth 294 m).

Salinity and temperature profiles were recorded with a Sea-Bird SBE 911 plus CTD unit. The chlorophyll *a* fluorescence profiles were recorded with fluorometer Minitracka II (Chelsea), which was installed on the transparency meter PUM-200. Photosynthetically active radiation (PAR) incident on the sea surface and its vertical profile were measured with a Li-COR set on the deck (LI-190SA) and with submersible device (LI-192). Water samples were collected at several depths (3–7 depths) using Niskin bottles attached to the CTD. The sampling depths were chosen based on the temperature, salinity, PAR, and fluorescence profiles.

The depth of the photosynthetic zone (Z_{eu}) was taken as the depth with 1% of the PAR incident on the sea surface. The concentrations of chlorophyll *a* and phaeopigments were determined spectrophotometrically with 90% acetone (Jeffrey and Humphrey 1975; Lorenzen 1967) using a dual-beam spectrophotometer Lambda 35 (PerkinElmer). Water samples were filtered immediately after collection under a low vacuum on Whatman GF/F glass fiber filters and stored in liquid nitrogen until analysis for particulate absorption and chlorophyll *a* concentration.

Particulate absorption was measured following the standard quantitative filter technique (QFT) in accordance with the current NASA protocol (Boss et al. 2019).

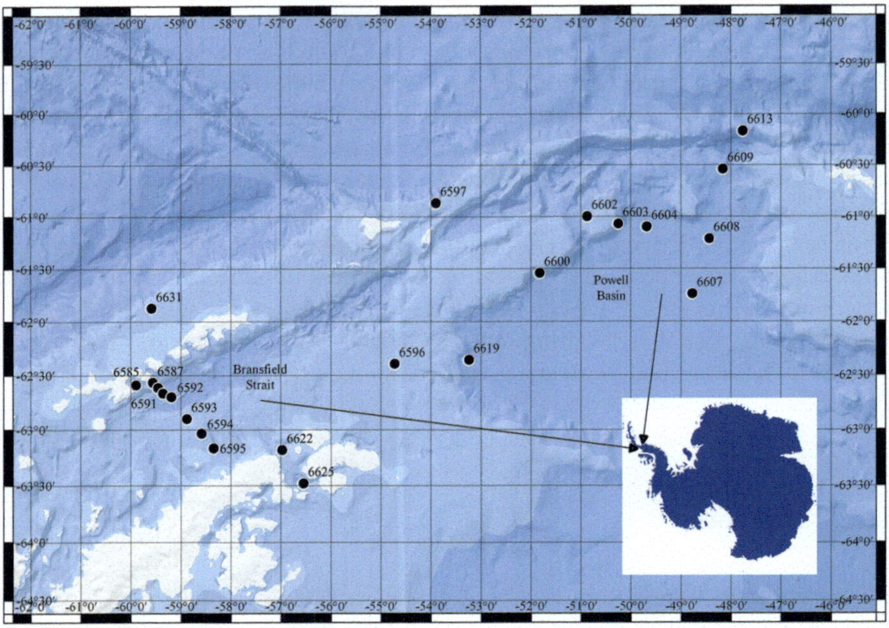

Fig. 16.1 Map of stations (●) of the bio-optical investigations carried out in the scientific cruise 79 of the R/V "Akademik Mstislav Keldysh" in the Bransfield Strait and in the Powell Basin from January 11 to February 4, 2020

To separate the phytoplankton pigments within the particulate matter from NAP, methanol extraction was performed (Kishino et al. 1985). Subsequent extraction with hot water was applied to remove phycobilipigments (Roesler and Perry 1995). The phytoplankton pigment absorption coefficient ($a_{ph}(\lambda)$) was estimated by subtracting bleached spectral absorption ($a_{NAP}(\lambda)$) from unbleached spectral absorption $a_p(\lambda)$. The chlorophyll a-specific light absorption coefficient of phytoplankton was obtained by dividing $a_{ph}(\lambda)$ by chlorophyll a concentration together with phaeopigments (TChl-a). The optical densities were corrected for path length amplification using beta (β) correction algorithm (Mitchell 1990). Optical density of the samples was measured from 350 to 750 nm with dual-beam spectrophotometer Lambda 35 (PerkinElmer) equipped with an integrating sphere.

The $a_{NAP}(\lambda)$ spectra were described by an exponential function (Babin et al. 2003):

$$a_{NAP}\left(\lambda\right) = a_{NAP}\left(\lambda_r\right) \cdot e^{-S_{NAP}\mathrm{n}\left(\lambda - \lambda_r\right)} \tag{16.3}$$

Wavelength λ_r is the reference wavelength, 440 nm; S_{NAP} represents the spectral slope for $a_{NAP}(\lambda)$. S_{NAP} was calculated using a nonlinear least square fitting from 400 to 700 nm.

The CDOM absorption was determined following the current NASA protocol (Boss et al. 2019). Discrete samples were filtered (under low vacuum) through

47 mm 0.2 mm nylon filters (Sartorius Nuclepore) (0.2 μm pore size, 47 mm diameter) using a prefiltration with GF/F (Whatman). Both filters were rinsed with deionized water before filtration. The CDOM absorption was measured with a dual-beam spectrophotometer Lambda 35 (PerkinElmer) using a 10 cm path length quartz cuvette and deionized water for reference. Parameterization of the $a_{CDOM}(\lambda)$ was done similarly to NAP (Eq. 16.3), but S_{CDOM}, which is the spectral slope for $a_{CDOM}(\lambda)$, was retrieved for a wavelength of 350–500 nm (Bricaud et al. 1981).

16.3 Results and Discussion

In the study site, hydrographical characteristics in the investigated layer (30–120 m) varied significantly at the stations (Fig. 16.2). The surface temperature varied from −0.47 °C to 3.19 °C, and the average value was 1.29 ± 1.4 °C in the BS. In the PB the surface temperature was lower than in the BS: it varied from −0.77 °C to 2.01 °C and was 0.72 ± 0.89 °C on average. The largest change in temperature with depth (between 0 m and 80 m) reached ~3 °C in the BS. Vertical profiles of temperature at some stations showed the existence of the upper mixed layer (Fig. 16.2). Conversely, at several stations (at two stations in the BS and at seven stations in the PB), the studied layer (up to 120 m) was almost uniform in temperature.

Salinity of the surface layer varied from 31.6 to 35.8 PSU, while the average value was 33.6 ± 1.0 PSU. In the salinity profile, there was a minimum in the layer 0 ~ 25 m. In this minimum region, the salinity was by 0.5–5.0 PSU lower in comparison with that in the surface layer. In a deeper water layer (> ~ 25 m), a salinity of ~34.5 PSU remained constant at all stations (Fig. 16.2).

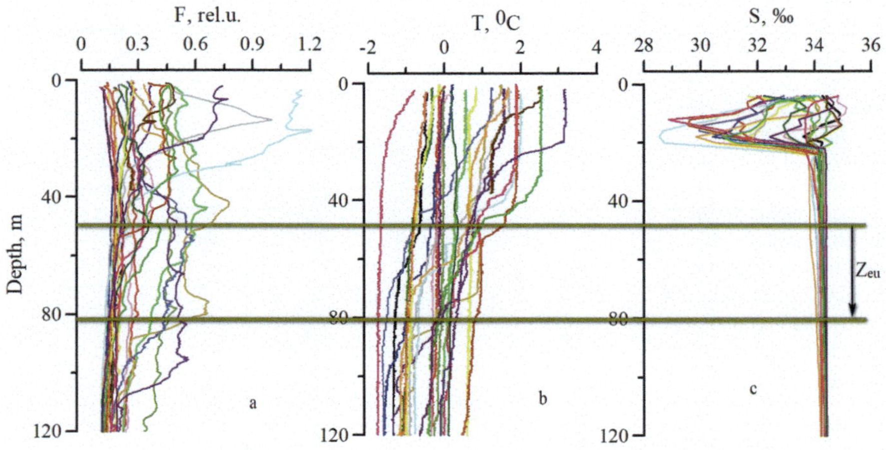

Fig. 16.2 Vertical profiles of the chlorophyll *a* fluorescence intensity (F) (**a**), temperature (T) (b), and salinity (S) (**c**) in the Bransfield Strait and in the Powell Basin from January 11 to February 4, 2020: green line shows the variability in euphotic depth (Z_{eu}) between stations

The euphotic zone varied between stations in a narrow range from 50 m at station 6614 to 83 m at station 6619 (Fig. 16.2), with the exception of two stations (6609 and 6613), where the euphotic layer was shallower (Z_{eu} = 28 and 35 m). The mean Z_{eu} was equal to 64 ± 12 m.

The variability in the profiles of hydrographical characteristics resulted in different vertical distributions of chlorophyll *a* fluorescence (F) (Fig. 16.2). At several stations, the values of F were distributed homogenously within the layer. But at other stations, depth-dependent variability in F was evident: F profiles are characterized by a maximum, and the intensity and location of maximum varied between stations (Fig. 16.2). The variability in the F profiles is likely to be caused by the different stability in the water column. A similar variability in the profiles of hydrographical characteristics and chlorophyll *a* fluorescence was found in previous research studies in the austral summer in the northwestern Antarctic Peninsula (Figueiras et al. 1999; Ferreira et al. 2017).

The TChl-*a* in the surface layer varied by more than one order of magnitude, from 0.25 to 4.44 mg m^{-3} (Fig. 16.3). The maximum values of TChl-*a* were found in the Powell Basin: 4.44 mg m^{-3} at station 6609 and 2.18 mg m^{-3} at station 6613. The mean value of TChl-*a* in the surface layer at all stations with the exception of these two stations (6609 and 6613) was 0.72 ± 0.35 mg m^{-3}. In the euphotic layer, TChl-*a* was in a range of 0.15–1.81 mg m^{-3} except for stations 6609 and 6613, where TChl-*a* reached 4.44 and 2.18 mg m^{-3}, respectively. In the layer of chlorophyll *a* maximum, which was observed at some stations, the TChl-*a* exceeded 1.3–1.5 times that in the surface layer. On average, in the euphotic layer, the TChl-*a* was equal to 0.46 ± 0.26 mg m^{-3}.

In the vertical distribution of TChl-*a*, three types of profiles could be distinguished (Fig. 16.4): (1) profiles with relatively uniform distribution of TChl-*a* (e.g., at st. 6602, Fig. 16.4), (2) profiles with a maximum near the bottom of euphotic zone (~ (1–0.1%) PAR) (e.g., at st. 6604, Fig. 16.4), and (3) profiles with TChl-*a* decreasing with depth (e.g., at st. 6614, Fig. 16.4). The third type of profile was

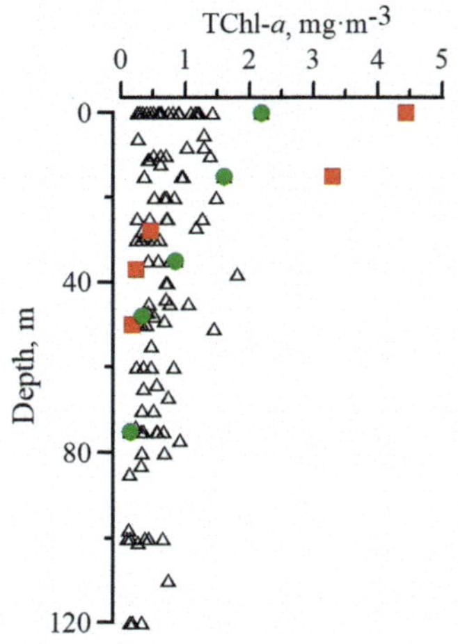

Fig. 16.3 Vertical distribution of the chlorophyll *a* concentration together with phaeopigments (TChl-*a*) in the Bransfield Strait and in the Powell Basin from January 11 to February 4, 2020: stations 6609 (■), 6613 (●), and other stations (Δ)

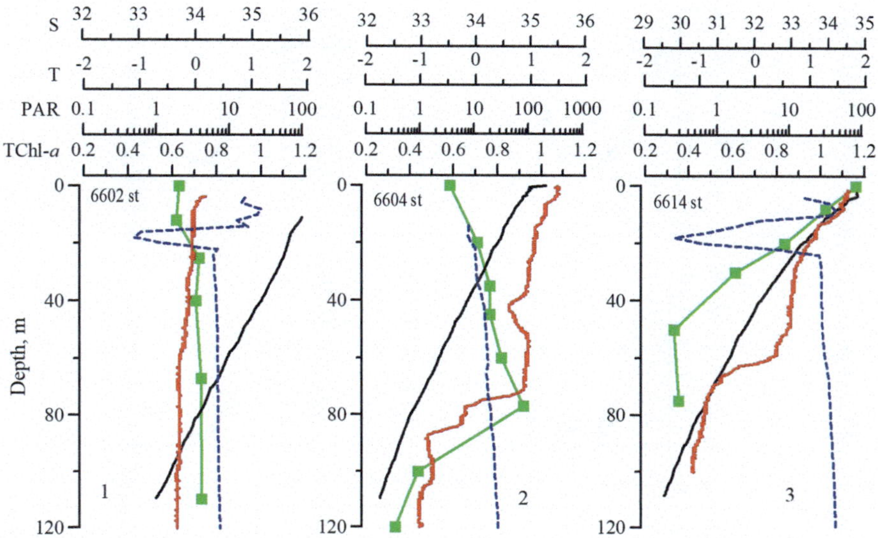

Fig. 16.4 Examples of typical vertical profiles of temperature (T, °C, red line), salinity (S, PSU, blue line), photosynthetically active radiation (PAR, μmol·m^{-2}·s^{-1}, black line), and chlorophyll *a* concentration (TChl-*a*, mg m^{-3} ■) at 6602 (1, first type), 6604 (2, second type), and 6614 (3, third type) stations in the Bransfield Strait and in the Powell Basin (January 11–February 4, 2020)

most common: 16 stations of a total of 26 stations. The obtained TChl-*a* range of variability was in a good agreement with the results in austral summer in the north-western Antarctic Peninsula (Ferreira et al. 2017, 2018).

High variability of TChl-*a* was accompanied by significant change (more than an order of magnitude) in the spectral light absorption coefficients of phytoplankton. The spectrum $a_{ph}(\lambda)$ has two main maxima: in the blue (at ~438 nm) and in the red (at ~678 nm) spectral ranges.

Light absorption coefficients at blue $a_{ph}(438)$ and red maximum $a_{ph}(678)$ varied within the euphotic layer from 0.005 to 0.29 m^{-1} and from 0.003 to 0.12 m^{-1}, respectively. The maximum values were recorded at stations with a high TChl-*a* (st. 6609 and 6613), the minimum — in the lower part of the euphotic layer.

Vertical profiles of $a_{ph}(\lambda)$ corresponded to different types of vertical distribution of TChl-*a* (Fig. 16.5). In the case of a uniform distribution of TChl-*a* (e.g., at station 6602, Fig. 16.5), the values of $a_{ph}(\lambda)$ were almost univariable with depth. The ratio between the coefficients in the blue and red peaks (R) did not change with depth (~ 2.0). Moreover, the chlorophyll *a*-specific absorption coefficient $a_{ph}^*(\lambda)$ did not vary within the euphotic layer. Morel and Bricaud (1981) reported that phytoplankton species composition, cell size, and cellular pigment complex did not change with depth. Such homogeneity in vertical distribution of pigments and functional characteristics of phytoplankton (in particular, specific light absorption) is likely the result of the mixing of waters and phytoplankton within the euphotic layer at the rate exceeding the rate of photoadaptive processes in phytoplankton (Cullen and Lewis 1988).

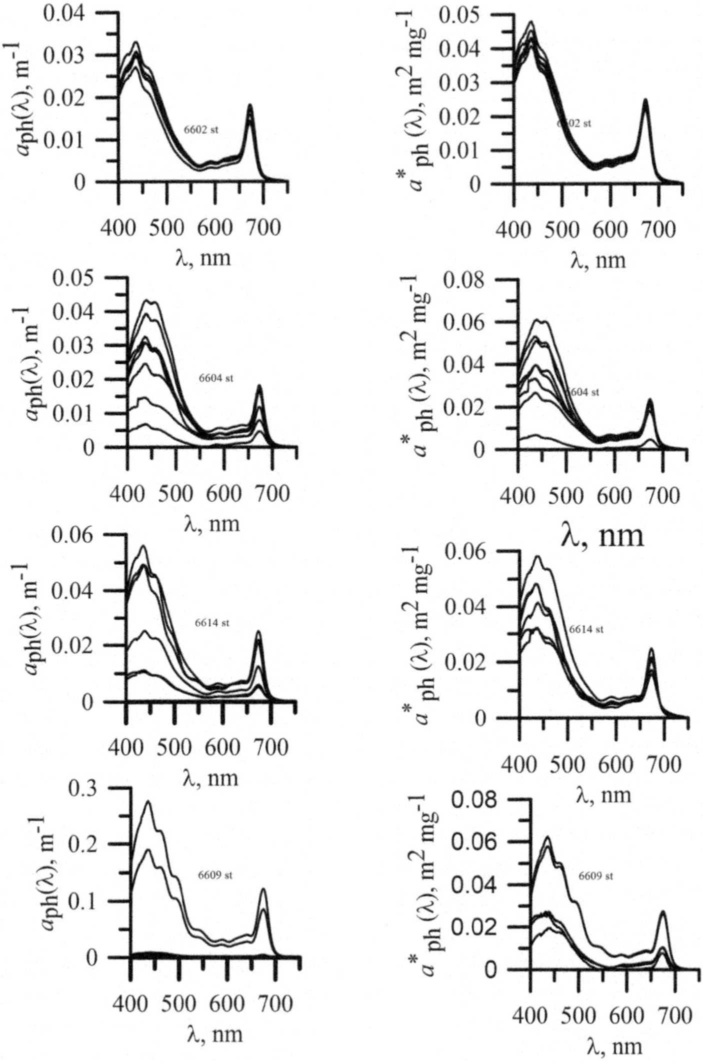

Fig. 16.5 Light absorption coefficient of phytoplankton ($a_{ph}(\lambda)$) and chlorophyll *a*-specific light absorption coefficient ($a_{ph}^*(\lambda)$) at stations in the Bransfield Strait and in the Powell Basin (January 11–February 4, 2020)

In the case of a depth-dependent vertical distribution of TChl-*a* (the second and third types of chlorophyll *a* fluorescence profiles), the $a_{ph}(\lambda)$ changed markedly with depth (Fig. 16.5). The R decreased from 2.8–2.4 in the surface layer to 1.9–1.5 in the lower part of the euphotic zone. The red maximum of the spectrum is formed due to the absorption of light by chlorophyll *a* and phaeopigments. In the blue part of the spectrum, not only chlorophyll *a* but also accessory pigments absorb light (Jeffrey et al. 1997). Consequently, the obtained decreasing of R (smoothing of spectra) is associated with an adaptive decrease in the proportion of accessory pigments relative to chlorophyll *a* and with a package effect (MacIntyre et al. 2002; Morel and Bricaud 1981). The values of $a_{ph}(678)$ decreased with depth from 0.025–0.023 m^2 mg^{-1} at the surface to 0.019–0.017 m^2 mg^{-1} near the bottom of the euphotic zone. The observed variations with depth (smoothing of the spectra and a decrease in chlorophyll *a*-specific absorption coefficients of phytoplankton) reflected the phytoplankton adjustment to the environmental conditions that varied with depth (Bricaud et al. 1995). Water stratification (formation of the upper layer with higher temperature/lower density) probably leads to an increase in the stability of the water column, which provides the conditions for the phytoplankton to adjust to the vertical environment gradient within the euphotic zone.

The $a_{ph}(\lambda)$ linked with TChl-*a*. A relationship between $a_{ph}(438)$), $a_{ph}(678)$, and TChl-*a* was revealed and described with the power function:

$$a_{ph}(438) = 0.046 \times C_a^{1.05}, r^2 = 0.84$$

$$a_{ph}(678) = 0.021 \times C_a^{1.04}, r^2 = 0.89.$$

In the equations, the power exponent is slightly higher than unity, which indicates a slight increase in $a_{ph}(\lambda)$ with an increase in the TChl-*a* in water. This reflects a decrease in the degree of pigments packing in phytoplankton cells (Morel and Bricaud 1981). This is probably due to the fact that in more trophic waters (where TChl-*a* values are higher), phytoplankton is represented by smaller-celled species. Comparison of the obtained equations with those revealed from a large dataset of ocean data (Bricaud et al. 1995) exhibited the difference mainly in terms of the power coefficient.

At station 6609, where TChl-*a* was relatively higher, at depths of 0 and 15 m, a local maximum was observed at a wavelength of ~545 nm, corresponding to the absorption band of phycoerythrin (Fig. 16.5). The appearance of this local maximum on the spectra is probably due to the relatively higher abundance of phycoerythrin-containing species in the community, in particular: cryptophyte and cyanobacteria (Moore et al. 1995; Novarino 2003; Heidenreich and Richardson 2020). At other depths, no local maximum at ~545 nm was observed.

The NAP absorption is determined by the absorption of light by suspended particles not associated with phytoplankton pigments such as bacteria and detritus (nonliving organic particles) and mineral particles (Morel and Ahn 1991; Stramski and Kiefer 1998). The $a_{NAP}(440)$ varied from 0.002 to 0.027 m^{-1}. The highest $a_{NAP}(440)$ values (0.018–0.027 m^{-1}) were recorded at the shallow-water (37 m) station 6585 near the Bransfield Strait and at stations 6609 and 6613, where the high TChl-*a* values were found. It was found that $a_{NAP}(440)$ increased with TChl-*a*. In the euphotic layer, the ratio between $a_{NAP}(440)$ and $a_p(440)$ varied from 0.05 to 0.82 and was equal to 0.25 ± 0.12 on average. The spectral slope coefficient (S_{NAP}) ranged from 0.006 to 0.016 nm^{-1} with mean equal to 0.010 ± 0.002 nm^{-1}. The mean S_{NAP} value is in good agreement with data published for multiple different regions of the global ocean (Bricaud et al. 1998; Babin et al. 2003; Churilova et al. 2017; Efimova et al. 2018).

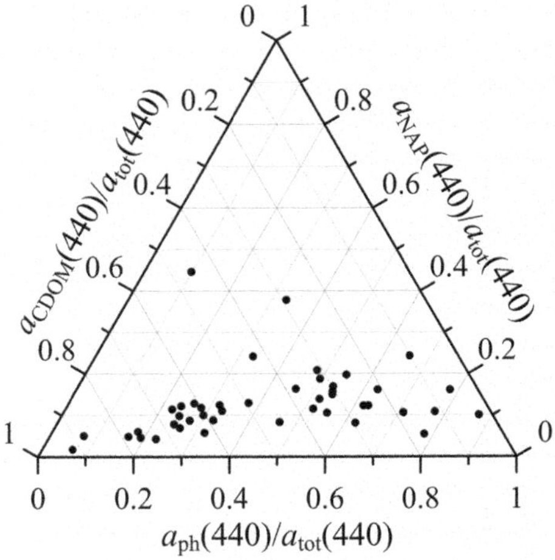

Fig. 16.6 Contribution of phytoplankton $a_{ph}(\lambda)$, non-algal particles $a_{NAP}(\lambda)$, and colored dissolved organic matter $a_{CDOM}(\lambda)$ to the total light absorption (without water) $a_{tot}(\lambda)$ at a wavelength of 438 nm in the first optical depth in the Bransfield Strait and in the Powell Basin from January 11 to February 4, 2020

The CDOM light absorption varied significantly. The $a_{CDOM}(440)$ was in a range of 0.016–0.19 m^{-1}. No correlation was found between $a_{CDOM}(440)$ and TChl-*a*. The spectral slope coefficient (S_{CDOM}) varied from 0.009 to 0.022 m^{-1} with inverse trend with $a_{CDOM}(440)$ as observed elsewhere (Nelson and Siegel 2013; Churilova et al. 2017). The mean of the S_{CDOM} was 0.013 ± 0.003 nm^{-1}. The S_{CDOM} values obtained in our studies in the BS and PB regions agree with the data in the northwestern Antarctic Peninsula (Ferreira et al. 2018). The S_{CDOM} values in the Antarctic waters were in the lower range of data for the open ocean (0.015–0.030 nm^{-1}) (Nelson and Siegel 2013) and in the European coastal waters 0.014–0.020 nm^{-1} (Babin et al. 2003).

The relative contribution of all optical components to the total (without water) light absorption at 440 nm was assessed in the upper layer whose thickness was assumed equal to the first optical depth (Fig. 16.6). The contribution of CDOM to the total light absorption varied from 3 to 92% (on average 45 ± 26%), and phytoplankton varied from 6 to 87% (on average 42 ± 21%). The NAP contribution was the smallest: from 2 to 44%, (on average 13 ± 8%). The absorption budget at 440 nm showed that the contribution of optically active components to light absorption in the first optical depth varied significantly and CDOM and phytoplankton were the main optically active components.

16.4 Conclusions

In this research, we examined variability and parameterized the spectral absorption coefficients of optically significant components in Antarctic waters (Bransfield Strait and Powell Basin) based on the in situ dataset in austral summer 2020. Analysis showed variability of more than an order of magnitude of the chlorophyll *a* concentration, spectral light absorption coefficients of phytoplankton, NAP, and CDOM. The results of parameterization of light absorption by all optically significant components in waters will provide the required scientific basis for the adjustment of the bio-optical algorithm for assessment of chlorophyll *a* and colored detrital matter absorption (Suslin and Churilova 2016) to the peculiarities of the bio-optical properties of the Antarctic waters (Bransfield Strait and Powell Basin).

Acknowledgments This work was carried out as part of state assignments of the IBSS on the topic "Study of the spatiotemporal organization of aquatic and terrestrial ecosystems in order to develop an operational monitoring system based on remote sensing data and GIS technologies" (state registration No. AAAA-A19-119061190081-9) and on the topic "Comprehensive studies of the current state of the ecosystem of the Atlantic sector of Antarctica" (state registration No. AAAA-A19-119100290162-0), as well as with the financial support of state assignment 0128-2019-0008. The analysis of optical data was carried out with the financial support of the RFBR grant no. 19-55-45024 IND-a.

References

Atkinson A, Siegel V, Pakhomov E, Rothery P (2004) Long-term decline in krill stock and increase in salps within the Southern Ocean. Nature 432:100–103

Babin M, Stramski D, Ferrari GM, Claustre H, Bricaud A, Obolensky G, Hoepffner N (2003) Variations in the light absorption coefficients of phytoplankton, non-algal particles, and dissolved organic matter in coastal waters around Europe. J Geophys Res 108(C7)

Boss E, Haëntjens N, Ackleson SG, Balch B, Chase A, Dall'Olmo G, Freeman S, Liu Y, Loftin J, Neary W, Nelson N (2019) IOCCG ocean optics and biogeochemistry protocols for satellite ocean colour sensor validation inherent optical property measurements and protocols: best practices for the collection and processing of ship-based underway flow-through optical data (v4. 0)

Bricaud A, Morel A, Prieur L (1981) Absorption by dissolved organic matter of the sea (yellow substance) in the UV and visible domains. Limnol Oceanogr 26:43–53

Bricaud A, Babin M, Morel A, Claustre H (1995) Variability in the chlorophyll-specific absorption coefficients of natural phytoplankton: analysis and parameterization. J Geophys Res Oceans 100(C7):13321–13332. https://doi.org/10.1029/95JC00463

Bricaud A, Morel A, Babin M, Allali K, Claustre H (1998) Variations of light absorption by suspended particles with chlorophyll a concentration in oceanic (case 1) waters: analysis and implications for bio-optical models. J Geophys Res 103:31033–31044

Churilova T, Moiseeva N, Efimova T, Suslin V, Krivenko O, Zemlianskaia E (2017) Annual variability in light absorption by particles and colored dissolved organic matter in coastal waters of Crimea (the Black Sea). 23rd International Symposium on Atmospheric and Ocean Optics: Atmospheric Physics 10466:104664B. https://doi.org/10.1117/12.2288339

Clarke A, Murphy EJ, Meredith MP, King JC, Peck LS, Barnes DK, Smith RC (2007) Climate change and the marine ecosystem of the western Antarctic peninsula. Philos Trans R Soc B Biol Sci 362:149–166

Cullen JJ, Lewis MR (1988) The kinetics of algal photoadaptation in the context of vertical mixing. J Plankton Res 10(5):1039–1063

Dierssen HM, Smith RC (2000) Bio-optical properties and remote sensing ocean color algorithms for Antarctic peninsula waters. J Geophys Res Oceans 105(11):26301–26312

Ducklow HW, Baker K, Martinson DG, Quetin LB, Ross RM, Smith RC, Stammerjohn SE, Vernet M, Fraser W (2007) Marine ecosystems: the West Antarctic peninsula. Philos Trans R Soc B Biol Sci 362:67–94

Efimova T, Churilova T, Moiseeva N, Zemlianskaia E, Dzhulay A, Krivenko O (2018) Dynamics in pigment concentration and light absorption by phytoplankton, non-algal particles and colored dissolved organic matter in the Black Sea coastal waters (near Sevastopol). In: Proceedings of

SPIE, 10833. 24th International Symposium on Atmospheric and Ocean Optics: Atmospheric Physics. 108336C. https://doi.org/10.1117/12.2504657

Ferreira A, Ciotti ÁM, Mendes CRB, Uitz J, Bricaud A (2017) Phytoplankton light absorption and the package effect in relation to photosynthetic and photoprotective pigments in the northern tip of Antarctic Peninsula. J Geophys Res Oceans 122(9):7344–7363

Ferreira A, Ciotti ÁM, Garcia CAE (2018) Bio-optical characterization of the northern Antarctic peninsula waters: absorption budget and insights on particulate backscattering. Deep-Sea Res II Top Stud Oceanogr 149:138–149

Figueiras FG, Arbones B, Estrada M (1999) Implications of bio-optical modeling of phytoplankton photosynthesis in Antarctic waters: further evidence of no light limitation in the Bransfield Strait. Limnol Oceanogr 44(7):1599–1608

Heidenreich KM, Richardson TL (2020) Photopigment, absorption, and growth responses of marine cryptophytes to varying spectral irradiance. J Phycol 56:507–520. https://doi.org/10.1111/jpy.12962

Jeffrey SW, Humphrey GF (1975) New spectrophotometric equations for determining chlorophylls a, b, c1 and c2 in higher plants, algae and natural phytoplankton. Biochem Physiol Pflanz 167(2):191–194. https://doi.org/10.1016/S0015-3796(17)30778-3

Jeffrey SW, Mantoura RFC, Wright SW (1997) Phytoplankton pigments in oceanography: guidelines to modern methods. UNESCO Publishing, Paris

Kerr R, Mata MM, Mendes CRB, Secchi ER (2018) Northern Antarctic peninsula: a marine climate hotspot of rapid changes on ecosystems and ocean dynamic. Deep-Sea Res II Top Stud Oceanogr 149:4–9

Kirk JTO (2011) Light and photosynthesis in aquatic ecosystems. Cambridge University Press, p 649

Kishino M, Takahashi N, Okami N, Ichimura S (1985) Estimation of the spectral absorption coefficients of phytoplankton in the sea. Bull Mar Sci 37:634–642

Lorenzen CJ (1967) Determination of chlorophyll and pheo-pigments: spectrophotometric equations. Limnol Oceanogr 12:343–346

MacIntyre HL, Kana TM, Anning J, Geider R (2002) Photoacclimation of photosynthesis irradiance response curves and photosynthetic pigments in microalgae and cyanobacteria. J Phycol 38(1):17–38

Mitchell BG (1990, September) Algorithms for determining the absorption coefficient for aquatic particulates using the quantitative filter technique. Ocean Optics X 1302:137–148

Moore LR, Georicke R, Chisholm SW (1995) Comparative physiology of *Synechococcus* and *Prochlorococcus*: influence of light and temperature on growth, pigments, fluorescence and absorptive properties. Mar Ecol Prog Ser 116:259–275

Morel A, Bricaud A (1981) Theoretical results concerning light absorption in a discrete medium and application to specific absorption of phytoplankton. Deep-Sea Res 28(11):1375–1393

Morel A, Ahn Y-H (1991) Optics of heterotrophic nanoflagellates and ciliates: a tentative assessment of their scattering role in oceanic waters compared to those of bacterial and algal cells. J Mar Res 49:177–202

Nelson NB, Siegel DA (2013) The global distribution and dynamics of chromophoric dissolved organic matter. Annu Rev Mar Sci 5:447–476

Novarino G (2003) A companion to the identification of cryptomonad flagellates (Cryptophyceae = Cryptomonadea). Hydrobiologia 502:225–270

Roesler CS, Perry MJ (1995) In situ phytoplankton absorption, fluorescence emission, and particulate backscattering spectra determined from reflectance. J Geophys Res 100:13279–13294

Stramski D, Kiefer DA (1998) Can heterotrophic bacteria be important to marine light absorption? J Plankton Res 20:1489–1500

Suslin V, Churilova T (2016) The Black Sea regional algorithm of separation of light absorption by phytoplankton and colored detrital matter using ocean color scanner's bands from 480-560 nm. Int J Remote Sens 37(18):4380–4400. https://doi.org/10.1080/01431161.2016.1211350

Chapter 17
Variability of Seawater Optical Properties in the Adjacent Water Basins of the Antarctic Peninsula in January and February 2020

Alexandr A. Latushkin, Vladimir A. Artemiev, Anton V. Garmashov, Pavel A. Salyuk, Inna V. Sahling, and Dmitry I. Glukhovets

Abstract Results of field bio-optical studies carried out in the South Atlantic (Falkland (Malvinas) Current) and in the northern part of the Antarctic Peninsula (Bransfield Strait and Antarctic Sound) are presented. It was found that bio-optical parameters in the main stream of the Falkland (Malvinas) Current are lower than in the surrounding waters. Upwelling was observed in the western periphery of the current, which contributes to the ascent of nutrients to the surface. Thus, intense development of phytoplankton exists in the surface layer, which is the main component in the suspended matter composition. A high correlation between remote sensing and field data was revealed in this region. Based on the results of research in the Bransfield Strait, the presence of two streams with different hydrological and bio-optical characteristics was found. Waters rich in nutrients were found in the northeastern part of the strait along the South Shetland Islands. Less productive waters from the Powell Basin were revealed in the eastern part of the strait. No influence of the alongshore current from the Drake Passage on bio-optical structure was observed at depths exceeding 100 m in the Bransfield Strait. The Antarctic Sound was characterized by the presence of clearer water. Only in the 50 m surface layer, were areas with different hydrological and bio-optical characteristics from the main water mass. Presumably, such a pattern of the subsurface layer was associated with the inflow of more turbid waters from the Bransfield Strait, caused by wind forcing.

A. A. Latushkin (✉) · A. V. Garmashov
Marine Hydrophysical Institute, Russian Academy of Science, Sevastopol, Russia

V. A. Artemiev · I. V. Sahling · D. I. Glukhovets
Shirshov Institute of Oceanology, Russian Academy of Sciences, Moscow, Russia

P. A. Salyuk
V.I. Ilichev Pacific Oceanological Institute, Far Eastern Branch, Russian Academy of Science, Vladivostok, Russia

E. G. Morozov et al. (eds.), *Antarctic Peninsula Region of the Southern Ocean*, Advances in Polar Ecology 6, https://doi.org/10.1007/978-3-030-78927-5_17

Keywords Bio-optics · Light beam attenuation coefficient · Fluorescence · Chlorophyll-a · Total suspended matter · Salinity · Temperature · Antarctic Peninsula · Southern Ocean · Weddell Sea · Bransfield Strait · Antarctic Sound

17.1 Introduction

At present much attention is focused on the research of variability of bio-optical characteristics, as indirect indicators of the ecological state of waters (Antoine et al. 2011; Schallenberg et al. 2019; Kerkar et al. 2020; Dias et al. 2020). These research studies are generally based on satellite data. Contact measurements of bio-optical characteristics in the seas and oceans are not numerous, and they are almost absent in the Antarctic. However, solution of the problems, related to the estimates of the current condition of Antarctic waters and predicting their state, is urgent today because the Antarctic ecosystem is most sensitive to natural climatic changes and to the anthropogenic impact. It is also known that the bio-optical properties of Antarctic waters differ significantly from those at midlatitudes (Mitchell and Holm-Hansen 1991; Dierssen and Smith 2000; Ferreira et al. 2018; Duarte et al. 2018). Therefore, remote sensing data collected with the help of a standard ocean color algorithm, probably, are not entirely correct when applied to the Southern Ocean, and this is the reason why field bio-optical data collection and accumulation for further verification and correction of remote sensing data is of high interest.

Conducting comprehensive hydrological and bio-optical researches is important because, first, optical measurements combined with hydrological data make it possible to get an idea of the peculiarities of formation of large-scale hydro-optical structure and its relation to the thermohaline structure of Antarctic waters and, second, synchronous measurements of the vertical structure of hydrological and hydro-optical fields in the coastal waters can provide unique information about the fine structure of these parameters in ocean active layer, which, in its turn, will make it possible to evaluate the relationship between these parameters and to reveal the spatial peculiarities of the vertical hydro-optical structure depending on hydrological processes.

Within the field research program of cruise 79 of the R/V "Akademik Mstislav Keldysh," in January–February 2020, integrated field studies were carried out in the subsurface layer (0–200 m) of the southern sector of the Atlantic Ocean (the Falkland (Malvinas) Islands) and in the northeastern part of the Antarctic Peninsula (Bransfield Strait and Antarctic Sound), including measurements of bio-optical and hydrological parameters. Locations of the stations are shown in Fig. 17.1.

The goal of our research is to identify the features of the formation of the spatial (vertical and horizontal) structure of seawater bio-optical fields in the adjacent waters of the Antarctic Peninsula in the summer season.

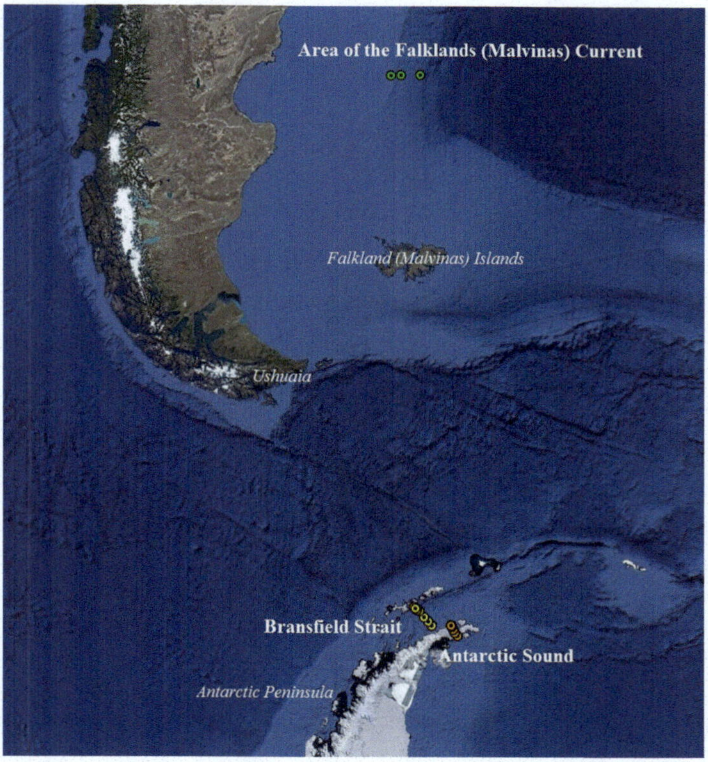

Fig. 17.1 Map of stations

17.2 Data and Methods

Hydrological parameters were measured using the Sea-Bird SBE-9p instrument. The following equipment were used for bio-optical measurements:

- An autonomous measuring complex PUM-200, which included a beam attenuation meter (Artemiev et al. 2008) and a chlorophyll-a fluorescence meter Minitracka II (Chelsea) (light radiation at 430 nm and recording at 685 nm). Measurements of the beam attenuation coefficient (BAC) were performed in the green spectrum range at a wavelength of 525 nm.
- An autonomous spectral meter of beam attenuation (BAM) (Latushkin and Martynov 2016). The measurements were carried out in the red spectrum range at a wavelength of 625 nm with a data readout frequency of 4 Hz.

The maps of remote measurements of chlorophyll-a concentration and particle backscattering coefficient (bbp) from the MODIS-Aqua and MODIS-Terra scanners were also used in the work. We used level 2 satellite data from ocean color scanners, available at http://oceancolor.gsfc.nasa.gov. The data were averaged over 16 satellite files using software developed at the Shirshov Institute of Oceanology of the Russian Academy of Sciences (SIO RAS) (Sheberstov 2015). A standard NASA algorithm was used to calculate the chlorophyll concentration; bbp was calculated using a SIO RAS algorithm (Burenkov and Sheberstov 2013).

17.3 Results

17.3.1 Region of the Falkland (Malvinas) Current

A latitudinal section of three stations along 46° S was made (Fig. 17.2) across the Falkland (Malvinas) Current; two stations were in the current and one on its western periphery in the shelf zone.

In the middle of the Falkland Current (Fig. 17.3, station 6575) in the subsurface layer, relatively clear waters were observed with chlorophyll-a (*Chl*) concentrations of 0.3–0.5μg/L (chlorophyll-a fluorescence – 0.3 r.u.), and values of the beam attenuation rate in the green spectrum area were ~0.5 m^{-1} as compared to the surrounding areas, where the values of chlorophyll-a concentrations exceeded 1μg/l and the values of the beam attenuation coefficient rate at 525 nm were greater than 1.5 m^{-1}, respectively. The water in the 10 m surface layer was warmer than the underlying layers. The water temperature reached 12.2 °C. The thermocline was located at a depth of 40 m. Development of phytoplankton cells in the main branch of the current occurred deeper than 25 m at several jumps in water density up to the depths of about 100 m, which is the main factor determining the variability of bio-optical characteristics. The presence of phytoplankton at such depths is possible due to the

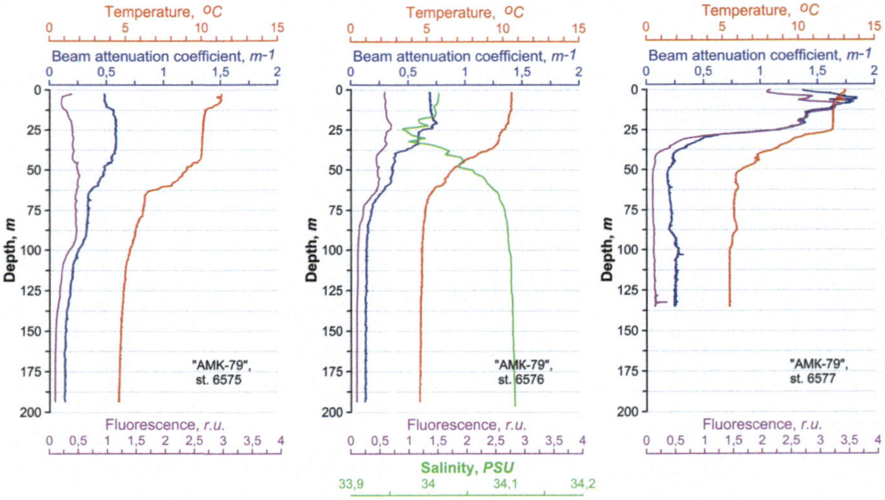

Fig. 17.2 Distributions of chlorophyll-a (Chl) concentration and particle backscattering coefficient (bbp) on 12.01.2020 based on the MODIS-Aqua and MODIS-Terra ocean color scanners

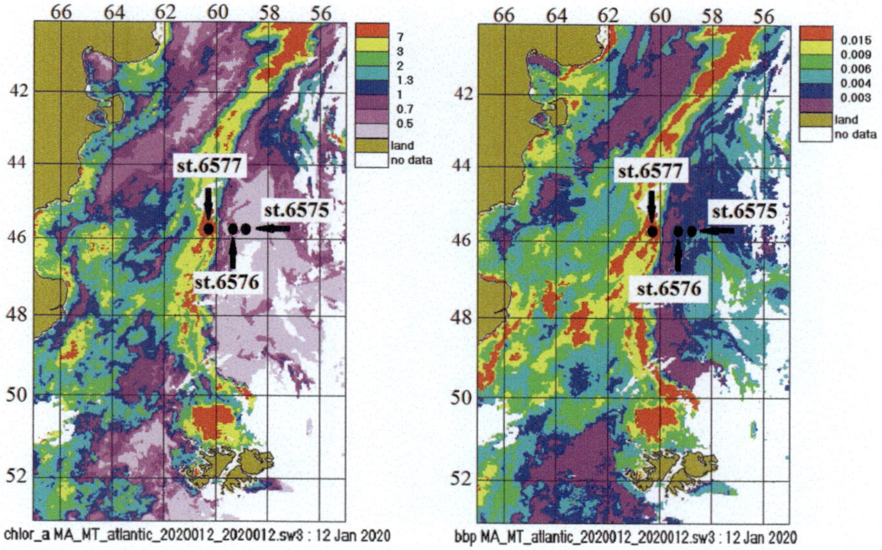

Fig. 17.3 Vertical distributions of hydrological and hydro-optical parameters along the section across the Falkland (Malvinas) Current

high transparency of water, which determines the deep occurrence of the euphotic zone (1% PAR). In accordance with the deepening of phytoplankton communities, they are poorly recorded with the help of satellite sounding of the sea color.

Ascent of phytoplankton to the surface and an increase in its concentration were observed from the center of the current to its periphery. A similar variation was observed in the distribution of the beam attenuation coefficient rate. Thus, at station 6576 (Fig. 17.3), which was located west of the center of the current, the highest phytoplankton content and higher BAC values were noted in the surface layer. Here, the phytoplankton concentration increased up to 2μg/l (fluorescence of chlorophyll-a was 0.6 relative units (r.u.) and the values of the beam attenuation rate at 525 nm were up to 0.75 m^{-1}).

On the western edge of the coastal branch of the current, near the shelf zone (Fig. 17.3, station 6577), an intense ascent of waters was observed (the thermocline was located at 25 m), which led to an increase in nutrients and to the corresponding development of phytoplankton communities. The measured bio-optical parameters in the surface layer increased sharply: the beam attenuation coefficient rate was two times greater (1.5 m^{-1}); the concentration of chlorophyll-a increased by more than three times up to 6–8μg/l. The maximum values were noted at 8–10 m, which intensifies the contribution of phytoplankton to the formation of the sea brightness spectrum and is interpreted as an additional increase in the remotely defined chlorophyll-a concentration.

At stations 6576 and 6577, there was a high correlation in the vertical distribution between the beam attenuation coefficient and chlorophyll-a fluorescence throughout the entire measured layer, which makes it possible to conclude that the main component of suspended matter is phytoplankton.

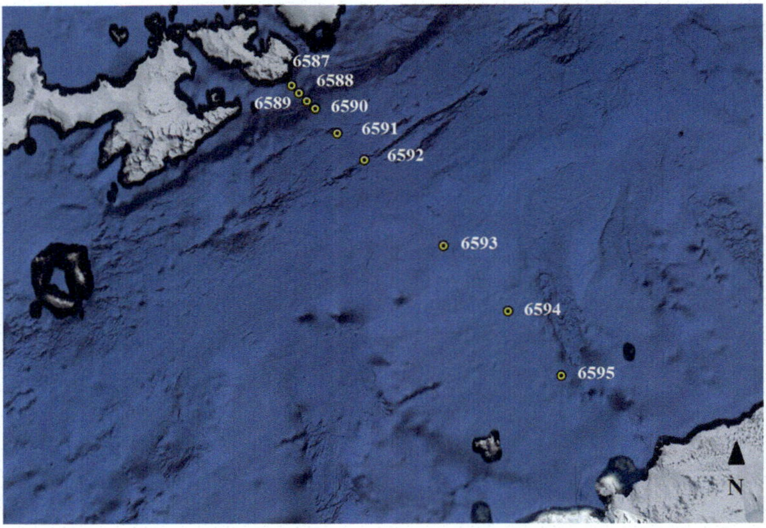

Fig. 17.4 Map of the location of stations in the Bransfield Strait

During the work on the section, scattered cloudy weather was observed, which made it possible to obtain satellite images gathering data on *Chl a* and b_{bp} that characterize the distribution of suspended matter.

As shown in Fig. 17.4, there was a high correlation in the spatial distribution of the concentration of *Chl a* and b_{bp}, which confirms the assumption that the composition of suspended matter here is determined by phytoplankton. The in situ data agree well with remote measurements.

17.3.2 Bransfield Strait

In the Bransfield Strait, measurements were made over a section of nine stations normal to the shore. The section was located between the South Shetland Islands and the Antarctic Peninsula (Fig. 17.4).

The maximum values of all measured bio-optical parameters were observed in the northwestern part of the strait, near the South Shetland Islands, in the surface layer (0–60 m) (Fig. 17.5). Here, the values of chlorophyll-a concentration reached 0.37μg/l (fluorescence of chlorophyll-a was 0.5 r.u.), and the beam attenuation coefficient reached 0.76 m^{-1} in the green (525 nm) and 0.95 m^{-1} in the red (625 nm) ranges of the spectrum. A trend for gradual decrease in the surface distribution of the beam attenuation coefficient at 525 nm and 625 nm was observed along the section toward the Antarctic Peninsula. Waters with higher values of the beam attenuation coefficient were found approximately in the middle of the strait. In the southeastern part of the strait, the water structure was more homogeneous with a

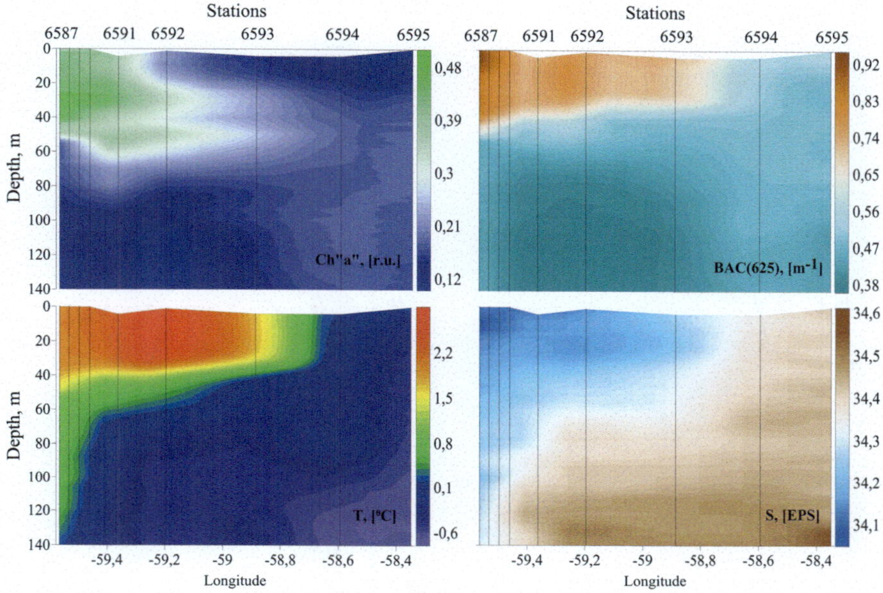

Fig. 17.5 Hydrological and bio-optical sections in the southeastern part (Bransfield Strait)

low beam attenuation coefficient (0.25–0.3 m^{-1} at 525 nm and 0.47–0.52 m^{-1} at 625 nm, respectively). In the vertical distribution, the turbid waters of the northwestern part of the section reached the pycnocline (~40 m), and below the pycnocline, the beam attenuation coefficient decreased sharply to the minimum values (0.13 m^{-1} at 525 nm and 0.36 m^{-1} at 625 nm, respectively). The highest content of chlorophyll-a also was observed off the shore of the South Shetland Islands; the maximum values of chlorophyll-a fluorescence were observed in the pycnocline. In contrast to the distribution of the attenuation coefficient in the surface layer (0–20 m), the concentration of chlorophyll-a decreased more significantly along the section in the southern direction. However, closer to the middle of the strait in the vertical distribution of chlorophyll-a, the layered structure was found (Fig. 17.6, stations 6591 and 6592) characterized by low values in the 0–20 m layer and the presence of two peaks at depths of ~35 m and 50 m. That type of hydro-optical structure in this area is explained by the motion of waters and changes in the species composition of phytoplankton. Waters from the Drake Passage rich in nutrients were found in the northwestern part of the strait, along the South Shetland Islands; in the southeastern part of the strait, less productive waters from the Powell Basin were recorded. The two water masses observed in the Bransfield Strait, different in hydro-optical features, are also different in the hydrological structure. The waters of the Drake Passage, in relation to the waters of the Powell Basin, are less saline but warmer.

Below 100 m, in the middle part of the strait, the beam attenuation coefficient, fluorescence, and temperature reached the level of quasi-permanent values, which

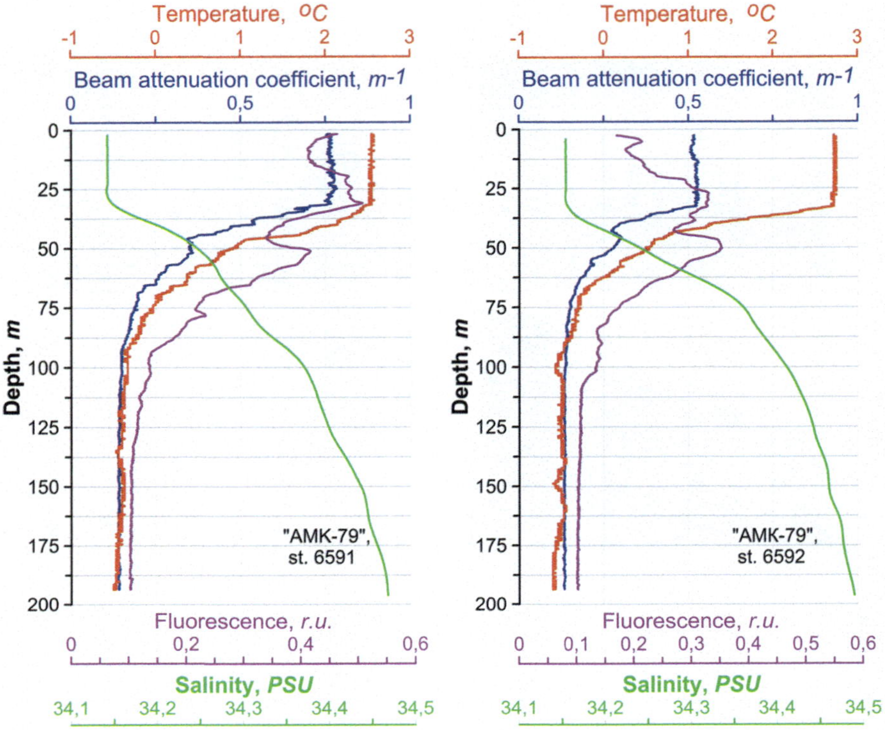

Fig. 17.6 Vertical distributions of hydrological and bio-optical parameters at stations 6591 and 6592

suggests that at these depths the influence of the alongshore current from the Drake Passage on the hydrological structure of waters was almost absent.

17.3.3 Antarctic Sound

Field measurements were carried out at five stations along the Antarctic Sound (Fig. 17.7).

In the Antarctic Sound, the main variability of bio-optical and hydrological parameters was traced only in the upper 50 m layer, where their distributions were almost identical (Fig. 17.8). The patchiness in the distributions of the beam attenuation coefficient and fluorescence of chlorophyll-a in this layer is associated with the changes in salinity and temperature: increased values of bio-optical values were observed in less saline waters with temperature above 0 °C; at stations 6623 and 6625, the values of the attenuation coefficient at 525 nm were 0.45 and 0.38 m^{-1}; the fluorescence of chlorophyll-a was 0.3 and 0.27 r.u., respectively. It can be assumed that such a distribution in the subsurface layer was associated with the water inflow

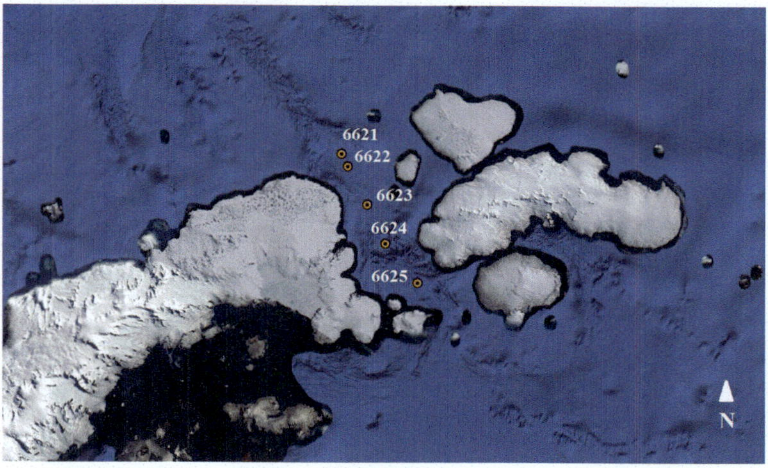

Fig. 17.7 Map of stations in the Antarctic Sound

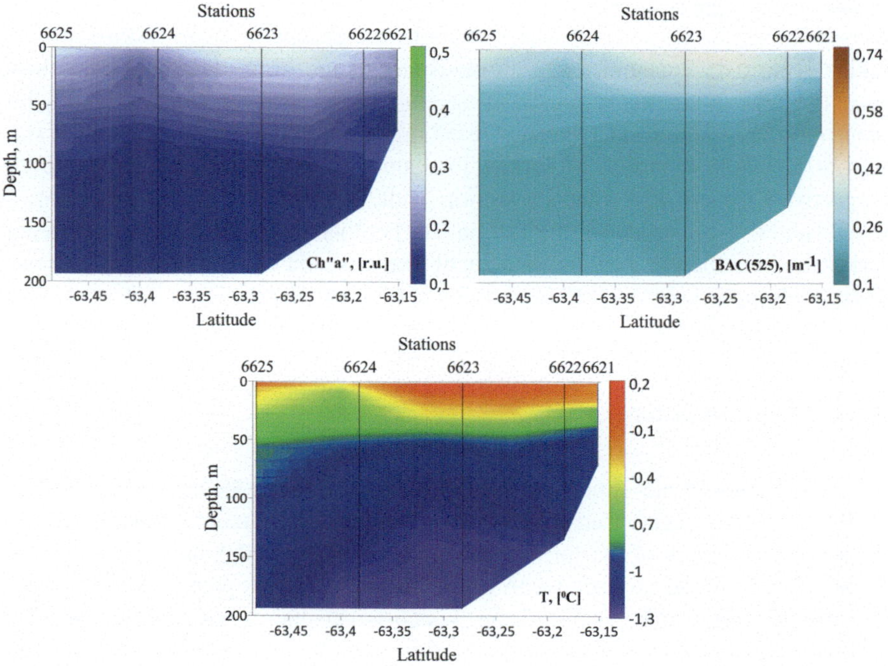

Fig. 17.8 Hydrological and bio-optical sections (Antarctic Sound)

from the Bransfield Strait, caused by wind forcing. The values of the measured quantities in the adjacent to the Antarctic Peninsula waters of this strait are the same. In general, water in the strait was clear. In the surface layer, the beam attenuation coefficient changed insignificantly: from 0.32 to 0.45 m^{-1}. The maximum

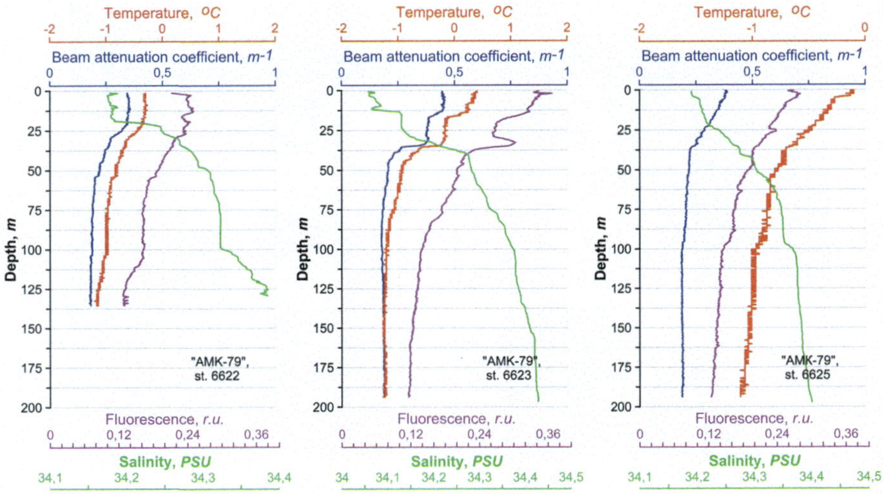

Fig. 17.9 Vertical distributions of hydrological and bio-optical parameters (Antarctic Sound). Stations are indicated in the figure

values were 0.45 at station 6623, in the center of the strait; the minimum values were 0.31 m^{-1} at the first station of the section (6621). The values of chlorophyll-a concentrations in the surface layer were 0.71 μg/l at station 6622 and 0.76μg/l at station 6625, while at depths of ~60 m, the values decreased to 0.41 and 0.34μg/l, respectively. Figure 17.9 shows a section of the vertical distributions of the measured values at the stations of the section. Deeper than 70 m, the values of the beam attenuation coefficient were the same at all stations, and close to 0.2 m^{-1}, the temperature dropped below −1 °C; salinity increased with depth.

17.4 Conclusions

1. Lower values of bio-optical parameters were recorded in the main stream of the Falkland (Malvinas) Current compared with the surrounding waters. On the western periphery of the current, upwelling was observed, which caused ascent of nutrients to the surface, due to which the intense development of phytoplankton took place in the surface layer, which is the main component of the suspended matter composition in the region. A high correlation between remote sensing and field data was found.
2. Two streams with different bio-optical and hydrological properties were identified in the Bransfield Strait. Waters from the Drake Passage rich in nutrients flow in the northwestern part of the strait, along the South Shetland Islands, while less productive waters from the Powell Basin were found in the southeastern part of the strait. No influence of the alongshore current from the Drake Passage on the

bio-optical structure was observed at depths exceeding 100 m in the Bransfield Strait.

3. The Antarctic Sound was characterized by the presence of clearer water in comparison with the Bransfield Strait. Only in the 50 m surface layer, waters with different hydrological and bio-optical characteristics from the main water mass of the strait were observed. Presumably, such a distribution in the subsurface layer is associated with the inflow of more turbid waters from the Bransfield Strait, caused by wind forcing.

Acknowledgments This work was carried out within the framework of state orders: Marine Hydrophysical Institute of the RAS No.0555-2019-0003 "Experimental studies of the variability of hydrophysical, hydrochemical and bio-optical fields at different spatiotemporal scales for identification of the peculiarities of climatic changes in oceanographic conditions in the Atlantic part of Antarctic"; IO RAS, No. 0128-2019-0008 "Assessment of the current state of natural complexes of the Atlantic sector of the Southern Ocean and their different periods of variability (ecosystems, bioproductivity, hydrophysics, hydro- and geochemistry)"; and POI FEB RAS, No. FWMM-2019-0007 "Integrated Environmental Studies of the Southern Ocean."

References

Antoine D, Siegel DA, Kostadinov T, Maritorena S, Nelson NB, Gentili B et al (2011) Variability in optical particle backscattering in contrasting bio-optical oceanic regimes. Limnol Oceanogr 56:955–973. https://doi.org/10.4319/lo.2011.56.3.0955

Artemiev AA et al. Transmissometer for sea water (in Russian). Patent RU 2341786 on 20.12.2008. Int.Cl. G01N 21/59

Burenkov VI, Sheberstov SV (2013) Modified algorithm for assessment of the particle backscattering coefficient from satellite data. Current problems in optics of natural waters:151–153

Dias A, Kurian S, Thyapurath S (2020) Influence of environmental parameters on bio-optical characteristics of colored dissolved organic matter in a complex tropical coastal and estuarine region. Estuar Coast Shelf Sci:106864

Dierssen HM, Smith RC (2000) Bio-optical properties and remote sensing ocean color algorithms for Antarctic Peninsula waters. J Geophys Res 105(11):26, 301–26, 312. https://doi.org/10.1029/1999JC000296

Duarte B, Cabrita MT, Vidal T, Pereira J, Pacheco M, Pereira P, Canario J, Goncalves F, Matos A, Rosa R, Marques J, Caçador I, Gameiro C (2018) Phytoplankton community-level bio-optical assessment in a naturally mercury contaminated Antarctic ecosystem (Deception Island). Mar Environ Res 140:412–421. https://doi.org/10.1016/j.marenvres.2018.07.014

Ferreira A, Ciottia AM, Garcia CAE (2018) Bio-optical characterization of the northern Antarctic Peninsula waters: Absorption budget and insights on particulate backscattering. Deep-Sea Res II Top Stud Oceanogr 149:138–149. https://doi.org/10.1016/j.dsr2.2017.09.007

Kerkar AU, Tripathy SC, Minu P et al (2020) Variability in primary productivity and bio-optical properties in the Indian sector of the Southern Ocean during an austral summer. Polar Biol 43:1469–1492. https://doi.org/10.1007/s00300-020-02722-2

Latushkin AA, Martynov OV. Method for determining the spectral attenuation of direct light in seawater in situ (in Russian). Patent RU 2605640 on 27.12.2016. Int.Cl. G01N2159

Mitchell BG, Holm-Hansen O (1991) Bio-optical properties of Antarctic Peninsula waters: differentiation from temperate ocean models. Deep Sea Res Part A Oceanogr Res Pap 38(8–9):1009–1028. https://doi.org/10.1016/0198-0149(91)90094-V

Schallenberg C, Harley JW, Jansen P, Davies DM, Trull TW (2019) Multi-year observations of fluorescence and backscatter at the Southern Ocean Time Series (SOTS) shed light on two distinct seasonal bio-optical regimes. Front Mar Sci 6:595. https://doi.org/10.3389/fmars.2019.00595
Sheberstov SV (2015) A system of batch processing of oceanological satellite data (in Russian). Curr Prob Remote Sens Earth Space 12(6):154–161

Chapter 18
Bio-Optical Models for Estimating Euphotic Zone Depth in the Western Atlantic Sector of the Southern Ocean in the Antarctic Summer

Pavel A. Salyuk, Vladimir A. Artemiev, Dmitry I. Glukhovets, Alexander N. Khrapko, Anatoly V. Grigoriev, Alexandr A. Latushkin, and Nadezda D. Romanova

Abstract The goal of this work was to develop two types of regional models for the western part of the Atlantic sector of the Southern Ocean in the Antarctic summer period to assess the photosynthetically active radiation (PAR) profile and the depth of the euphotic layer. We used the data of underway flow-through optical measurements and ocean color data, which could be measured remotely.

The measured values of the euphotic zone depth were in the range of 30–90 m. In some areas with a significant development of phycoerythrin-containing algae in the upper layer of the sea, the depth of the euphotic zone depth decreased to 30–45 m. Because of the pronounced stratification in the distribution of the light attenuation coefficient, with a maximum in the upper 20–30 m layer, the PAR attenuation coefficient changes with depth.

The data of calibrated flow-through laser fluorimeter and a transparency meter were used to estimate the euphotic zone depth from the surface measurements. The best results in assessing the euphotic zone depth were provided by in situ reflectance measurements, which could be compared to those remotely sensed; we also used a model to apply the satellite ocean color data.

P. A. Salyuk (✉)
V.I. Ilichev Pacific Oceanological Institute, Far Eastern Branch, Russian Academy of Science, Vladivostok, Russia

V. A. Artemiev · D. I. Glukhovets · A. N. Khrapko · A. V. Grigoriev · N. D. Romanova
Shirshov Institute of Oceanology, Russian Academy of Sciences, Moscow, Russia

A. A. Latushkin
Marine Hydrophysical Institute, Russian Academy of Science, Sevastopol, Russia

E. G. Morozov et al. (eds.), *Antarctic Peninsula Region of the Southern Ocean*, Advances in Polar Ecology 6, https://doi.org/10.1007/978-3-030-78927-5_18

Keywords Photosynthetically active radiation · Euphotic zone depth · Model · Chlorophyll-a · Light attenuation coefficient · Remote sensing reflectance · Ocean color · Southern Ocean · Antarctica

18.1 Introduction

The photosynthetically active radiation (PAR) level (Mitchell et al. 1991; Tilzer et al. 1994) and iron content (Behrenfeld and Kolber 1999) are the main limiting factors for the development of phytoplankton cells in the Southern Ocean, despite the fact that usually, the concentration of nutrients is sufficient in this region (Minas and Minas 1992). Express estimation of the vertical distribution of PAR is an important task that is necessary for the correct choice of sampling depths. In addition, it is needed for primary production (PP) incubation experiments, as well as for calculating the integral PP in the seawater column, both based on contact and remote sensing measurements.

The practice of marine field research shows that part of the vertical sampling of seawater must be performed at night for the optimal use of the expedition time when direct in situ measurements of the PAR are not available. Therefore, it is necessary to develop such models that make it possible to quickly calculate the PAR profile and estimate various depths of light penetration on the basis of alternative underwater in situ data or near-surface measurements.

In addition, an important task is to assess the PAR profile to calculate integral PP values in the water column only from the data of near-surface measurements or from the remote sensing data. Corresponding models make it possible to estimate different light penetration depths during underway shipboard measurements or from satellite ocean color data, which is necessary for the correct work of satellite bio-optical models to calculate the PP (Soppa et al. 2013).

Thus, the aim of this work was to develop two types of regional models for the western part of the Atlantic sector of the Southern Ocean to assess the PAR profile and the depth of the euphotic layer (Z_{eu}): (1) from the data of optical measurements in the flow-through system while the vessel is underway, (2) from the ocean color data, which could be remotely measured. The importance of this aim arises from the fact that in the study site (especially in the Weddell Sea), the amount of satellite data is limited due to frequent cloudiness and ice cover. Moreover, the peculiarities of regional circulation prevent use of BioARGO buoys in the Weddell Sea.

18.2 Research Region

The optical characteristics of the western part of the Atlantic sector of the Southern Ocean are determined mainly by the functioning of phytoplankton. The contribution of suspended matter (SM) and colored dissolved organic matter (CDOM) is relatively small, as it in the entire Southern Ocean (Tilzer et al. 1994; Szeto et al. 2011; Ortega-Retuerta et al. 2010; Kieber et al. 2009). However, two different cases can be distinguished in the phytoplankton communities of the study site based on our measurements: (1) only chlorophyll-a (chl-a)–containing algae; (2) chl-a- and phycoerythrin (phy)–containing algae. The relatively small contribution of SM and CDOM to the total light-attenuation coefficient compared to the pigments of phytoplankton and differences in the pigment composition of phytoplankton communities are the main reasons that determine the regional features of the models in the process of their development.

18.3 Measurements and Methods

The PAR values were determined using the LI-COR device (https://www.licor.com/env/products/light/quantum_underwater.html) consisting of two hermetic photodiode sensors LI-192 for measuring upwelling and downwelling underwater irradiance and a deck sensor LI-190SA for monitoring the level of incident surface radiation.

We measured the beam-attenuation coefficient, temperature, and Chl fluorescence intensity using a portable transparency meter (Artemiev et al. 2011) connected with a Chelsea Minitracka II miniature fluorimeter. The transparency meter operates at 530 nm in the range of 0.050–1.0 m^{-1} with an error of 0.005 m^{-1}; the excitation wavelength of the fluorimeter is 430 nm and the recording one is 685 nm, while the measurement range is 0–100 μg/l.

We used a PRO-1 floating spectroradiometer developed at the SIO RAS (Artemiev et al. 2000) for measurements of the spectral upwelling radiance just beneath the sea surface. It also measures the spectral surface irradiance, which makes it possible to calculate the spectral remote sensing reflectance $R_{rs}(\lambda)$. The measurements are carried out at drift stations, and the device floats off the ship at a distance of about 50 m to avoid shading and reflectance from the ship. Depending on weather conditions, 20–50 spectral scans are averaged during processing, using a specially developed algorithm.

The operative chlorophyll-a concentration measurements were carried out by a flow-through laser hyperspectral fluorimeter that provides recording seawater fluorescence in the range of 400–750 nm with 1-nm resolution when excited at wavelengths of 355 and 532 nm (Nagorniy et al. 2014).

In addition, direct determination of the Chl concentration was carried out by the fluorometric method after filtration of seawater samples and extraction (JGOFS 1994) according to the formulas prescribed by Holm-Hansen and Riemann (Holm-Hansen and Riemann 1978). The laboratory spectrofluorimeter was calibrated using the spectrophotometric equipment according to the method described in the work of Demidov et al. (Demidov et al. 2019).

18.4 Results

18.4.1 Analysis of the Vertical Distribution of PAR and Calculation of Light Penetration Depths

Figure 18.1 shows an example of the vertical profile of PAR in the water column: initial values are in μmol/m²/s units (Fig. 18.1a) and normalized on a logarithmic scale (Fig.18.1b). Figure 18.1c also shows vertical profiles of light attenuation coefficient at 532 nm *c*(532) and chlorophyll-a fluorescence intensity at station 6604 (61° 05.98' S, 49° 41.22' W).

In the upper 5-m layer, measurements are highly variable due to the influence of the rough sea surface and the board of the vessel. In addition, in the upper 20–30 m, there is a greater amount of optically active components, which leads to a greater attenuation of the PAR in comparison with the underlying layers. In the case of unstratified sea, the PAR attenuation coefficient (K_{PAR}) would be constant, and we expected to observe a linear dependence on the logarithm of the PAR; however this was not the case. In order to take into account the increase in K_{PAR} in the upper layers, the following model was used, which is the sum of linear and an exponential dependence:

$$\text{Model «p1e1»}: \ln\left(PAR(z)/PAR(0)\right) = k_1 * \exp\left(-k_2 * z\right) + k_3 * z - k_1 \quad (18.1)$$

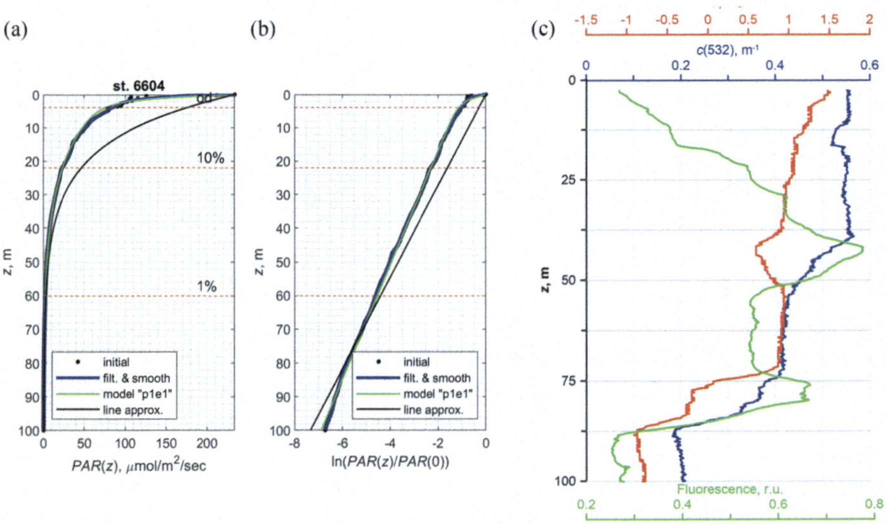

Fig. 18.1 Vertical profiles of PAR in μmol/m²/s (**a**); normalized and logarithmic PAR values (**b**); light attenuation coefficient *c*(532), temperature, and chlorophyll-a fluorescence intensity in the water column (**c**). Dotted red lines indicate characteristic light penetration depths

Coefficient k_1 at the exponent is the same as the constant term in this model. It is necessary for obligatory inclusion of the point [0, 0]. The presence of an exponent in Eq. (1) makes it possible to take into account a sharper decrease in PAR in the upper layers, which is apparently associated with a high content of phytoplankton. The approximation result is shown in Fig. 18.1a and b with a green line. Smoothed values of PAR and log (PAR) are shown in blue in Fig. 18.1a and b. The dashed red lines show the optical depth (decrease in PAR values in e^1 (or 2.72) times), in which the main formation of the ocean color occurs. The depth of 10% attenuation of surface PAR values is also shown, where there is still a statistically significant effect on the ocean color. Euphotic zone depth corresponding to 1% of surface PAR values is shown, where the majority of PP is produced. For comparison, linear approximation is shown with a black line. It would be the correct choice in the absence of stratification in the phytoplankton content in the water column, or if the influence of phytoplankton and other optically active components on the attenuation of the PAR is small.

The performance of the "p1e1" model for the euphotic zone depth (Z_{eu}) calculations is shown in Fig. 18.2a. The same, but for a linear approximation, is shown in Fig. 18.2b. These figures show that the use of a line to approximate the decrease in PAR on a logarithmic scale does not allow us to approximate the vertical profile of the PAR correctly and assess the euphotic zone depth, while Eq. (1) gives satisfactory results.

The results of the spatial distribution of the euphotic zone depth in the study site are shown in Fig. 18.3, where the blue arrows schematically show the Falkland (Malvinas) and Antarctic Circumpolar Current (ACC).

It can be seen that the values of Z_{eu} range from 30 to 90 m. The northernmost point at 46° S is retained in the dataset, because it was measured in the middle of the Falkland Current, which transports the ACC waters northward (Friocourt et al. 2005). Hence, by its origin, these waters belong to the study region.

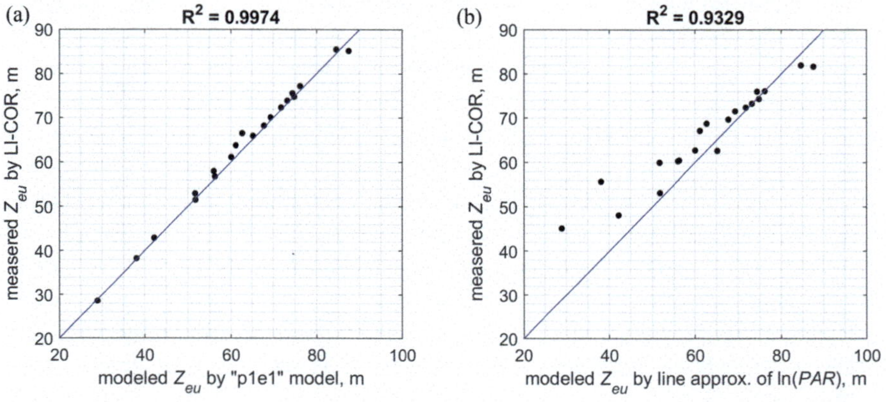

Fig. 18.2 Comparison of the measured and calculated euphotic zone depth based on the data: of the "p1e1" model (**a**); and using a line approximation of the PAR profile on a logarithmic scale (**b**)

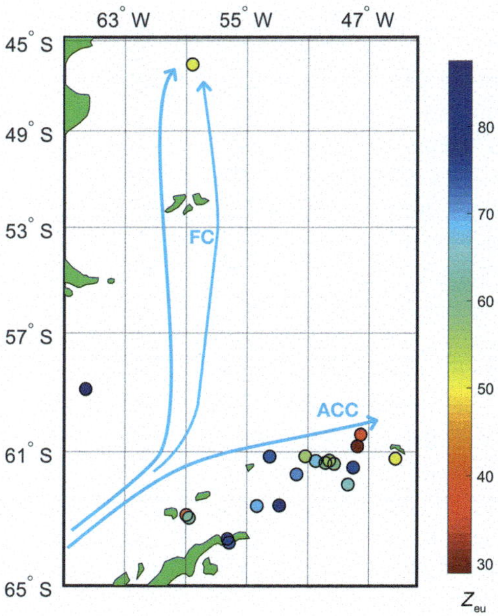

Fig. 18.3 Map of the distribution of euphotic zone depth measurements in the western Atlantic sector of the Southern Ocean. ACC and FC denote the Antarctic Circumpolar Current and Falkland (Malvinas) Current, respectively

18.4.2 Estimation of Euphotic-Zone Depth from the Surface and Ocean Color Data

Table 18.1 presents the results of estimating the euphotic zone depth from the surface measurements and $R_{rs}(\lambda)$ data. The first column contains a list of variables from which the Z_{eu} values were calculated. The second column presents the coefficient of determination R^2 relative to the 1:1 line between the measured and model values of Z_{eu}, which were calculated using linear regression "poly1" for one variable and "poly11" for two variables. The third column gives R^2 values calculated with the help of exponential model "exp1" for one variable and "exp11" for two variables:

$$\text{"poly1"}: Z_{eu}(x_1) = k_1 \cdot x_1 + k_2 \tag{18.2}$$

$$\text{"poly11"}: Z_{eu}(x_1,x_2) = k_1 \cdot x_1 + k_2 \cdot x_2 + k_3 \tag{18.3}$$

$$\text{"exp1"}: Z_{eu}(x_1) = k_1 \cdot \exp(k_2 \cdot x_1) + k_3 \tag{18.4}$$

$$\text{"exp11"}: Z_{eu}(x_1,x_2) = k_1 \cdot \exp(k_2 \cdot x_1 + k_3 \cdot x_3) + k_4 \tag{18.5}$$

Here is a list of input variables x_1, x_2: Chl_{FL} is chl-a concentration measured using a flow-through laser fluorimeter (in μg/L); Phy_{FL} is phycoerythrin fluorescence intensity measured by a flow-through laser fluorimeter (in Raman units, R.U.); $c_{surf}(532)$ is the seawater light-attenuation coefficient at 532 nm in the surface layer, measured with a PUM-200 transparency meter (in m^{-1}), and the ocean color index

Table 18.1 Results of euphotic zone depth calculation from various types of alternative measurements

Input parameters (x_1, x_2)	R^2 for "poly1" or "poly11", number of points	R^2 for "exp1" or "exp11", number of points
Chl_{FL}	0.58, n = 16 (0.76, n = 15)	0.59, n = 16 (0.82, n = 15)
Chl_{FL}, Phy_{FL}	0.64, n = 16 (0.78, n = 15)	0.66, n = 16 (0.82, n = 15)
$c_{surf}(532)$	0.73, n = 16 (0.77, n = 15)	0.83, n = 16 (0.86, n = 15)
$c_{surf}(532)$, Phy_{FL}	0.77, n = 16 (0.81, n = 15)	0.81, n = 16 (0.85, n = 15)
$Rrs(490)/Rrs(555)$	0.94, n = 11	0.93, n = 11
$c_{surf}(532)$ for R_{rs} selection[a]	0.78, n = 11	0.82, n = 11

[a]limited $c_{surf}(532)$ dataset was selected based on the radiometer data for correct comparison of the approximation results

$R_{rs}(490)/R_{rs}(555)$ is the blue-green ratio measured with a floating hyperspectral radiometer (dimensionless).

Figure 18.4 shows the results of the best models for a flow-through laser fluorimeter (Fig. 18.4a, b), a PUM-200 transparency meter (Fig. 18.4c), and a floating radiometer (Fig. 18.4d), respectively:

$$Z_{eu}\left(Chl_{FL}\right) = 105.4 \cdot \exp\left(-0.923 \cdot Chl_{FL}\right) + 20.3 \tag{18.6}$$

$$Z_{eu}\left(Chl_{FL}, Phy_{FL}\right) = 102.4 \cdot \exp\left(-0.233 \cdot Chl_{FL} - 0.226 * Phy_{FL}\right) - 12.5 \tag{18.7}$$

$$Z_{eu}\left(c_{surf}\left(532\right)\right) = 102.7 \cdot \exp\left(-2.407 \cdot c_{surf}\left(532\right)\right) + 29.1 \tag{18.8}$$

$$Z_{eu}\left(Rrs\left(490\right), Rrs\left(555\right)\right) = 25.63 \cdot Rrs\left(490\right) / Rrs\left(555\right) - 3.9 \tag{18.9}$$

Noteworthy is the fact that the use of the exponential model "exp1" or "exp11" usually leads to an improvement in the result of measurements using a flow-through fluorimeter and a transparency meter. This is related to the fact that the exponent partially takes into account the variability of the vertical K_{PAR} profile in the sea. At the same time, the use of the exponent in the measurements with a radiometer did not lead to an improvement of the result. This can be interpreted so that reflectance measurements in the study site contain information on the vertical distribution of optically active components within the optical depth, and partially take into account the K_{PAR} variability with depth.

The additional use of phycoerythrin fluorescence intensity in the estimation of euphotic zone depth resulted in improved linear models for the fluorimeter and transparency meter but did not produce a statistically significant improvement when using exponential fit. However, the only outlier in the Z_{eu} scatter plot and modeled Z_{eu} (Chl_{FL}), corresponding to the Drake Passage region, was partially corrected by additionally taking into account the phycoerythrin intensity in the Z_{eu} scatter plot and modeled Z_{eu} (Chl_{FL}, Phy_{FL}). The true value of Z_{eu} at the station in the Drake

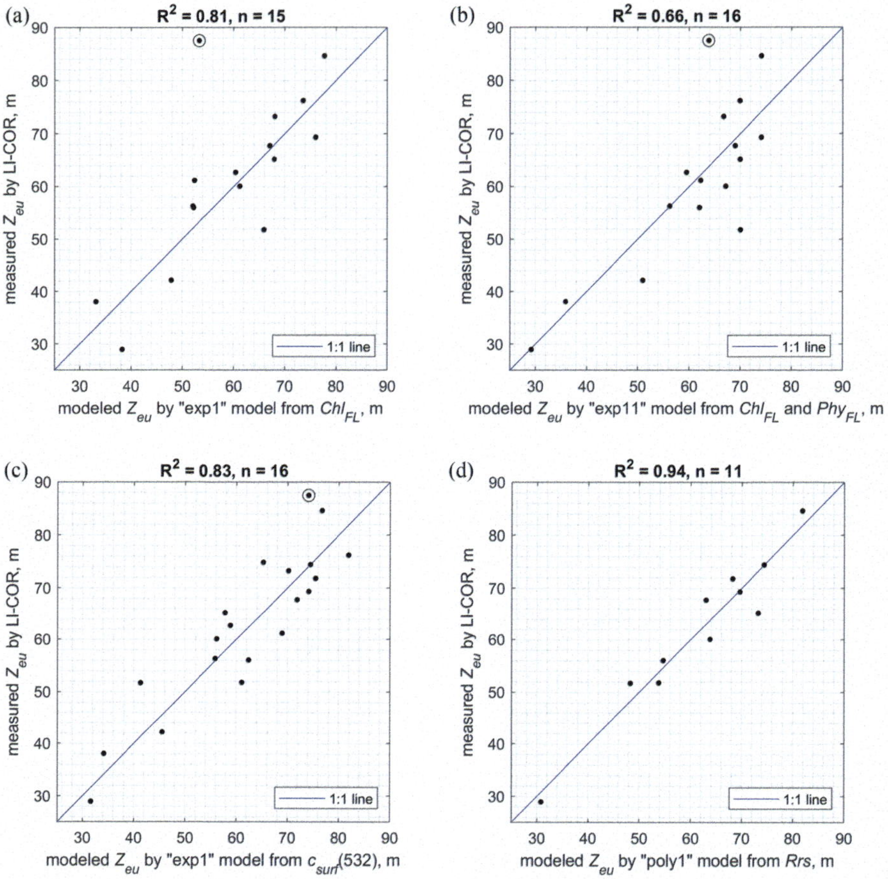

Fig. 18.4 Results of euphotic zone depth Z_{eu} estimation from surface measurements: chlorophyll-a concentration (**a**), chlorophyll-a concentration and phycoerythrin fluorescence (**b**), light attenuation coefficient at 532 nm (**c**), remotely sensed reflectance (**d**)

Passage was 87 m, and the model estimate of Z_{eu}, taking into account only the content of Chl-a, was equal to 54 m, while taking into account the content of Chl-a and phycoerythrin, the value changed to 64 m. In general, the improvement in the result of linear models with additional phycoerythrin accounting is explained by the fact that algae containing phycoerythrin are found in the upper layers of seawater. They significantly increase the K_{PAR} values in the upper layers (approximately at depths of 0–20 m) relative to underlying layers. The addition of an extra parameter allows us to partially consider this feature.

The best result for estimating the euphotic zone depth was obtained in radiometric ocean color measurements and calculating the color index $R_{rs}(490)/R_{rs}(555)$. Most likely, this is due to the fact that the sea's upwelling radiation is formed not at one depth, but at the range of depths, and this is sufficient to better estimate the Z_{eu} values.

18.5 Conclusions

The euphotic zone depths in the study site were in the range of 30–90 m. In some areas, with a significant development of phycoerythrin-containing algae in the upper layer of the sea, the depth of the euphotic zone decreased to 30–45 m. In the western part of the Atlantic sector of the Southern Ocean, in summer, due to the small amount of dissolved and suspended substances, light penetrates deep enough to allow phytoplankton communities to develop even at greater depths of about 100 m.

The PAR attenuation coefficient changes with depth due to the pronounced stratification in the distribution of the light-attenuation coefficient, the maximum of which is in the upper 20–30 m. Its vertical profile is well approximated by the sum of linear and exponential functions according to Eq. (1).

The data of a calibrated flow-through laser fluorimeter and a transparency meter give adequate results to estimate the euphotic zone depth from the surface measurements, which is seen from Eqs. (18.6) and (18.8), respectively. These results can be used during underway measurements, including at night, in rain, and in stormy weather.

The best results in assessing the euphotic zone depth were provided by in situ reflectance measurements, which can be obtained only during daylight hours under good weather conditions with Eq. (18.9). This type of measurement can be organized from the ship automatically. In addition, this model is applicable to the satellite ocean color data.

Acknowledgments This work was carried out within the framework of State Orders: POI FEB RAS No. FWMM-2019-0007 “Integrated Environmental Studies of the Southern Ocean”; SIO RAS No. 0128-2019-0008 “Assessment of the current state of natural complexes of the Atlantic sector of the Southern Ocean and their different periods of variability (ecosystems, bioproductivity, hydrophysics, hydro- and geochemistry)”; MHI of RAS No.0555-2019-0003 “Experimental studies of the variability of hydrophysical, hydrochemical and bio-optical fields at different spatio-temporal scales for identification of the peculiarities of climatic changes in oceanographic conditions in the Atlantic part of Antarctic”.

References

Artemiev VA, Burenkov VI, Vortman MI, Grigoriev AV, Kopelevich OV, Khrapko AN (2000) Sea-truth measurements of ocean color: a new floating spectroradiometer and its metrology. Oceanology 40:139–145

Artemiev VA, Grigoriev AV, Vortman MI (2011) Advanced meter of the seawater beam attenuation coefficient. Current problems in optics of natural waters (ONW`2011). St-Petersburg, Russia, September 8-11, Proceedings 173-178

Behrenfeld MJ, Kolber ZS (1999) Widespread Iron limitation of phytoplankton in the South Pacific Ocean. Science 283(5403):840–843. https://doi.org/10.1126/science.283.5403.840

Demidov AB, Mosharov SA, Gagarin VI et al (2019) Spatial variability of phytoplankton primary production characteristics in the North Atlantic in summer 2013. Oceanology 59(2):223–234. https://doi.org/10.1134/S000143701902005X

Friocourt Y, Drijfhout S, Blanke B, Speich S (2005) Water mass export from Drake Passage to the Atlantic, Indian, and Pacific oceans: a Lagrangian model analysis. J Phys Oceanogr 35(7):1206–1222

Holm-Hansen O, Riemann B (1978) Chlorophyll a determination: improvements in methodology. Oikos 30:438–447

JGOFS (Joint Global Ocean Flux Study Protocols) (1994) Protocols for the joint Global Ocean flux study protocols (JGOFS). Core Meas, Manual Guides 119–122

Kieber DJ, Toole DA, Kiene RP (2009) Chromophoric dissolved organic mattercycling during a Ross Sea Phaeocystis Antarcticabloom. In: Krupnik I, Lang MA, Miller SE (eds) Smithsonian at the poles: contributions to international polar year science – a Smithsonian contribution to knowledge. Smithsonian Institute Scholarly Press, Washington DC, pp 319–334

Le Traon PY, D'Ortenzio F, Babin M, Leymarie E, Marec C, Pouliquen S, Thierry V, Cabanes C, Claustre H, Desbruyères D, Lacour L (2020) Preparing the new phase of Argo: scientific achievements of the NAOS project. Front Mar Sci 7:838

Minas HJ, Minas M (1992) Net community production in "high nutrient-low chlorophyll" waters of the tropical and Antarctic Oceans: Grazing vs. iron hypothesis. Oceanol Acta 15:145–162

Mitchell BG, Brody EA, Holm-Hansen O, McClain C, Bishop J (1991) Light limitation of phytoplankton biomass and macronutrient utilization in the Southern Ocean. Limnol Oceanogr 36(8):1662–1677. https://doi.org/10.4319/lo.1991.36.8.1662

Nagornyi IG, Maior AY, Salyuk PA, Doroshenkov IM (2014) A mobile complex for on-line studying water areas and surface atmosphere. Instrum Exp Tech 57(1):68–71. https://doi.org/10.1134/S0020441214010175

Soppa MA, Dinter T, Taylor BB, Bracher A (2013) Satellite derived euphotic depth in the Southern Ocean: implications for primary production modelling. Remote Sens Environ 137:198–211

Szeto M, Werdell PJ, Moore TS, Campbell JW (2011) Are the world's oceans optically different? J Geophys Res 116(C7). https://doi.org/10.1029/2011JC007230

Tilzer MM, Gieskes WW, Heusel R, Fenton N (1994) The impact of phytoplankton on spectral water transparency in the Southern Ocean: implications for primary productivity. Polar Biol 14(2):127–136. https://doi.org/10.1007/bf00234975

Chapter 19
Phycoerythrin Pigment Distribution in the Upper Water Layer Across the Weddell-Scotia Confluence Zone and Drake Passage

Pavel A. Salyuk, Dmitry I. Glukhovets, Alexander Yu. Mayor, Natalia A. Moiseeva, Vladimir A. Artemiev, and Alexander N. Khrapko

Abstract The presence of phycoerythrin (Phy) pigment in seawater was detected in the Atlantic sector of the Southern Ocean, mainly in the Weddell-Scotia Confluence zone, in January and early February 2020, by optical contact and remote sensing methods. Phy significantly influenced the seawater fluorescence and light absorption coefficient and, correspondingly, the remote sensing reflectance spectrum. The spatial distribution of Phy fluorescence intensity and chlorophyll-a (Chl) concentration and their fluorescence ratio (Phy/Chl) was investigated. The ratio might be an indicator of changes in the species composition of phytoplankton communities. In addition, a comparative analysis with the map of gap-free Chl concentrations interpolated from satellite radiometer VIIRS measurements is provided.

In general, phycoerythrin-containing algae were located in the Weddell-Scotia Confluence zone near the Antarctic Circumpolar Current (ACC) at the end of January 2020. At the same time, phycoerythrin was completely absent in the ACC in the Drake Passage. Phycoerythrin-containing algae dominated in the South American shelf region on the Drake Passage side in early February 2020.

P. A. Salyuk (✉)
V.I. Ilichev Pacific Oceanological Institute, Far Eastern Branch, Russian Academy of Science, Vladivostok, Russia

D. I. Glukhovets · V. A. Artemiev · A. N. Khrapko
Shirshov Institute of Oceanology, Russian Academy of Sciences, Moscow, Russia

A. Yu. Mayor
Far Eastern Federal University, Vladivostok, Russia

N. A. Moiseeva
A.O. Kovalevsky Institute of Biology of the Southern Seas, Russian Academy of Sciences, Sevastopol, Russia

E. G. Morozov et al. (eds.), *Antarctic Peninsula Region of the Southern Ocean*, Advances in Polar Ecology 6, https://doi.org/10.1007/978-3-030-78927-5_19

Keywords Phycoerythrin · Southern Ocean · Antarctic Peninsula · Fluorescence · Absorbance · Remote sensing reflectance · Sun-induced · Fluorimeter · Gap-free satellite image

19.1 Introduction

The presence of the phycoerythrin (Phy) pigment was detected in the Atlantic sector of the Southern Ocean during cruise 79 of the R/V "Akademik Mstislav Keldysh" at the end of January 2020 mainly in the Weddell-Scotia Confluence Zone. In this region Phy is contained in cryptophytes, such as *Geminigera cryophila* (Sapozhnikov et al. 2020; van den Hoff et al. 2020), whose growth could be stimulated by the relatively high irradiance and ocean acidification (Trimborn et al. 2019) in Antarctic summer. Throughout the year, low-light conditions are observed in the study site because of frequent cloudiness, the presence of ice, and the polar winter (Mitchell 1992; Thomalla et al. 2011). Phy pigment significantly affected the optical characteristics of the photic zone because it has specific optical features (Hoge and Swift 1990; Hoge et al. 1999; Churilova et al. 2019). In this research we show the change in hydro-optical characteristics of seawater caused by the presence of Phy in the water basins around the Antarctic Peninsula, where Phy is quite common (García-Muñoz et al. 2013) in January and early February 2020. The spatial distribution of phycoerythrin fluorescence intensity and chlorophyll-a concentration and their ratio are also presented, which may be an indicator of changes in the species composition of phytoplankton communities. This research is devoted to the results of the stage of the expedition in January. The results of the transatlantic part of the cruise are presented in Glukhovets et al. (2021).

19.2 Methods and Data

The following measuring instruments and methods were used in the work:

- Flow-through laser hyperspectral fluorimeter that provides measurements of seawater fluorescence in the range of 560–750 nm when excited at wavelengths of 532 nm (Nagornyi et al. 2016) during the ship's motion.
- Light absorption by particulate matter ($a_p(\lambda)$) was determined in accordance with the NASA protocol (Boss et al. 2019). The measurement was performed on a two-beam Lambda 35 spectrophotometer (PerkinElmer) with an integrating sphere in the wavelength range from 350 to 750 nm with a step of 1 nm. The values of $a_p(\lambda)$ were separated into light absorption by phytoplankton pigments (a_{ph} (λ)) and by non-algal particles (a_{NAP} (λ)) described in Kishino (1985). Absorption correction (β-correction) was performed according to Mitchell

(1990). The correction for nonspecific absorption was made by "zeroing" the average value for 700–750 nm.

- PRO-1 floating spectroradiometer (Artemiev et al. 2000). The device measures upwelling radiance spectra directly below the sea surface and the spectral irradiance of the radiation incident on the surface, which makes it possible to calculate the spectral remotely sensed reflectance of the sea surface in the 390–700 nm range. Measurements were made at the stations.
- Satellite data were used to study the bio-optical characteristics of the surface layer and the processes occurring in it over a wide spatial range. The NOAA service (NOAA MSL12 Ocean Color) launched at the end of 2019 was selected as the data source. This service provides interpolated time and space data on chlorophyll concentration calculated using the method of empirical orthogonal functions (Liu and Wang 2019). The interpolation is based on the combined data of the VIIRS ocean color scanners installed on the SNPP and NOAA-20 satellites. The result is daily data with a spatial resolution of 9 km. The relative error of the Chl concentration interpolation algorithm is about 20%, which corresponds to the accuracy of the standard satellite algorithms. The interpolated data is best suited for our study because frequent dense clouds prevent obtaining actual data from satellite ocean color scanners.

19.3 Results and Discussion

The main data set used for the analysis was obtained using a flow fluorimeter. The examples of fluorescence spectra are plotted in Fig. 19.1. A significant peak of fluorescence of phycoerythrin (Phy) at 580 nm is seen.

To compare the magnitude of fluorescence intensity values, it is necessary to normalize them by the intensity of Raman scattering (RS) by seawater and to analyze Phy in "Raman" units (R.u.). Chlorophyll-a (Chl) fluorescence at 680 nm was estimated, of which intensities in R.u. are calibrated for spectrophotometric Chl.

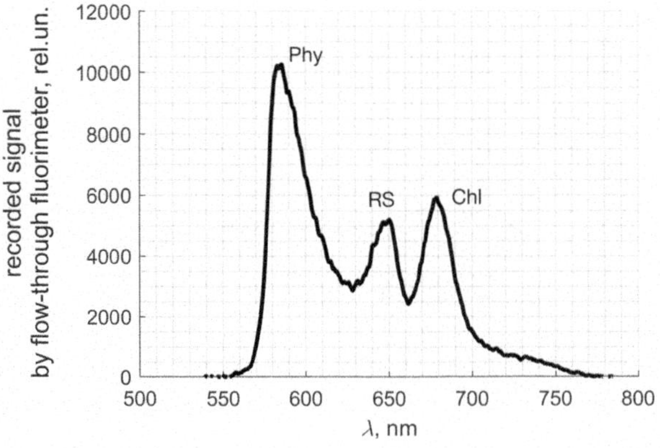

Fig. 19.1 Example of fluorescence spectra of seawater when excited by laser radiation with a wavelength of 532 nm obtained by laser flow-through fluorimeter: Phy is phycoerythrin fluorescence; Chl is chlorophyll-a fluorescence; RS is Raman scattering of seawater

The device enabled us to measure seawater fluorescence every 60 s (with a spatial resolution of 50–100 m) in automatic mode; thus, our large data array was collected, making it possible to identify the boundaries of the regions with algae containing Phy. A similar result was obtained in Babichenko et al. (1993), but the intensity of the recorded fluorescence in RS units measured in this work was much lower.

In addition, the results of measurements at stations were used. The data at station 6607 (61°44.43′ S, 48°46.73′ W, January 27, 2020) were analyzed; no Phy was identified within the sensitivity of the optical devices. The highest Phy content during the expedition was recorded at station 6609 (60°32.58′ S, 48°09.91′ W, January 28, 2020). Station 6607 was used as the background data in the further analysis. A more detailed analysis of the results in terms of the effect of the detected phycoerythrin on the optical characteristics of seawater is presented in Salyuk et al. (2020).

Figure 19.2 shows the spectra of the seawater absorption coefficient of phytoplankton particles at stations 6607 and 6609, where the features related to the absorption of Phy were seen. Additional absorption peaks in the range of 480–500 nm and 550–560 nm associated with the presence of Phy (Six et al. 2007) were seen at station 6609.

The presence of Phy at the station was so great that it affected the remote sensing reflectance (Rrs) spectra $\rho(\lambda)$, which are presented in Fig. 19.3 for the same stations. Additional Phy absorption at station 6609 resulted in the corresponding decrease in the $\rho(\lambda)$ spectra. In addition, in the range of 575–590 nm, the fluorescence of Phy induced by solar radiation appears.

Such an impact by Phy on the $\rho(\lambda)$ spectrum can lead to errors in determining the concentrations of Chl estimated from the remotely sensed sea color data. This effect appears due to the fact that standard satellite algorithms like OCx (O'Reilly and Werdell 2019) use a color index, the denominator of which includes Rrs in the green range of the spectrum and the numerator – Rrs from the blue spectral range. Each range is affected by the absorption of Phy, and, under certain conditions, this can

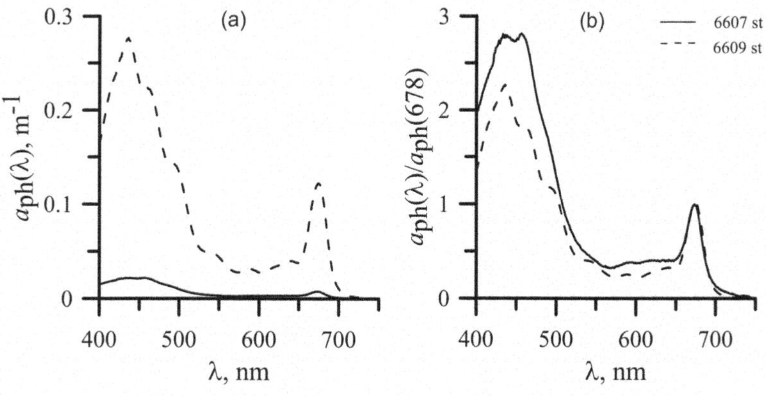

Fig. 19.2 Absorption spectra by phytoplankton at the background station 6607 and station with phycoerythrin 6609, Atlantic sector of the Southern Ocean, January 28–29, 2020: initial spectra (**a**) and normalized spectra (**b**)

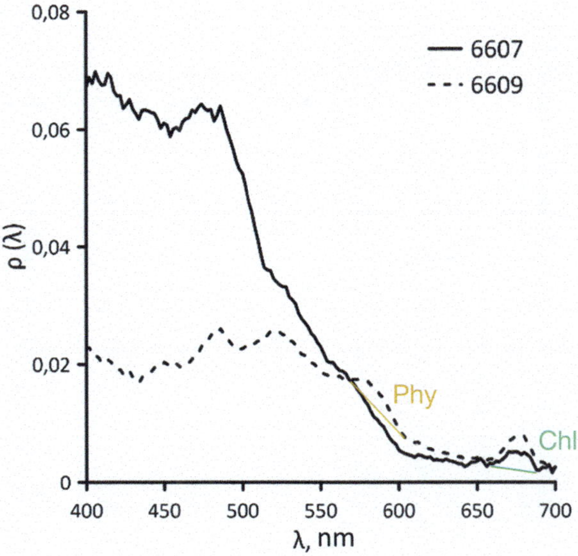

Fig. 19.3 Remote sensing reflectance spectra at background station 6607 and at station 6609 with phycoerythrin, Atlantic sector of the Southern Ocean, January 28–29, 2020

lead to additional uncertainty in determining the concentration of Chl. However, this effect may not be revealed since Phy strongly absorbs both in the ranges of 490–500 and 550–560 nm and this effect may be eliminated during division.

The corresponding vertical distributions of chlorophyll-a fluorescence, total light attenuation coefficient, and seawater temperature at the background station 6607 (left) and at the station with Phy 6609 (right) are plotted in Fig. 19.4. While direct measurements of Phy are not presented, it manifests itself in a signal of total attenuation of light in the upper layer to the depths of about 20 m. Thus, it is present almost in the entire layer that causes the color of the sea surface. At depths greater than 20 m, only chlorophyll-a fluoresces. The main conclusion that can be made from this figure is that the attenuation coefficient in the upper 15 m is constant, and thus it can be expected that variations in stratification would not affect the formation of the satellite ocean color image.

The spatial distributions of Phy fluorescence intensity, Chl concentration, and ratio of Phy fluorescence to Chl fluorescence are shown in Fig. 19.5. The same figures indicate the positions of stations 6607 and 6609. The red color in Fig. 19.6c corresponds to the cases where the fluorescence intensity of Phy was comparable or greater than Chl. It can be seen that, in general, phycoerythrin-containing algae are located across the Weddell-Scotia Confluence zone south of the Antarctic Circumpolar Current (ACC). At the same time, phycoerythrin is completely absent in the ACC region; generally, it is an area of pure low-productive waters. A small amount of Phy was also detected in the upwellings near the Falkland (Malvinas) Current at 46°N and 52°N, but these are the waters dominated by chlorophyll-a-containing algae. Another area where there was a significant relative contribution of phycoerythrin-containing algae is the South American shelf in the Drake Passage, where measurements were made in early February 2020.

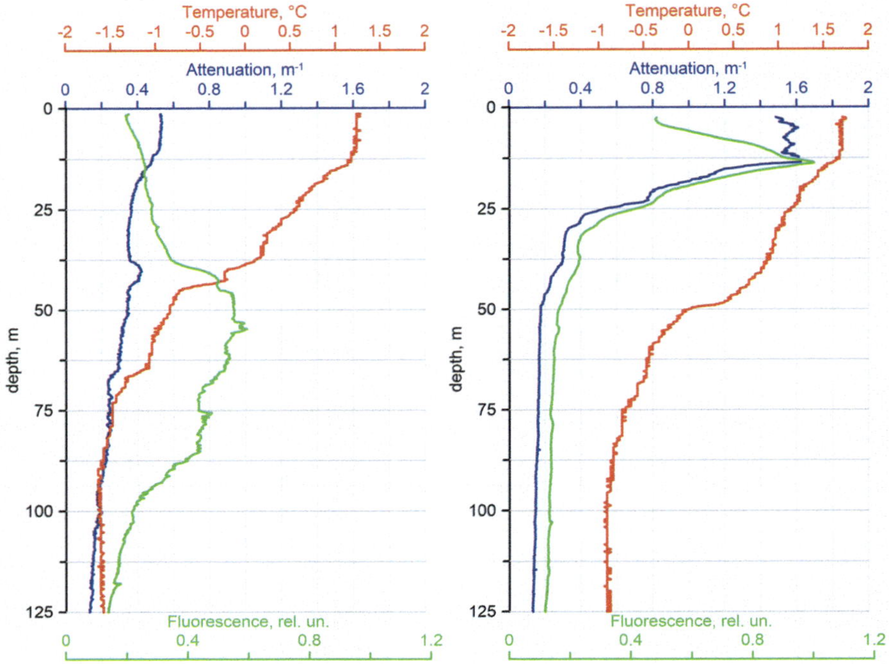

Fig. 19.4 Vertical profiles of chlorophyll-a fluorescence, total light attenuation coefficient, and seawater temperatures at background station 6607 (left) and at station 6609 with Phy (right)

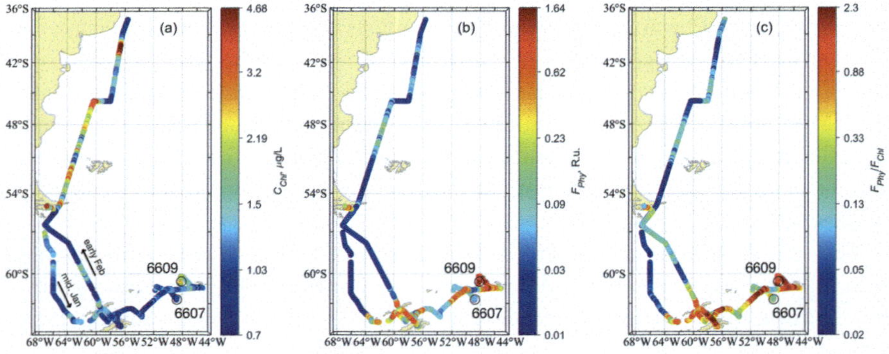

Fig. 19.5 Spatial distribution of chlorophyll-a concentration in µg/l (**a**), phycoerythrin fluorescence intensity in R.u. (**b**), and the ratio of phycoerythrin and chlorophyll-a fluorescence intensities (**c**)

For comparison, Fig. 19.6 shows "gap-free" satellite data. In general, there is good agreement between the data of direct shipboard measurements and that interpolated from satellites: lower water productivity in the Drake Passage and higher values of Chl concentration are recorded in the Weddell-Scotia Confluence zone.

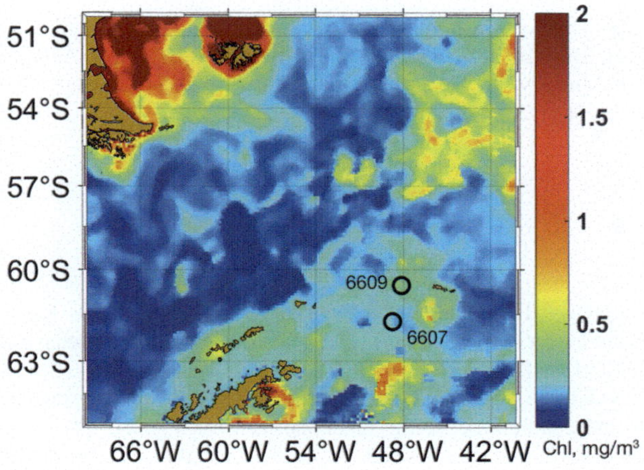

Fig. 19.6 "Gap-free" Chl concentration based on the interpolated satellite radiometer VIIRS data on January 27, 2020. Black circles indicate position of stations 6609 and 6607

The Chl concentration at stations 6607 and 6609 were 0.44 and 0.61 mg/m^3, respectively, which qualitatively agrees with the results just presented. However, the quantitative differences in bio-optical characteristics (Figs. 19.2, 19.3, 19.4, and 19.5) at these stations were pronounced stronger. One possible explanation is related to the difference in the spatial resolution of the compared methods: shipboard measurements were made at a point, while satellite data were interpolated to a grid with a resolution of 9 km. In addition, frequent dense clouds over the study region lead to a decrease in the amount of satellite images, which consequently leads to an increase in interpolation errors. Another possible reason is the effect of phycoerythrin pigment on the standard satellite algorithm for estimating chlorophyll concentration, leading to underestimations of the result (Salyuk et al. 2020).

19.4 Conclusions

A high content of phycoerythrin in the surface layer was recorded in the Atlantic sector of the Southern Ocean in January 2020. It strongly influenced seawater fluorescence and the light absorption coefficient and, correspondingly, remote sensing reflectance spectrum.

In general, phycoerythrin-containing algae were found in the Weddell-Scotia Confluence zone south of the Antarctic Circumpolar Current (ACC) in the end of January 2020. At the same time, phycoerythrin was completely absent in the ACC in the Drake Passage. Phycoerythrin-containing algae dominated in the South American shelf on the Drake Passage side in early February 2020.

Acknowledgments This work was carried out as part of Russian State task (POI FEB RAS № FWMM-2019-0007; SIO RAS № 0128-2019-0008; IBSS RAS № AAAA-A19-119100290162-0) and grant of the Ministry of Science and Higher Education of the Russian Federation FZNS-2020-0003, № 0657-2020-0003.

References

Artemiev VA, Burenkov VI, Vortman MI, Grigoriev AV, Kopelevich OV, Khrapko AN (2000) Sea-truth measurements of ocean color: a new floating spectroradiometer and its metrology. Oceanology 40(1):139–145

Babichenko S, Poryvkina L, Arikese V, Kaitala S, Kuosa H (1993) Remote sensing of phytoplankton using laser-induced fluorescence. Remote Sens Environ 45(1):43–50

Boss E et al (2019) IOCCG ocean optics and biogeochemistry protocols for satellite ocean colour sensor validation inherent optical property measurements and protocols: best practices for the collection and processing of ship-based underway flow-through optical data (v4. 0)

Churilova T, Suslin V, Sosik H, Efimova T, Moiseeva N, Moncheva S, Krivenko O (2019) Phytoplankton light absorption in the deep chlorophyll maximum layer of the Black Sea. Eur J Remote Sens 52(1):123–136

García-Muñoz C, Lubián LM, García CM et al (2013) A mesoscale study of phytoplankton assemblages around the South Shetland Islands (Antarctica). Polar Biol 36:1107–1123. https://doi.org/10.1007/s00300-013-1333-5

Glukhovets DI, Salyuk PA, Artemiev VA, Shtraikhert EA, Zakharkov SP (2021) Variability of the bio-optical characteristics of the water surface layer during the transatlantic transect in 2019–2020. Oceanology 41(1) (in press)

Hoge FE, Swift RN (1990) Photosynthetic accessory pigments: Evidence for the influence of phycoerythrin on the submarine light field. Remote Sens Environ 34(1):19–35

Hoge FE, Wright CW, Lyon PE, Swift RN, Yungel JK (1999) Satellite retrieval of the absorption coefficient of phytoplankton phycoerythrin pigment: theory and feasibility status. Appl Opt 38:7431–7441

Kishino M, Takahashi N, Okami N, Ichimura S (1985) Estimation of the spectral absorption coefficients of phytoplankton in the sea. Bull Mar Sci 37:634–642

Liu X, Wang M (2019) Filling the gaps of missing data in the merged VIIRS SNPP/NOAA-20 ocean color product using the DINEOF method. Remote Sens 11(2):178

Mitchell BG (1990) Algorithms for determining the absorption coefficient for aquatic particulates using the quantitative filter technique. SPIE Ocean Optics X 1302:137–148

Mitchell BG (1992) Predictive bio-optical relationships for polar oceans and marginal ice zones. J Mar Syst 3:91–105

Nagornyi IG, Maior AY, Salyuk PA, Doroshenkov IM (2016) A mobile complex for on-line studying water areas and surface atmosphere. Instrum Exp Tech 57(1):68–71

NOAA MSL12 Ocean Color, science quality, VIIRS multi-sensor (SNPP + NOAA-20), chlorophyll gap-filled analysis: https://coastwatch.noaa.gov/cw/satellite-data-products/ocean-color/science-quality/viirs-multi-sensor-gap-filled-chlorophyll.html

O'Reilly JE, Werdell PJ (2019) Chlorophyll algorithms for ocean color sensors - OC4, OC5 & OC6. Remote Sens Environ 229:32–47. https://doi.org/10.1016/j.rse.2019.04.021

Salyuk PA, Glukhovets DI, Moiseeva NA, Artemiev VA, Mayor AY, Khrapko AN (2020) Phycoerythrin influence on the optical characteristics of seawater in the Atlantic sector of the Southern Ocean. Proc. SPIE 11560, 26th International Symposium on Atmospheric and Ocean Optics, Atmospheric Physics, 1156056 (12 November 2020). https://doi.org/10.1117/12.2575813

Sapozhnikov PhV, Kalinina OYu, Morozova TV (2020) Phytopelagic communities of the Powell Basin in the summer of 2020. (This book)

Six C, Thomas JC, Garczarek L, Ostrowski M, Dufresne A, Blot N, Partensky F (2007) Diversity and evolution of phycobilisomes in marine Synechococcus spp.: a comparative genomics study. Genome Biol 8(2):R259

Thomalla SJ, Fauchereau N, Swart S, Monteiro PMS (2011) Regional scale characteristics of the seasonal cycle of chlorophyll in the Southern Ocean. Biogeosciences 8(10):2849–2866. https://doi.org/10.5194/bg-8-2849-2011

Trimborn S, Thoms S, Karitter P, Bischof K (2019) Ocean acidification and high irradiance stimulate the photo-physiological fitness, growth and carbon production of the Antarctic cryptophyte Geminigera cryophila. Biogeosciences 16(15):2997–3008. https://doi.org/10.5194/bg-16-2997-2019

van den Hoff J, Bell E, Whittock L (2020) Dimorphism in the Antarctic cryptophyte Geminigera cryophila (Cryptophyceae). J Phycol 56(4):1028–1038. https://doi.org/10.1111/jpy.13004

Chapter 20
Nanophytoplankton in the Bransfield Strait: Contribution of Cryptophyta to the Community Abundance and Biomass During Austral Summer

Vladimir S. Mukhanov, Evgeny G. Sakhon, Aleksander A. Polukhin, and Vladimir A. Artemiev

Abstract The chapter describes the flow cytometry data on the abundance and biomass of nanophytoplankton in the central Bransfield Strait, and their distribution across the strait in relation to hydrology and hydrochemistry of the two major water masses: the warmer and less saline Transitional Zonal Water with Bellingshausen Sea influence (TBW), and the cold and salty Transitional Zonal Water with Weddell Sea influence (TWW). Two nanophytoplankton clusters were distinguished and enumerated in the cytograms: larger cells (about 9 μm) of cryptophytes with bright orange fluorescence, and smaller cells (3 μm) of other nanophytoplankton (NP). It was shown that both assemblages develop to higher abundances and biomasses in the warmer and less saltine TBW. This biotope was characterized by a much more diverse nanophytoplankton community with a pronounced dominance of Cryptophyta in terms of biomass. Temperature did not appear to be a key factor controlling NP growth. The results support the idea that the increasing melt-water input can potentially increase the spatial and temporal extent of cryptophytes. The replacement of large diatoms with small cryptophytes leads to a significant shift in trophic processes in favor of the consumers like salps that are able to graze on smaller prey. If this trend persists, the food web shifts will eventually impact biogeochemical cycling in Antarctic coastal waters.

Keywords Nanophytoplankton · cryptophytes · phytoplankton bloom · Bransfield Strait · Antarctic Peninsula · flow cytometry

V. S. Mukhanov (✉) · E. G. Sakhon
A. O. Kovalevsky Institute of Biology of the Southern Seas, Russian Academy of Sciences, Sevastopol, Russia

A. A. Polukhin · V. A. Artemiev
Shirshov Institute of Oceanology, Russian Academy of Sciences, Moscow, Russia

E. G. Morozov et al. (eds.), *Antarctic Peninsula Region of the Southern Ocean*, Advances in Polar Ecology 6, https://doi.org/10.1007/978-3-030-78927-5_20

20.1 Introduction

Diatoms, haptophytes (predominantly *Phaeocystis antarctica*), and cryptophytes represent the major taxonomic groups of phytoplankton in the Southern Ocean (Rodriguez et al. 2002; Garibotti et al. 2003, 2005; Kozlowski et al. 2011; Mendes et al. 2012). Summer phytoplankton blooms in coastal waters of the Antarctic Peninsula are usually associated with a shallow upper mixed layer (UML) that keeps phytoplankton under favorable light conditions and a better supply of dissolved iron (e.g., Prezelin et al. 2000). As a rule, diatoms and/or *P. antarctica* make the largest contribution to the blooms, but it was reported that the role of cryptophytes increased, especially in the region of melting glaciers (Moline and Prezelin 1996; Moline et al. 2004; Gonçalves-Araujo et al., 2015). The mechanisms behind Cryptophyta bloom formation are still poorly understood. Lower water temperatures in the surface layer, narrowing of the UML, and a decrease in water salinity appear to favor the development of the phenomenon (Mendes et al. 2013; Gonçalves-Araujo et al. 2015).

Publicly available data provide evidence of transformation of the Antarctic phytoplankton community as a probable consequence of global warming (Mendes et al. 2013). The transformation covers not only taxonomic composition but also the size structure of the community. The replacement of large diatoms with small cryptophytes leads to a significant shift in trophic processes in favor of the consumers able to graze on smaller prey. Since Antarctic krill is not one of them (Haberman et al. 2003), such a trend gives rise to a well-founded concern. A decrease in the food supply of krill and a substitution of the latter by other zooplankton (like salps) implies a reduction of the carbon pool available for higher trophic levels by almost 70% (Moline et al. 2001).

It has been shown in a number of experimental studies that Antarctic krill feed on small prey (about 10 μm in size) with low efficiency (McClatchie and Boyd 1983; Boyd et al. 1984), while the optimal size of its food objects is about 50 μm (Quentin and Ross 1985) due to the features of the filtration apparatus of euphausiids (McClatchie and Boyd 1983). This well explains why krill exhibit positive selectivity for diatoms and avoid cryptophytes when feeding on complex prey mixture (Haberman et al. 2003).

On the contrary, salps, a food competitor of Antarctic krill, are satisfied with a wide range of taxonomic and size composition of phytoplankton prey, thus demonstrating much lower feeding selectivity (Haberman et al. 2003). Against the background of their ongoing expansion to higher latitudes (Moline et al. 2004; Montes-Hugo et al. 2009), their potential ability to utilize the biomass of cryptophytes during phytoplankton-bloom periods poses additional threats to Antarctic krill populations, and hence, also for higher consumers including penguins, seals, and whales, which feed on krill (Loeb et al. 1997). Thus, an increasing role of Cryptophyta in Antarctic waters implies restrictions in food supply for krill and strengthening of its competitor's position.

The phenomenon of increased cryptophyte abundances in the Antarctic waters has been initially explained by advection (Moline and Prézelin 1996), sedimentation of large diatoms (Castro et al. 2002), grazing (Garibotti et al. 2003), and tolerance of cryptophytes to lower salinity waters (Moline et al. 2004). Recently, Mendes et al. (2018), who studied the prevalence of cryptophytes in the coastal waters of the western Antarctic Peninsula, hypothesized that the phenomenon can be due to their pigment protection capability and, as a result, an ability to successfully grow under high irradiance exposure in strongly stratified shallow mixing surface layers. In this context, our objective was to characterize the nanophytoplankton assemblages by flow cytometry (FCM), identify their Cryptophyta component, and analyze their distribution in relation to hydrological and hydrochemical conditions in the coastal waters of the Antarctic Peninsula.

20.2 Material and Methods

Sampling Sites and Methods The oceanographic and microbiological data were collected on January 21, 2020, over a transect across the central Bransfield Strait (BS) (seven stations, Fig. 20.1). The water area differs sufficiently from the rest of the western Antarctic Peninsula shelf due to the inflow of cold and saline Weddell Sea Shelf Water (Mendes et al. 2018). According to Tokarczyk's (1987) classification, the water masses in the BS are the Transitional Zonal Water with Bellingshausen Sea influence (TBW; water temperature $T > 1$ °C, salinity $S < 34.1$ psu), and the Transitional Zonal Water with Weddell Sea influence (TWW; $T < 1$ °C, $S > 34.1$ psu). At the surface, TBW covers most of the BS (from station 6587 to 6593, Fig. 20.1). The two water masses are separated from each other in the surface layer (between stations 6593 and 6594, Fig. 20.2) and below the surface by the Peninsula Front and Bransfield Front, respectively (Garcıa et al. 1994; Sangra et al. 2011), which are identified using the 1 °C isotherm. The latter did not deepen below 50 m (station 6587) and corresponded to the maximum temperature gradient at the boundary between TBW and TWW (marked in Fig. 20.2 with short horizontal lines).

The Bransfield Front is linked to a baroclinic jet called the Bransfield Current (Niiler et al. 1991; Garcıa et al. 1994; Zhou et al. 2006) directed from the southwest to the northeast and transporting the TBW along the southern slope of the Southern Shetland Islands. The highest current velocity (0.45 m s^{-1}) was observed at station 6590 in the surface layer. The inflow (< 0.1 m s^{-1}) of cold and saline Weddell Sea Shelf Water was recorded only at station 6594 over the entire range of water depths (0 to 200 m) with a maximum at about 100 m.

Flow cytometry. Nanophytoplankton samples were frozen on board in liquid nitrogen for their storage and transportation to a stationary laboratory before flow cytometric analysis. Two nanophytoplankton clusters were distinguished and enumerated using a Cytomics™ FC 500 flow cytometry system (Beckman Coulter Inc., USA): cryptophytes (CPs) and other nanophytoplankton (NP). Cryptophytes

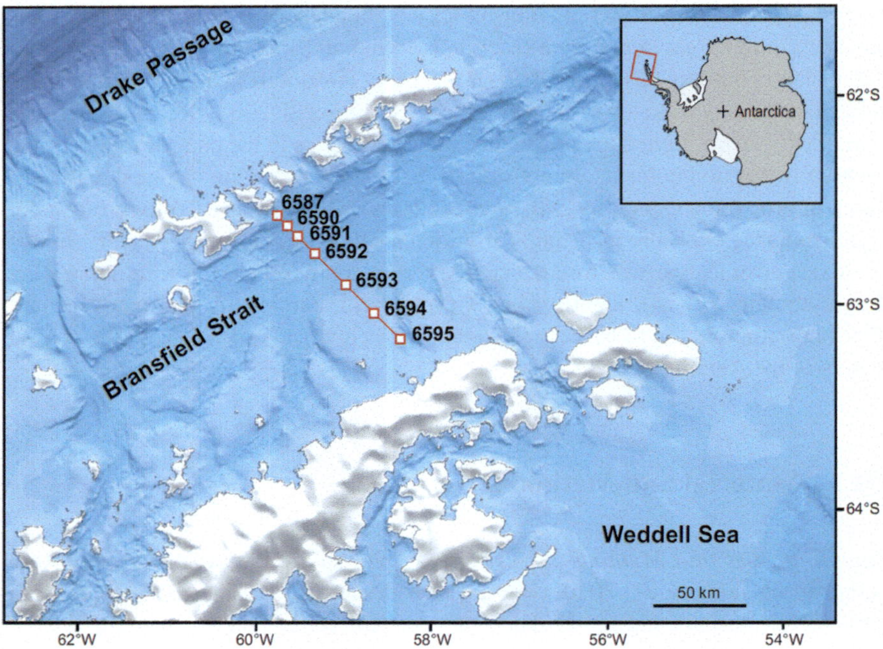

Fig. 20.1 Location of oceanographic stations in the Bransfield Strait (cruise 79 of the R/V *Akademik Mstislav Keldysh*) where nanophytoplankton samples were collected at 5–7 depths in austral mid-summer, January 2020

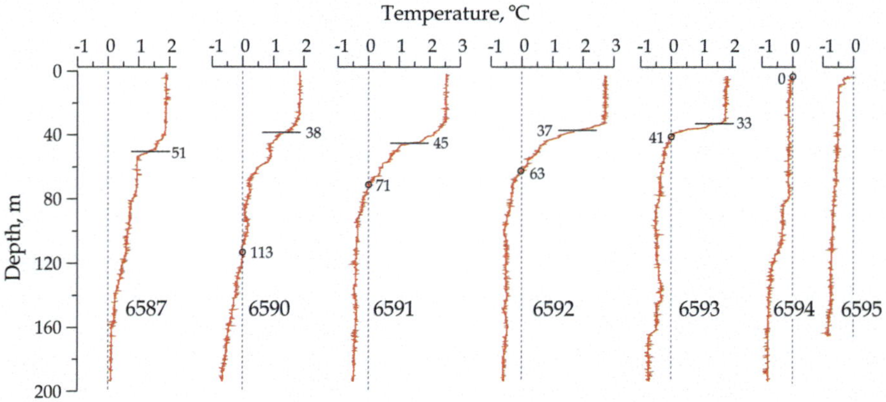

Fig. 20.2 Water temperature profiles at the stations (6587–6595) along the transect in the Bransfield Strait, January 2020. The depths of zero temperature are marked with empty circles. The depths of the maximum temperature gradient (horizontal lines) corresponded to the Bransfield Front (1 °C isotherm)

contain a distinctive set of pigments, the orange fluorescing phycobilins (although there are also colorless members (Hoef-Emden et al. 2002). These cryptomonads were identified by their bright orange fluorescence while NP did not produce it (Fig. 20.3). Light microscopy observation of cryptomonad cells confirmed the flow-cytometry data. Cytometric gating and sizing of the cells were performed using Flowing Software v. 2.5.0 (P. Terho, Turku Centre for Biotechnology, University of Turku, Finland, www.flowingsoftware.com). The standard forward light scatter (FSC), orange fluorescence (FL2, 575 nm), and red fluorescence (FL4, 675 nm) channels were used for the acquisition and analysis.

Cell-size measurements were calibrated in terms of equivalent spherical diameters (ESD, μm) using a set of standard polystyrene beads (Polysciences, Inc.) in the size range between 0.5 and 10 μm; calibration curves were obtained for each flow-cytometry protocol by plotting the ESD as a function of the FSC signal (coefficient of determination $r^2 > 0.8$ in all cases). Average ESD and cell volume were calculated for every cell population from the calibration curves (Olson et al. 1993). To allow for a better comparison with earlier published data, nanophytoplankton abundances and biovolumes were converted to cellular carbon (C). Carbon conversion factors have been shown to vary depending on taxonomy, cell size, and growth-regulating factors such as light and nutrient availability (Alvarez et al. 2017). In this study, C content of each nanophytoplankton population was estimated from the average ESD using a conversion factor of 196.5 fg C μm^{-3} (Garrison et al. 2000).

Statistics. Multivariate statistical analysis (cluster analysis and nonmetric multidimensional scaling, nMDS) of microbiological and hydrological–hydrochemical data obtained in the upper 200-m layer was carried out using the PAST3 software

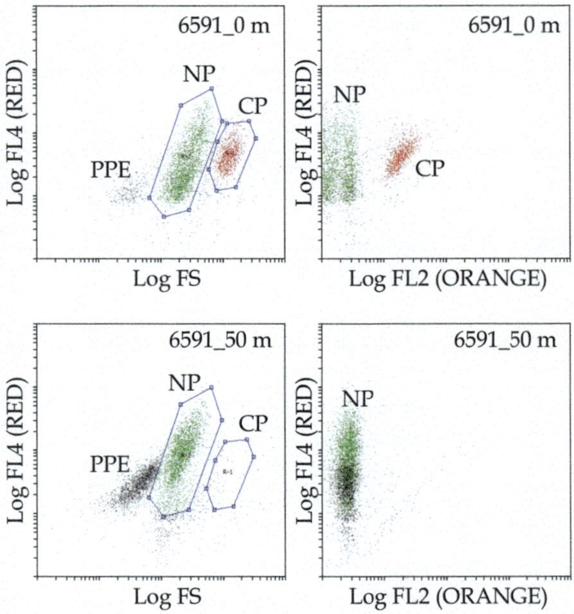

Fig. 20.3 Gating cryptophytes (CPs) and other nanophytoplankton (NP) in the space of forward light scattering (FS, cell size), red autofluorescence (FL4, chlorophyll), and orange autofluorescence (FL2, phycoerythrin) in two samples (station 6591, at 0 and 50 m). PPE is eukaryotic picophytoplankton

(Hammer et al. 2001). Mapping the spatial distribution of the variables and graph plotting were made using Surfer v. 10 and Grapher v. 8 (Golden Software, Inc.), respectively.

In total, 21 biological, hydrological, and hydrochemical variables were measured in the study site and investigated by the methods of multivariate statistical analysis (Table 20.1). The 12 nanophytoplanktonic variables included in this list characterized both the structure (*CPN, CPN%, CPB%, NPN, NPB, CP-ESD, NP-ESD*) and some functional characteristics of the communities, namely, intracellular content of photosynthetic pigments in the photoautotrophic cells (*CP-CHL, CP-PE, NP-CHL*).

20.3 Results and Discussion

Vertical distribution of water temperature and salinity across the BS (Fig. 20.4) corresponded to the typical BS circulatory system, with the Southern BS Current carrying waters advected from the Weddell Sea and the Gerlache Strait Current carrying waters from the southwest (Jiang et al. 2013). The Bransfield Front was located

Table 20.1 Symbols of variables used in this study with their definitions, units, and descriptive statistics (n = 39). R. u. is relative units; CP is cryptophyte; NP is nanoplankton; ESD is equivalent spherical diameter; SD is standard deviation

Symbol	Definition, units	Min	Max	Mean ± SD
T	Water temperature, °C	−0.79	2.77	0.59 ± 1.19
S	Water salinity, psu	34.1	34.4	34.3 ± 0.1
PO4	Phosphates, μM	1.56	4.37	3.09 ± 0.58
Si	Silicates, μM	67.9	112.3	81.3 ± 10.1
NO3	Nitrates, μM	22.6	35.4	29.4 ± 3.7
NO2	Nitrites, μM	0.11	0.23	0.17 ± 0.04
NH4	Ammonium, μM	1.32	3.81	2.13 ± 0.58
DO	Dissolved oxygen, ml l^{-1}	6.44	7.98	7.44 ± 0.36
FLUOR	Chl-a fluorescence, r. u.	0.10	0.49	0.24 ± 0.12
CPN	Cryptophyte abundance, 10^6 cells l^{-1}	0.00	1.55	0.35 ± 0.46
CP-ESD	Mean CP ESD, μm	6.70	11.59	9.42 ± 1.77
CPB	CP biomass, μg C l^{-1}	0	179	33 ± 45
CP-CHL	Intracellular chlorophyll content in CP, r. u. cell $^{-1}$	0	163	69 ± 39
CP-PE	Intracellular phycoerythrin content in CP, r. u. cell $^{-1}$	0	206	31 ± 39
NPN	NP abundance, 10^6 cells l^{-1}	0.19	3.95	1.71 ± 1.07
NP-ESD	Mean NP ESD, μm	2.14	3.67	2.88 ± 0.38
NPB	NP biomass, μg C l^{-1}	0.4	15.7	4.3 ± 3.2
NP-CHL	Intracellular chlorophyll content in NP, r. u. cell $^{-1}$	8	220	74 ± 52
TNPB	Total NP + CP biomass, μg C l^{-1}	0.4	194.2	37.3 ± 47.7
CPN%	Portion of CP in the total abundance, %	0	49	14 ± 12
CPB%	Portion of CP in the total biomass, %	0	96	74 ± 21

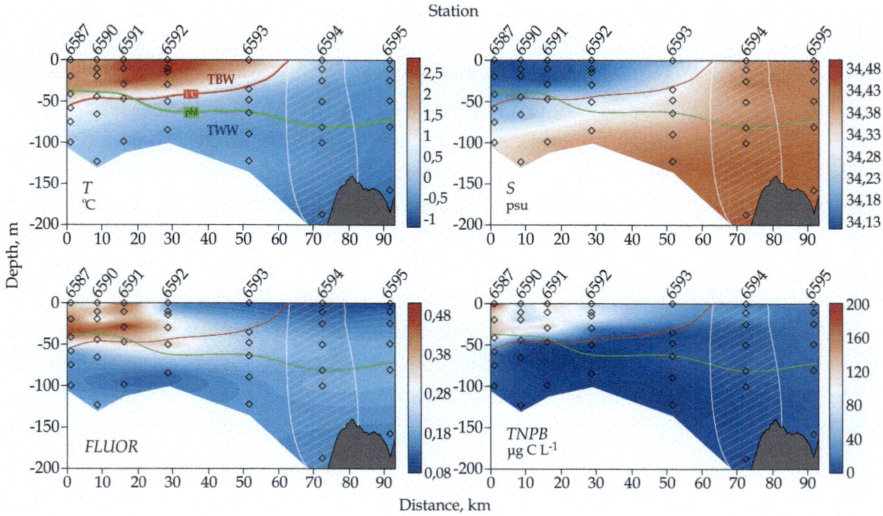

Fig. 20.4 Vertical distribution of water temperature (*T*), salinity (*S*), Chl-a fluorescence (*FLUOR*), and total nanophytoplankton biomass (*TNPB*) along the transect in the Bransfield Strait during austral mid-summer, January 2020. Lines represent the bottom of the photic layer (phl) and 1 °C isotherm (as border between TBW and TWW; explanations are in the text). Shaded area shows location of inflow of Weddell Sea shelf water (station 6594) based on direct measurements

between stations 6593 and 6594; the boundary between the TBW and colder and saltier TWW was recognized according to the 1 °C isotherm position (Fig. 20.4).

The highest Chl-a fluorescence values were measured at the northern stations of the transect (6587, 6590, 6591) at the depths between 0 and 50 m, i.e., within TBW (Fig. 20.4). The distribution pattern of the total biomass of nanophytoplankton estimated by flow cytometry was the same, with a patch of high values at the same stations and depths. This provided evidence that nanophytoplankton was a major, if not dominant, component of phytoplankton in the study site.

In general, CPs were significantly less abundant ($0.35 \pm 0.46 \times 10^6$ cells l^{-1}; ± SD is presented here and further) than NP ($1.71 \pm 1.07 \times 10^6$ cells l^{-1}); however, their cells were much larger (about 9.5 µm versus 3 µm in NP). Hence, their average carbon biomass (33 ± 45 µg C l^{-1}) sufficiently exceeded those of NP (4.3 ± 3.2 µg C l^{-1}) (Table 20.1). The CP average ESD measured by flow cytometry (9.4 ± 1.8 µm) slightly exceeded the estimates (8 ± 2 µm) obtained earlier microscopically for Antarctic cryptophytes (McMinn and Hodgson 1993). The contribution of Cryptophyta to the total nanophytoplankton abundance and biomass reached 49% and 96%, respectively, with the average values of 14 ± 12% and 74 ± 21% (Table 20.1).

The Cryptophyta biomass maximum (about 180 µg C l^{-1}) observed in TBW in the northern part of the transect (station 6587) is an outstanding characteristic of this phytoplankton component. The importance of cryptophytes in multiple basins of the Southern Ocean was reported earlier (Jacques and Panouse 1991; Kopczynska

1991; Buma et al. 1992; Kang and Lee 1995; Rodrıguez et al. 2002; Garibotti et al. 2003) with the maximum concentrations being much lower (between 0.3 and 6.0 × 10^6 cells l^{-1}) than our estimates. Garibotti et al. (2005) have surprisingly revealed peak concentrations in the region of Anvers Island, reaching values up to 15.7 × 10^6 cells l^{-1} and 369 μg C l^{-1}. The estimates obtained in this study proved to be close to this peak in terms of biomass. Therefore, our results widen the area of Cryptophyta blooms to include the central Bransfield Strait and support the conclusion of the cited authors that cryptophytes are a distinctive and annually recurrent component of Antarctic phytoplankton, exhibiting blooms, which can be considered a peculiarity of the continental shelf west of the Antarctic Peninsula.

Patches of the highest CP abundances and biomasses were observed in the photic layer around the jet of the Bransfield Current (station 6590, surface layer), while in TWW (i.e., below the 1 °C isotherm), cryptophytes were scarce or undetectable, especially in the deeper layers (Fig. 20.5). The NP demonstrated similar distribution

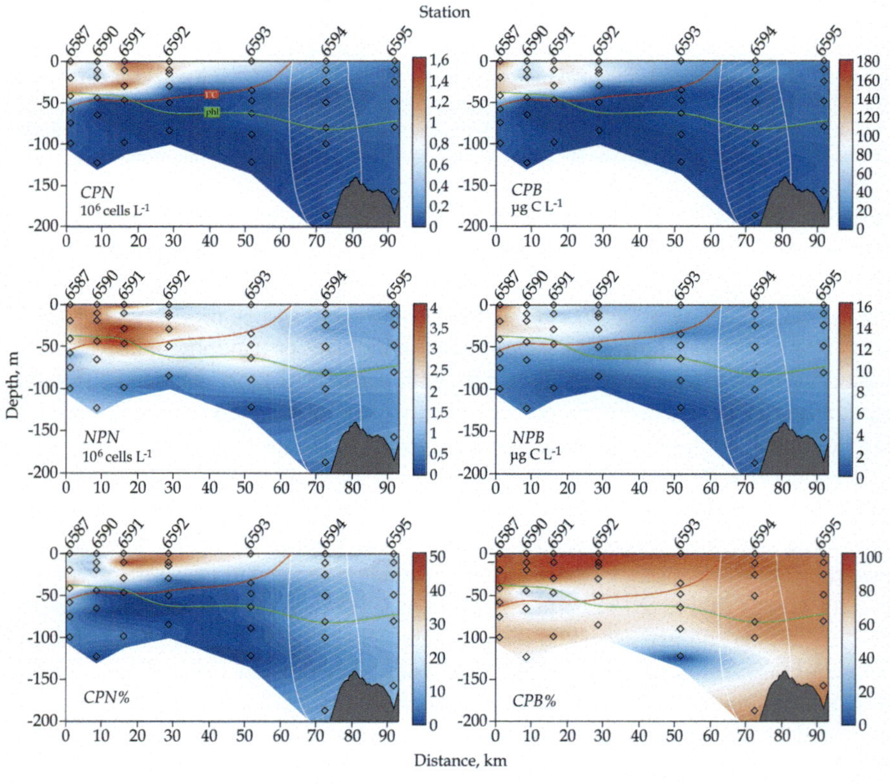

Fig. 20.5 Vertical distribution of cryptophytes and other nanophytoplankton along the transect in the Bransfield Strait during austral mid-summer, January 2020. Lines represent the bottom of the photic layer (phl) and 1 °C isotherm (as the boundary between TBW and TWW water masses). Shaded area shows location of inflow of Weddell Sea Shelf Water (station 6594) based on direct measurements. Description of the variables is given in Table 20.1

pattern with the highest abundances at depths of 35–50 m at station 6591 (Fig. 20.5). The maximum contribution of CP to the total nanophytoplankton biomass was recorded farther south in TBW (6591 and 6592) at shallow depths (about 10 m) where NP were not abundant (Fig. 20.5).

All quantitative measurements of nanophytoplankton (*CPN, NPN, CPB, NPB, TNPB*) revealed significant positive correlations with water temperature (*T*), dissolved oxygen (*DO*), and Chl-a fluorescence (*FLUOR*), while negative correlations were found with salinity (*S*) and nitrates (*NO3*) (Table 20.2). These environmental variables varied gradually with depth and horizontally in the transition from TBW to TWW, which explained well the dependencies. Thus, both CP and NP preferred to develop in warmer and less saline waters of TBW, which supported earlier published observations (Holm-Hansen et al. 1989; Smith and Sakshaug 1990; Mitchell and Holm-Hansen 1991; Kang and Lee 1995; Garibotti et al. 2003). Portions of CP in the total nanophytoplankton abundance and biomass appeared the highest within TBW (Fig. 20.5).

Although Chl-a fluorescence significantly correlated with the total nanophytoplankton biomass, the linkage with NP concentrations was more pronounced (Table 20.2). This was likely due to higher NP abundance (*NPN*) and intracellular chlorophyll content (*NP-CHL*) compared with the same characteristics in CP (Table 20.1). Intracellular chlorophyll values in both CP and NP had a positive relationship with ammonium concentration. Significant correlations have also been revealed between *NP-CHL* and silicates (0.44), *CP-CHL* and phosphates (−0.32), *CP-CHL* and nitrates (−0.41) (Table 20.2).

It should be noted that the strong relationship between the microorganism abundance and water temperature was found over the entire set of data (see regressions in Fig. 20.6a and b). However, within each of the biotopes, TBW and TWW, the nature of the relationship was different. The *CPN* maintained strong dependence ($r^2 = 0.65$) on temperature within TBW, but the other links were lost (see the

Table 20.2 Spearman's correlation between the studied variables. Statistically significant ($p < 0.05$) indicators are in bold. The description of the variables is in Table 20.1

Variable	*T*	*S*	*PO4*	*Si*	*NO3*	*NO2*	*NH4*	*DO*	*FLUOR*
CPN	**0.59**	**−0.53**	−0.22	−0.22	**−0.45**	0.29	0.20	**0.78**	**0.37**
CP-ESD	−0.24	0.02	−0.14	0.28	0.29	−0.18	0.27	−0.15	−0.11
CPB	**0.54**	**−0.52**	−0.23	−0.16	**−0.40**	0.25	0.21	**0.74**	**0.36**
CP-CHL	−0.12	−0.06	**−0.32**	0.27	0.09	**−0.41**	**0.62**	0.04	−0.04
CP-PE	0.10	−0.06	−0.01	−0.13	0.03	0.00	0.28	0.11	−0.03
NPN	**0.74**	**−0.81**	0.32	0.24	**−0.59**	**0.56**	0.32	**0.60**	**0.73**
NP-ESD	−0.02	0.01	−0.32	−0.17	−0.10	−0.26	0.08	0.31	0.06
NPB	**0.68**	**−0.75**	0.12	0.10	**−0.63**	**0.42**	0.25	**0.73**	**0.73**
NP-CHL	0.01	−0.14	0.04	**0.44**	−0.16	−0.12	**0.34**	−0.15	0.12
TNPB	**0.59**	**−0.57**	−0.19	−0.14	**−0.45**	0.27	0.22	**0.77**	**0.42**
CPN%	**0.43**	**−0.33**	−0.31	**−0.32**	−0.30	0.15	0.07	**0.65**	0.15
CPB%	**0.47**	**−0.41**	−0.30	−0.25	−0.30	0.25	0.15	**0.65**	0.19

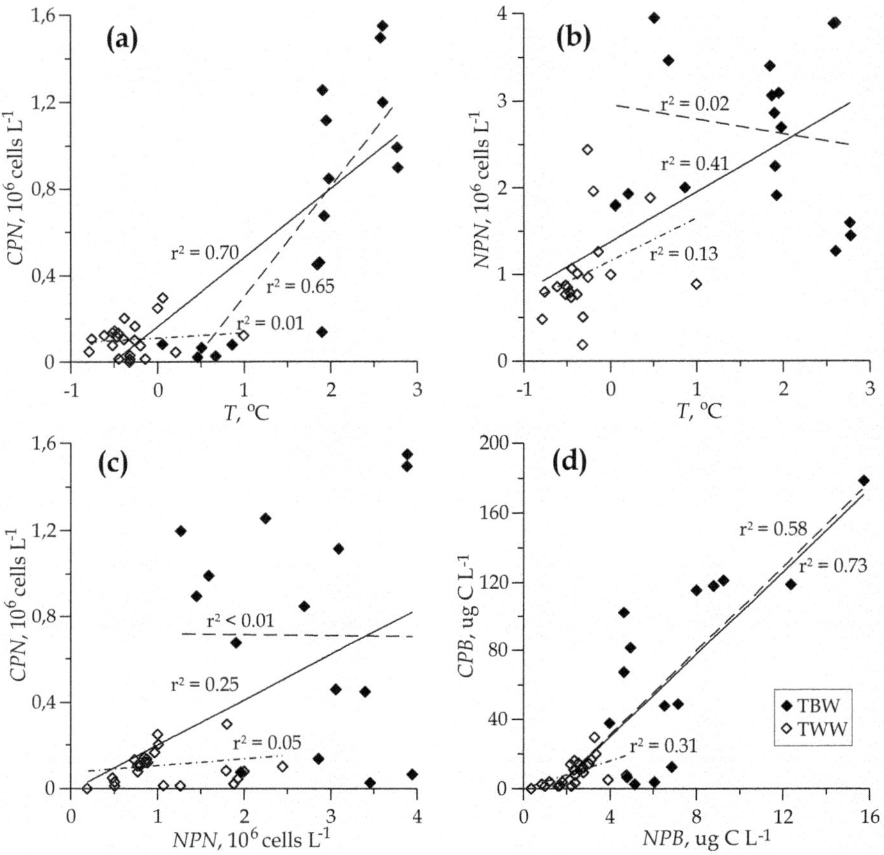

Fig. 20.6 Relationships among nanophytoplankton abundance (*CPN, NPN*), biomass (*CPB, NPB*), and water temperature (*T*). Samples from TBW (above 1 °C isotherm) and TWW (below 1 °C isotherm) water masses are represented with filled and empty symbols, respectively. Solid lines are regressions for the entire set of data, dashed ones are for individual biotopes. Description of variables is given in Table 20.1

regressions represented with dashed lines in Fig. 20.6a and b). Temperature did not appear to be a key factor controlling NP growth.

The CP and NP exhibited significant intercorrelation in terms of biomass ($r^2 = 0.73$) but not of abundance ($r^2 = 0.25$) (Fig. 20.6c and d) that could provide evidence of their competition for the same resources. Yet again, the linkage between *CPB* and *NPB* remained significant ($r^2 = 0.65$) within TBW, while the other relationships were completely lost within the individual biotopes.

Cluster analysis allowed us to classify the nanophytoplankton samples, depending on the community structure (Fig. 20.7a). Four sample groups were identified, three of which (clusters 2–4) were within the TBW biotope (see the scheme of cluster distribution in the water column in Fig. 20.7b). The sample collected in the surface layer at station 6587 formed individual cluster 4, owing to the outstanding

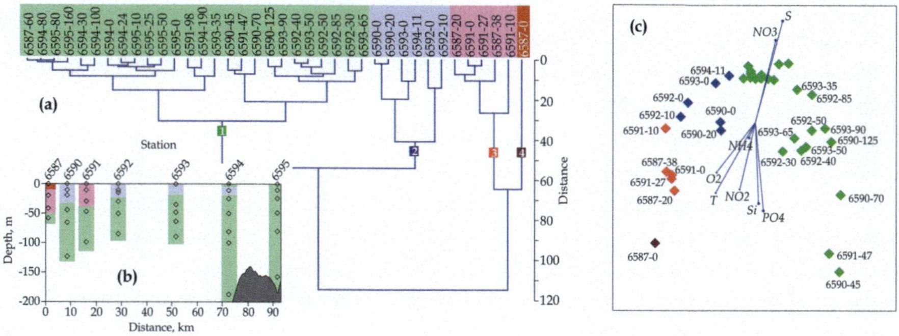

Fig. 20.7 Cluster and nonmetric multidimensional scaling (nMDS) analyses performed on the variables characterizing the nanophytoplankton community in the Bransfield Strait during austral summer, January 2020. (**a**) Dendrogram of a hierarchical cluster analysis with Euclidean distance measure, and four identified clusters; single samples are marked with a code indicating the station and depth of sampling. (**b**) Locations of nanophytoplankton assemblages identified by cluster analysis; background colors correspond to cluster numbers in (**a**). (**c**) nMDS ordination of nanophytoplankton samples with environmental variables overlaid; symbol colors correspond to cluster numbers in (**a**); single samples are marked with a code indicating the station and depth of sampling

microbial characteristics: the highest biomasses and average ESD of CP (178.5 μg C l^{-1}; 11.6 μm) and NP (15.7 μg C l^{-1}; 3.7 μm). Cluster 3 included "patches" of nanophytoplankton at stations 6587 and 6591 within the TBW water masses (upper 50 m), while in cluster 2, CP abundance and portion in the nanophytoplankton decreased sufficiently. In the TWW biotope, the community was represented by cluster 1 with low or undetectable microbial abundances (Fig. 20.7b). Thus, hydrographic and chemical properties of different water masses deeply influenced nanophytoplankton structure. The TBW biotope was characterized by a much more abundant and diverse community with a pronounced dominance of Cryptophyta in terms of biomass.

Ordination of the nanophytoplankton samples by nMDS proved to be mutually consistent with the cluster analysis (Fig. 20.7c); however, the analysis has not revealed a clear distinction between the clusters as the samples, which formed a continuous sequence of dots from cluster 1 to cluster 4. The clouds of dots corresponding to both biotopes, TBW (clusters 2–4) and TWW (cluster 1), were located along the environmental vectors, which indicated a strong environmental control of nanophytoplankton. Distribution of TBW and TWW fronts, anticyclonic eddies and their thermohaline structures, and nutrients are among the factors that can control the community across the BS (García-Muñoz et al. 2013; Mura et al. 1995; Sangrà et al. 2014).

In general, the spatial distribution patterns of chlorophyll-a (Chl-a) fluorescence that we have revealed in the central BS agree well with the earlier long-term observations: the highest Chl-a is commonly located close to the South Shetland Islands (SSI) with the upper mixed layer depth (UMLD) being negatively related with Chl-a (c9); at well-stratified sites (TBW), with euphotic layers varying from 40 to 50 m, the highest Chl-a is at or near the surface, while weakly stratified sites (TWW) are associated with low Chl-a and deeper euphotic zones (Holm-Hansen et al. 1997).

However, localization of cryptophyte maxima in the BS water masses is controversial. In an eastern sector of the BS, Gonçalves-Araujo et al. (2015) observed the dominance of microplanktonic diatoms and higher Chl-a within shallow upper mixed layers in the TBW, close to the South Shetland Islands (SSI), while TWW was primarily characterized by lower Chl-a within a well-mixed water column. Nanoplanktonic flagellates (including haptophytes and cryptophytes) dominated in this water. This contradicts our observations of coincident maxima of Chl-a fluorescence and cryptophyte concentration within the TBW surface layer with low salinities and relatively high temperatures (>1 °C).

According to Schofield et al. (2017), peak cryptophyte populations were observed in the coastal areas of the West Antarctic Peninsula with lower salinity (33.75 to 32.5 psu) and colder water (−1 to 1 °C), while they were absent at higher temperatures (> 1 °C) and salinities (> 33.75 psu). On the contrary, Mendes et al. (2018) provided different data on the distribution of cryptophyte biomass in the Gerlache Strait, which was mostly associated with the sea surface temperature rather than salinity. The authors found cryptophytes to be strongly associated with high temperature, water-column stability, and Chl-a, and negatively associated with nutrient concentrations, UMLD, and salinity, which is in perfect agreement with our observations and earlier published data (Holm-Hansen et al. 1989; Smith and Sakshaug 1990; Mitchell and Holm-Hansen 1991; Kang and Lee 1995; Garibotti et al. 2003).

An explanation of the phenomenon could consist of an early retreat of sea ice, which, together with the increase in sea surface temperature, leads to the development of a shallow mixing layer and strong vertical water column stratification. The latter confines phytoplankton near the surface under high irradiance (Moreau et al. 2010). Mendes et al. (2018) hypothesized that the dominance of cryptophytes in coastal waters of the AP is associated with their ability to grow under high irradiance exposure due to their pigment protection capability; however, this idea has yet to be tested.

Thus, the increase in melt-water input would potentially increase the spatial and temporal extent of cryptophytes (Moline et al. 2004). Our study supports this suggestion, providing evidence of extremely high cryptophyte biomass within the low-salinity, high-temperature TBW. Historical data also show an emerging dominance of cryptophytes and other phytoflagellates in low-salinity coastal waters (Ferreyra and Tomo 1979; Whitaker 1982; Krebs 1983; Buma et al. 1992; Kopczynska 1992; McMinn and Hodgson 1993; Kang and Lee 1995). As a consequence of these processes, the zooplankton community structure undergoes cascade changes (Loeb et al. 1997; Walsh et al. 2001), leading to an increase in salp abundance and concurrent variations in penguin migration patterns (Ainley et al. 2003). If this trend persists for a long time, the food web shifts will eventually impact biogeochemical cycling in Antarctic coastal waters.The research was conducted in the frame of the Russian state assignment No. AAAA-A19-119100290162-0.

20.4 Conclusions

Two nanophytoplankton clusters were distinguished and enumerated by flow cytometry including larger cells (about 9.4 μm) of cryptophytes (CP) with bright orange fluorescence, and smaller cells of (3 μm) other nanophytoplankton (NP). Despite lower CP abundance ($0.35 \pm 0.46 \times 10^6$ cells l^{-1}) compared with NP ($1.71 \pm 1.07 \times 10^6$ cells l^{-1}), CP sufficiently exceeded NP in terms of biomass (33 ± 45 versus 4.3 ± 3.2 μg C l^{-1}). Hydrographic and chemical properties of the water masses deeply influenced nanophytoplankton structure, while temperature did not appear to be a key factor controlling NP growth. The TBW biotope was characterized by a much more abundant and diverse community with a pronounced dominance of Cryptophyta in terms of biomass. The results support the idea that the increasing melt-water input can potentially increase the spatial and temporal extent of cryptophytes in Antarctic coastal waters. The replacement of large diatoms with small cryptophytes leads to a significant shift in trophic processes in favor of the consumers, like salps, that are able to graze on smaller prey.

References

Ainley DG, Ballard G, Emslie SD et al (2003) Adélie penguins and environmental change. Science 300:429–430

Alvarez E, Nogueira E, Lypez-Urrutia B (2017) In vivo single-cell fluorescence and size scaling of phytoplankton chlorophyll content. Appl Environ Microbiol 83:1–16. https://doi.org/10.1128/aem03317-16

Boyd CM, Heyraud M, Boyd CN (1984) Feeding of the Antarctic krill *Euphausia superba*. J Crustac Biol 4(5):123–141

Buma AGJ, Gieskes WWC, Thomsen HA (1992) Abundance of Cryptophyceae and chlorophyll b-containing organisms in the Weddell-Scotia Confluence area in the spring of 1988. In: Weddell Sea ecology. Springer, Berlin, Heidelberg, pp 43–52

Castro CG, Ríos AF, Doval MD, Pérez FF (2002) Nutrient utilisation and chlorophyll distribution in the Atlantic sector of the Southern Ocean during Austral summer 1995–96. Deep-Sea Res II 49:623–641

Ferreyra GA, Tomo AP (1979) Variacion estational de las diatomeas planctonicas en Puerto Paraiso-I. Contribucion del Instituto Antartico Argentio 264:149–184

Garibotti IA, Vernet M, Ferrario ME, Smith RC, Ross RM, Quetin LB (2003) Phytoplankton spatial distribution patterns along the western Antarctic peninsula (Southern Ocean). Mar Ecol Prog Ser 261:21–39

Garibotti IA, Vernet M, Ferrario ME (2005) Annually recurrent phytoplanktonic assemblages during summer in the seasonal ice zone west of the Antarctic peninsula (Southern Ocean). Deep-Sea Res I Oceanogr Res Pap 52(10):1823–1841

Garcia MA, López O, Sospedra J, Espino M, Gracia V, Morrison G et al (1994) Mesoscale variability in the Bransfield Strait region (Antarctica) during Austral summer. Ann Geophys 12:856–867

García-Muñoz C, Lubián LM, García CM, Marrero-Díaz Á, Sangrà P, Vernet M (2013) A mesoscale study of phytoplankton assemblages around the South Shetland Islands (Antarctica). Polar Biol 36(8):1107–1123

Garrison D, Gowing M, Hughes M et al (2000) Microbial food web structure in the Arabian Sea: a US JGOFS study. Deep Sea Res Part II 47:1387–1422. https://doi.org/10.1016/S0967-0645(99)00148-4

Gonçalves-Araujo R, de Souza MS, Tavano VM, Garcia CAE (2015) Influence of oceanographic features on spatial and interannual variability of phytoplankton in the Bransfield Strait, Antarctica. J Mar Syst 142:1–15

Jacques G, Panouse M (1991) Biomass and composition of size fractionated phytoplankton in the Weddell-Scotia confluence area. Polar Biol 5:315–328

Haberman KL, Ross RM, Quetin LB (2003) Diet of the Antarctic krill (*Euphausia superba* Dana): II selective grazing in mixed phytoplankton assemblages. J Exper Mar Biol and Ecol 283(1–2):97–113

Hammer O, Harper DAT, Ryan PD (2001) PAST: paleontological statistics software package for education and data analysis. Paleontologia Electronica 4:1–9

Hoef-Emden K, Marin B, Melkonian M (2002) Nuclear and nucleomorph SSU rDNA phylogeny in the Cryptophyta and the evolution of cryptophyte diversity. J Mol Evol 55:161–179

Holm-Hansen O, Mitchell BG, Hewes CD, Karl DM (1989) Phytoplankton blooms in the vicinity of Palmer Station, Antarctica. Polar Biol 10:49–57

Holm-Hansen O, Hewes CD, Villafane VE, Helbling EW, Silva N, Amos T (1997) Distribution of phytoplankton and nutrients in relation to different water masses in the area around Elephant Island, Antarctica. Polar Biol 18(2):145–153

Jiang M, Charette MA, Measures CI, Zhu Y, Zhou M (2013) Seasonal cycle of circulation in the Antarctic Peninsula and the off-shelf transport of shelf waters into southern Drake Passage and Scotia Sea. Deep Sea Res Part II Top Stud Oceanogr 90:15–30

Kang S-H, Lee S (1995) Antarctic phytoplankton assemblage in the western Bransfield Strait region, February 1993: composition, biomass, and mesoscale distributions. Mar Ecol Prog Ser 129:253–267

Kopczynska EE (1991) Distribution of microflagellates and diatoms in the sea-ice zone between Elephant Island and the South Orkney Islands (December 1988–January 1989). Polish Polar Res 12:515–528

Kopczynska EE (1992) Dominance of microflagellates over diatoms in the Antarctic areas of deep vertical mixing and krill concentrations. J Plankton Res 14:1031–1054

Kozlowski WA, Deutschman D, Garibotti I, Trees C, Vernet M (2011) An evaluation of the application of CHEMTAX to Antarctic coastal pigment data. Deep-Sea Res I Oceanogr Res Pap 58(4):350–364

Krebs WN (1983) Ecology of neritic marine diatoms, Arthur Harbor, Antarctica. Micropaleontology 29:267–297

Loeb V, Siegel V, Holm-Hansen O, Hewitt R, Fraser W, Trivelpiece W, Trivelpiece S (1997) Effects of sea-ice extent and krill or salp dominance on the Antarctic food web. Nature 387(6636):897–900

McClatchie S, Boyd CM (1983) Morphological study of sieve efficiencies and mandibular surfaces in the Antarctic krill, *Euphausia superba*. Canadian J Fish Aquatic Sci 40(7):955–967

McMinn A, Hodgson D (1993) Summer phytoplankton succession in Ellis Fjord, eastern Antarctica. J Plankton Res 15(8):925–938

Mendes CRB, de Souza MS, Garcia VMT, Leal MC, Brotas V, Garcia CAE (2012) Dynamics of phytoplankton communities during late summer around the tip of the Antarctic peninsula. Deep-Sea Res I Oceanogr Res Pap 65:1–14

Mendes CRB, Tavano VM, Leal MC, de Souza MS, Brotas V, Garcia CAE (2013) Shifts in the dominance between diatoms and cryptophytes during three late summers in the Bransfield Strait (Antarctic peninsula). Polar Biol 36(4):537–547

Mendes CRB, Tavano VM, Dotto TS, Kerr R, De Souza MS, Garcia CAE, Secchi ER (2018) New insights on the dominance of cryptophytes in Antarctic coastal waters: a case study in Gerlache Strait. Deep-Sea Res II Top Stud Oceanogr 149:161–170

Mitchell BG, Holm-Hansen O (1991) Observations of modeling of the Antarctic phytoplankton crop in relation to mixing depth. Deep Sea Research Part A Oceanographic Res Papers 38(8–9):981–1007

Moline MA, Prezelin BB (1996) Long-term monitoring and analyses of physical factors regulating variability in coastal Antarctic phytoplankton biomass, *in situ* productivity and taxonomic composition over subseasonal, seasonal and interannual time scales. Mar Ecol Prog Ser 145:143–160

Moline MA, Claustre H, Frazer TK, Grzymski J, Vernet M (2001) Changes in phytoplankton assemblages along the Antarctic peninsula and potential implications for the Antarctic food web. Antarctic ecosystems: models for wider ecological understanding. W Davidson, C Howard-Williams and P Broady. Christchurch, New Zealand

Moline MA, Claustre H, Frazer TK, Schofield O, Vernet M (2004) Alteration of the food web along the Antarctic peninsula in response to a regional warming trend. Glob Chang Biol 10(12):1973–1980

Montes-Hugo M, Doney SC, Ducklow HW, Fraser W, Martinson D, Stammerjohn SE, Schofield O (2009) Recent changes in phytoplankton communities associated with rapid regional climate change along the western Antarctic peninsula. Science 323(5920):1470–1473

Moreau S, Ferreyra GA, Mercier B, Lemarchand K, Lionard M, Roy S, Mostajir B, Roy, van Hardenberg B, Demers S (2010) Variability of the microbial community in the western Antarctic peninsula from late fall to spring during a low ice cover year. Polar Biol 33:1599–1614

Mura MP, Satta MP, Agustí S (1995) Water-mass influences on summer Antarctic phytoplankton biomass and community structure. Polar Biol 15(1):15–20

Olson RJ, Zettler ER, DuRand MD (1993) Phytoplankton analysis using flow cytometry. Handbook of methods in aquatic microbial ecology 175–186

Niiler PP, Amos A, Hu J-H (1991) Water masses and 200 m relative geostrophic circulation in the western Bransfield Strait region. Deep Sea Res 38:943–959

Prézelin BB, Hofmann EE, Mengelt C, Klinck JM (2000) The linkage between upper circumpolar deep water (UCDW) and phytoplankton assemblages on the West Antarctic peninsula continental shelf. J Mar Res 58(2):165–202

Ross RM, Quetin LB (1985) The effect of pressure on the sinking rates of the embryos of the Antarctic krill, *Euphausia superba*. Deep Sea Res Part A Oceanographic Res Papers 32(7):799–807

Rodriguez F, Varela M, Zapata M (2002) Phytoplankton assemblages in the Gerlache and Bransfield Straits (Antarctic Peninsula) determined by light microscopy and CHEMTAX analysis of HPLC pigment data. Deep Sea Research Part II Topical Stud Oceanography 49(4–5):723–747

Sangrà P, Gordo C, Hernández-Arencibia M, Marrero-Díaz A, Rodríguez-Santana A, Stegner A, Pichon T (2011) The Bransfield current system. Deep Sea Res I 58:390–402. https://doi.org/10.1016/jdsr201101011

Sangrà P, García-Muñoz C, García CM, Marrero-Díaz Á, Sobrino C, Mouriño-Carballido B, Hernández-Arencibia M (2014) Coupling between upper ocean layer variability and size-fractionated phytoplankton in a non-nutrient-limited environment. Mar Ecol Prog Ser 499:35–46

Schofield O, Saba G, Coleman K, Carvalho F, Couto N, Ducklow H, Montes-Hugo M (2017) Decadal variability in coastal phytoplankton community composition in a changing West Antarctic Peninsula. Deep-Sea Res I Oceanogr Res Pap 124:42–54

Smith WO Jr, Sakshaug E (1990) Polar phytoplankton. Polar Oceanography, Part B WO Smith Jr:477–525

Tokarczyk R (1987) Classification of water masses in the Bransfield Strait and southern part of the Drake Passage using a method of statistical multidimensional analysis. Pol Res 8:333–366

Walsh JJ, Dieterle DA, Lenes J (2001) A numerical analysis of carbon dynamics of the Southern Ocean phytoplankton community: the roles of light and grazing in effecting both sequestration of atmospheric CO_2 and food availability to larval krill. Deep-Sea Res 48:1–48
Whitaker TM (1982) Primary production of phytoplankton off Signy Island, South Orkneys, the Antarctic. Proc R Soc Lond 214:169–189
Zhou M, Niiler PP, Zhu Y, Dorland RD (2006) The western boundary current in the Bransfield Strait, Antarctica. Deep Sea Res I 53:1244–1252. https://doi.org/10.1016/jdsr200604003

Part III
Section Marine Ecosystems and Their Oceanographical Background

Chapter 21
Phytopelagic Communities of the Powell Basin in the Summer of 2020

Philipp V. Sapozhnikov, Olga Yu. Kalinina, and Tatiana V. Morozova

Abstract Microphytic communities in the pelagic zone of the Powell Basin (Weddell Sea, Southern Ocean) were studied based on an extensive network of stations during two observation periods in 2020 (late January to early February and mid-to-late February). The surveys conducted in these periods showed a high diversity of cenoses and their significant spatial heterogeneity. In the first observation period, eight cenotic groups were identified (of which seven were in the open waters of the basin and one in the Antarctic Sound); in the second period, four groups were found, all in open waters. Substantial rearrangements in the structure of the communities by the late summer and a significant growth of their area were recorded. The complex structures that are formed on elements of tests of *Corethron*, *Chaetoceros*, and *Thalassiothrix* were described. They were based on the matrix of bacterial colonies that gradually colonize its surface and deeper layers with a variety of microphyte species, mainly from the ice and/or benthic flora, and also with planktonic species in some regions. These structures that we referred to as "aggregates," having specific genesis and hierarchical structural patterns, proliferated most intensely by the late summer. In the first period of observations, *Corethron pennatum* dominated free-floating microphytes everywhere; in the second period, *Thalassiothrix antarctica*, *Chaetoceros criophillum*, and *Ch. convolutus* were among the dominant species over a vast part of the study site. Heterotrophic flagellates abundantly proliferated within aggregates in the southern regions; benthic diatoms from the genera *Amphora*, *Halamphora*, and *Nitzschia* were also abundant in the northern regions.

P. V. Sapozhnikov (✉)
Shirshov Institute of Oceanology, Russian Academy of Sciences, Moscow, Russia

O. Yu. Kalinina
Faculty of Geography, Lomonosov Moscow State University, Moscow, Russia

T. V. Morozova
A.V. Zhirmunsky National Scientific Center of Marine Biology, Far Eastern Branch, Russian Academy of Sciences, Vladivostok, Russia

E. G. Morozov et al. (eds.), *Antarctic Peninsula Region of the Southern Ocean*, Advances in Polar Ecology 6, https://doi.org/10.1007/978-3-030-78927-5_21

Keywords Phytoplankton · Microphytes · Phytopelagic · Diatoms · Weddell Sea · Powell Basin · Antarctica · Southern Ocean

21.1 Introduction

The seas of the Antarctic region are to a greater extent subject to global changes such as climate trends that have become particularly pronounced recently. On the one hand, the ecosystems of these seas are complex and diverse and, on the other hand, very fragile. Communities of the Southern Ocean are sensitive to overharvesting of fish and other commercial resources due to their very low rate of recovery. Furthermore, these marine areas are highly productive; some of them have served as a kind of sanctuary for natural Antarctic ecosystems for many decades, while others are still subject to heavy fishing pressure from, e.g., Antarctic krill fisheries.

The previously untypical processes that have occurred in recent years due to the climate change affecting high-latitude regions (e.g., greater, than previously recorded, seasonal warming of surface waters, melting of ice, and breaking off of giant icebergs) necessitate expanding the area of research to cover these seas. This will provide not only assessment and understanding of the ecosystem rearrangements, but also prediction of their further trends, including those affecting the potential use of resources. The biological studies of the pelagic zone of the Bransfield Strait and the Powell Basin, carried out during the first and second stages of research in January and February 2020, were aimed at assessing the recovery potential of the Antarctic krill population in these waters. The main food supply for krill is pelagic communities of microalgae (microphytes) (Cleary et al. 2018; Schmidt and Atkinson 2016). Therefore, the study of the structure of these communities in the region and, in particular, manifestations of the forms of their development in terms of abundance and spatial organization is of high practical significance. The observations that we conducted in the first and second stages of research in January and February 2020 have revealed a number of currently important features of pelagic microphytic cenoses in the Powell Basin and, partly, in the Bransfield Strait, including the facts and phenomena not observed previously.

In this study, we consider the groups of communities identified on the basis of structural parameters of microphytic flora at stations using statistical methods. The presented results show the general diversity of the microphytic flora of phytopelagic communities in the study site, as well as the α-diversity and some other integral characteristics of cenoses within the selected cenotic groups. We also answer the question of why the communities under study should be considered as phytopelagic communities rather than phytoplankton in the common sense of the term. Finally, we provide a brief overview of the free-floating microphytic flora of cenoses, as well as the flora that forms structures provisionally referred to as "aggregates" here, for each of the cenotic groups. This has been done for each of the observation periods, since the study revealed a significant difference between the communities that inhabited the studied water areas in the first and second stages of research in January and February 2020.

While considering the studies on macrophyte communities inhabiting the euphotic zone in the water column of the Antarctic seas, we, first, paid attention to the difference between concepts such as phytoplankton and phytopelagic community. According to the Encyclopedia Britannica, "phytoplankton is a flora of freely floating, often minute organisms that drift with water currents; like land vegetation, phytoplankton uses carbon dioxide, releases oxygen, and converts minerals to a form animals can use" (Britannica.com: Phytoplankton). We have no reason to distrust this definition, which is applicable to all free-floating phototrophic and mixotrophic organisms, as well as to some of their heterotrophic relatives in phylogenetic groups.

Furthermore, pelagic ecosystems are inhabited by other photo-, mixotrophic organisms and their close heterotrophic relatives that migrate in the water column not freely, but either in an attached form like epibionts, or as part of more complex structures formed with the involvement of bacterial colonies, phytoplankton cells, elements of their membranes, as well as remnants of membranes and/or tests of zooplankton organisms. The self-organization of such structures can proceed "from smaller to greater," by increasing the number of live and dead structural elements. In the process of self-assembly, they frequently become hierarchically organized and generally have special forms of spatial regularity. Moreover, the composition and structure of such formations will differ between various parts of the sea, depending on the habitat conditions. Phytopelagic communities (cenoses) are distinguished from phytoplankton by the presence of aggregates, phytoplankton (single and colonial planktonic microphytes), and epiphytic/basiphytic complexes formed by young colonies of the prymnesiophyte *Phaeocystis antarctica* in the early stages of fission of cells that live solely on the surface of planktonic microphyte cells (Assmy 2009).

We propose to introduce the term "phytopelagic" by analogy with the commonly accepted term "phytobenthic" (Wilkinson 2001) that denotes the entire range of forms of live photo- and mixotrophic organisms (as well as heterotrophs from the same phylogenetic groups, which are their closest relatives) in bottom habitats.

21.2 Materials and Methods

The phytoplankton of the Powell Basin (Weddell Sea, Southern Ocean) was studied based on the materials collected during two stages of cruise 79 of the R/V *Akademik Mstislav Keldysh:* from January 21 to February 4 and from February 12 to February 24, 2020. In the first stage, samples were taken at 20 stations; in the second stage, at 19 stations (Table 21.1).

Phytoplankton samples were collected with a nylon mesh conical plankton net (50 cm in diameter, 1 m in length, mesh size of 37μm). At the stations, the net was lowered to a depth of 50 m (relative to the upper ring) and then hauled up at a speed of 0.5 m/s with an electric winch. The samples were drained into 250-ml PET bottles. At each station, three samples were taken by vertical hauls.

Table 21.1 Stations of phytoplankton sampling during cruise 79 of the R/V *Akademik Mstislav Keldysh*

Study period	Date	Station	Latitude	Longitude
Stage 1	January 22, 2020	6596	62°23.618’ S	54°44.024’ W
	January 23, 2020	6597	60°52.284’ S	53°54.161’ W
	January 24, 2020	6598	61°06.008’ S	53°11.734’ W
	January 24, 2020	6599	61°25.855’ S	52°9.168’ W
	January 25, 2020	6600	61°32.582’ S	51°49.8914’ W
	January 25, 2020	6601	60°51.728’ S	51°34.417’ W
	January 25, 2020	6602	61°00.518’ S	50°53.261’ W
	January 25, 2020	6603	61°04.277’ S	50°15.224’ W
	January 26, 2020	6604	61°06.044’ S	49°41.173’ W
	January 27, 2020	6607	61°44.474’ S	48°47.010’ W
	January 28, 2020	6608	61°12.924’ S	48°26.335’ W
	January 28, 2020	6609	60°32.657’ S	48°09.950’ W
	January 29, 2020	6613	60°10.141’ S	47°56.08’ W
	January 29, 2020	6614	60°56.232’ S	45°39.206’ W
	February 01, 2020	6619	62°22.666’ S	53°16.11’ W
	February 02, 2020	6620	62°32.374’ S	53°34.771’ W
	February 03, 2020	6622	63°11.039’ S	55°58.687’ W
	February 03, 2020	6625	63°28.952’ S	56°32.969’ W
	February 03, 2020	6627	63°22.82’ S	56°39.62’ W
	February 04, 2020	6631	61°52.476’ S	59°35.394’ W
Stage 2	February 12, 2020	6646	63° 16.6938’ S	56° 51.2365’ W
	February 13, 2020	6648	63° 29.4985’ S	56° 32.5229’ W
	February 14, 2020	6650	62° 23.0859’ S	54° 43.0859’ W
	February 14, 2020	6651	62° 32.2974’ S	53° 33.7168’ W
	February 15, 2020	6652	63° 17.3950’ S	53° 36.1108’ W
	February 16, 2020	6653	62° 27.2275’ S	52° 15.0781’ W
	February 17, 2020	6656	62° 17.1559’ S	50° 51.9994’ W
	February 18, 2020	6657	61° 42.2896’ S	49° 58.2266’ W
	February 19, 2020	6659	60° 59.8052’ S	50° 37.2458’ W
	February 19, 2020	6662	60° 42.0928’ S	50° 36.4595’ W
	February 19, 2020	6665	60° 53.9434’ S	50° 23.9878’ W
	February 20, 2020	6669	60° 47.9707’ S	50° 12.2907’ W
	February 20, 2020	6671	60° 35.9282’ S	50° 12.0400’ W
	February 20, 2020	6674	60° 35.8701’ S	50° 48.1538’ W
	February 21, 2020	6676	60° 46.7275’ S	50° 48.0513’ W
	February 21, 2020	6679	61° 06.1321’ S	50° 48.4384’ W
	February 21, 2020	6682	61° 05.4316’ S	50° 08.2642’ W
	February 23, 2020	6689	60° 32.0290’ S	49° 03.2497’ W
	February 24, 2020	6691	59° 45.2524’ S	50° 30.7070’ W

The first of the samples was fixed with a mixture of a 70% ethanol solution and a 2.5% formaldehyde solution and then placed in a dark box for settling during four days. After that, the upper layer of the supernatant liquid was carefully removed from the sample using a disposable PE pipette so that a layer of at least 3 cm of the fixative solution remained above the sediment surface.

The second sample was thoroughly mixed and viewed live under a Carl Zeiss Primo Star Axio Zoom light microscope at magnifications of 200x, 400x, and 1000x. This visual analysis was performed directly in the onboard laboratory, while the vessel was en route between stations. Live phytoplankton was photographed using the digital camera of a BlackView 5800 smartphone (with a resolution of 13 MP) and a Levenguk photo camera built directly into the eyepiece tube (with a resolution of 5 MP). The images (about 100 per each station) were later used for the taxonomic identification of a number of microphytic species, the assessment of the relative abundance of certain species, and the determination of the spatial organization of compact formations in phytopelagic communities found in the samples and provisionally referred to as "aggregates." The third sample was fixed with a 5% Lugol's iodine solution.

In addition, phytoplankton samples were collected from the discharge hose of the onboard flow-through system during the ship's travel between stations. The water intake by this system from a depth of 2.5–3.0 m was continuous as the vessel was en route. In particular, such samples were taken between stations 6688 and 6689 (February 23, 2020, a 15 km segment – trench #1), as well as at the entrance to the Bransfield Strait (February 26, 2020, a 14 km segment between the points with coordinates 62°07'48.3421" S, 56°40'37.9536" W and 62°10'39.2279" S, 56°56'54.0312" W – trench #2) and in the northeastern part of the strait (February 26, 2020, a 17 km segment between the points with coordinates 62°26'45.4455" S, 58°32'17.7678" W and 62°21'15.6420" S, 58°49'09.6120" W – trench #3). These samples were collected on a nylon filter with a mesh size of 37μm installed immediately at the outlet of the dischar+ge hose. The sampling was completed when the filter clogged and water stopped flowing through. The flow rate of the system at the hose outlet was approximately 60 L of seawater per 1 km of the vessel's travel. Thus, the volume of filtered water was ≈900 l for the first, ≈840 l for the second, and ≈1020 l for the third sample.

The samples were washed off the nylon filter with 250 ml of filtered seawater into PET flasks, thoroughly mixed, and viewed live under a microscope with photographing the components of communities. The analysis of these samples was aimed both at identifying the composition and structure of near-surface phytoplankton at the sampling sites and studying the structure of aggregates.

All ethanol-fixed samples were then transported to a laboratory at the Shirshov Institute of Oceanology RAS (Moscow) and additionally examined under Leica DMLS and Leica DM 2500 light microscopes at magnifications of 400x and 1000x for a more detailed identification of a number of microphytic species and analysis of the species structure of the community (relative abundance of species). During the analysis of this material, a number of species were also photographed with the built-in digital camera of the Leica DM 2500 microscope.

The Lugol's-fixed samples were transported to the A.V. Zhirmunsky National Scientific Center of Marine Biology (NSCMB FEB RAS, Vladivostok). There the material was photographed by scanning electron microscopy (SEM). For SEM, cells were placed on 2-μm nuclepore filters in a Swinnex filter-holder. The cells were washed twice with distilled water and passed through a graded series of alcohol baths (25%, 50%, 75%, and 96%). The filters with cells were air-dried and sputter-coated with chrome on an Edwards Auto 306 vacuum coater (Crawley, Great Britain). Observations were carried out using a Sigma 300 VP electron microscope (Carl Zeiss, Cambridge, Great Britain) at the Center for Collective Use "Far East Center of Electron Microscopy," NSCMB FEB RAS.

Statistical processing of the data on relative abundance of multiple microphytic species in the communities at the stations was performed using the algorithms of the program for environmental data analysis available in the Primer-6 software. For taxonomic identification of various microphytic species, we used the latest and historical atlases and keys, as well as scientific papers (containing species diagnoses and illustrative material) on the phytoplankton of the Antarctic region and those parts of the World Ocean where the same species as in Antarctica were described (Stickley et al. 2006; Rytter Hasle and Lange 1992; Johansen and Fryxell 1985; Fryxell and Prasad 1990; Dodge and Hermes 1981; Cefarelli et al. 2010, 2015; Al-Handal, Wulf 2008; Park et al. 2016; Scott and Marchant 2005; Gilbert 1991; Hasle nad Syvertsen 1996; Hasle and de Mendiola 1967; Dietrich and Arndt 2004; Tong et al. 1997). The species that made up more than 30% of the total abundance of microphytes were considered dominant.

21.3 Results

21.3.1 Phytopelagic Cenoses during the First Stage of the Expedition (January 21, 2020–February 4, 2020)

A total of 60 species and subspecies of eukaryotic microphytes were found at the integrated research stations where phytopelagic cenoses were studied. Photographs of some of them are shown in Fig. 21.1. They formed communities of various types in various combinations. Of them, 37 species were recorded exclusively as free-floating, represented by colonial or solitary forms (Table 21.2); 17 species and subspecies were found only in aggregates (Table 21.2), including the diatom *Chaetoceros socialis* represented only by resting cells (spores). Massive tree-like colonies of vegetative *C. socialis* cells were recorded only from the stations in the Antarctic Sound, whereas spores of this species were absent in aggregates from this region.

Six diatoms species, *Pseudo-nitzschia turgiduloides*, *P. calliantha, Nitzschia* sp. 2, *Haslea trompii, Actinocyclus actinochilus*, and *Fragilariopsis nana* were found not only in aggregates, but also in free-living form. However, it should be noted that *A. actinochilus* in the vast majority of cases was observed in the form of single

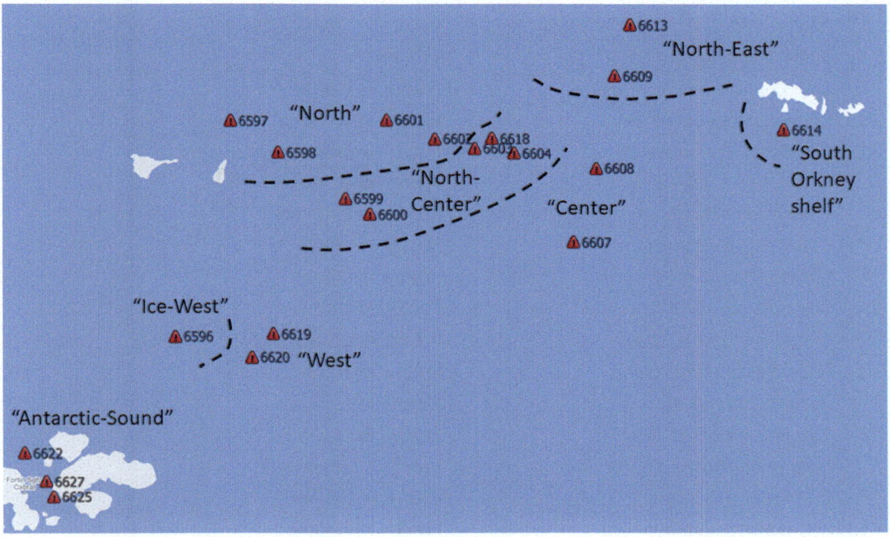

Fig. 21.1 Areas of distribution of cenotic groups in the first stage of research (late January to early February)

Table 21.2 Species composition of eukaryotic microphytes from phytopelagic cenoses of the Powell Basin from late January to early February, 2020

No.	Species and subspecies	Life forms
1	*Cafeteria roenbergensis*	A part of aggregates
2	Cysts of Chrysophyta	A part of aggregates
3	*Fragilariopsis curta*	A part of aggregates
4	*Fragilariopsis cylindrus*	A part of aggregates
5	*Fragilariopsis pseudonana*	A part of aggregates
6	*Fragilariopsis sublinearis*	A part of aggregates
7	*Licmophora gracilis*	A part of aggregates
8	*Navicula directa*	A part of aggregates
9	*Nitzschia* cf. *acicularis*	A part of aggregates
10	*Nitzschia lecointei*	A part of aggregates
11	*Odontella litigiosa*	A part of aggregates
12	*Paraphysomonas* sp. 1	A part of aggregates
13	*Pinnularia quadratarea* var. *soederlundii*	A part of aggregates
14	*Ploeotia* sp.	A part of aggregates
15	*Seminavis* sp. 1	A part of aggregates
16	Spores of *Chaetoceros socialis*	A part of aggregates
17	*Thalassiosira gracilis*	A part of aggregates
18	*Fragilariopsis nana*	A part of aggregates and freely
19	*Phaeocystis antarctica* (early stages of colony)	Epiphytic
20	*Chaetoceros atlanticus*	Freely, in colonies

(continued)

Table 21.2 (continued)

No.	Species and subspecies	Life forms
21	*Chaetoceros concavicornis*	Freely, in colonies
22	*Chaetoceros convolutus*	Freely, in colonies
23	*Chaetoceros criophilum*	Freely, in colonies
24	*Chaetoceros debilis*	Freely, in colonies
25	*Chaetoceros dichaeta*	Freely, in colonies
26	*Chaetoceros flexuosus*	Freely, in colonies
27	*Chaetoceros hendeyi*	Freely, in colonies
28	*Chaetoceros socialis*	Freely, in colonies
29	*Chaetoceros tortissimus*	Freely, in colonies
30	*Dactylosolen antarcticus*	Freely, in colonies
31	*Eucampia antarctica*	Freely, in colonies
32	*Fragilaria* cf. *islandica*	Freely, in colonies
33	*Leptocylindrus mediterraneus*	Freely, in colonies
34	*Odontella veissflogii*	Freely, in colonies
35	*Pseudo-nitzschia heimii*	Freely, in colonies
36	*Pseudo-nitzschia lineola*	Freely, in colonies
37	*Pseudo-nitzschia subcurvata*	Freely, in colonies
38	*Pseudostaurosira brevistriata*	Freely, in colonies
39	*Thalassiosira gravida*	Freely, in colonies
40	*Thalassiosira tumida*	Freely, in colonies
41	*Corethron pennatum*	Mostly free, single
42	*Coscinodiscus radiatus*	Mostly free, single
43	*Dictyocha speculum*	Mostly free, single
44	*Diplopsalis lenticula*	Mostly free, single
45	*Entomoneis kjellmannii*	Mostly free, single
46	*Geminigera cryophila*	Mostly free, single
47	*Nitzschia australis*	Mostly free, single
48	*Ochromonas* sp.	Mostly free, single
49	*Proboscia alata*	Mostly free, single
50	*Proboscia* cf. *truncata*	Mostly free, single
51	*Proboscia inermis*	Mostly free, single
52	*Protoperidinium archiovatum*	Mostly free, single
53	*Rhizosolenia antennta* f. *semispina*	Mostly free, single
54	*Stellarima microtrias*	Mostly free, single
55	*Thalassiothrix antarctica*	Mostly free, single
56	*Trichotoxon reinboldii*	Mostly free, single
57	*Pseudo-nitzschia calliantha*	Mostly free, single and colonial, and often in aggregates
58	*Pseudo-nitzschia turgiduloides*	Mostly free, single and colonial, and often in aggregates
59	*Actinocyclus actinochilus*	Mostly free, single and often in aggregates
60	*Haslea trompii*	Mostly free, single and often in aggregates
61	*Nitzschia* sp. 2	Mostly free, single and often in aggregates

free-floating cells, and only once as part of an aggregate. In contrast, *F. nana* was recorded almost always in aggregates and only once in free-floating form, such as a small colony.

We found prymnesiophyte *Phaeocystis antarctica* at 14 stations out of 20, but only at the stage of initial cells of future colonies, as well as at the stages of their fission with the formation of a large number of cells enclosed under a common spherical shell. These stages of the *Ph. antarctica* life cycle were only found as epiphytes, on test elements of *Corethron*, and, occasionally, on setae of *Chaetoceros*.

Based on the data on the relative abundance of species in cenoses at the stations, we assessed the degree of similarity and difference between these communities in their quantitative structure. Using the Bray-Curtis similarity index (Primer 6 (PRIMER E)) and successively applying the procedures of *Multidimensional scaling* (MDS) and *Cluster* analysis, as well as the ANOSIM-1 algorithm, we identified eight cenotic groups (Global $R = 0.943$; $p = 0.1$). Each lived in a certain region of the study site. The general list of groups is provided in Table 21.3, which indicates the stations where their communities were recorded. This table shows the total number of species for each group, as well as the number of species found as free floating, the number of species in aggregates, and the number of epiphytic species. The approximate location of the areas occupied by the cenotic groups in the study area is shown in Fig. 21.1.

The West, North-Center, and Antarctic Sound groups were the richest in species. Furthermore, while having a significantly lower total richness of the flora, the cenoses in the fibrous aggregates floating at the surface among the ice fields, i.e., the Ice-West group (station 6596), proved to be exceptionally rich in species.

Table 21.3 Groups of cenoses in the waters of the Powell Basin in the first stage of the expedition (January 21–February 4, 2020)

Names of groups of cenoses	North-East	South Orkney shelf	West	North	North-Center	Center	Ice-West	Antarctic Sound
Stations	6609, 6613	6614	6619, 6620	6601, 6602, 6597, 6598	6603, 6604, 6618, 6599, 6600	6607, 6608	6596	6622, 6625, 6627
Number of free floating species	6	1	17	4	21	6	2	23
Number of species in aggregates	0	6	9	2	6	2	11	3
Epiphytic species	0	1	1	1	1	1	0	0
Total number of species in the group	6	7	24	5	24	8	13	26

The species forming each group of cenoses are listed in Table 21.4 indicating the life form of their presence in the pelagic zone and their relative quantitative level. The table shows that the planktonic diatom *Corethron pennatum* is not only the most widely distributed species in the study sites, but also dominates everywhere. None of the other species were found simultaneously in all the cenotic groups.

When considering the similarity and difference of the identified groups (using the Bray-Curtis similarity index), we obtained an average similarity between them at a level of 38.26%, which is a relatively low level. The groups differed from each other significantly in their structure. This means that both the free-living microphytic flora and that developing within aggregates formed community patterns that significantly varied between various parts of the study site with the exception of the ubiquitous dominance of *C. pennatum*.

Below, we consider the structure of aggregates found in the area of each cenotic group more in detail. In the communities of the North-East group, no aggregates were recorded: here we observed a mass proliferation of free-living forms, with the mixotrophic cryptophyte *Geminigera cryophila* reaching a significant abundance.

Within cenoses of the South Orkney shelf group, the aggregates formed on the basis of abundant bacterial colonies: *Fragilariopsis nana* and *F. pseudonana* (members of the ice flora) proliferated to large numbers in the matrix, while the heterotrophic flagellate *Cafeteria roenbergensis* was frequently found on the surface. Cells of *C. roenbergensis* attached to the surface of the occupied substrate with one flagellum and used the other one to create an active flow of pico-particles toward their mouthparts and move to short distances around the point of attachment.

A structural feature of aggregates of the West group was the mass proliferation of not only the ice-associated species such as *F. nana* and *F. curta*. Along with them, two planktonic species of *Nitzschia* (*N. lecontei* and *Nitzschia* sp. 2), as well as a member of the typically benthic genus *Seminavis* sp., were abundant here.

Aggregates of the North group cenoses were organized through the mass proliferation of cells of the colonial diatom *Pseudonitzschia calliantha*. The structure of their colonies became completely destroyed, and the cells lived singly and in groups on the surface of a relatively thin bacterial matrix that developed on setae of *Corethron*.

In the communities of the North-Center group, the organization of aggregates was somewhat more complex and diverse: abundant, often mixed colonial aggregations of *P. calliantha* and *P. turgiduloides* with a small portion of *F. nana* and *F. curta* grew on a relatively thin layer of the bacterial matrix that enveloped setae and valves of some of *Corethron* cells.

In deep layers and on the surface of the matrix of aggregates in the Center group, only *F. nana* cells were found frequently.

In the structure of cord aggregates of the Ice-West group recorded from ice fields in the west of the basin, a very specific pattern of the ice-associated flora community developed on the plicate biopolymer substrate of empty sheaths of *Berckeleya adeliensis*. The absolutely dominant species here was *F. nana* mostly represented by single cells relatively evenly distributed over folds of the main substrate and on the matrix of bacterial colonies. Common species were *F. curta* and cysts of flagellates from the Chrysophyta group. Compact groups of *Fragilariopsis cylindrus* and

Table 21.4 Species structure of the cenotic groups in the Powell Basin from late January to early February 2020

Life forms	Species and subspecies	Relative quantitative level of species in the cenotic groups							
		North-East	South Orkney shelf	West	North	North-Center	Center	Ice-West	Antarctic Sound
A part of aggregates	*Actinocyclus actinochilus*			Single					
A part of aggregates	*Cafeteria roenbergensis*		Frequently						
A part of aggregates	*Cysts of Chrysophyta*							Abundant	
A part of aggregates	*Fragilariopsis curta*			Abundant		Rarely		Abundant	
A part of aggregates	*Fragilariopsis cylindrus*							Frequently	
A part of aggregates	*Fragilariopsis nana*		Abundant	Abundant		Rarely	Frequently	Abundant & dominant	
A part of aggregates	*Fragilariopsis pseudonana*		Abundant						
A part of aggregates	*Fragilariopsis sublinearis*							Rarely	
A part of aggregates	*Haslea trompii*							Rarely	
A part of aggregates	*Licmophora gracilis*								Rarely
A part of aggregates	*Navicula directa*							Rarely	
A part of aggregates	*Nitzschia* cf. *acicularis*			Single					

(continued)

Table 21.4 (continued)

Life forms	Species and subspecies	Relative quantitative level of species in the cenotic groups							
		North-East	South Orkney shelf	West	North	North-Center	Center	Ice-West	Antarctic Sound
A part of aggregates	*Nitzschia lecointei*			Abundant				Frequently	
A part of aggregates	*Nitzschia* sp. 2			Abundant					
A part of aggregates	*Odontella litigiosa*								Rarely
A part of aggregates	*Paraphysomonas* sp. 1							Frequently	
A part of aggregates	*Pinnularia quadratarea* var. *soederlundii*							Rarely	
A part of aggregates	*Ploeotia sp.*		Rarely						
A part of aggregates	*Pseudo-nitzschia calliantha*				Abundant	Abundant			
A part of aggregates	*Pseudo-nitzschia turgiduloides*					Abundant			
A part of aggregates	*Seminavis* sp. 1			Abundant				Rarely	
A part of aggregates	Spores of *Chaetoceros socialis*			Single					
A part of aggregates	*Thalassiosira gracilis*		Rarely			Single			Rarely
Epiphytic	*Phaeocystis antarctica*		Single	Single	Rarely	Frequently	Frequently		
Freely, in colonies	*Chaetoceros atlanticus*			Rarely					

Freely, in colonies	*Chaetoceros concavicornis*								Abundant
Freely, in colonies	*Chaetoceros convolutus*			Abundant		Frequently			Abundant
Freely, in colonies	*Chaetoceros criophilum*			Frequently		Frequently	Abundant		Abundant
Freely, in colonies	*Chaetoceros debilis*			Rarely					
Freely, in colonies	*Chaetoceros dichaeta*			Rarely	Frequently	Frequently	Frequently		Frequently
Freely, in colonies	*Chaetoceros flexuosus*						Rarely		Frequently
Freely, in colonies	*Chaetoceros hendeyi*								Frequently
Freely, in colonies	*Chaetoceros tortissimus*								Abundant
Freely, in colonies	*Dactylosolen antarcticus*					Singularly			
Freely, in colonies	*Eucampia antarctica*								Frequently
Freely, in colonies	*Fragilaria* cf. *islandica*			Rarely		Rarely			Abundant
Freely, in colonies	*Fragilariopsis nana*	Single							
Freely, in colonies	*Leptocylindrus mediterraneus*						Single		
Freely, in colonies	*Odontella veissflogii*			Rarely					Abundant

(continued)

Table 21.4 (continued)

Life forms	Species and subspecies	Relative quantitative level of species in the cenotic groups							
		North-East	South Orkney shelf	West	North	North-Center	Center	Ice-West	Antarctic Sound
Freely, in colonies	*Pseudo-nitzschia calliantha*				Abundant & dominant	Abundant & dominant			
Freely, in colonies	*Pseudo-nitzschia heimii*					Rarely			
Freely, in colonies	*Pseudo-nitzschia lineola*			Rarely					Abundant
Freely, in colonies	*Pseudo-nitzschia subcurvata*					Frequently			
Freely, in colonies	*Pseudo-nitzschia turgiduloides*					Abundant			Frequently
Freely, in colonies	*Pseudostaurosira brevistriata*								Rarely
Freely, in colonies	*Thalassiosira gravida*					Rarely			Frequently
Freely, in colonies	*Thalassiosira tumida*			Single		Single			Abundant
Freely, in colonies	*Chaetoceros socialis*								Abundant & dominant
Mostly free, single	*Actinocyclus actinochilus*	Rarely		Abundant		Single			
Mostly free, single	*Corethron pennatum*	Abundant & dominant	Abundant & dominant	Abundant & dominant	Abundant & dominant	Abundant & dominant	Abundant & dominant	Abundant & dominant	Abundant & dominant
Mostly free, single	*Coscinodiscus radiatus*								Frequently
Mostly free, single	*Dictyocha octonaria*					Rarely			

Mostly free, single	*Diplopsalis lenticula*	Rarely				Rarely			Rarely
Mostly free, single	*Entomoneis kjellmannii*			Single					Rarely
Mostly free, single	*Geminigera cryophila*	Abundant & dominant							
Mostly free, single	*Haslea trompii*					Single			
Mostly free, single	*Nitzschia australis*								Rarely
Mostly free, single	*Nitzschia* sp. 2			Single					
Mostly free, single	*Ochromonas* sp.	Single							
Mostly free, single	*Proboscia alata*			Single					
Mostly free, single	*Proboscia* cf. *truncata*								Frequently
Mostly free, single	*Proboscia inermis*					Rarely			
Mostly free, single	*Protoperidinium archiovatum*								Single
Mostly free, single	*Rhizosolenia antennta* f. *semispina*					Single			
Mostly free, single	*Stellarima microtrias*			Abundant		Single		Rarely	Abundant
Mostly free, single	*Thalassiothrix antarctica*			Single	Single	Single	Frequently		
Mostly free, single	*Trichotoxon reinboldii*			Single		Single			

Note: a blank cell in the table means that the species was not recorded at that location

N. lecontei, vegetative cells of the flagellate *Paraphysomonas* sp. 1, and, occasionally, compact groups of large cells of the ice-associated diatom *Pinnularia quadratarea* var. *soederlundii* and, separately from them, single cells of *Seminavis* sp. were also frequently found.

This list of aggregate structures is not complete.

21.3.2 Phytopelagic Cenoses in the Second Stage of the Expedition (February 12–24, 2020)

During the second stage of studies, the pattern of structural features of phytopelagic cenoses in the Powell Basin changed substantially. First, only 54 species and subspecies of microphytes were identified within the communities in the entire study site. However, the water area covered by the network of stations was slightly larger than in the first stage, but did not include the Antarctic Sound. Second, as many as 22 species were identified within aggregates, and three as epiphytic forms. Thus, in addition to *Ph. antarctica*, two small-sized species of *Fragilariopsis* were also recorded: *F. nana* and *F. pseudonana*.

By applying the same statistical methods to the analysis of similarities and differences of communities at the stations as those in the first stage of the research, we identified only four groups of cenoses that were unevenly distributed over multiple parts of the basin. Furthermore, we found that intergroup differences substantially prevailed over intragroup ones (Global $R = 0.987$) on a general level of differences, which was close to highly significant ($p = 0.1\%$). Table 21.5 shows these new cenotic groups that divided the study site into a new pattern. According to the table, the number of species generally increased both in the entire flora and in the flora in aggregates. This fits into the concept of complication of species compositions of microphytes by late summer that occurs due to successional processes. The approximate location of cenotic groups in the water area is shown in Fig. 21.2.

Table 21.5 Groups of cenoses in waters of the Powell Basin in the second stage of the expedition (February 12–24, 2020)

Names of groups of cenoses	Center	North	Group from the trench	South
Stations	6656, 6659, 6662, 6665, 6669, 6679, 6682	6669, 6671, 6672, 6674, 6676, 6684	Trench #1, trench #2, trench #3	6650, 6651, 6652, 6653, 6657
Number of free floating species	12	19	14	11
Number of species in aggregates	10	14	16	6
Epiphytic species	3	3	3	1
Total number of species in the group	21	32	29	16

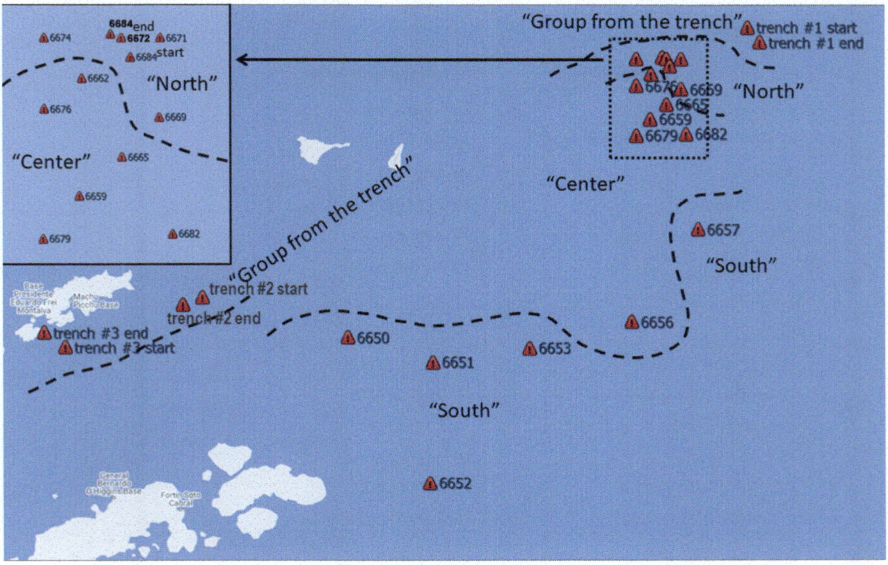

Fig. 21.2 Areas of distribution of cenotic groups in the water area at the second stage of research (mid-to-late February)

A complete list of macrophyte species (not only eukaryotic, but also with the presence of the cyanoprokaryotic species *Anabaena* sp. 1), with indicating their life forms, is provided in Table 21.6. The table shows that many species that had not been recorded in the first stage of observations appeared in the flora of the basin. This especially applies to aggregates: their structural diversity increased significantly.

Tables 21.7, 21.8, 21.9 and 21.10 below successively show the cenotic groups identified in various parts of the study site during the second stage of the research indicating the stations corresponding to the group. Species within the groups are listed in accordance with their significance in the community structure. For some of the species, ranges of their statuses in the structure of cenoses are indicated. In the first stage of observations, there were many groups, and each included from one to five stations; in the third stage, however, each group covered a larger number of stations. It should also be noted that in the second stage of the expedition, the average similarity in structure between the communities was 20.72%, which was almost twice as low as that in the first stage.

The aggregates that developed within the communities of the identified groups also differed significantly in their structure. Thus, small, ribbon-like colonies of *Fragilariopsis kerguelensis*, with rare presence of the periphytic species *Licmophora gracilis*, cysts of the dinoflagellates *Polarella glacialis* and *Pseudogomphonema kamtschaticum*, represented occasionally by single cells, frequently developed in aggregates of the southern cenotic group (Table 21.7). This group and its aggregates were mostly found close to the boundaries of ice fields.

Table 21.6 Species composition of microphytes from phytopelagic cenoses of the Powell Basin in late February 2020

No.	Species and subspecies	Life forms
1	*Amphora* sp. 1 (large)	In aggregates
2	*Amphora* sp. 2 (small, groups)	In aggregates
3	*Anabaena* sp. 1	In aggregates
4	*Bodo designis*	In aggregates
5	*Cafeteria roenbergensis*	In aggregates
6	*Fragilariopsis curta*	In aggregates
7	*Fragilariopsis kerguelensis*	In aggregates
8	*Halamphora* sp. 1 (large)	In aggregates
9	*Licmophora antarctica*	In aggregates
10	*Licmophora gracilis*	In aggregates
11	*Navicula* cf. *gelida* var. *parvula*	In aggregates
12	*Navicula* sp. 1	In aggregates
13	*Nitzschia* cf. *decipiens*	In aggregates
14	*Nitzschia* sp. 4 (groups)	In aggregates
15	*Polarella glacialis*	In aggregates
16	*Pseudogomphonema kamtschaticum*	In aggregates
17	*Pterosperma parallelum*	In aggregates
18	*Fragilariopsis nana*	In aggregates and epiphytic
19	*Fragilariopsis pseudonana*	In aggregates and epiphytic
20	*Phaeocystis antarctica*	Epiphytic and freely, in colonies
21	*Chaetoceros convolutus*	Freely, in colonies
22	*Chaetoceros criophilus*	Freely, in colonies
23	*Chaetoceros debilis*	Freely, in colonies
24	*Chaetoceros socialis*	Freely, in colonies
25	*Eucampia antarctica*	Freely, in colonies
26	*Pseudo-nitzschia* cf. *lineola*	Freely, in colonies
27	*Pseudo-nitzschia heimii*	Freely, in colonies
28	*Thalassiosira frenguellii*	Freely, in colonies
29	*Thalassiosira gracilis*	Freely, in colonies
30	*Thalassiosira gravida*	Freely, in colonies
31	*Thalassiosira tumida*	Freely, in colonies
32	*Actinocyclus actinochilus*	Mostly free, single
33	*Ceratium extensum*	Mostly free, single
34	*Corethron pennatum*	Mostly free, single
35	*Coscinodiscus asteromphalus*	Mostly free, single
36	*Dictyocha octonaria*	Mostly free, single
37	*Dictyocha speculum*	Mostly free, single
38	*Geminigera cryophila*	Mostly free, single
39	*Gymnodinium* sp. 1	Mostly free, single

(continued)

Table 21.6 (continued)

No.	Species and subspecies	Life forms
40	*Gyrodinium lachryma*	Mostly free, single
41	*Hantzschia* sp. 1	Mostly free, single
42	*Navicula* cf. *bergstromiana*	Mostly free, single
43	*Navicula* cf. *cristata*	Mostly free, single
44	*Nitzschia lecontei*	Mostly free, single
45	*Nitzschia longissima*	Mostly free, single
46	*Porosira glacialis*	Mostly free, single
47	*Protoperidinium defectum*	Mostly free, single
48	*Stellarima microtrias*	Mostly free, single
49	*Talassiothrix Antarctica*	Mostly free, single
50	*Trichotoxon reinboldii*	Mostly free, single
51	*Tropidoneis* sp. 1	Mostly free, single
52	*Pseudo-nitzschia calliantha*	Mostly free, single and colonial, and often in aggregates
53	*Pseudo-nitzschia subcurvata*	Mostly free, single and colonial, and often ins aggregates
54	*Pseudo-nitzschia turgiduloides*	Mostly free, single and colonial, and often in aggregates

Table 21.7 Southern cenotic group: stations 6650, 6651, 6652, 6653, 6657

Species and subspecies	Relative quantitative level of species in the cenotic groups
Chaetoceros convolutus	Abundant & dominant
Chaetoceros criophilus	Abundant & dominant
Corethron pennatum	Abundant & dominant
Talassiothrix Antarctica	Rarely → abundant
Chaetoceros socialis	Not recorded → abundant
Fragilariopsis kerguelensis (in aggregates)	Not recorded → frequently
Eucampia antarctica	Not recorded → frequently
Phaeocystis antarctica (epiphytically)	Single → rarely
Licmophora gracilis (in aggregates)	Not recorded → rarely
Polarella glacialis (in aggregates)	Not recorded → rarely
Coscinodiscus wailesii	Not recorded → rarely
Pseudo-nitzschia subcurvata	Not recorded → rarely
Pseudo-nitzschia turgiduloides	Not recorded → rarely
Actinocyclus actinochilus	Not recorded → single
Trichotoxon reinboldii	Not recorded → single
Pseudogomphonema kamtschaticum (in aggregates)	Not recorded → single

Table 21.8 Central group: 6656, 6659, 6662, 6665, 6669, 6679, 6682

Species and subspecies	Relative quantitative level of species in the cenotic groups
Talassiothrix Antarctica	Abundant & dominant
Cafeteria roenbergensis (in aggregates)	Abundant & dominant
Fragilariopsis nana (in aggregates and epiphytically)	Rarely → abundant & dominant
Fragilariopsis pseudonana (in aggregates and epiphytically)	Rarely → abundant & dominant
Corethron pennatum	Frequently → abundant
Chaetoceros criophilus	Rarely → abundant
Chaetoceros convolutus	Not recorded → abundant
Bodo designis (in aggregates)	Not recorded → abundant
Pseudo-nitzschia calliantha	Not recorded → abundant
Pseudo-nitzschia heimii	Not recorded → abundant
Pseudo-nitzschia turgiduloides	Not recorded → abundant
Phaeocystis antarctica (epiphytically)	Not recorded → frequently
Licmophora gracilis (in aggregates)	Not recorded → frequently
Thalassiosira gracilis	Not recorded → frequently
Stellarima microstrias	Not recorded → rarely
Polarella glacialis (in aggregates)	Not recorded → rarely
Pterosperma parallelum (in aggregates)	Not recorded → rarely
Ceratium extensum	Not recorded → single
Nitzschia longissima	Not recorded → single
Dictyocha speculum	Not recorded → single
Fragilariopsis curta (in aggregates)	Not recorded → single

Table 21.9 Northern group: 6669, 6671, 6672, 6674, 6676, 6684

Species and subspecies	Relative quantitative level of species in the cenotic groups
Pseudo-nitzschia calliantha	Abundant & dominant
Cafeteria roenbergensis (in aggregates)	Frequently → abundant & dominant
Bodo designis (in aggregates)	Rarely → abundant & dominant
Talassiothrix Antarctica	Abundant
Pseudo-nitzschia turgiduloides	Abundant
Corethron pennatum	Frequently → abundant
Licmophora gracilis (in aggregates)	Rarely → abundant
Amphora sp. 2 (small-sized, in groups, in aggregates)	Singularly → abundant
Nitzschia sp. 4 (in aggregates, in groups)	Not recorded → abundant
Protoperidinium defectum	Frequently
Phaeocystis antarctica (epiphytic)	Rarely → frequently
Fragilariopsis nana (in aggregates and epiphytically)	Rarely → frequently

(continued)

Table 21.9 (continued)

Species and subspecies	Relative quantitative level of species in the cenotic groups
Fragilariopsis pseudonana (in aggregates and epiphytically)	Rarely → frequently
Pseudo-nitzschia heimii	Singularly → frequently
Cryptomonas sp. 1	Not recorded → frequently
Chaetoceros convolutus	Rarely
Pterosperma parallelum (in aggregates)	Single → rarely
Chaetoceros debilis	Not recorded → rarely
Thalassiosira gracilis	Not recorded → rarely
Amphora sp. 1 (large-sized, in aggregates)	Not recorded → rarely
Gymnodinium sp. 1	Not recorded → rarely
Nitzschia cf. *decipiens* (in aggregates)	Not recorded → rarely
Phaeocystis antarctica (colonies)	Not recorded → rarely
Tropidoneis sp. 1	Not recorded → rarely
Pseudo-nitzschia cf. *lineola*	Not recorded → single
Dictyocha octonaria	Not recorded → single
Dictyocha speculum	Not recorded → single
Gyrodinium lachryma	Not recorded → single
Pseudo-nitzschia cf. *lineola*	Not recorded → single
Hantzschia sp. 1	Not recorded → single
Navicula cf. *gelida* var. *parvula* (in aggregates)	Not recorded → single
Nitzschia lecontei	Not recorded → single

Table 21.10 The group of the trench: three samplings from the discharge hose of the onboard hydrological flow-through system

Species and subspecies	Relative quantitative level of species in the cenotic groups
Amphora sp. 2 (small-sized, in groups, in aggregates)	Abundant & dominant
Corethron pennatum	Frequently → abundant & dominant
Nitzschia sp. 4 (in aggregates, in groups)	Not recorded → abundant & dominant
Licmophora gracilis (in aggregates)	Rarely → abundant
Halamphora sp. 1 (large-sized, in aggregates)	Singularly → abundant
Cafeteria roenbergensis (in aggregates)	Singularly → abundant
Bodo designis (in aggregates)	Not recorded → abundant
Talassiothrix Antarctica	Not recorded → abundant
Thalassiosira tumida	Not recorded → abundant
Thalassiosira gravida	Not recorded → abundant
Pseudo-nitzschia calliantha	Not recorded → frequently
Pseudo-nitzschia turgiduloides	Not recorded → frequently
Protoperidinium defectum	Not recorded → frequently
Thalassiosira frenguellii	Not recorded → frequently

(continued)

Table 21.10 (continued)

Species and subspecies	Relative quantitative level of species in the cenotic groups
Fragilariopsis nana (in aggregates and epiphytically)	Rarely
Navicula cf. *gelida* var. *parvula* (in aggregates)	Single → rarely
Thalassiosira gracilis	Not recorded → rarely
Anabaena sp. 1 (in aggregates)	Not recorded → rarely
Chaetoceros debilis	Not recorded → rarely
Fragilariopsis kerguelensis (in aggregates)	Not recorded → rarely
Fragilariopsis pseudonana (in aggregates and epiphytically)	Not recorded → rarely
Porosira glacialis	Not recorded → rarely
Navicula cf. *bergstromiana*	Not recorded → rarely
Navicula glaciei (in aggregates)	Not recorded → single
Amphora sp. 1 (large-sized, in aggregates)	Not recorded → single
Pseudo-nitzschia heimii	Not recorded → single
Navicula cf. *cristata*	Not recorded → single
Nitzschia cf. *decipiens* (in aggregates)	Not recorded → single

Within the Center cenotic group (Table 21.8), aggregates formed on the basis of matrix of bacterial colonies stretched over long and thin tests of *Thalassiothrix antarctica* (the dominant of communities) like sails. In many cases, such matrix combined several cells of this diatom into bundles and also bound different bundles into larger and branched structures with nodes. Attached flagellates *Cafeteria roenbergensis* lived in high abundance on the surface of the matrix, with flagellates *Bodo designis* moving among them freely and in large numbers. *F. nana* and *F. pseudonana* were abundant (these species could develop epiphytically, directly on cells of *Th. antarctica*), and cells and small colonies of *Licmophora gracilis* were also frequently found. The three other species occurred rarely or singly.

Aggregates of the North cenotic group (Table 21.9) differed significantly from all aggregates that developed more southerly. There the attached (*C. roenbergensis*) and freely moving flagellates (*B. designis*) occurred in large numbers with the general dominance of colonies and single cells of *Pseudo-nitzschia calliantha*. However, compact, but very abundant, colonial aggregations of the small-size benthic species *Amphora* sp. 2 and *Nitzschia* sp. 4 could also actively proliferate in the bacterial matrix. The ice-associated *F. nana* and *F. pseudonana* were sometimes frequently found here, both in the matrix and epiphytically. A few other species, including the large-sized benthic *Amphora* sp. 1 and *Nitzschia* cf. *decipiens*, could live within these aggregates, which developed on cells and dead tests of *Th. antarctica*, but as rare and single cells.

Aggregates of the Group from the trench (Table 21.10) represented very specific structures: large-sized, well developed, and very diverse in species. The communities where they developed were recorded from the area of the trench that extends in the northern Powell Basin and enters the Bransfield Strait. In these cenoses, the small-sized benthic *Amphora* sp. 2 clearly dominated in number, forming very large, up to several hundreds or even thousands of cells, colonies in aggregates. It was often accompanied by the small-size *Nitzschia* sp. 4, which showed equally high proliferation. These two species competed in abundance with *C. pennatum,* which dominated the free-living forms here, but were often found in aggregates as a reinforcing element of their structure. *Licmophora graclis* was also abundant and inhabited, among other substrates, the surface of small filaments of brown algae found in aggregates. The previously mentioned heterotrophic flagellates, as well as the large benthic *Halamphora* sp. 1, also reached a high abundance. Ice-associated species were also encountered occasionally: *F. kerguelensis*, as well as *F. nana* and *F. pseudonana* in aggregates and epiphytically. Here, trichomes of the cyanobacterium *Anabaena* sp. 1 and rare cells of the small benthic diatom *Navicula glaciei*, in addition to several other benthic species, were observed for the first time. Aggregates could be considered here as completely independent forms of communities in the photic zone, which were not inferior to free-living phytoplankton in their species diversity and richness.

It is also possible that we found not all the minor species, especially those that developed in the earlier periods of the summer of 2020. Figure 21.3 shows the images (SEM) of some species of microphytes which were often found at certain

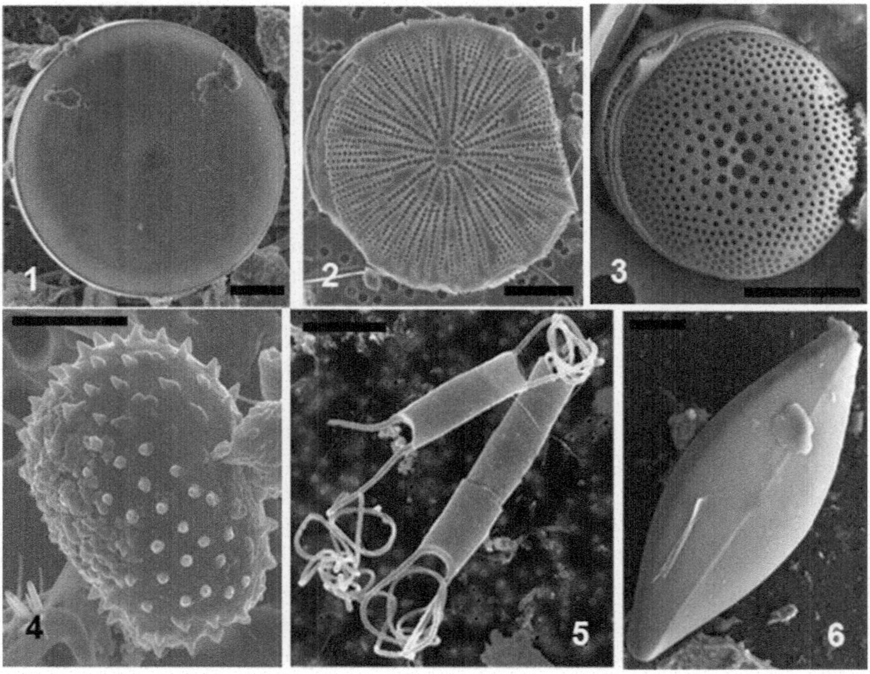

Fig. 21.3 SEM photographs of the species found at some of the stations. (1) *Stellarima microtrias*, (2) *Actinocyclus actinochilus*, (3) *Thalassiosira gracilis*, and (4) cyst of *Polarella glacialis* were found frequently, while (5) *Attheya flexuosa*, (6) *Parlibellus* cf. *hamulifer* were found rarely, in few samples and only as dead valves. Scale bar (black line): 10μm

stations, as well as photographs of valves of some species that were not encountered in a live state in the samples.

21.4 Discussion

The results of our study of microphytic cenoses in the pelagic ecosystems of the Powell Basin showed the presence of typically planktonic microphytes having complex adaptations to inhabit the water column in vegetative and generative states during many generations. Benthic microphytes, including those traditionally considered as phytoperiphytic, epipsammic, and epipelic (intropsammic) that lacked complex adaptations for living and breeding in the water column, also occurred there in large quantities. Many members of the ice or cryopelagic flora well adapted to live and reproduce in pore channels permeating the ice cover were also found in these communities.

It could be assumed that the benthic flora was brought to the pelagic zone by storms from areas of coastal shoals and carried by currents to open waters over hundreds of nautical miles that separate the sites where these organisms were found in the upper layers of the sea, successfully reproducing and forming colonies.

It would also be relevant to suggest that the presence of ice flora in the euphotic zone of the sea is a "plume" left by moving and melting icebergs, from which it dispersed, if the hypothetical icebergs had passed by the sampling sites two or three days before, and if the distance to the nearest icebergs would not be hundreds of nautical miles by the sampling time.

Nevertheless, we found structures forming in various parts of the Powell Basin with the involvement of ice flora, benthic microphyte species, and, in some cases, colonies of planktonic microalgae and spores of some of them, as well as colonies of heterotrophic bacteria. The polysaccharide secretions of the latter, forming "clouds," served as a kind of matrix, a relatively transparent cementing basis to provide habitat for all others. Heterotrophic flagellates that built their colonial settlements on the surface of the matrix of bacterial colonies were found here sometimes in large numbers.

It should be noted that all these complex structures built with the involvement of species that are not adapted to independent life in the water column of open sea began their development on the surface of the test elements of large planktonic diatoms of the genera *Corethron*, *Chaetoceros*, and *Thalassiothrix*, which are common members of phytoplankton in the Southern Ocean. With the increasing volume of the structure, abundance of eukaryotes in their composition, and their diversity, they eventually could include those cells of planktonic diatoms on which they began to build, or at least "tear" them off from the elements of the test on which they grew, and make a transition to independent floating in water. By the end of the second stage of the expedition, we observed very large such structures, measured thousands of μm in length, which continued to develop in streams of the upper pelagic currents (from a depth of 2.0–3.5 m to 50 m) and also included live and dead cells of these planktonic diatoms. The latter now performed the functions (among others) of internal reinforcing elements that can give mechanical rigidness to the structures.

The second type of macrostructure was algobacterial communities formed on the basis of empty polysaccharide sheaths of macrocolonies of the ice colonial diatom *Berckeleya adeliensis*. In the fields of ice floes (near station 6596), we found macrostructures in the form of pink-orange cords of a few to dozens of centimeters in length, which easily disintegrated from the motion of water created by a paddle. These formations were densely filled with gas bubbles, which allowed them to stay afloat.

A microscopic examination showed that those were dispersing sheaths of *B. adeliensis* which remained on the surface when macrocolonies were freed from ice-pore channels after ice melting, were abandoned by the majority of cells of the species that had built them, and then repopulated by the diverse ice flora of diatoms, as well as by heterotrophic flagellates and bacterial colonies. We found only remnants of dead *B. adeliensis* cells in these structures.

We combined all these macrostructures under a common term "aggregates." Our observations showed that in various regions of the pelagic zone and, moreover, in different periods, their composition and pattern of spatial organization could vary significantly.

Below, we summarize the results of observations on the composition, abundance structure, and spatial organization of phytopelagic cenoses in the Powell Basin during the middle and late biological summer of 2020:

1. The macrophytic cenoses in the pelagic zone of various regions of the Powell Basin differed in composition and structure in both stages of observations. In the middle of the biological summer (late January to early February), we could reliably identify eight cenotic groups (one of them in the Antarctic Sound, which was not surveyed in the second stage) and then, by the late summer, we could identify only four. By that time, we recorded significant structural rearrangements of cenoses, their enrichment in species and, in general, expansion over larger parts of the study site. By the late summer, cenotic groups became somewhat more heterogeneous within themselves (in terms of abundance of rare species at certain stations) than in the mid-summer. However, they became more clearly formed by that time, first of all, in terms of abundant and dominant species.
2. In mid-summer, the large-size planktonic diatom *Corethron pennatum* played the dominant role at most stations in the study site; by the late summer, other dominants showed their influence in different areas. Those were, e.g., *Chaetoceros convolutus* and *Ch. criophilus* in the structure of the South cenotic group, located most closely to ice fields, while *C. pennatum* already proved to be the third dominant. *Thalassiothrix antarctica* became the first dominant of the Center group. By that time, the North cenotic group, located north of the oceanographic front, had a more significant proportion of the complex of *Pseudonitzschia* spp. species with a dominance of *P. calliantha*, while *C. pennatum* already moved to the category of frequently occurring or common species.

3. The mass proliferation of *P. calliantha* in the northern regions of the Powell Basin, recorded in various forms in both stages of the expedition, is a new, previously undescribed event in the life of phytopelagic communities in the Powell Basin.
4. The algo-bacterial structures, often complex and hierarchically organized, developing on setae or other elements of tests of planktonic diatoms (*Corethron, Thalassiothrix* and *Chaetoceros*), have been described from the study site for the first time. Within these structures, members of the ice-associated algal flora developed, located at a distance of dozens and hundreds of miles from edges of ice fields and the zones of concentration of large icebergs. Some species of benthic microphytes can also develop here, at considerable distances from the coast and shallow-water zones. These structures do not sink to the bottom when live and inhabit mainly the photic layer of the water column, because the matrix of bacterial colonies contains large amounts of gas (presumably oxygen released by microphytes) frequently in the form of numerous microbubbles. We referred to such structures as "aggregates."
5. By the late summer, aggregates included a larger number of species in their structure in various regions, characterized by discrete cenotic groups, and stable complexes of common and dominant species were formed. Thus, due to the mass formation of aggregates, heterotrophic flagellates (the Center and North cenotic groups) and benthic diatoms (the Group from the trench) began playing the first roles in communities.
6. The formation of aggregates containing large quantities of benthic and/or ice-associated macrophytic species in the euphotic zone of open waters in the study site may answer the question of where these species, found in bottom sediments that are rich in summer species of Antarctic phytoplankton, came from. Motions of icebergs that disperse the ice-associated flora, as well as the spread of large ice floes from shallow coastal to open waters, over deep-sea regions, also serve as a source of diatom tests from ice and the upper subtidal and intertidal zones that settle onto bottom sediments. Nevertheless, the ice-associated flora on the lower side of iceberg proliferates in pore channels, to a maximum of several meters inside (authors' personal observations). Furthermore, the formation of aggregates containing the ice-associated flora covers at least the upper 50-m layer of water column. Thus, the scales of settlement of tests of these species on bottom sediments are quite comparable. This means that the detection of ice-associated flora on the bottom of deep waters in the study area no longer indicates boundaries of distribution of ice fields.
7. On the other hand, the structure of aggregates on the northernmost boundaries of the Powell Basin suggests that benthic diatoms played the leading role here in the late summer. In this region, ice melts earlier than in the other parts of the basin, and developed, well-structured aggregates form by late summer. Thus, the ice flora in the structure of these aggregates is secondary. The balance of benthic (intertidal and upper subtidal) species and members of the ice-associated flora in the bottom sediment microlayers may also indicate that these ice-associated species bred beyond pore channels of icebergs.

8. It can also be assumed that the colonization of young ice in open deep waters occurs with the direct involvement of ice-associated species that make a transition from aggregates onto forming ice.
9. The presence of young thalli of brown macrophytic algae within large (up to 2–3 mm), well-formed aggregates of the Group from the trench suggests that these macrophytic species can disperse with the involvement of algo-bacterial structures.

Acknowledgments The study was conducted within the framework of the State assignment no. 0128-2019-0008 (Shirshov Institute of Oceanology, Russian Academy of Sciences) and no. 0208-2019-0013 (A.V. Zhirmunsky National Scientific Center of Marine Biology, Far Eastern Branch, Russian Academy of Sciences).

References

Al-Handal AY, Wulf A (2008) Marine benthic diatoms from potter cove, King George Island. Antarctica Bot Mar 51:51–68

Assmy P (2009) In: Schaechter M (ed) Algal blooms. Encyclopedia of microbiology. Elsevier, Oxford, pp 27–41

Cefarelli AO, Ferrario ME, Almandoz GO et al (2010) Diversity of the diatom genus *Fragilariopsis* in the argentine sea and Antarctic waters: morphology, distribution and abundance. Polar Biol 33:1463–1484

Cefarelli A, Ferrario M, Vernet M (2015) Diatoms (Bacillariophyceae) associated with free-drifting Antarctic icebergs: taxonomy and distribution. Polar Biol 39:443–459

Cleary A, Durbin E, Casas M (2018) Feeding by Antarctic krill Euphausia superba in the West Antarctic peninsula: differences between fjords and open waters. Mar Ecol Prog Ser 595:39–54

Dietrich D, Arndt H (2004) Benthic heterotrophic flagellates in an Antarctic melt water stream. Hydrobiologia 529:59–70

Dodge JD, Hermes HB (1981) A revision of the *Diplopsalis* group of dinoflagellates (Dinophyceae) based on material from the British Isles. Botanical J Linnean Soc 83:15–26

Fryxell GA, Prasad AKSK (1990) *Eucampia antarctica* var. *recta* (Mangin) stat. Nov. (Biddulphiaceae, Bacillariophyceae): life stages at the Weddell Sea ice edge. Phycologia 29:27–38

Gilbert NS (1991) Microphytobenthic seasonality in nearshore marine sediments at Signy Island, South Orkney Islands, Antarctica. Estuar Coast Shelf Sci 33:89–104

Hasle GR, de Mendiola BRE (1967) The fine structure of some *Thalassionema* and *Thalassiothrix* species. Phycologia 6(2/3):107–126

Hasle GR, Syvertsen EE (1996) Marine diatoms. In: Tomas CR (ed) Identifying marine phytoplankton. Academic Press, San Diego, pp 5–385

Johansen JR, Fryxell GA (1985) The genus *Thalassiosira* (Bacillariophyceae): studies on species occurring south of the Antarctic convergence zone. Phycologia 24:155–179

Park JS, Jung SW, Lee JH, Yun SM, Lee JH (2016) Species diversity of the genus *Thalassiosira* (Thalassiosirales, Bacillariophyta) in South Korea and its biogeographical distribution in the world. Phycologia 55:403–423

Rytter Hasle G, Lange CB (1992) Morphology and distribution of *Coscinodiscus* species from the Oslofjord, Norway, and the Skagerrak, North Atlantic. Diatom Res 7(1):37–68

Schmidt K, Atkinson A (2016) Feeding and food processing in Antarctic krill (Euphausia superba Dana). In Biology and Ecology of Antarctic Krill (Springer) 175–224 https://www.britannica.com/, https://www.britannica.com/science/phytoplankton

Scott FJ, Marchant HJ (Ed.) (2005) Antarctic marine protists. Australian Biological Resources Study: Canberra 563

Stickley CE, Pike J, Leventer A (2006) Productivity events of the marine diatom *Thalassiosira tumida* (Janisch) Hasle recorded in deglacial varves from the East Antarctic margin. Marine Micropaleontol 59:184–196

Tong S, Vørs N, Patterson D (1997) Heterotrophic flagellates, centrohelid heliozoa and filose amoebae from marine and freshwater sites in the Antarctic. Polar Biol 18:91–106

Wilkinson M (2001) What is Phytobenthos. In: Steele JH, Turekian KK, Thorpe SA (eds) Encyclopaedia of ocean sciences. Academic Press, London, pp 2172–2179

Chapter 22
Bioluminescence in the Atlantic Sector of the Southern Ocean Based on the Field Observations and Sounding Data

Aleksandr V. Melnik, Viktor V. Melnikov, Lidiya A. Melnik, Olga V. Mashukova, and Sergei V. Kapranov

Abstract In this work, the results of studies on seawater bioluminescence variation in the Atlantic sector of the Antarctic Ocean (Weddell Sea) are presented. The data were collected using the hydrobiophysical system *Salpa-M*, in February 2020 (122 soundings at 18 stations in cruise 79 of the R/V *Academic Mstislav Keldysh*). The bioluminescence studies were coupled with the simultaneous measurements of temperature, electrical conductivity, and photosynthetically active radiation; the measurements were compared to the data of plankton sample processing. As a result, it was found that the glow of the Antarctic waters in the photic layer of this region occurs within the range of $8.4 \cdot 10^{-12}$–$104.42 \cdot 10^{-12}$ $W \cdot cm^{-2} \cdot l^{-1}$. Bioluminescence peaks (up to 104 $pW \cdot cm^{-2} \cdot l^{-1}$) were recorded below the thermocline at a depth of 45 m in the areas of concentration of salps *Salpa thompsoni* Foxton, 1961, close to the hydrological front at a distance of about 6–7 miles on each side.

Keywords Bioluminescence intensity · Atlantic sector of Antarctica · Euphotic zone · Plankton

22.1 Introduction

Krill (*Euphausia superba* Dana, 1852) is a crucial component of the Southern Ocean ecosystem, acting as the main link between primary production and upper trophic levels with an annual predator demand of about 470 mln t (Hunt et al. 2017). This species is the most abundant in the circumpolar belt between the Antarctic continent and the polar front (Nicol et al. 2000; Nicol and Foster 2016), with an approximate biomass of 379 mln t (Mt) and postlarval production of 342–536 Mt/

A. V. Melnik (✉) · V. V. Melnikov · L. A. Melnik · O. V. Mashukova · S. V. Kapranov
A.O. Kovalevsky Institute of Biology of the Southern Seas (IBSS), Russian Academy of Sciences, Sevastopol, Russia

E. G. Morozov et al. (eds.), *Antarctic Peninsula Region of the Southern Ocean*, Advances in Polar Ecology 6, https://doi.org/10.1007/978-3-030-78927-5_22

year (Atkinson et al. 2009; Hunt et al. 2017). To date, assessing the status of krill communities is one of the high-priority research fields in the Antarctic sector of the Atlantic Ocean (Spiridonov and Uryupova 2009; Sprong and Schalk 1992).

Krill swarms glow because of bioluminescent photophores located on the body of this crustacean: one pair is at the eyestalks; another one is on the hips of the second and seventh thoracopods and singular organs on the pleon segments. These organs periodically emit yellow-green light for 2–3 s. Such a glow of living organisms, clearly visible in the dark, is called bioluminescence. It is visible electromagnetic radiation, kinetic regularities of which are closely related to the mechanisms of chemical reactions and metabolic processes that generate it (Harvey 1957). A large number of luminous species of aquatic organisms are known to date: dinoflagellate microalgae, radiolarians, fungi, and motile multicellular animals of different types, from polyps, jellyfishes, and comb jellies to squids, crustaceans, and fishes (Labas and Gordeeva 2003).

The highest intensity and duration of the glow (up to 22 s) occurs in various krill species (Tokarev et al. 2006). The habitat of Antarctic krill is separated from other regions of the oceans by the Antarctic polar front, which is an efficient barrier to the spreading of marine organisms, making the Southern Ocean a largely isolated ecosystem (Petrov 2016). Therefore, this almost completely isolated ecosystem can be easily destroyed both as a result of direct anthropogenic impact and indirectly due to climate change. In particular, overfishing of krill (*E. superba*) can lead to negative consequences for many species of birds, fish and mammals.

Another dominant luminous organism in Antarctic waters is salps. In the Southern Ocean, krill (*E. superba*) and salps (*Salpa thompsoni)* are both found in dense aggregations. They can, on occasion, coexist in the same waters (Woodd-Walker et al. 2003), but salps tend to occupy warmer water than krill do and prefer oceanic regions with lower food concentrations (Atkinson et al. 2004). The temporal and spatial variability in the abundance of salps and krill is commonly assigned to water temperature, timing of sea-ice advance, duration of sea-ice cover, and food availability (Loeb et al. 1997; Pakhomov et al. 2006).

In this work, for the first time, the results of a mesoscale survey of bioluminescence conducted in February 2020 in the photic zone in the northern part of the Weddell Sea (Powell Basin) are presented.

22.2 Material and Methods

22.2.1 Material

The fieldwork was carried out at 18 stations during cruise 79 of the R/V "Akademik Mstislav Keldysh." At each station, 5–10 bioluminescence profile soundings were performed using the hydrobiophysical system *Salpa-M* (Tokarev et al. 2016), which were coupled with the simultaneous measurements of environmental characteristics

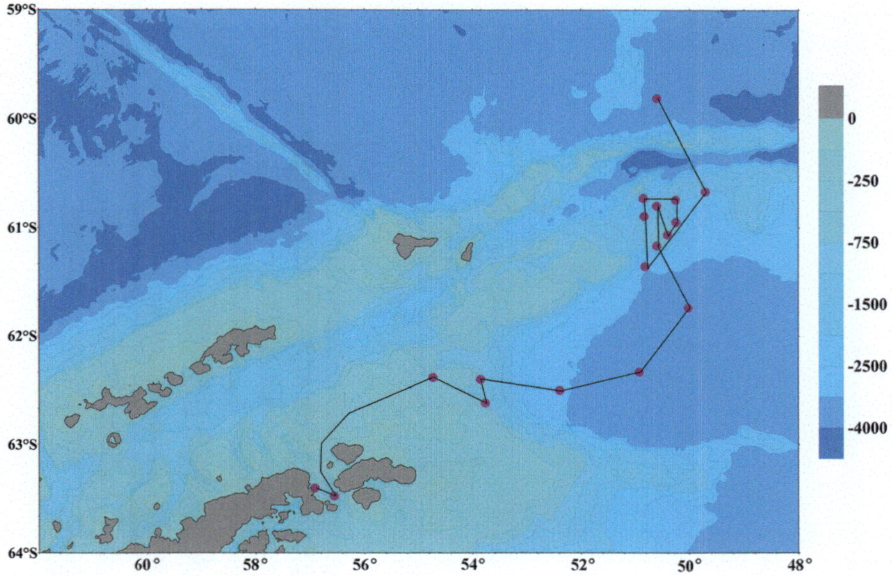

Fig. 22.1 Map of the location of stations

(temperature, electrical conductivity, and photosynthetically active radiation) and juxtaposed with the results of plankton sampling (Morozov et al. 2020). The survey was performed in the region of an oceanic front that manifests itself in sharp horizontal gradients of the thermohaline and chemical parameters (Morozov et al. 2020). A total of 188 soundings were performed in the layer 0–65 m (Fig. 22.1).

22.2.2 Biophysical System **Salpa-M**

The multiparameter system *Salpa-M* was used (Fig. 22.2).

Structurally and functionally, the *Salpa-M* system comprises six measuring and service channels (Fig. 22.3):

- Bioluminescence channel (measurement range: 10^{-12}–10^{-8} W·cm^{-2}·l^{-1}).
- Temperature channel (measurement range: -2–35 °C).
- Pressure channel (measurement range: 0–2 MPa).
- Photosynthetically active radiation (PAR) measurement channel.
- Information and remote-control channel.
- Control and signaling channel.

Channel for Measuring Bioluminescent Radiation Intensity The channel consists of several functionally related parts: a measurement chamber, light collector, photodetector (PMT-71), and control and interface devices. The interface device is

Fig. 22.2 Submersible unit of the *Salpa-M* system

used to measure temperature and pressure. The activation of bioluminescence, after which the organisms start emitting light, is performed mechanically. A measurement chamber is used for this purpose; it consists of eight bladed impellers, which are located in the chamber in two axially spaced groups, with four impellers in each.

To reduce the influence of sunlight, four rows of black impellers consisting of two groups of rows mutually perpendicular to the incidence angles make up a moving light trap. With a negligible resistance to the incoming flow of the water medium, they ensure attenuation of light energy by a factor of $2{\cdot}10^7$. When the

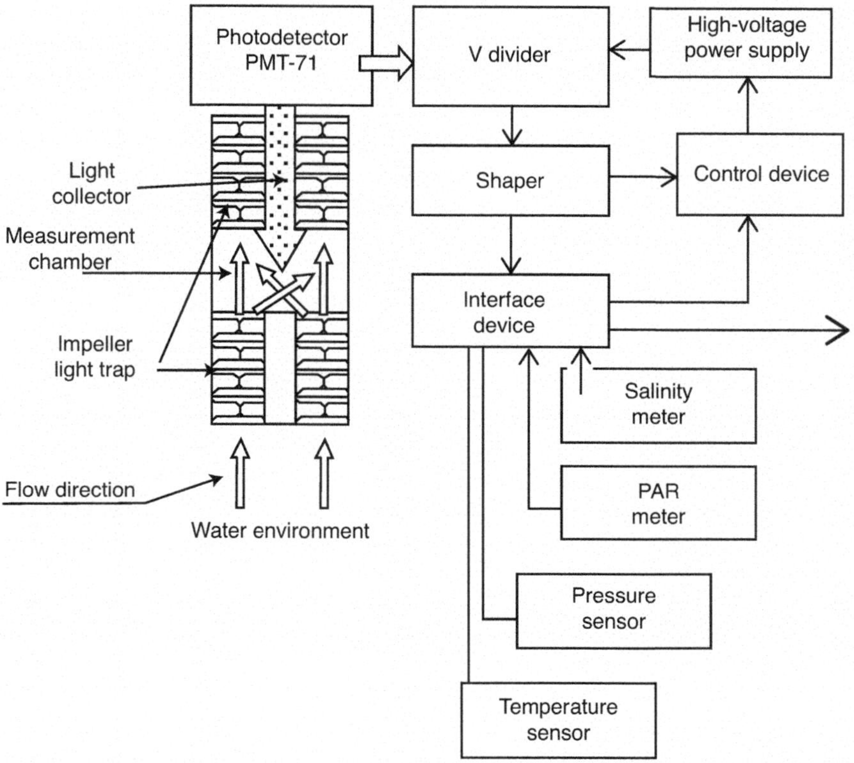

Fig. 22.3 Structural and functional diagram of the submersible unit of *Salpa-M*

bioluminescence sensor moves axially, the incoming flow of the water medium under study enters the measurement chamber. Passing four rows of bladed impellers, the water medium is vigorously agitated, and the bioluminescence in it is excited. The lag time (the time from the agitation to the onset of the bioluminescent glow) is determined by the species composition. For measuring the integral bioluminescence intensity, the residence time of excited bioluminescence in the flow chamber must exceed the lag time and glow duration. The bioluminescence time is also determined by the species composition.

The residence time of microorganisms in the measurement chamber after the mechanical action is determined by the flow of the water medium through the chamber. When solving the problem of bioluminescence recording, a preliminary analysis of the hydrodynamic flow of the water medium through the measurement chamber reveals the need for the formation of a turbulent flow in the middle of the chamber. It is worthwhile to note that a special light guide design forming a cone radiation pattern ensures the transfer of bioluminescence energy of microorganisms from the entire volume of the measurement chamber to the photodetector.

Photodetector Bioluminescence is represented by pulses of low-intensity light energy. The measurement of the light energy is performed using a photomultiplier tube PMT-71, which has high anode sensitivity of 1000 A·lm^{-1} at a supply voltage no higher than 1000 V. The photodetector is powered by a stabilized high-voltage converter, whose pulse switch and multiplication circuits are structurally located in the immediate vicinity of the PMT voltage divider (V divider). The control and stabilization circuits for the stabilized high-voltage converter are located on the control board. Upon sending the *sensitivity* signal "1," "2," "3," or "4" from the *Salpa-T,P* software shell for PC, the control unit generates the required high-voltage power supply for the PMT corresponding to the sensitivities of 1000, 100, 10, or 1 A·lm^{-1}, respectively.

The photomultiplier measuring the signal is fed through a follower and a subsequent amplifier to the interface, an analogue-to-digital converter. Filtering of the information signal with a time constant of 1–5 s is controlled via the *Salpa-M* software shell. The channels for measuring the bioluminescent radiation intensity, electrical conductivity, temperature, and hydrostatic pressure (BCTP) are central to the system. The hydrostatic pressure measurement channel consists of a strain gauge-based pressure transducer and a measuring transducer located on the interface board of the BCTP unit. The temperature measurement channel consists of a temperature sensor ST4–16 and a measuring transducer located on the interface board of the BCTP unit.

22.2.3 Bioluminescent Field and Its Characteristics

The range of Antarctic krill includes its hyper population that forms a continuous distribution field for this species along the Antarctic Circumpolar Current and Antarctic continent. The net light field produced by krill is called the bioluminescence field (BF). Besides, other luminous crustaceans make a significant contribution to the formation of the BF. As a physical field, it is characterized by its light intensity, energy, and frequency spectrum. Since this field is generated by biological objects, it is also a function of the biological characteristics, namely, the number of flashes of individual organisms and the nonuniform structure (heterogeneity) of their distribution in time and space.

The optical spectrum and bioluminescent capacity are important characteristics of the BF. Bioluminescent capacity is the time-integrated radiant flux emitted by planktonic organisms per unit volume in which excitation of their bioluminescence was provided (Tokarev 2006):

$$B = \int_0^t B(t')\,dt$$

where *B* is the bioluminescent capacity of 1-m-thick layer, *t* is the time of the bathyphotometer passing through this layer, and $B(t')$ is the time-dependent radiant flux of bioluminescence in the dark chamber of the bathyphotometer.

In the study of the bioluminescence of Antarctic waters, a method of multiple (5–10) soundings of the euphotic layer of the pelagic zone (1–80 m) was used. The advantages of this method over other techniques of studying marine bioluminescence are obvious:

- it is possible to study the mosaic nature of the spatial distribution of the BF simultaneously with the background characteristics of the environment;
- the influence on the environment and the stimulation of luminous organisms are constant;
- a detailed study of the vertical structure of bioluminescent populations is feasible;
- there is no influence of surface waves on the recorded signal.

22.3 Results

22.3.1 Hydrological Characteristics of the Region

In the mesoscale study site in the Weddell Sea, a hydrological front was found (Morozov et al. 2020), which was formed between the warmer waters with cold subsurface layer and the colder advective waters entering the study site from the western shallower part of Powell Basin (northwestern part of the Weddell Sea). Studies have shown that the northern part of Powell Basin contains relatively warm waters. The frontal zone extended from southwest to northeast in the latitude range 58°–61°S (Fig. 22.4). The hydrological front was especially pronounced in the photic zone, where the temperature difference in the upper layer reached 2 °C (Morozov et al. 2020).

22.3.2 Bioluminescence

During the study period, the planktonic abundance was characterized by a seasonal outbreak in the luminous gelatinous species *S. thompsoni*, which absolutely dominated the composition of zooplankton in the zone of the southern branch of the Antarctic Circumpolar Current, Bransfield Strait, coastal waters of the Antarctic Peninsula, and bottom elevations in the northwestern part of the Powell Basin (Morozov et al. 2020). The background biomass of euphausiids (which also glow), including Antarctic krill, estimated by the catches of the Bongo net in a layer of 200–0 m was two orders of magnitude lower (than salps *S. thompsoni*), and that of the other groups was several orders of magnitude lower. These data can be

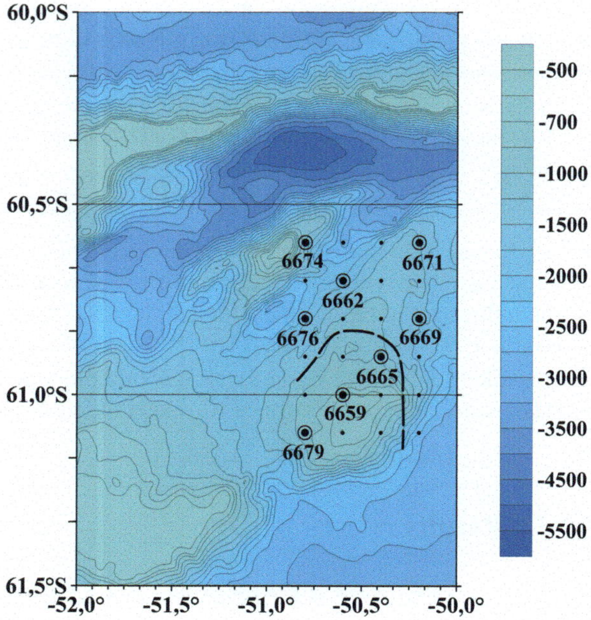

Fig. 22.4 Mesoscale fieldwork. The dashed black line schematically shows the hydrological front line

interpreted as the manifestation of the negative impact of global warming on krill abundance with an increase in the abundance of salps (Driscoll et al. 2015; Flores et al. 2012).

In the mesoscale study site, the highest abundance of *S. thompsoni* was recorded in the communities located on both sides of the hydrological front, with the biomass concentration of 0.66 and 1.62 g·m^{-3}. This is accordingly reflected in the bioluminescence profiles (Figs. 22.5 and 22.6). The bioluminescence peaks (stations 6669 and 6679) were observed at a depth of about 45 m in the areas of accumulation of salps (Driscoll et al. 2015). This is especially clearly seen in the spatial section across stations 6679, 6659, 6665, and 6669: the bioluminescence peaks are located on both sides of the front at a depth of 45 m (Fig. 22.6). Intense flashes on both sides of the hydrological front (with low krill abundance) can be unambiguously associated with the presence of salps *S. thompsoni* capable of generating bursts of light of such intensity (Fig. 22.7).

Table 22.1 shows the average amplitude values of the BF at 18 stations in the region under study in 2020. The highest bioluminescence was recorded at station 6679 (in the southern part of the Powell Basin) in the area of accumulation of salps, where it reached 104.42·10^{-12} W·cm^{-2}·l^{-1}.

Within a single-maximum vertical bioluminescence structure, a layer with increased bioluminescence was found at a depth of 40–50 m. At station 6676 the bioluminescence level lowered to 25.66·10^{-12} W·cm^{-2}·l^{-1}. The layer with increased bioluminescence was formed here at a depth of 15–20 m. At the northern station 6673, the bioluminescence was 19.49·10^{-12} W·cm^{-2}·l^{-1}. The increased bioluminescence layer was recorded here at a depth of 10 m. The vertical structure of

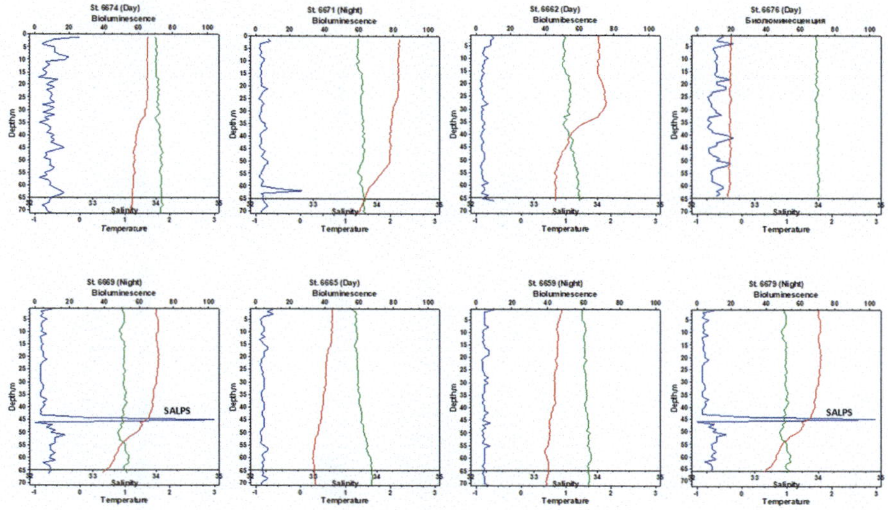

Fig. 22.5 Bioluminescence (pW·cm^{-2}·l^{-1}), temperature (°C), and salinity (PSU) profiles

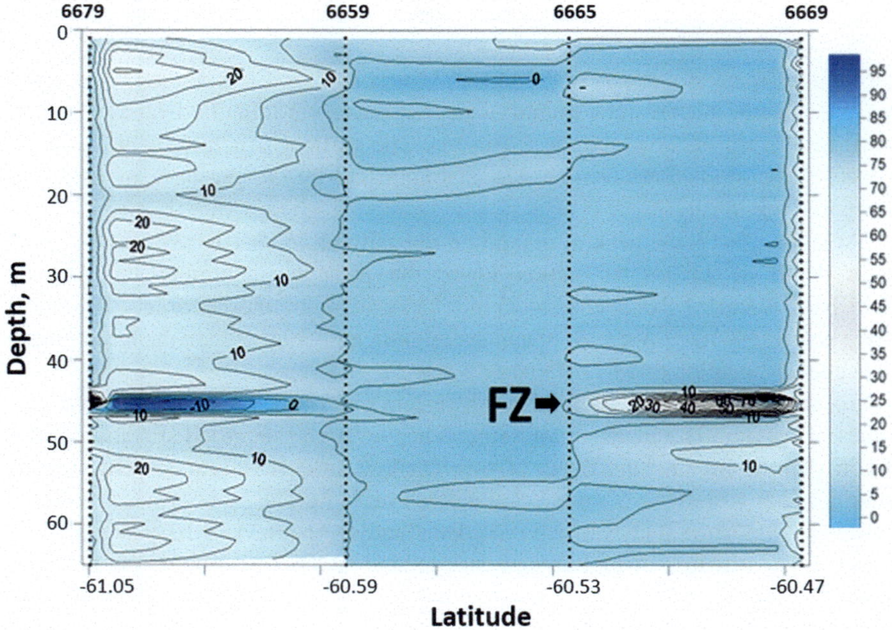

Fig. 22.6 Bioluminescence: spatial section across stations 6679, 6659, 6665, and 6669. FZ is frontal zone. Scale of bioluminescence intensity (W·cm^{-2}·l^{-1}) is shown on the right

Fig. 22.7 Salps (*S. thompsoni*, marked by arrows) on the probe of *Salpa-M* after boarding at station 6679

bioluminescence at stations 6676 and 6673 was similar. Stations 6647 and 6648 were located in the Antarctic Sound, which separates the Joinville Islands from the northeastern tip of the Antarctic Peninsula. The vertical structure of bioluminescence of the studied water area was distinguished by the presence of peaks of increased luminosity.

At station 6647, the peak of maximum intensity of $25.93 \cdot 10^{-12}$ W·cm^{-2}·l^{-1} was at a depth of 11 m. At station 6648, the peak of maximum intensity of $5.75 \cdot 10^{-12}$ W·cm^{-2}·l^{-1} deepened to 45 m. The mean bioluminescence level at station 6647 was $38.98 \cdot 10^{-12}$ W·cm^{-2}·l^{-1} and was significantly higher than the bioluminescence level at station 6648 $17.52 \cdot 10^{-12}$ W·cm^{-2}·l^{-1}.

Six stations were located in the northwestern part of the Weddell Sea at various times of the day. The highest bioluminescence level was recorded at station 6650: $12.56 \cdot 10^{-12}$ W·cm^{-2}·l^{-1} at a depth of 8 m. The lowest-level bioluminescence was at station 6652: $4.51 \cdot 10^{-12}$ W·cm^{-2}·l^{-1} at a depth of 58 m. The highest mean bioluminescence level in the study site was recorded at station 6653: $25.78 \cdot 10^{-12}$ W·cm^{-2}·l^{-1}. The vertical structure of bioluminescence at stations 6650 and 6653 had multiple peaks. At station 6652, a single-maximum vertical bioluminescence structure was observed.

The results of our research revealed that differences in the structure of waters on either side of the hydrological front are also manifested in the characteristics of the integral bioluminescence profiles. This corresponds to significant differences in

Table 22.1 Mean ($\bar{x}$) and standard deviation (δ) of bioluminescence, temperature, and salinity at several stations in the region under study in 2020

Station	Characteristic	Bioluminescence 10^{-12} W·cm^{-2}·l^{-1} Layers, m. All layers	1–10	11–20	21–30	31–40	41–50	51–60	61–70	71–80	Max. BP	Level max. BP	Temperature (layer 1–100 m)	Salinity (layer 1–100 m)
6647	$\bar{x}$	25.93	27.22	27.58	23.11	25.66	–	–	–	–	38.98	11	0.69	34.29
	δ	7.85	6.75	5.49	2.61	7.85	–	–	–	–			0.05	0.06
6648	$\bar{x}$	5.75	3.24	3.29	4.3	7.73	12.84	–	–	–	17.52	46	1.65	35.5
	δ	4.41	3.79	3.9	4.89	2.92	3.43	–	–	–			0.58	0.69
6650	$\bar{x}$	12.56	16.22	13.11	11.37	11.59	11.83	11.57	11.86	–	25.65	8	0.22	33.94
	δ	5.06	6.4	5.07	4.39	5.48	4.65	3.68	4.22	–			0.03	0.03
6651	$\bar{x}$	5.14	4.87	5.02	5.13	5.26	5.04	5.15	6.08	–	17.52	35	0.62	33.72
	δ	2.51	2.36	1.94	2.54	2.8	2.56	2.36	3.43	–			0.15	0.1
6652	$\bar{x}$	4.51	4.63	4.30	4.35	4.32	4.38	4.49	4.93	4.68	15.99	58	0.49	33.9
	δ	1.09	1.07	0.81	0.77	0.83	0.86	1.34	1.76	2.09			0.3	0.13
6653	$\bar{x}$	6.76	8.49	7.01	6.22	6.25	6	6.16	6.55	6.32	25.78	3	0.39	33.56
	δ	4.02	5.34	3.26	3.44	3.53	3.36	3.21	4.03	4.2			0.42	0.45
6656	$\bar{x}$	4.61	5.02	4.54	4.35	4.52	4.47	4.58	5.11	–	10.25	65	0.77	33.25
	δ	1.17	1.49	1.05	0.96	1.03	1.06	1.14	1.5	–			0.13	0.55
6657	$\bar{x}$	4.79	5.47	4.48	4.53	4.52	4.5	4.79	6.09	–	10.25	63	0.58	33.14
	δ	1.58	1.7	1.35	1.45	1.32	1.46	1.77	1.81	–			0.37	0.53
6659	$\bar{x}$	4.75	4.8	4.98	4.69	4.41	4.63	4.89	4.85	–	10.25	4	0.69	33.81
	δ	1.42	1.85	1.42	1.32	1.12	1.22	1.38	1.63	–			0.16	0.07
6662	$\bar{x}$	4.61	4.77	4.41	4.36	4.2	4.49	4.59	5.11	–	8.4	67	1.25	33.7
	δ	1.34	1.72	1.35	0.96	0.94	1.11	1.15	1.75	–			0.45	0.1
6665	$\bar{x}$	6.97	7.65	7.06	7.17	7.17	6.99	6.45	7.3	5.77	21.03	24	0.4	33.85
	δ	3.88	4.35	3.96	4.32	4.11	4.09	2.99	3.93	2.75			0.13	0.09

(continued)

Table 22.1 (continued)

Station	Characteristic	Bioluminescence 10^{-12} W·cm^{-2}·l^{-1}											Temperature (layer 1–100 m)	Salinity (layer 1–100 m)
		Layers, m.												
		All layers	1–10	11–20	21–30	31–40	41–50	51–60	61–70	71–80	Max. BP	Level max. BP		
6669	$\bar{x}$	11.41	12.43	10.61	10.63	10.34	11.79	10.95	11.97	11.44	103.4	45	1.15	33.61
	δ	5.29	4.83	3.64	3.87	3.67	10.02	3.5	4.14	4.75			0.53	0.08
6671	$\bar{x}$	4.41	4.47	4.08	4.05	4.17	4.36	4.28	4.96	4.53	12.04	61	1.9	33.79
	δ	1.81	1.66	0.97	1.11	1.5	1.66	1.3	3.55	1.71			0.29	0.07
6673	$\bar{x}$	8.05	9.39	6.9	6.75	8.08	7.46	7.46	8.97	8.61	19.49	10	1.36	34
	δ	3.9	5.26	3.89	3.38	3.71	3.52	2.98	3.79	3.54			0.16	0.07
6676	$\bar{x}$	11.44	11.79	11.91	10.83	10.91	11.45	11.52	11.23	–	25.66	18	0.22	33.95
	δ	4.11	4.14	4.41	4.08	4.65	3.9	3.58	3.83	–			0.03	0.03
6679	$\bar{x}$	11.13	11.48	10.85	10.1	10.3	11.68	11.46	12.01	10.38	104.42	45	1.15	33.62
	δ	5.29	4.72	3.81	3.96	3.68	9.98	3.71	4.14	3.47			0.52	0.08
6689	$\bar{x}$	2.71	1.89	2.84	2.91	2.8	2.46	2.66	2.55	3.59	13.77	26	1.97	33.24
	δ	2.83	1.96	3.09	2.73	2.3	2.24	2.88	2.21	2.56			0.7	0.06
6691	$\bar{x}$	1.85	1.21	1.76	1.79	2.17	1.8	1.75	2.18	1.91	8.4	12	2.13	33.32
	δ	2.14	1.12	2.25	1.93	2.24	1.58	1.25	1.15	2.37			0.48	0.08

taxonomic composition, structure, and spatial distribution of zooplankton from the south of the hydrological front to its north (Morozov et al. 2020). In the region under study, glowing krill did not form dense clusters, and therefore bioluminescence was due mainly to salps. These gelatinous organisms are typical indicators of frontal zones. The increase in salps in Antarctic waters is currently believed to be related to ocean warming (Flores et al. 2012) with particular impact in the areas accessible by predators and used by fisheries (Hill et al. 2013). This increase is also attributed to the ocean acidification (Kawaguchi et al. 2013).

The results of the presented research show that the bioluminescence peaks (about $100 \cdot 10^{-12}$ W·cm^{-2}·l^{-1}) were located in the immediate vicinity of the frontal zone, at a distance of about 6–8 miles on both sides below the thermocline at a depth of about 45 m.

Acknowledgments The authors express their sincere gratitude to the crew and captain of the R/V *Akademik Mstislav Keldysh* for the help in carrying out this study. This work received financial support of the Russian Academy of Science (projects no. AAAA-A19-119100290162-0, no. AAAA-A18-118021490093-4, no. 121040600178-6 and no. AAAA-A18-118021350003-6).

References

Atkinson A, Siegel V, Pakhomov E, Rhotery P (2004) Long-term decline in krill stock and increase in salps within the Southern Ocean. Nature 432:100–103

Atkinson A, Siegel V, Pakhomov EA, Jessopp MJ, Loeb V (2009) A reappraisal of the total biomass and annual production of Antarctic krill. Deep Sea Res Part 1 Oceanogr Res Pap 56(5):727–740

Driscoll RM, Reiss CS, Hentschel BT (2015) Temperature-dependent growth of *Thysanoessa macrura*: inter-annual and spatial variability around Elephant Island. Antarctica Mar Ecol Prog Ser 529:49–61

Flores H, Atkinson A, Kawaguchi S, Krafft BA, Milinevsky G, Nicol S (2012) Impact of climate change on Antarctic krill. Mar Ecol Prog Ser 458:1–19. https://doi.org/10.3354/meps09831

Harvey EN (1957) A history of luminescence from the earliest times until 1900. Mem Am Philos Soc Phila 44:692

Hill SL, Phillips T, Atkinson A (2013) Potential climate change effects on the habitat of Antarctic krill in the weddell quadrant of the Southern Ocean. PLoS One 8(8):e72246

Hunt BJ, Özkaya Ö, Davies NJ, Gaten E, Seear P, Kyriacou CP, Tarling G, Rosato E (2017) The Euphausia superba transcriptome database, Superba SE: An online, open resource for researchers. Ecol Evol Aug 7(16):6060–6077. https://pubmed.ncbi.nlm.nih.gov/30094004/

Kawaguchi S, Ishida A, King R, Raymond B, Waller N, Constable A, Ishimatsu A (2013) Risk maps for Antarctic krill under projected Southern Ocean acidification. Nat Clim Chang 3(9):843–847

Labas Yu A, Gordeeva AV (2003) Light and color of living organisms. Nature 2:25–31

Loeb V, Siegel V, Holm-Hansen O, Hewitt R, Fraser W, Trivelpiece W, Trivelpiece S (1997) Effects of sea-ice extent and krill or salp dominance on the Antarctic food web. Nature 387:897–900. https://doi.org/10.1038/43174

Morozov EG, Frey DI, Polukhin AA, Krechik VA, Artemiev VA, Gavrikov AV, Kasian VV, Sapozhnikov FV, Gordeeva NV, Kobylyansky SG (2020) Mesoscale Variability of the Ocean in the Northern Part of the Weddell Sea. Oceanology 60(5):663–679

Nicol S, Constable AJ, Pauly T (2000) Estimates of circumpolar abundance of Antarctic krill based on recent acoustic density measurements. CCAMLR Science 7:87–99. https://www.ccamlr.org/ru/node/72892

Nicol S, Foster J (2016) The fishery for Antarctic krill: Its current status and management regime. In: Siegel V (ed) Biology and ecology of Antarctic krill. Springer, Cham, pp 387–421. https://link.springer.com/chapter/10.1007/978-3-319-29279-3_11

Pakhomov E, Dubischar C, Strass V, Brichta M, Bathmann U (2006) The tunicate Salpa thompsoni ecology in the Southern Ocean. I. Distribution, biomass, demography and feeding ecophysiology. Mar Biol 149:609–623

Petrov AF (2016) On state of understanding for bioresources of Antarctic, their status and commercial using. Izv TINRO 184:41–53

Spiridonov V, Uryupova E (2009) Antarctica and the Southern Ocean today: 50 years of the Antarctic Treaty. State of Russia in the Surrounding World, MNEPU, pp 125–150

Sprong I, Schalk PH (1992) Acoustic observations on krill spring-summer migration and patchiness in the northern Weddell Sea. In: Hempel G (ed) Weddell Sea Ecology. Springer, Berlin/Heidelberg, pp 261–268. https://doi.org/10.1007/978-3-642-77595-6_30

Tokarev Y, Vasilenko V, Piontkovski S (2006) Peculiarities of the bioluminescence distribution in the Atlantic sector of the Antarctic Ocean. Luminescence 21:294–294

Tokarev Yu N (2006) Fundamentals of biophysical ecology of aquatic organisms. ECOSI-Hydrophysics, Sevastopol, p 342

Tokarev Yu N, Evstigneev PV, Mashukova OV (2016) Planktonic bioluminescent of the World Ocean: species diversity, characteristics of light emission in normal conditions and under anthropogenic impact. N. Orianda, Simferopol, p 347

Woodd-Walker RS, Walkins JL, Brierley AS (2003) Identification of Southern Ocean acoustic targets using aggregation backscatterer and shape characteristics. ICES J Mar Sci 60:641–649. https://doi.org/10.1016/S1054-3139(03)00062-6

Chapter 23
Parasites as an Inseparable Part of Antarctic and Subantarctic Marine Biodiversity

Tatyana A. Polyakova and Ilya I. Gordeev

Abstract The biodiversity of parasites of marine invertebrates and vertebrates in the Antarctic and Subantarctic is high. Parasitic organisms are represented by 377 species belonging to 18 taxa. The vast majority of parasites in this region have a complex life cycle, which includes intermediate/paratenic and definitive hosts of various taxonomic groups and ecological levels. The dominant group among the identified parasites in the invertebrates and vertebrates is the endoparasites (271 species), whose share amounts to over 72% of the total parasitic fauna. Poor knowledge and a fragmentary degree of research are characteristic of the hosts of parasites among invertebrates. The greatest attention of the researchers of parasites of this region has been paid to the parasites of the fish, which is the dominant group of vertebrate hosts. Currently, 189 parasite species have been found in 135 fish species, i.e., only 50% of the ichthyofauna has been studied for infection. An uneven distribution of parasites of the various taxa among fish families was noted, which indicates that the study of parasites in widespread and large families of Antarctic and Subantarctic fish has been performed, while rare and living at great depths fish remained either not studied in terms of parasites or the studies were fragmentary. The parasite fauna of Antarctica and Subantarctica is characterized by a high level of endemism.

Keywords Biodiversity · Parasites · Invertebrates · Vertebrates · Endemism · Antarctic · Subantarctic

T. A. Polyakova (✉)
A.O. Kovalevsky Institute of Biology of the Southern Seas, Russian Academy of Sciences, Sevastopol, Russia

I. I. Gordeev
Russian Federal Research Institute of Fisheries and Oceanography, Moscow, Russia

Faculty of Biology, Lomonosov Moscow State University, Moscow, Russia

E. G. Morozov et al. (eds.), *Antarctic Peninsula Region of the Southern Ocean*, Advances in Polar Ecology 6, https://doi.org/10.1007/978-3-030-78927-5_23

23.1 Introduction

Parasitic organisms are an integral component of any bio-community (Beklemishev 1956; Fonseca et al. 2010; Marcogliese 2004). They exist in a huge variety that includes over 75,000 species, which is significantly higher than the number of free-living species (~45,000 species) (Poulin and Morand 2000, 2004). According to some estimates, the total number of parasitic species is thought to be as high as 300 thousand species (de Meeus and Renaud 2002; Dobson et al. 2008).

In various ways parasites can have a major impact on life within the ecosystems they are a part of. They can regulate their host population sizes (Albon et al. 2002), as well as affect the composition, structure (Mouritsen and Poulin 2002), and functioning of the community as a whole (Thomas et al. 2005). In their turn, climate change and human impact can become factors affecting the parasites themselves. For example, ocean warming can lead to an increase in numbers and habitat expansion of parasitic species (Poulin and Mouritsen 2006; Studer et al. 2010), while their introduction into a non-native body of water through human activity can have catastrophic consequences for potential hosts. Therefore, an assessment of the parasite species diversity is especially relevant in ecosystems impacted by anthropogenic pressures and located in an area undergoing climate-induced changes. Such assessments have become an important trend in modern biology and ecology.

Antarctica is a unique and relatively underexplored region inhabited by a large number of endemic species. The unique characteristics of the Antarctic waters, such as the consistently low temperatures and, as a result, good oxygen solubility, the narrow continental shelf (with the exception of the Ross and Amundsen seas), and the seasonal variations in light have all contributed to the isolation of the Antarctic ecosystem and resulted in unique adaptations and a high degree of endemism both in the indigenous free-living species and in their parasites (Kock 1992; Eastman 1993; Bargelloni et al. 1994; Rocka 2006; Klimpel and Palm 2011). Due to the adverse weather conditions during most of the year, the marine fauna of the Antarctic remains not fully understood in many aspects, including its involvement in the parasite life cycles.

In this chapter we have attempted to give an extensive overview of all the groups of parasitic organisms and include all the major works on marine parasites in the Antarctic and Subantarctic, as well as to assess parasite diversity in various taxonomic groups of hosts. In total, we have assessed the systematic position of 377 parasite species belonging to 162 genera, 84 families, 39 orders, and 18 taxa (Tables 23.1 and 23.2), as well as their invertebrate and vertebrate hosts, as listed in WoRMS (http://www.marinespecies.org). We have attached an check-list of parasitic species (Table 23.1), indicating their host species and reference. We have put in the check-list well-discovered parasitic taxons without listing parasitic species, but including quantity of found parasitic species, the name of their host-species class and references. Such taxons are: monogeneans on fish (see table 4.1, Klapper et al. 2017), trematodes on fish (see table 1, Oğuz et al. 2015; table 5.3, Faltýnková et al. 2017), cestodes on fish and birds (see table 1, Oğuz et al. 2015; Rocka 2017), nematodes

Table 23.1 Distribution of parasites by hosts in the Antarctica and Subantarctic

Parasite	Host	References
Protista Haeckel, 1866		
Coccidiasina *Goussia cascosa* Lom, Dyková, 1982	*Macrourus berglax*	Rodyuk (1988)
G. clupearum (Thélohan, 1894) Labbé, 1896		
Goussia sp.1	*Merluccius hubbsi*	
Goussia sp.2		
Desseria nototheniae Barber, Mills Westerman, Storoz, 1987 (syn. *Haemogregarina nototheniae*)	*Notothenia rossii*, *N. coriiceps*	Barber et al. (1987) and Barber and Westermann (1988)
Haemogregarina georgianae Barber, Westermann, 1988	*Parachaenichthys georgianus*	
Eimeria pygosceli Golemansky, 2003	*Pygoscelis antarcticus*, *P. papua*, *P. adeliae*	Golemansky (2011)
Eimeria sp.		
Isospora sp.		
Hepatozoon albatrossi Peirce, Prince, 1980	*Diomedea chrysostoma*, *D. exulans*, *D. melanophris*	Peirce and Prince (1980)
Gregarinasina *Cephaloidophora indica* Avdeev, Avdeeva, 1989	*Euphausia superba*	Avdeev (1985) and Avdeev and Avdeeva (1989)
C. pacifica Avdeev, Avdeeva, 1989		Avdeev (1985), Avdeev and Avdeeva (1989), Takahashi et al. (2008), Gómez-Gutiérrez and Morales-Ávila (2016) and Wallis et al. (2017)
C. vulgaris Avdeev, Avdeeva, 1989	*Vibilia antarctica*	Avdeev and Avdeeva (1989)
C. macropa Avdeev, Avdeeva, 1989	*Primno macropa*	
C. hyperiellae Avdeev, Avdeeva, 1989	*Hyperiella antarctica*	
C. antarctica Avdeev, Avdeeva, 1989	*Euphausia superba*, *Thysanoessa macrura*	
C. lumbricalis Avdeev, Avdeeva, 1989	*Themisto gaudichaudii*	
C. thysanoessae Avdeev, Avdeeva, 1989	*Thysanoessa macrura*	Avdeev and Avdeeva (1989) and Wallis et al. (2017)
Cephaloidophora sp.	*Euphausia triacantha*, *E. valentini*	Wallis et al. (2017)
Thalicola sp.	*Salpa thompsoni*	Wallis et al. (2017)
Schistodeutomeritus kammi Avdeev, Avdeeva, 1989	*Boreomysis* sp.	Avdeev and Avdeeva (1989)

(continued)

Table 23.1 (continued)

Parasite	Host	References
Microsporidia *Glugea merluccii*	*Merluccius hubbsi*	Rodyuk (1988)
G. cholecystia	*Macrourus carinatus*	
Eumetazoa Butschli, 1910		
Dicyemida *Dicyemennea antarcticensis* Short, Hochberg, 1970	*Pareledone charcoti*, *Pareledone turqueti*	Short and Hochberg (1970), Hochberg (1990) and Furuya and Hochberg (2002)
D. eltanini Short, Powell, 1969	*Pareledone charcoti*	Short and Hochberg (1970) and Furuya and Hochberg (2002)
D. bathybenthum Furuya, Hochberg, 2002	*Graneledone antarctica*	Furuya and Hochberg (2002)
D. dorycephalum Furuya, Hochberg, 2002	*Graneledone macrotyla*	
D. littlei Hochberg, Short, 1970	*Benthoctopus magellanicus*	Hochberg and Short (1970)
D. discocephalum Hochberg, Short, 1983	*Cirroctopus glacialis*	Short and Hochberg (1970) and Furuya and Hochberg (2002)
Dicyemennea sp.	*Graneledone* cf. *gonzalezi*	Furuya and Hochberg (2002)
Dicyema benthoctopi Hochberg, Short, 1970	*Benthoctopus magellanicus*	Hochberg and Short (1970)
Kantharella antarctica Czaker, 1994	*Pareledone turqueti*	Czaker (1994) and Furuya and Hochberg (2002)
Myxosporea *Zschokkella globulosa* Davis, 1917	*Muraenolepis microps*	Kovaleva et al. (2002)
Z. meglitschi Moser et Noble, 1977	*Dissostichus eleginoides*	
Z. australis Kovaleva, Rodjuk, Grudnev, 2002	*Champsocephalus gunnari*	
Myxodavisia nototheniae (Kovaleva, Rodjuk, Grudnev, 2002) (syn. *Davisia nototheniae*)	*Nototheniops larseni*	
Ceratomyxa orthospora Kovaleva, Rodjuk, Grudnev, 2002	*Chaenodraco wilsoni*, *Dissostichus eleginoides*	
C. ellipsoidea Kovaleva, Rodjuk, Grudnev, 2002	*Lepidonotothen squamifrons*	
C. antarctica Kovaleva, Rodjuk, Grudnev, 2002	*L. squamifrons*, *C. wilsoni*, *Champsocephalus gunnari*	
C. ampla (Kovaleva, Rodjuk, Grudnev, 2002) (syn. *Leptotheca ampla*)	*Lepidonothothen* sp.	
Kudoa vesica Kovaleva, Gaevskaya, 1984	*Icichthys australis*	Kovaleva et al. (2002) and Eiras et al. (2014)

(continued)

Table 23.1 (continued)

Parasite	Host	References
K. alliaria Kovaleva, Shulman, Yakovlev, 1979	*Micromesistius australis*	Kovaleva et al. (2002)
Pseudalataspora meridionalis Kovaleva, Rodjuk, Grudnev, 2002	*Notothenia rossii*	
P. squamifrons Kovaleva, Rodjuk, Grudnev, 2002	*Lepidonothothen* sp.	
Neoparvicapsula ovalis Gaevskaya, Kovaleva, Shulman, 1982	*Gobionotothen gibberifrons*	
Monogenea 23 species	Fishes	Klapper et al. (2017)
Cestoda 21 species	Fishes	Rocka (2017)
25 species	Birds	Rocka (2017, 2019)
Cestoda in pinnipeds and cetaceans: *Diphyllobothrium archeri* (Leiper, Atkinson, 1914) Meggitt, 1924	*Leptonychotes weddellii*	Yurakhno (1990) and Yurakhno and Maltsev (1997)
D. lashleyi (Leiper, Atkinson, 1914) Meggitt, 1924	*L. weddellii, Ommatophoca rossii, Mirounga leonina, Hydrurga leptonyx*	
D. lobodoni Yurakhno, Maltsev, 1994	*Lobodon carcinophagus*	
D. mobile (Rennie, Reid, 1912) Meggitt, 1924 (syn. *Diphyllobothrium wilsoni*)	*Le. weddellii, O. rossii, H. leptonyx*	Yurakhno (1990) and Yurakhno and Maltsev (1997) and Maltsev (2000)
D. quadratum (Linstow, 1892) Railliet, Henry, 1912	*Le. weddellii, O. rossii, H. leptonyx, Lobodon carcinophagus*	
D. scoticum (Rennie, Reid, 1912) Meggitt, 1924	*Le. weddellii, H. leptonyx*	Yurakhno (1990) and Yurakhno and Maltsev (1997)
D. balaenopterae (Lönnberg, 1892) Waeschenbach, Brabec, Scholz, Littlewood, Kuchta, 2017 (syn. *Diplogonoporus balaenopterae*)	*Balaenoptera borealis*	Delyamure (1955)
Baylisia baylisi Markowski, 1952	*Lobodon carcinophagus*	Yurakhno (1990) and Yurakhno and Maltsev (1997)
B. supergonoporis Yurakhno, 1989		
Baylisiella tecta (Linstow, 1892)	*Mirounga leonina*	
Flexobothrium microovatum Yurakhno, 1989	*M. leonina, Lo. carcinophagus*	Yurakhno (1990), Yurakhno and Maltsev (1997) and Maltsev (2000)
Dibothriocephalus coatsi Rennie, Reid, 1912	*Hydrurga leptonyx*	Rennie and Reid (1912) and Yurakhno (1990)
Glandicephalus antarcticus (Baird, 1853) Fuhrmann, 1921 (syn. *Diphyllobothrium antarcticum*)	*O. rossii, H. leptonyx*	Delyamure (1955), Yurakhno (1990) and Yurakhno and Maltsev (1997)
G. perifoliatus (Railliet, Henry, 1912) Markowski, 1952	*Leptonychotes weddellii*	Yurakhno (1990) and Yurakhno and Maltsev (1997)

(continued)

Table 23.1 (continued)

Parasite	Host	References
Tetrabothrius (*Oriana*) *affinis* (Lönnberg, 1891) Lönnberg, 1892	*Balaenoptera borealis*	Delyamure (1955)
T. (*Oriana*) *wilsoni* Leiper, Atkinson, 1914	*B. borealis*, *Physeter macrocephalus*	
T. (*Oriana*) *arsenyevi* Delyamure, 1955	Cetaceans	
Tetrabothrius (*Tetrabothrius*) *curilensis* Gubanov in Delyamure, 1955	*B. borealis*, *P. macrocephalus*	
Priapocephalus grandis Nybelin, 1922		
Monorygma grimaldii (Moniez, 1899) Meggitt, 1924 l.	*Physeter macrocephalus*	
Phyllobothrium delphini (Bosc, 1802) Gervais, 1885 l.	*Balaenoptera physalus*	Skryabin (1972)
	Physeter macrocephalus	Dailey and Vogelbein (1991)
	Leptonychotes weddellii	Yurakhno (1990)
Trematoda 65 species	Fishes	Faltýnková et al. (2017)
Trematoda in crustaceans, pinnipeds, cetaceans *Neolebouria antarctica* (Szidat, Graefe, 1967) Zdzitowiecki, 1990 mtc	*Antarctomysis maxima*	Gaevskaya (1982), Rocka (2006) and Santoro et al. (2014)
Lepidapedidae gen. sp. mtc	Mysides	Gaevskaya (1982)
Ogmogaster antarctica Johnston, 1931	*Lo. carcinophagus*, *Le. weddlli*, *H. leptonyx*	Yurakhno (1990)
	Balaenoptera borealis	Dailey and Vogelbein (1991)
	B. musculus, *B. physalus*	Delyamure (1955)
O. plicata (Creplin, 1829) Jägerskiöld, 1891	*B. musculus*, *B. physalus*, *B. acutorostrata*	
	Balaenoptera borealis	Dailey and Vogelbein (1991)
Orthosplanchnus antarcticus Kurochkin, Nikol'skii, 1972	*Leptonychotes weddlli*	Yurakhno (1990)
Brachycladium goliath (Van Beneden, 1858) Fraija-Fernández, Aznar, Raga, Gibson, Fernández, 2014	*Balaenoptera borealis*	Dailey and Vogelbein (1991)
Nematoda 19 species	Fishes and birds	Mattiucci et al. (2017) and Rocka (2017)
Nematoda in asteroides, polychaetes, crustaceans, pinnipeds, cetaceans		
Ananus asteroides Rubtzov, 1977 l.	*Diplopteraster peregrinator*	Rubtsov (1977) and Chesunov and Spiridonov (1985)
Aborjinia eulagiscae (Tchesunov, Spiridonov, 1985) Özdikmen, 2010 larva	*Eulagisca gigantea*	Chesunov and Spiridonov (1985)

(continued)

Table 23.1 (continued)

Parasite	Host	References
Harpagonchoides crassus Platonova, Potin, 1972	*Hemipodus digitifera*	Platonova and Potin (1972) and Venekey et al. (2019)
Harpagonchus averincevi Platonova, Potin, 1972	*Aglapohamus macroura*	
H. simillis Platonova, Potin, 1972		
Stegophorus macronectes (Johnston, Mawson, 1942) larvae	*Eupausia superba*	Cleary et al. (2019)
Contracaecum mirounga Nikolskij, 1974	*Mirounga leonina*	Delyamure (1955), Yurakhno (1990) and Mattiucci et al. (2017)
C. osculatum D Orecchia, Mattiucci, D'Amelio, Paggi, Plotz, Cianchi, Nascetti, Arduino, Bullini, 1994	*Lo. carcinophagus, H. leptonyx*	Yurakhno (1990)
	Leptonychotes weddellii	Mattiucci et al. (2017)
C. osculatum D larvae	*Euphasia crystallorophias* presumably	Mattiucci et al. (2017)
C. osculatum E Orecchia, Mattiucci, D'Amelio, Paggi, Plotz, Cianchi, Nascetti, Arduino, Bullini, 1994	*Leptonychotes weddellii*	Delyamure (1955), Yurakhno (1990) and Mattiucci et al. (2017)
	Lobodon carcinophagus	Yurakhno (1990)
C. osculatum E larvae	Amphipod/polychaete/isopod?	Mattiucci et al. (2017)
C. radiatum (Linstow, 1907) Baylis, 1920	*Lo. carcinophagus, H. leptonyx, O. rossii*	Yurakhno (1990)
	Leptonychotes weddellii	Mattiucci et al. (2017)
C. radiatum larvae	Crustaceans?	Santoro et al. (2014)
C. stenocephalum (Railliet, Henry, 1907)	*H. leptonyx, Le. weddellii*	Delyamure (1955), Yurakhno (1990) and Felix (2013)
C. rectangulum (von Linstow, 1907)	*Le. weddellii, H. leptonyx, Lo. carcinophagus*	Delyamure (1955) and Yurakhno (1990)
Pseudoterranova decipiens (Krabbe, 1878) Gibson, 1983	*Leptonychotes weddellii*	Delyamure (1955) and Yurakhno (1990)
	Ommatophoca rossii	Yurakhno (1990) and Mattiucci et al. (2017)
Anisakis pegreffii Campana-Rouget, Biocca, 1955	*B. borealis, B. acutorostrata, Physeter catodon*	Dailey and Vogelbein (1991)
	Globicephala melas	Mattiucci et al. (2017)
Anisakis physeteris (Baylis, 1923)	*B. borealis, B. acutorostrata, P. catodon*	Dailey and Vogelbein (1991)
	Physeter catodon	Delyamure (1955)

(continued)

Table 23.1 (continued)

Parasite	Host	References
A. simplex (Rudolphi, 1809)	*B. borealis*, *B. acutorostrata*, *P. catodon*	Dailey and Vogelbein (1991)
	O. rossii, *H. leptonyx*	Yurakhno (1990)
	Mirounga leonina	Delyamure (1955)
	Hydrurga leptonyx	Felix (2013)
Anisakis berlandi Mattiucci, Cipriani, Webb, Paoletti, Marcer, Bellisario, Gibson, Nascetti, 2014	*Globicephala melas*	Mattiucci et al. (2017)
Anisakis sp. *Crassicauda delamureana* Skrjabin, 1966 *C. carssicauda* (Creplin, 1929) Leiper, Atkinson, 1914	*Balaenoptera acutorostrata* *Balaenoptera borealis*	Dailey and Vogelbein (1991)
Filaria sp. *Uncinaria hamiltoni* Baylis, 1933	*Mirounga leonina*	Yurakhno (1990)
Parafilaroides hydrurgae Mawson, 1953	*Hydrurga leptonyx*	Delyamure (1955); Yurakhno (1990)
Acanthocephala 32 species	Crustaceans, fishes, pinnipeds, cetaceans	Stryukov (2000, 2001, 2002) and Laskowski and Zdzitowiecki (2017)
Annelida 20 species	Crustaceans, pycnogonids, fishes	Utevsky (1995, 2005, 2007), Utevsky and Gordeev (2015), Utevsky and Utevsky (2018) and Parket et al. (2020)
Ascothoracida *Ascothorax bulbosus* Heegaard, 1951	Amphiura	Grygier (1981) and Gryger and Salvet (1984)
A. gigas Wagin, 1968	*Ophionotus victoriae*	
Dendrogaster usarporum Grygier, 1987	*Porania antarctica grabra*	Grygier (1987)
D. antarctica Grygier, 1980	*Acodontaster conspicuus*	Grygier (1981, 1987) and Gryger and Fratt (1984)
Gongylophysema asetosum Grygier, 1987	*Odontaster validus*	Grygier (1987)
Copepoda *Cephalorhiza flaccida* Boxshall, Harrison, 1988	*Ilyarachna antarctica*	Boxshall and Harrison (1988)
Tripartisoma trapezoidalis Avdeev, 1983	*Pareledone harrisoni*	Avdeev (1983) and Bresciani and Lützen (1994)
T. ovalis Avdeev, 1983	*P. turqulti*, *P. chorcoti*, *P. harrisoni*	
Avdeevia antarctica Bresciani, Lützen, 1994	*Megaleledone senoi*	Bresciani and Lützen (1994)

(continued)

Table 23.1 (continued)

Parasite	Host	References
Cancerilla ampla Heegaard, 1951	*Ophiacantha vivipara*	Heegard (1951)
C. alata Heegaard, 1951	*O. disjuncta*	
Ophioika tenuibrachia Heegaard, 1951	*O. vivipara, O. disjuncta*	
Codoba discoveryi Heegaard, 1951	*Ophiura meridionalis*	
Ascothorax bulbosus Heegaard, 1951	*Amphiura belgicae*	
Lernaeosaccus ophiacanthae Heegaard, 1951	*O. disjuncta*	Heegard (1951) and Bartsch (1996)
Asterocheres hirsutus Bandera, Conradi, López-González, 2005	*Rossella* sp.	Bandera et al. (2005)
Lophoura szidati Stadler, 1978	*Macrourus holotrachys, M. whitsoni, M. caml*	Gordeev (2015)
	M. whitsoni	Walter et al. (2002)
Paeonocanthus antarcticensis (Hewitt, 1965)	*Bathylagus antarcticus*	Ho et al. (2003)
Clavellopsis tomoi Stadler, 1973	*Trematomus bernacchii*	Moser and Cowen (1991)
Clavella adunca (Strøm, 1762)	*Trematimus loennbergi, M. whitsoni, M. holotrachus*	Walter et al. (2002)
Chondracanthodes tuberofurcatus Kabata, Gusev, 1966	*M. whitsoni*	
Eubrachiella antarctica (Quidor, 1906)	*Notothenia gibberifrons,*	Rokicki and Scora (1987)
	N. rossii	Rokicki and Zdzitowiecki (1991)
	Dissostichus eleginoides	Brickle et al. (2006)
E. gaini (Quidor, 1913)	*Chionodraco hamatus*	Santoro et al. (2014)
Pennella balaenoptera Koren, Danielssen, 1877	*Balaenoptera borealis, B. acutorostrata*	Dailey and Vogelbein (1991)
Zanclopus antarcticus Gravier, 1912	*Cephaloidiscus* sp.	Gravier (1912d)
Isidicola antarctica Gravier, 1914	Isididae	Gravier (1914)
Selioides tardus Gravier, 1912	Polyniida	Gravier (1912a)
Serpulidicoloides cystopomati (Gravier, 1912)	Serpulidae	Gravier (1912b)
Thylacoides sarsi Gravier, 1912	Syllidae	Gravier (1912c)
Herpyllobius rotundus Lützen, Jones B., 1976	Harmothae	Lützen and Jones (1976) and Boxshall et al. (2019)
H. hartmanae Lützen, Jones B., 1976	*Laetmonice producta*	Lützen and Jones (1976)
H. antarcticus Vanhöffen, 1913	*Polyeunoa laevis, Harmothoe fullo*	López-González et al. (2006)
H. stocki López-González, Bresciani, Conradi, 2000	*Austrolaenilla antarctica*	Boxshall et al. (2019)

(continued)

Table 23.1 (continued)

Parasite	Host	References
H. polarsterni López-González, Bresciani, Conradi, 2000	*Eulagisca gigantea*	López-González et al. (2006)
Eurysilenium australis López-González, Bresciani, Conradi, 2006	*Polyeunoa* sp.	
Gottoniella antarctica López-González, Bresciani, Conradi, 2006	*Harmotho* sp.	
G. andeepi López-González, Bresciani, Conradi, 2006		
Amphipoda *Cyamus erraticus* Roussel de Vauzème, 1834	*Eubalaena australis*, *Megaptera novaeangliae*	Berzin and Vlasova (1982) and De Broyer et al. (2007)
	Balaena australis	Margolis (1955)
C. bahamondei Buzeta, 1963	*Physeter macrocephalus*, *Balaenoptera acutorostrata*	Berzin and Vlasova (1982)
C. catodontis Margolis, 1954	*B. acutorostrata P. macrocephalus*	Berzin and Vlasova (1982) and De Broyer et al. (2007)
C. boopis Lütken, 1870	*M. novaeangliae*, *Physeter catodon*, *P. macrocephalus*	
C. gracilis Roussel de Vauzème, 1834	*Eubalaena australis*	
C. ovalis Roussel de Vauzème, 1834	*E. australis*, *Balaena* sp.	De Broyer et al. (2007)
Isocyamus antarcticensis Vlasova in Berzin, Vlasova, 1982	*Orcinus orca*	Berzin and Vlasova (1982) and De Broyer et al. (2007)
Balaenocyamus balaenopterae (K.H. Barnard, 1931)	*Balaenoptera physalus*, *B. acutorostrata*	
Isopoda *Arcturocheres gaussicola* Schultz, 1980	*Dolichiscus gaussianus*	Schultz (1980)
Caecognathia calva (Vanhöffen, 1914)	*Chionodraco hamatus*	Kononenko (2009) and Santoro et al. (2014)
Aegiochus antarctica (Hodgson, 1910)	*Notothenia coriiceps*	Palm et al. (1998)
Phthiraptera *Austrogoniodes* 13 species	Birds	Diaz et al. (2017)
Echinophthirius horridus (von Olfers, 1816)	*Lobodon carcinophagus*, *Ommatophoca rossii*, *Leptonychotes weddellii*	Mehlhorn and Mehlhorn (2017)
Lepidophthirus macrorhini Enderlein, 1904	*Mirounga leonina*	Leonardi et al. (2015) and Mehlhorn and Mehlhorn (2017)
Proechinophthirus zumpti Werneck, 1955	*Arctocephalus gazella*	Leonardi et al. (2015)

(continued)

Table 23.1 (continued)

Parasite	Host	References
Antarctophthirus lobodontis Enderlein, 1909	*Lobodon carcinophagus*	Leonardi et al. (2014), Leonardi et al. (2015) and Mehlhorn and Mehlhorn (2017)
A. ogmorhini Enderlein, 1906	*Hydrurga leptonyx*	
A. mawsoni Harrison, 1937	*Ommatophoca rossii*	
A. carlinii Leonardi, Poljak, Carlini, Galliari, Bobinac, Santos, Márquez, Negrete, 2014	*Leptonychotes weddellii* *Hydrurga leptonyx*	
Siphonaptera 3 species	Birds	Diaz et al. (2017)
Acari 5 species	Birds	

on fish and birds (Rocka 2017), acanthocephalans on crustaceans, fish, birds and mammals (see checklist p. 143, Laskowski and Zdzitowiecki 2017), leeches on invertebrates and fish (Utevsky 2005, 2007), as well as lice, fleas and ticks on birds (see table 9.1, Diaz et al. 2017).

23.2 Systematic Review of Parasites of the Antarctic and Subantarctic

Protista Haeckel, 1866 Parasitic protozoa of the Antarctic and Subantarctic are represented by 11 species of gregarines, coccidia, and microsporidian (Table 23.2).

Phylum Apicomplexa Levine, 1970. In total, coccidia are represented by ten species from the genera of *Goussia* Labbé, 1896; *Eimeria* Schneider, 1875; *Desseria* Siddall, 1995; *Haemogregarina* Danilewsky, 1885; *Isospora* Schneider, 1881; and *Hepatozoon* Miller, 1908 (Tables 23.1 and 23.2). Thus, the five examined species of fish were found to host coccidia (Table 23.2) of the genera *Goussia* (four species) (Rodyuk 1988), *Desseria* (one), and *Haemogregarina* (one) (Barber and Westermann 1988). Four species of coccidia were detected in birds: in penguins, *Eimeria* (two) and *Isospora* (one) (Golemansky 2011); in albatrosses, *Hepatozoon* (one) (Peirce and Prince 1980). In total in the Pacific and Indian (eastern part) sectors of the Antarctic, 11 species of gregarines (Tables 23.1 and 23.2) from the genus of *Cephaloidophora* Mawrodiadi, 1908 (nine species); *Thalicola* Ormières, 1965 (one); and *Schistodeutomeritus* Avdeev, Avdeeva, 1989 (one) (Avdeev 1985; Avdeev and Avdeeva 1989; Takahadhi et al. 2008; Wallis et al. 2017; Cleary et al. 2019), were found. Due to a variety of reasons, the systematic position of these gregarine species remains inconclusive. Also, gregarines were detected in nine species of crustaceans (one species of mysids, four amphipods, and four euphausiids) as well as in one species of salps (Table 23.1).

Phylum Microsporidia Balbiani, 1882. Microsporidia of the Antarctic and Subantarctic are represented by two species from the genera of *Glugea* Thélohan, 1891 (Tables 23.1 and 23.2), which were detected in fish (Rodyuk 1988).

Table 23.2 Distribution of taxonomic diversity of parasite with various life cycles by groups of their hosts in Antarctic and Subantarctic

Parasites	Number of species	Species with a simple life-cycle	Species with a complex life-cycle	Parasites species terminating development in intermediate/definitive hosts			
				Invertebrata	Pisces	Mammalia	Aves
Protista	**23**	**23**	**–**	**11**	**8**	**–**	**4**
Coccidia	10	10	–	–	6	–	4
Gregarinasina	11	11	–	11	–	–	–
Microsporidia	2	2	–	–	2	–	–
Eumetazoa	**354**	**151**	**203**	**58**	**181**	**74**	**63**
Dicyemida	9	9	–	9	–	–	–
Myxosporea	13	13	–	–	13	–	–
Monogenea	23	23	–	–	23	–	–
Cestoda	67	–	67	–	21	21	25
mature	–	–	–	–	17	19	25
plerocercoid	–	–	–	–	4†	2	–
Trematoda	69	–	69	1	65	4	–
marita		–	–	–	65	4	–
metacercaria		–	–	1	1	–	–
Nematoda	35	–	35	9‡‡	14	19	5
adult		–	–	–	7	19	5
larval		–	–	6	8	–	–
Acanthocephala	32	–	32	5	19	14	8
adult		–	–	–	13	–	5
cystacanth		–	–	5	6	–	–
Hirudinea	20‡	20	–	5	16	–	–
Ascothoracidae	5	5	–	5	–	–	–
Copepoda	32	32	–	24	7	1	–
Amphipoda	8	8	–	–	–	8	–
Izopoda	11	11	–	–	3††	–	–
Phthiraptera	22	22	–	–	–	7	17
Siphonaptera	3	3	–	–	–	–	3
Acari	5	5	–	–	–	–	5
Total	**377**	**174**	**203**	**69**	**189**	**74**	**67**

† – taxa of cestode larvae, identified to the species level are given without 51 forms of larvae of unclear species affiliation;

†† – of eleven known species of isopods, only three are found on aquatic organisms;

‡ – of 20 species of leeches, one is found in invertebrate and vertebrate hosts.

‡‡ – species of nematode larvae out of five, found only in invertebrates.

Eumetazoa Butschli, 1910 Parasitic eumetazoa in the Antarctic waters are represented by: dicyemids, myxosporidians, monogeneans, trematodes, cestodes, nematodes, acanthocephalans, leeches, ascothoracida, amphipods, parasitic copepods and isopods, chewing lice, ticks, and fleas (Table 23.2).

Phylum Dicyemida Van Beneden, 1876. Dicyemida are parasites specific to cephalopod mollusks; they live inside the renal appendages and pericardial cavity of the octopus and cuttlefish. In the Antarctic and Subantarctic, nine species have been found (Tables 23.1 and 23.2), representing the genera *Dicyemennea* Whitman, 1883 (seven species); *Dicyema* von Kölliker, 1849 (one); and *Kantharella* Czaker, 1994 (one) (Hochberg and Short 1970; Short and Hochberg 1970; Furuya and Hochberg 2002; Furuya 2008).

Phylum Cnidaria Hatscher, 1888, subclass Myxosporea Bütschli, 1881. Myxosporea in fish in the Antarctic and Subantarctic are represented by 13 species (Tables 23.1 and 23.2): *Ceratomyxa* Thélohan, 1892 (four species); *Zschokkella* Auerbach, 1909 (three); *Kudoa* Meglitsch, 1947 (two); *Pseudalataspora* Kovaleva, Gaevskaya, 1983 (two); *Myxodavisia* Zhao, Zhou, Kent, Whipps, 2008 (syn. *Davisia*) (one); and *Neoparvicapsula* Gaevskaya, Kovaleva, Schulman, 1982 (one) (Kovaleva et al. 2002).

Phylum Platyhelminthes Claus, 1887. Some of the most commonly found parasites of the hydrobionts in the Antarctic are represented by monogeneans, trematodes, and cestodes (Table 23.2).

Class Monogenea Carus, 1863. Monogeneans are ectoparasites of all major fresh- and saltwater fish groups, including teleosts and elasmobranches; they are also found in amphibians. In the Antarctic and Subantarctic waters, only fishes have been found as their hosts. At the time of writing, a total of 23 monogenean species have been found, belonging to 2 taxonomic groups (Table 23.2): Monopisthocotylea (Gyrodactylidae Cobbold, 1864 (ten species); Tetraonchoididae Bychowsky, 1951 (seven); Capsalidae Baird, 1853 (three)) and Polyopisthocotylea (Diclidophoridae Cerfontaine, 1895 (one); Mazocraeidae Price, 1936 (one); Hexabothriidae Price, 1942 (one)), which are detected in 25 fish species (Table 4.1, Klapper et al. 2017). Notably, the highest species richness is exhibited by the species of Monopisthocotylea. Both groups are ambisexual with a monoxenous life cycle, reproduce asexually; most are ovigerous, with the exception of live-bearers from the genus of *Gyrodactylus* von Nordmann, 1832.

Class Cestoda. All cestodes are parasitic, usually with complex life cycles switching between two or more hosts. Mature cestodes are found in fish, amphibians, reptiles, birds, and mammals, as well as in humans, while plerocercoids live mostly in the bodies of various arthropods and other invertebrates (Palm 2004). The fish, birds, and mammals of the Antarctic and Subantarctic are known to host 67 cestode species (Table 23.2), excluding larvae, where species cannot be identified with certainty (Lyadov 1981; Lyadov et al. 1981; Parukhin and Lyadov 1981, 1982; Parukhin 1986; Rodyuk 1988; Gaevskaya and Rodyuk 1988; Gaevskaya et al. 1990; Wojciechowska 1990a, b, 1991a, b; Rocka and Zdzitowiecki 1998; Rocka 2003, 2006, 2017, 2019; Laskowski et al. 2005; Laskowski et al. 2007; Laskowski and

Rocka 2014; Oğuz et al. 2015). The highest species richness of cestodes (25 species) is found in migratory and nonmigratory birds (Rocka 2017, 2019), mostly from the family of Tetrabothriidae Baer, 1954 (18 species): *Tetrabothrius* (*Oriana*) Leiper, Atkinson, 1914 (six); *Tetrabothrius* (*Culmenamniculus*) Murav'eva, 1975 (four); *Tetrabothrius* (*Neotetrabothrius*) Nybelin, 1929 (four); *Tetrabothrius* (*Tetrabothrius*) Rudolphi, 1819 (two); and *Tetrabothrius* Rudolphi, 1819 (two), and the lowest species richness in birds from the family of Dilepididae Railliet, Henry, 1909, from the genera of *Parorchites* Fuhrmann, 1932 (one); *Alcataenia* Spasskaya, 1971 (one); *Nototaenia* Jones, Williams, 1967 (one); and *Reticulotaenia* Hoberg, 1985), as well as in birds from the family of Hymenolepididae Ariola, 1899 (three): *Microsomacanthus* Lopez-Neyra, 1942 (two) and *Branchiopodataenia* Bondarenko, Kontrimavichus, 2004.

In the mammals of the Antarctic and Subantarctic, 21 cestode species (Table 23.1) have been found (Delyamure 1955; Skryabin 1972; Yurakhno 1990; Dailey and Vogelbein 1991; Yurakhno and Maltsev 1997; Maltsev 2000). Cestodes in pinnipeds (14 species) are represented by the genera of *Diphyllobothrium* Cobbold, 1858 (six species); *Glandicephalus* Fuhrmann, 1921 (two); *Baylisia* Markowski, 1952 (two); *Baylisiella* Markowski, 1952 (one); *Dibothriocephalus* Lühe, 1899 (one); *Phyllobothrium* Van Beneden, 1850 (one); and *Flexobothrium* Yurakhno, 1989 (one) (Table 23.1) (Rennie and Reid 1912; Delyamure 1955; Yurakhno 1990; Yurakhno and Maltsev 1997; Maltsev 2000). Cestodes in cetaceans are represented by six species of mature cestodes from the subgenera of *Tetrabothrius* (*Oriana*) Leiper, Atkinson, 1914 (three species); *Tetrabothrius* (*Tetrabothrius*) Rudolphi, 1819 (one); *Priapocephalus* Nybelin, 1922 (one); and *Diphyllobothrium* (one), as well as two species of larvae from the family of Phyllobothriidae Braun, 1900 (Skryabin 1972; Dailey and Vogelbein 1991). Larvae *Phyllobothrium delphini* (Bosc 1802) Gervais, 1885, have been found both in pinnipeds and cetaceans. Both of these mammal groups serve as reservoir hosts for phyllobothriids and get infected with them via teleosts, while the definitive hosts of these cestodes are sharks (Skryabin 1972).

A total of 21 cestode species have been found in fish. Out of these, elasmobranches act as definitive hosts for 12 species from the genera of *Phyllobothrium* (four species); *Guidus* Ivanov, 2006 (two); *Pseudanthobothrium* Baer, 1956 (two); *Dinobothrium* Van Beneden, 1889 (one); *Onchobothrium* de Blainville, 1828 (one); *Notomegarhynchus* Ivanov, Campbell, 2002 (one); and *Echinobothrium* Van Beneden, 1849 (one) (Lyadov 1985; Gaevskaya and Rodyuk 1988; Gaevskaya et al. 1990; Wojciechowska 1990a, 1990b, 1991a, b; Wojciechowska et al. 1995; Rocka and Zdzitowiecki 1998; Walter et al. 2002; Rocka 2003, 2006, 2017; Laskowski and Rocka 2014; Oğuz et al. 2015). Five species from the genera of *Bothriocephalus* Rudolphi, 1808 (two species); *Parabothriocephalus* Yamaguti, 1934 (two); and *Clestobothrium* Lühe, 1899 (one) (Wojciechowska 1993a, b; Wojciechowska et al. 1994; Oğuz et al. 2015; Rocka 2017) reach maturity while hosted by teleosts. Also, in teleosts, 51 forms of plerocercoids have been found, 14 of which were identified to genus level and 37 to order level. Only four were identified as species:

Onchobothrium antarcticum, *Grillotia* (*Grillotia*) *erinaceus*, *Lacistorhynchus tenuis*, and *Hepatoxylon trichiuri* (Table 23.1, Oğuz et al. 2015; Rocka 2017).

Class Trematoda Rudolphi, 1808. Most trematodes have leaf-shaped unsegmented bodies with two suckers. This class of flatworm has a complex life cycle, characterized by alteration of sexually reproductive (marita) and parthenogenetic generations (sporocysts and redia). Vast majority of families use gastropods or bivalve mollusks as first intermediate hosts, while various vertebrates act as definitive hosts. Both invertebrates and vertebrates can play the part of second intermediate hosts, depending on the group of trematodes. Sixty-nine trematode species have been found in the hydrobionts of the Antarctic and Subantarctic (Zdzitowiecki 1991, 1997, 2001a; Zdzitowiecki et al. 1997, 1998; Zdzitowiecki and Zadrozny 1999; Zdzitowiecki and Laskowski 2004; Jeżewski et al. 2014; Oğuz et al. 2015; Gordeev and Socolov 2016a, b; Faltýnková et al. 2017). Metacercariae *Neolebouria antarctica* (Szidat, Graefe, 1967) Zdzitowiecki, 1990, and metacercariae from the family of Lepidapedidae Yamaguti, 1958 (Table 23.1) (Gaevskaya 1982), are hosted by mysids. Out of 65 trematode species found in fish, the majority belongs to the families of Lepidapedidae (18 species) and Opecoelidae (16) (Table 5.3, Faltýnková et al. 2017). Pinnipeds and cetaceans play host to four species (Table 23.1) from the genera of *Ogmogaster* Jägerskiöld, 1891 (two); *Orthosplanchnus* Odhner, 1905 (one); and *Brachycladium* Looss, 1899 (one) (Delyamure 1955; Yurakhno 1990; Dailey and Vogelbein 1991).

Phylum Nematoda Rudolphi, 1808. Among nematodes there are both free-living species inhabiting different environments and parasites hosted by animals and plants. In the hydrobionts of the Antarctic and Subantarctic, 35 parasitic species of nematodes are found (Tables 23.1 and 23.2) (Delyamure 1955; Chesunov and Spiridonov 1985; Yurakhno 1990; Rocka 2006, 2017; Felix 2013; Santoro et al. 2014; Oğuz et al. 2015; Mattiucci et al. 2017). The definitive hosts of parasitic nematodes are vertebrates (fish, birds, mammals including humans), while their intermediate/paratenic hosts are invertebrates (crustaceans, cephalopods, polychaetes, sea stars, and others). Nematode development process is through a complex life cycle, involving switching of intermediate/paratenic hosts belonging to different taxa. Certain nematode species, for instance, from the orders Marimermilhida Rubzov, 1980, and Chromadorida Chitwood, 1933, exist as parasites inside the body of the host (polychaetes, sea stars) before reaching sexual maturity, when they exit their host's body to complete their development and reproduce outside. In three polychaete species, four nematode species (Table 23.1) of the genera *Aborjinia* Özdikmen, 2010 (one species); *Harpagonchoides* Platonova, Potin, 1972 (one); and *Harpagonchus* Platonova, Potin, 1972 (two) (Platonova and Potin 1972; Chesunov and Spiridonov 1985; Venekey et al. 2019), were detected. A sea star *Diplopteraster peregrinator* (Sladen 1882) was found to host the nematode *Ananus asteroides* Rubtzov, 1977 (Rubtsov 1977; Chesunov and Spiridonov 1985). Larvae of four nematode species, *Stegophorus macronectes* (Johnston and Mawson 1942) (Cleary et al. 2019), *Contracaecum osculatum* sp. D, *C. osculatum* sp. E, and *C. radiatum* (Linstow, 1907) Baylis, 1920 (Santoro et al. 2014; Mattiucci et al. 2017), have been found in crustaceans. Larvae of the first nematode species in its mature state are found in birds (*Macronectes giganteus*, *Stercorarius skua loennbergi*, *Chionis alba*, *Pygoscelis adeliae*, *P. antarctica*, *P. papua*), the other species in seals (Santoro et al. 2014). Larvae of the nematode *Anisakis pegreffii* Campana-Rouget, Biocca, 1955; *A. berlandi* Mattiucci, Cipriani, Webb, Paoletti, Marcer, Bellisario, Gibson, Nascetti, 2014; and *A. nascettii* Mattiucci, Paoletti, Webb, 2009, were found in squid to the north of the Subantarctic by the coasts of South Africa, New Zealand, and in the Tasman Sea (Table 7.2, Mattiucci et al. 2017). Of these three species, only the larvae of *A. pegreffii* and *A. berlandi* were found in fish in the Antarctic and Subantarctic, as they complete their development in long-finned pilot whales *Globicephala melas* Traill, 1809 (Mattiucci et al. 2017). Fourteen nematode species are found in fish (Rocka 2006, 2017; Oğuz et al. 2015). Among them, the family Anisakidae Railliet, Henry, 1912, has the highest diversity (eight species), Capillariidae Railliet, 1915 (*Capillaria* sp.), and Cucullanidae Cobbold, 1864 (*Dichelyne* (*Cucullanellus*) *fraseri* (Baylis 1929)), the lowest (by one species). Out of the nematode species hosted in fish in the Antarctic and Subantarctic, only six were found to be at the stage of sexual maturity: *Ascarophis nototheniae* Johnston, Mawson, 1945; *Hysterothylacium aduncum* (Rudolphi 1802); *Dichelyne* (*Cucullanellus*) *fraseri* (Baylis, 1929); *Paranisakiopsis weddelliensis* Rocka, 2002; *Comephoronema beatriceinsleyae* (Holloway and Klewer 1969); and *Capillaria* sp. (female with eggs) (Mattiucci et al. 2017; Rocka 2017). Eight out of 14 nematode species found in fish complete their development in pinnipeds and cetaceans: *Contracaecum* (three species), *Pseudoterranova* (two), and *Anisakis* (three). Pinnipeds were found to host 12 nematode species (Table 23.1) from the genera: *Contracaecum* Railliet, Henry, 1912 (six species); *Pseudoterranova* Mozgovoi, 1951 (two); *Parafilaroides* Dougherly, 1946 (one); *Anisakis* Dujardin, 1845 (one); *Uncinaria* Frölich, 1789 (one); and *Filaria* Mueller, 1787 (one) (Delyamure 1955; Yurakhno 1990; Felix 2013; Mattiucci et al. 2017). Eight nematode species (Table 23.1) from the genera *Anisakis* Dujardin, 1845 (five species); *Crassicauda* Leiper, Atkinson, 1914 (two); and *Placentonema* Gubanov, 1951 (one) (Delyamure 1955; Dailey and Vogelbein 1991; Felix 2013; Mattiucci et al. 2017), were detected in cetaceans. The species *Anisakis simplex* (Rudolphi, 1809) is found in both cetaceans and pinnipeds. Five nematode species from the genera *Stegophorus* Wehr, 1934 (two species); *Paracuaria* Krishna Rao, 1951 (one); *Tetrameres* Creplin, 1846 (one); and *Eulimdana* Founikoff, 1934 (one) (Rocka 2017), were found in birds.

Phylum Acanthocephala Kölr., 1771. A wide range of marine and freshwater inhabitants (amphibians, reptiles, fish, birds, mammals), terrestrial vertebrates, and, in rare cases, humans can all act as definitive hosts for acanthocephalans. Their intermediate hosts are arthropods – crustaceans, millipedes, and insects (Shultz and Gvozdev 1972). A total of 32 acanthocephalan species from the orders Echinorhynchida Petrochenko, 1956 (four families, five genera), and Polymorphida Petrochenko, 1956 (one family, 4 genera) (Table 23.2) (Dailey and Vogelbein 1991; Laskowski and Zdzitowiecki 2017), have been found in the Antarctic and Subantarctic. The complete lists of acanthocephalan species and their definitive/parapostcyclic hosts are given in the checklist by Laskowski and Zdzitowiecki (2017). Acanthocephalans from the order Echinorhynchida use Antarctic fish as their definitive hosts, while those from the order Polymorphida use them as their parapostcyclic hosts. At the moment of writing, eight species of Antarctic amphipods, the first intermediate hosts of acanthocephalans, have been found to host cyctacanths of five species from the genera *Aspersentis* Van Cleave, 1929 (one species); *Metacanthocephalus* Yamaguti, 1959 (one); and from Polymorphida: *Corynosoma* Lühe, 1904 (three) (Zdzitowiecki 2001b; Laskowski and Zdzitowiecki 2017). Thirteen acanthocephalan species at the final stage of their development have been found in fish, who also act as parapostcyclic hosts for six species, which then complete their development in birds and mammals. Mammals are currently known to host 14 species from the genera *Corynosoma* Lühe, 1904 (eight species), and *Bolbosoma* Porta, 1908 (six) (Laskowski and Zdzitowiecki 2017). Eight species from the genera *Profilicollis* Meyer, 1931 (two species); *Andracantha* Schmidt, 1975 (two); and *Corynosoma* (four) (Laskowski and Zdzitowiecki 2017) have been detected in birds. Birds were also found to host immature forms *Corynosoma bullosum*, *C. hamanni*, and *C. pseudohamanni*, which then complete their development in pinnipeds. Since the infection of birds evidently occurs through fish, the birds in this case are considered aberrant hosts (Laskowski and Zdzitowiecki 2017).

Phylum Annelida Lamarck, 1809, subclass Hirudinea Lamarck, 1818. A total of 20 leech species have been found in invertebrates and vertebrates of the Antarctic and Subantarctic (Sawyer and White 1969; Sawyer 1972; Utevsky 1995, 2007; Bielecki et al. 2008; Utevsky and Gordeev 2015; Utevsky and Utevsky 2018; Parket et al. 2020). Sea spiders (Pantopoda) are found to host three species from the genera *Pleurobdella* A. Utevsky, 1995 (two species), and *Mooreobdellina* Epshtein, 1972 (one) (Utevsky 1995, 2005, 2007). Leeches *Glyptonotobdella antarctica* (Sawyer and White 1969) and *Trachelobdellina glabra* Moore, 1957 (Sawyer and White 1969; Utevsky 2005), were detected on the giant Antarctic isopod *Glyptonotus antarcticus* Eights, 1852. Sixteen leech species from nine genera of the family Piscicolidae Johnston, 1865, have been found in teleosts and elasmobranchs (Sawyer 1972; Utevsky 2005, 2007; Bielecki et al. 2008; Utevsky and Gordeev 2015; Utevsky and Utevsky 2018). The leech species *Glyptonotobdella antarctica* has been found in both isopods and fishes *Chionodraco* sp. and *Chaenocephalus aceratus* (Lönnberg, 1906) (Utevsky 2005).

Phylum Arthropoda Siebold, 1848. In the Antarctic and Subantarctic waters, vertebrates are found to host 82 arthropod species belonging to seven taxa: ascothoracida, copepods, amphipods, isopods, chewing lice, ticks, and fleas.

Infraclass Ascothoracida Lacaze-Duthiers, 1880. Ascothoracida are widespread and hosted by coelenterata and echinoderms. Five species (Tables 23.1 and 23.2) from the genera *Ascothorax* Djakonov, 1914 (two species); *Dendrogaster* Knipovich, 1890 (two); and *Gongylophysema* Grygier, 1987 (one), have been found in sea stars (Grygier 1981, 1987; Gryger and Fratt 1984; Gryger and Salvet 1984).

Subclass Copepoda Milne-Edwards, 1840. A total of 32 parasitic copepod species (Tables 23.1, 23.2) have been detected on the hydrobionts of the Antarctic (Rodyuk 1988; Avdeev 1983; Boxshall and Harrison 1988; Bresciani and Lützen 1994; Gordeev 2015; Heegard 1951; Ho et al. 2003). Polychaetes have been found to host 11 copepod species from the genera *Herpyllobius* Steenstrup, Lütken, 1861 (five species); *Gottoniella* López-González, Bresciani, Conradi, 2006 (two); *Eurysilenium* Sars M., 1870 (one); *Selioides* Levinsen, 1878 (one); *Serpulidicoloides* Boxshall, Halsey, 2004 (one); and *Thylacoides* Gravier, 1912 (one) (Gravier 1912a, b, c, d; Lützen and Jones 1976; López-González et al. 2006; Boxshall et al. 2019). Six species from the genera *Cancerilla* Dalyell, 1851 (two species); *Ophioika* Stephensen, 1933 (one); *Lernaeosaccus* Heegaard, 1951 (one); *Codoba* Heegaard, 1951 (one); and *Ascothorax* Djakonov, 1914 (one), have been found in ophiuroids (Heegard 1951; Bartsch 1996). Octopus plays host to three species from the genera *Tripartisoma* Avdeev, 1983 (two species), and *Avdeevia* Bresciani, Lützen, 1994 (one) (Avdeev 1983; Bresciani and Lützen 1994). *Isidicola antarctica* Gravier, 1914, were found in anthozoa (Gravier 1914), and *Zanclopus antarcticus* Gravier, 1912, on pterobranchia (Gravier 1912d). *Asterocheres hirsutus* Bandera, Conradi, López-González, 2005, were found to hexactinellid sponges *Rossella* sp. (Bandera et al. 2005). Of all the copepods, only one species, *Cephalorhiza flaccida* Boxshall, Harrison, 1988, was found in the isopod *Ilyarachna antarctica* Vanhöffen, 1914 (Boxshall and Harrison 1988). Seven copepod species from the genera *Eubrachiella* Wilson C.B., 1915 (two species), *Lophoura* Kölliker in Gegenbaur, Kölliker, Müller, 1853 (one); *Paeonocanthus* Kabata, 1965 (one); *Clavellopsis* Wilson C.B., 1915 (one); *Clavella* Oken, 1815 (one); and *Chondracanthodes* Wilson C.B., 1932 (one), have been detected on fish (Moser and Cowen 1991; Ho et al. 2003; Gordeev 2015). One of the largest copepod species from the family Pennellidae Burmeister, 1835 – *Pennella balaenoptera* Koren, Danielssen, 1877 – is found in the fins of whales (Dailey and Vogelbein 1991).

Order Amphipoda Latreille, 1816. Only a small number of parasitic forms of amphipods have been described, such as whale lice and commensal crustaceans from the family Cyamidae Rafinesque, 1815. They are ectoparasites and are found in skin lesions, genital folds, nostrils, and the eyes of marine mammals in the cetacean order. A total of eight whale lice species (Table 23.1) from the genera *Cyamus* Latreille, 1796 (six species); *Isocyamus* Gervais, Van Beneden, 1859 (one); and *Balaenocyamus* Iwasa-Arai, Serejo, 2018 (one), have been found in the Antarctic (Berzin and Vlasova 1982; De Broyer et al. 2007).

Order Isopoda Latreille, 1817. Out of 11 facultative and obligate parasitic isopod species of the Antarctic and Subantarctic, only 3 species were found in hosts (Table 23.1). The isopod *Arcturocheres gaussicola* Schultz 1980, was found in the isopod *Dolichiscus gaussianus* (Schultz 1980) and other isopods *Caecognathia calva* (Vanhöffen 1914) on fish *Chionodraco hamatus* (Lönnberg 1905) (Kononenko 2009; Santoro et al. 2014) and *Aegiochus antarctica* (Hodgson 1910) on *Notothenia coriiceps* Richardson, 1844 (Palm et al. 1998). The hosts of the other ten isopod species in this region remain unknown.

Order Phthiraptera Haeckel, 1896. Members of the order of hematophagous parasitic insects with incomplete metamorphosis are small obligate, highly host-specific ectoparasites. In total, birds and mammals of the Antarctic and Subantarctic are known to host 22 chewing lice species (Table 9.1, Diaz et al. 2017). *Nesiotinus demersus* Kellogg, 1903, and *Naubates prioni* (Enderlein 1909) and 13 species of the genera *Austrogoniodes* Harrison, 1915, are found in birds (Table 9.1, Diaz et al.

2017), while seals play host to seven species of the genera *Antarctophthirius* Enderlein, 1906 (four species), *Echinophthirius* (one), *Lepidophthirus* (one), and *Proechinophthirus* (one) (Leonardi et al. 2014, 2015; Mehlhorn and Mehlhorn 2017).

Order Siphonaptera Latreille, 1825. Fleas in the Antarctic and Subantarctic are only found in birds and represented by three species of the genus *Parapsyllus* Enderlein, 1903 (Table 9.1, Diaz et al. 2017).

Subclass Acari Leach, 1817. Ticks in the Antarctic and Subantarctic are found in birds and represented by five species of the genera *Rhinonyssus* Trouessart, 1894 (one); *Ayersacarus* Hunter, 1964 (one); *Gamasellus* Berlese, 1892 (one); and *Gamasolaelaps* Berlese, 1904 (one), including scale ticks *Ixodes uriae* White, 1852 (Table 9.1, Diaz et al. 2017).

23.3 Faunistic Review of Parasites by Host Groups

23.3.1 Invertebrates

Invertebrata, as the potential parasite hosts, are represented in the Antarctic and Subantarctic by the 6 highest taxa: 2628 arthropod species (2366 species are crustaceans, 262 species are sea spiders); sponges, 400; echinoderms, 235; polychaetes, 588; cephalopods, 32; and anthozoa – not more than 30 (De Broyer et al. 2014).

Arthropods have been found to host parasites belonging to 6 taxa: 11 gregarine species from the families Cephaloidophoridae Kamm, 1922 (nine species), and Actinocephalidae Léger, 1892 (one); family position of *Schistodeutomeritus kami* Avdeev, Avdeeva, 1989, is not identified (Avdeev 1985; Avdeev and Avdeeva 1989; Takahadhi et al. 2008; Wallis et al. 2017; Cleary et al. 2019); metacercariae of trematodes *Neolebouria antarctica* and Lepidapedidae gen. sp. (Gaevskaya 1982); cyctacanths of five acanthocephalan species from the families Polymorphidae Meyer, 1931 (three species); Heteracanthocephalidae Petrochenko, 1956 (one); and Rhadinorhynchidae Lühe, 1912 (one) (Zdzitowiecki 2001b; Laskowski and Zdzitowiecki 2017); larvae of four nematode species from the family Anisakidae Railliet, Henry, 1912 (Santoro et al. 2014; Cleary et al. 2019); five leech species from the family Piscicolidae Johnston, 1865 (Sawyer and White 1969; Utevsky 1955, 2007); and one copepod species from the family Nicothoidae Dana, 1852 (Boxshall and Harrison 1988). In total, 27 parasite species were found in 20 arthropod host species.

Nine dicyemid species from the families Dicyemidae Van Beneden, 1882 (eight species), and Kantharellidae Czaker, 1994 (one) (Hochberg and Short 1970; Short and Hochberg 1970; Furuya and Hochberg 2002; Furuya 2008), and three copepod species from the family Tisbidae (Avdeev 1983; Bresciani and Lützen 1994) have been found in eight cephalopod species.

Among the great variety of Antarctic sponges (Janussen and Downey 2014), only one species has been found to host parasitic copepods of the family Asterocheridae Giesbrecht, 1899 (Bandera et al. 2005).

Sea stars serve as hosts to larvae of one nematode species of the family Marimermithidae Rubtsov, Platonova, 1974 (Rubtsov 1977; Chesunov and Spiridonov 1985) and five ascothoracida species from the families Ascothoracidae Grygier, 1987 (two); Dendrogastridae Gruvel, 1905 (two); and Ctenosculidae Thiele, 1925 (one) (Grygier 1981). Six copepod species from the families Cancerillidae Giesbrecht, 1897 (two species); Chordeumiidae Boxshall, 1988 (two); Codobidae Boxshall, Ohtsuka, 2001 (one); and Ascothoracidae Grygier, 1987 (one), have been found in ophiuroids (Heegard 1951). A total of 12 parasite species has been found in nine echinoderm species.

Fourteen polychaete species have been found to host parasitic larvae of 4 nematode species from the families Chromadoridae Filipjev, 1917 (three), and Marimermithidae (one) (Platonova and Potin 1972; Chesunov and Spiridonov 1985; Venekey et al. 2019); 11 copepod species from the families Herpyllobiidae Hansen, 1892 (nine species); Nereicolidae Claus, 1875 (one); and Serpulidicolidae Stock, 1979 (one) (Gravier 1912a, b, c, d; Lützen and Jones 1976; López-González et al. 2006; Boxshall et al. 2019).

One gregarine species of the family Actinocephalidae has been found in one of the salp species (Wallis et al. 2017).

One copepod species of the family Lamippidae Joliet, 1882, has been found in an anthozoa species (Gravier 1914).

One species of pterobranchia has been found to host a copepod species of the family Enterognathidae Illg, Dudley, 1980 (Gravier 1912d).

To sum up all of the numbers, 69 parasite species have been found in 43 species of invertebrate hosts. Assuming the presence of parasites in various stages of their life cycle in all invertebrates, we can therefore conclude that less than 1 % of the total has been researched at this point.

23.3.2 *Vertebrates*

Fishes Out of 374 fish species in the Antarctic and Subantarctic waters (Duhamel et al. 2014; Froese and Pauly 2019), 135 species have been found to host 189 parasite species, belonging to 11 taxa (Table 23.1). Therefore, parasitic fauna has been studied in less than 36 % of all the fish fauna in the region.

Protozoans are represented by six coccidia species of the families Eimeriidae Minchin, 1903 (four species), and Haemogregarinidae Léger, 1911 (two), as well as two microsporidian species of the family Glugeidae Thélohan, 1892 (Rodyuk 1988). Protozoa are found in six fish species which amounts to less than 1% of all the fish fauna.

Among myxosporidia 13 species of the families Ceratomyxidae Doflein, 1899 (four species); Myxidiidae Thélohan, 1892 (three); Kudoidae Meglitsch, 1960 (two); Alatasporidae Shulman, Kovaleva, Dubina, 1979 (two); Sinuolineidae Shulman, 1959 (one); and Parvicapsulidae Shulman, 1953 (one), have been identified (Kovaleva et al. 2002); these were found in ten fish species (3% of the fish fauna).

The most compositionally diverse group of helminths found in fish is trematodes, which are represented by 65 species of 28 genera and 10 families (Oğuz et al. 2015; Faltýnková et al. 2017). These helminths are found in 79 fish species which amounts to 21% of the fish fauna. Other platyhelminths, monogeneans are represented by 23 species of 11 genera and 7 families, detected in 25 fish species, amounting to 7% of fish fauna (Oğuz et al. 2015; Klapper et al. 2017). There are 21 cestode species, present in different stages of their development in 59 fish species (16% of the fish fauna) (Oğuz et al. 2015; Rocka 2017). Forty-two fish species (11% of fish fauna) have been found to host 13 nematode species (Rocka 2006, 2017; Oğuz et al. 2015). Acanthocephalans are represented by 13 species, found in 34 fish species (9% of fish fauna) (Oğuz et al. 2015; Laskowski and Zdzitowiecki 2017).

Among leeches, 16 species of the family Piscicolidae have been identified, hosted by 27 fish species (7% of fish fauna) (Utevsky 2007; Bielecki et al. 2008; Utevsky and Gordeev 2015; Parker et al. 2020).

Crustaceans are represented by seven copepod species of three families Lernaeopodidae Milne Edwards, 1840 (four species); Chondracanthidae Milne Edwards, 1840 (one); Sphyriidae Wilson C.B., 1919 (two); and two isopod species of the families Gnathiidae Leach, 1814, and Aegidae White, 1850 (Rodyuk 1988; Bresciani and Lützen 1994; Ho et al. 2003; Gordeev 2015), which have been found in seven fish species (2% of fish fauna).

Waterbirds There are 148 bird species in the Antarctic and Subantarctic, both resident or migratory (Ropert-Coudert et al. 2014), and 34 of these, amounting to 23% of the known avian fauna, are known to host 64 parasite species belonging to eight taxa. Birds play host to endoparasites represented by 3 coccidia species of the family Eimeriidae Minchin, 1903 (Golemansky 2011); 25 cestode species of the families Tetrabothriidea Baer, 1954 (18 species); Dilepididae Railliet, Henry, 1909 (four); and Hymenolepididae Ariola, 1899 (three) (Rocka 2017, 2019); 8 acanthocephalan species of the family Polymorphidae Meyer, 1931 (Laskowski and Zdzitowiecki 2017); 5 nematode species of the subfamily Lemdaninae Seurat, 1917 (one species), and families Acuariidae Railliet, Henry, Sisoff, 1912 (three), and Tetrameridae Travassos, 1914 (one) (Rocka 2017). Among ectoparasites that are known in birds are 15 species of chewing lice of the family Philopteridae Burmeister, 1838; 5 tick species of the families Rhinonyssidae Trouessart, 1895; Dermanyssidae Kolenati, 1859; Ologamasidae Ryke, 1962; Veigaiidae Oudemans, 1939; and Ixodidae C. L. Koch, 1844; and 3 flea species of the family Rhopalopsyllidae Oudemans, 1909 (Diaz et al. 2017). Out of 64 parasitic species found in/on birds, 38 species of helminths use saltwater fish and invertebrates as their intermediate hosts.

Marine Mammals Out of 31 marine mammal species of the Antarctic and Subantarctic (Ropert-Coudert et al. 2014), 7 cetacean species and 7 pinniped species are found to host 76 parasite types belonging to 7 taxa: 4 trematode species from 2 families Notocotylidae Lühe, 1909 (two species), and Brachycladiidae Odhner, 1905 (two) (Delyamure 1955; Yurakhno 1990; Dailey and Vogelbein 1991); 21 cestode species of the families Diphyllobothriidae Lühe, 1910 (14 species); Tetrabothriidea Baer, 1954 (five); and Phyllobothriidea (two) (Delyamure 1955; Skryabin 1972; Yurakhno 1990; Dailey and Vogelbein 1991; Yurakhno and Maltsev 1997; Maltsev 2000); 14 acanthocephalan species of the family Polymorphidae (Laskowski and Zdzitowiecki 2017); 19 nematode species of the families Anisakidae Railliet, Henry, 1912 (13 species); Tetrameridae Travassos, 1914 (three); Filaroididae Schulz, 1951 (one); Filariidae Chabaud, Anderson, 1959 (one); and Ancylostomatidae (one) (Delyamure 1955; Yurakhno 1990; Dailey and Vogelbein 1991; Felix 2013; Mattiucci et al. 2017); 1 copepod species of the family Pennellidae Burmeister, 1835 (Dailey and Vogelbein 1991); 8 amphipod species of the family Cyamidae Rafinesque, 1815 (Berzin and Vlasova 1982; De Broyer et al. 2007); and 7 species of chewing lice of the family Echinophthiriidae Enderlein, 1904 (Leonardi et al. 2014; Leonardi et al. 2015; Mehlhorn and Mehlhorn 2017).

Parasitic fauna of the vertebrates has been studied in 183 species out of 553, fewer than 33% of their total: 40% of all fish fauna, 23% of known avian fauna, and 45% of the mammal fauna.

Therefore, taking into the account species diversity of the potential parasite hosts, invertebrates and vertebrates (4466 species) of the Antarctic and Subantarctic, we can note that their parasitic fauna, with only 377 recorded parasite species (5%), remains mostly unexplored.

23.4 Assessment of Parasite Species Diversity in the Antarctic and Subantarctic

Marine ecosystems are sensitive to external pressures, including human impact, and their biological components are fragile and vulnerable. Parasitological conditions are an integral part of its environmental state. Due to the dual nature of their environment (external environment and the body of the host), parasites constitute a special level in the ecosystem structure. In order to evaluate the role played by parasite communities in biocenoses, we require not only the quantitative and qualitative analyses of the parasitic fauna but also the understanding of the mechanisms of parasite interaction with the environmental factors and free-living components of the biota on every level of ecosystem organization. Therefore, in order to preserve the biodiversity of the biosphere, it is important to understand not only the function of parasitic communities within biocenoses but also all the aspects that affect successful completion of parasite life cycle.

At the start of the twenty-first century, the Antarctic ecosystem is under constant pressure due to climate change, rapid development of commercial fishing, and the general increase in human traffic (Kennicutt et al. 2015). The impact of these processes on the future conservation status of Antarctica is a field of keen interest (Parker et al. 2020). However, the lack of summarized data on parasitic communities of the Antarctic precludes their full integration into the biodiversity forecasts for the Southern Ocean (Bielecki et al. 2008).

Regrettably, the harsh weather conditions and limited accessibility of the Antarctic preclude continuous monitoring of parasitological conditions in this region. The majority of studies concerned with the parasitic fauna of the hydrobionts of the region is fragmentary in nature and, therefore, makes it difficult to appreciate the significant contribution parasites make to biodiversity in the communities of the Antarctic and Subantarctic. One of the latest updates on the parasite biodiversity of the Southern Ocean is the work of Klimpel et al. (2017), which presents accumulated data on five helminth taxa (monogeneans, trematodes, cestodes, nematodes, and acanthocephalans) and three ectoparasite taxa (chewing lice, fleas, and ticks) found in birds and mammals.

Among the invertebrates and vertebrates of the Antarctic and Subantarctic, the highest species diversity (271 species) is found in endoparasites (Coccidia, Gregarinasina, Microsporidia, Dicyemida, Myxosporea, Cestoda, Trematoda, Nematoda, Acanthocephala), which amounts to 72% of the total parasitic fauna of the region. Notably, the majority (203 species) have complex life cycles (Table 23.2), which can involve many host species of different taxonomic groups and ecological levels.

At the time of writing, 377 parasitic species belonging to 8 taxa have been found in the hydrobionts of the region, with the highest number of genera and families recorded among trematodes (32/14) and cestodes (31/12), and the lowest among fleas (one/one) and microsporidian (one/one) (Table 23.3). The highest number of orders has been found among cestodes (nine orders), nematodes (five), monogeneans (four), and copepods (four). More than half of parasitic taxa in the region are represented by one order and over 20% by two orders.

The Southern Ocean is unusual in that the bulk of teleost diversity and abundance is dominated by the suborder of Notothenioidei, including eight families (Eastman 2000). The members of this family include both the fish targeted by commercial fishing (Kock 1992; Delord et al. 2009) and the species acting as a link in the Antarctic food chain between the lower trophic levels and the consumers at the higher levels, such as whales, seals, and penguins (Targett 1981). Elasmobranchs of the Antarctic and Subantarctic are represented by sharks (five species) and rays (ten species). Sharks, which tend to prefer warmer waters, are mostly found in the northern part of the Southern Ocean. Up until now, the single finding of a shark in the southern part of the Antarctic occurred in the Ross Sea. All Antarctic rays belong to two families Rajidae: *Amblyraja* Malm, 1877 (two species), and *Bathyraja* Ishiyama, 1958 (seven), and Rhinobatidae: *Trygonorrhina* Müller, Henle, 1838 (one).

The species richness of parasites found in fish (189 species) of the Antarctic and Subantarctic is quite large (Table 23.2), and the dominant group of parasites are helminths; 142 species (75%) have been found (excluding larvae of cestodes where the species is unidentified) in 135 fish species. A thoroughly researched and,

Table 23.3 Numbers of parasitic species/endemic, genera, families, and order in the Antarctic and Subantarctic regions

Parasites	Species/endemic	Genera	Families	Order
Protista	**23/10**	**6**	**3**	**3**
Coccidia	10/2?	6	4	1
Gregarinasina	11/8	2	1	1
Microsporidia	2/?	1	1	1
Eumetazoa	**354/187**	**156**	**81**	**36**
Dicyemida	9/3?	3	2	1
Myxosporea	13/8	6	6	2
Monogenea	23/10?	11	7	4
Cestoda	67/26	31	12	9
Trematoda	69/63	32	14	2
Nematoda	35/6	22	13	5
Acanthocephala	32/29?	9	5	2
Hirudinea	20/16	13	1	1
Ascothoracidae	5/3	3	3	1
Copepoda	32/23?	24	17	4
Amphipoda	8/?	3	1	1
Izopoda	11/?	2	2	1
Phthiraptera	22/?	4	2	1
Siphonaptera	3/?	1	1	1
Acari	5/?	5	5	2
Total	**377/197**	**162**	**84**	**39**

evidently, dominant helminth group is the trematodes (65 species), which amounts to 46% of the total helminth fauna found in fish, while the fewest number of species (two each) is found in microsporidia and isopods (Table 23.2).

Since fish are the most investigated parasite hosts among the marine animals of the Antarctic and Subantarctic, the incidence analysis of all known parasites by fish families has shown the highest species richness in fish of three families: Macrouridae Bonaparte, 1831, was found to host parasites of nine different taxa; Channichthyidae T. N. Gill, 1861, of eight taxa; and Nototheniidae Günther, 1861, of seven (Table 23.4). A lesser degree of parasite diversity (five taxa) was found in fish of the family Bathydraconidae Regan, 1914; Harpagiferidae T. N. Gill, 1861; Merlucciidae T. N. Gill, 1884; and Muraenolepididae Regan, 1903. The uneven distribution of various parasite taxa identified in fish families in a given body of water points to the fact that the parasitic fauna has been most thoroughly researched among the widespread and most populous Antarctic fish families, while the rare deep-sea fish remains unexplored or underexplored in terms of parasites.

The level of endemism among the parasitic fauna of the Antarctic and Subantarctic has not been fully assessed. Out of the 18 parasite taxa identified in the invertebrates and vertebrates of the region, verified or estimated data on endemic species is available in 12 taxa (Table 23.3). Among protozoa, these are eight gregarine species of

Table 23.4 Distribution of parasites of different taxa by families of fish living in Antarctic and Subantarctic

Parasitic families of fish	Coccidiomorpha	Microsporidia	Myxosporea	Monogenea	Trematoda	Cestoda	Nematoda	Acanthocephala	Hirudinea	Copepoda	Isopoda
Achiropsettidae	–	–	–	–	1/1[a]	–	–	1/1	–	–	–
Arhynchobatidae	–	–	–	–	1/3	–	–	–	–	–	–
Artedidraconidae	–	–	–	–	1/9	4/2	1/4	2/1	–	–	–
Bathydraconidae	–	–	–	–	16/9	5/5	5/6	9/2	3/1	–	–
Bathylagidae	–	–	–	–	1/1	–	–	–	–	2/2	
Bovichtidae	–	–	–	–	2/1	–	–	–	–	–	–
Centrolophidae	–	–	1/1	–		–	–	–	–	–	–
Channichthyidae	–	–	3/2	–	18/10	13/10	6/11	12/8	5/6	1/1	1/1
Congiopodidae	–	–	–	–	1/1	–	–	–	–	–	–
Harpagiferidae	–	–	–	–	1/2	1/1	3/2	3/1	1/1	–	–
Gadidae	–	–	1/1	–	–	–	2/1	–	–	–	–
Gempylidae	–	–	–	–	–	1/1	–	–	–	–	–
Liparidae	–	–	–	–	5/3	–	–	1/1	–	–	–
Macrouridae	1/1	1/1	–	2/2	12/2	7/4	4/2	1/1	1/1	3/3	–
Merlucciidae	3/1	1/1	–	–		2/1	2/2	1/1	–	–	–
Muraenolepididae	–	–	1/1	–	17/2	6/2	4/1	–	1/2	–	–
Myctophidae	–	–	–	–	1/1	4/4	2/3	–	–	–	–
Nototheniidae	–	–	9/5	16/21	38/29	18/25	7/24	12/22	8/5	1/1	1/1
Ophidiidae	–	–	–	–	–	–	–	–	1/1	–	–
Rajidae	–	–	–	–	1/2	11/3	–	–	2/2	–	–
Rhinobatidae	–	–	–	–	–	–	–	–	1/1	–	–
Zoarcidae	–	–	–	3/2	7/4	–	2/3	–	–	–	–

[a]number of parasite species/number of species of infected hosts

the genus *Cephaloidophora* (Avdeev 1985; Avdeev and Avdeeva 1989; Takahashi et al. 2008; Gómez-Gutiérrez and Morales-Ávila 2016; Wallis et al. 2017) and two coccidia species *Haemogregarina georgianae* and *Desseria nototheniae* (Barber and Westermann 1988). Among dicyemids *Dicyemennea antarcticensis*, *D. bathybenthum*, and *Kantharella antarctica* are likely to be endemic (Short and Hochberg 1970; Furuya and Hochberg 2002). Eight endemic species have been found among myxosporidia – *Kudoa vesica*, *K. alliaria*, *Ceratomyxa antarctica*, *Zschokkella australis*, *Z. meglitschi*, *Z. globulosa*, *Neoparvicapsula ovalis*, and *Myxodavisia nototheniae* (Kovaleva et al. 2002; Eiras et al. 2014).

Among helminths the highest number of endemic species has been identified in trematodes. All the trematode species hosted by fish (61 species) are endemic, with the exception of cosmopolitan and bipolar species – *Otodistomum cestoides* (Beneden 1871); *Derogenes varicus* (Müller 1784); *Gonocerca phycidis* Manter, 1925; and *Glomericirrus macrouri* (Gaevskaya, 1975) (Rocka 2006), while in mammals only two endemic trematode species are identified – *Ogmogaster antarctica* Johnston, 1931, and *Orthosplanchnus antarcticus* Kurochkin, Nikol'skii, 1972 (Yurakhno 1990). Among the cestodes found in fish, 15 species are endemic: 12 cestode species in elasmobranchs and 3 in teleosts (Rocka 2006). Also, 11 cestode species have been identified in mammals (Yurakhno 1990). Of all the multitude of nematodes (35 species), four endemic species were found in fish *Ascarophis nototheniae* Johnston, Mawson, 1945; *Dichelyne (Cucullanellus) fraseri* (Baylis 1929); *Paranisakiopsis weddelliensis* Rocka, 2002; and *Comephoronema beatriceinsleyae* (Holloway and Klewer 1969) (Rocka 2006), and two species in mammals *Contracaecum mirounga* and *C. rectangulum* (Yurakhno 1990). The vast majority of acanthocephalans (approx. 29 species) of the Antarctic and Subantarctic are not found in the northern hemisphere (Rocka 2006), with the exception of the bipolar species *Corynosoma bulosum* and two bizonal species *C. australe* and *C. evae* (Yurakhno 1990). There is incomplete data on endemic monogeneans, but ten species are likely to be hosted by fish (Klapper et al. 2017).

Of the 20 leech species, only 4 are found north of the Subantarctic: *Pleurobdella australis* (Epstein, 1970); *Stibarobdella macrothela* (Schmarda, 1861); *Galatheabdella bruuni* Richardson, Meyer, 1974; and *Austrobdella translucens* (Badham, 1916) (Utevsky 1995, 2005, 2007; Utevsky and Gordeev 2015; Utevsky and Utevsky 2018; Parket et al. 2020). Two out of five ascothoracida species are thought to be endemic: *Dendrogaster antarctica* Grygier, 1980, and *Dendrogaster usarporum* Grygier, 1987 (Grygier 1987). Out of 32 copepod species, there is data confirming the endemic nature of 23 species: in fish, *Paeonocanthus antarcticensis*, *Lophoura szidati*, *Chondracanthodes tuberofurcatus*, and *Eubrachiella antarctica*; in polychaetes, *Herpyllobius antarcticus*, *H. polarsterni*, *Eurysilenium australis*, *Gottoniella antarctica*, *G. andeepi*, *Selioides tardus*, *Serpulidicoloides cystopomati*, and *Thylacoides sarsi*; in octopus, *Avdeevia antarctica, Tripartisoma trapezoidalis*, and *T. ovalis*; while in ophiuroids all the six found copepod species are represented (Table 23.1).

The biodiversity of parasites of marine invertebrates and vertebrates in the Antarctic and Subantarctic can be characterized as of a high level. The greatest attention of researchers of parasites of this region has been paid to parasites of the dominant group of vertebrate hosts, fish. Out of 377 parasite species recorded in hydrobionts of this region, 189 species were recorded in/on fish, i.e., fish are hosts for half of the entire fauna of parasites. An uneven distribution of parasites of different taxonomy among fish families was noted, which indicates an active study of parasites in widespread and well-investigated families of Antarctic fish, while rare and living at great depths, fish species remained either not studied in terms of parasites or the studies were fragmentary. According to both verified and estimated data, out of 377 parasite species found in marine life of the Antarctic and Subantarctic, 197 are endemic to the region, which indicative of the high level of endemism.

23.5 Concluding Remarks

In the conclusion of this chapter, it has to be noted that faunistic and taxonomic studies of Antarctic parasites remain sporadic, despite their considerable diversity in the Antarctic and Subantarctic waters. Parasitic organisms are represented by 18 taxa, the vast majority with the complex life cycle including intermediate/paratenic and definitive hosts in various taxonomic groups and ecological levels. The dominant group among the identified parasites in the invertebrates and vertebrates is endoparasites (271 species), whose share amounts to over 72% of the total parasitic fauna. Particularly underexplored, with only fragmentary studies available, are invertebrate hosts. The parasitic fauna of the Antarctic and Subantarctic is characterized by a high degree of endemism. Over half of the known species of the parasitic taxa are not found north of the Subantarctic. The highest diversity in parasitic fauna among the vertebrate hosts is found in fish, and the highest species richness among them is found in trematodes and cestodes.

Acknowledgment This work was supported by the Russian Academy of Sciences number AAAA-A18-118020890074-2 and AAAA-A19-119100290162-0. PTA is deeply grateful to Ph.D. Dmitrieva E.V. and Ph.D. Korniychuk Yu.M. (IBSS) for their help in discussing the results and recommendations.

References

Albon SD, Stien A, Irvine RJ, Langvatn R, Ropstad E, Halvorsen O (2002) The role of parasites in the dynamics of a reindeer population. Proc R Soc Lond B 269:1625–1632

Avdeev GV (1983) New harpacticoid copepods (Tisbidae), parasites of octopuses in the Ross Sea. Zool Zhurnal 62:1775–1785

Avdeev VV (1985) New species of gregarines of genus *Cephaloidophora* parasites of Euphausia superba. Parisitologiay 1:1–6 (In Russian)

Avdeev VV, Avdeeva NV (1989) On gregarine fauna from planktonic crustaceans from Antarctica. In: Lebadev BI (ed) Parasites of animals and plants: collection papers. Far East Division of Russia Academy of Science, Vladivostok, pp 40–44 (In Russian)

Bandera ME, Conradi M, Lopez-Gonzalez PJ (2005) *Asterocheres hirsutus*, a new species of parasitic copepod (Siphonostomatoida: Asterocheridae) associated with an Antarctic hexactinellid sponge. Helgol Mar Res 59:315–322

Barber DL, Westermann JEM (1988) *Haemogregarina georgianae* n. sp., a new haemogregarine parasite from the Antarctic teleost *Parachaenichthys georgianus*. Syst Parasitol 11:65–71

Barber DL, Westermann JEM, Storoz P (1987) *Haemogregarina nototheniae* n. sp. from the blood of Antarctic nototheniids. Syst Parasitol 10:135–147

Bargelloni L, Ritchie PA, Patarnello T et al (1994) Molecular evolution at subzero temperatures: mitochondrial and nuclear phylogenies of fishes from Antarctica (suborder Notothenioidei), and the evolution of antifreeze glycopeptides. Mol Biol Evol 11:854–863

Bartsch I (1996) Parasites of the Antarctic brittle star *Ophiacantha disjuncta* (Ophiacanthidae, Ophiuroidea): redescription of the copepod *Lernaeosaccus ophiacanthae* Heegard, 1951. Mitt Zool StInst Hamb 93:63–72

Beklemishev VN (1956) Biocenoses of the river and river valley as part of the living cover of the Earth. Proceedings of the All-Union Hydrobiological Society VII:340 (In Russian)

Berzin AA, Vlasova LP (1982) Fauna of the Cetacea Cyamidae (Amphipoda) of the World ocean. Investigations on Cetacea 13:149–164

Bielecki A, Rokicka M, Ropelewska E, Dziekońska-Rynko J (2008) Leeches (Hirudinida: Piscicolidae) – parasites of Antarctic fish from Channichthyidae family. Wiadomooeci Parazytologiczne 54(4):345–348

Boxshall GA, Harrison K (1988) New nicothoid copepods (Copepoda, Siphonostomatoida) from an amphipod and from deep-sea isopods. Bull Br Mus Nat Hist (Zool) 54:285–299

Boxshall GA, O'Reilly M, Sikorski A, Summerfield R (2019) Mesoparasitic copepods (Copepoda: Cyclopoida) associated with polychaete worms in European seas. Zootaxa 4579(1):001–069

Bresciani J, Lützen J (1994) Morphology and anatomy of *Avdeevia antarctica* new genus, new species (Copepoda, Harpacticoida, Tisbidae) parasitic on an Antarctic cephalopod. J Crust Biol 14:744–751

Brickle P, MacKenzie K, Pike A (2006) Variations in the parasite fauna of the patagonian tootfish (*Dissostichus eleginoides* Smitt, 1898), with length, season, and depth of habitat around the Falkland Islands. J Parasitol 92(2):282–291. https://doi.org/10.1645/GE-539R.1

Cleary AC, Casas MC, Durbin EG, Gómez-Gutiérrez J (2019) Parasites in Antarctic krill guts inferred from DNA sequences. Antarct Sci 31(1):16–22. https://doi.org/10.1017/S0954102018000469

Chesunov AV, Spiridonov SE (1985) *Australonema eulagiscae* gen. et sp. n. (Nematoda, Marimemithidae) – polychaete parasite from Antarctica. Bul Zool 2:16–21 (In Russian)

Czaker R (1994) *Kantharella antarctica*, a new and unusual dicyemid mesozoan from the Antarctic. Zoologischer Anzeiger 232:151–158

Dailey MD, Vogelbein WK (1991) Parasite fauna of 3 species of Antarctic whales with reference to their use as potential stock indicators. Fish B-NOAA 89:355–365

De Broyer C, Lowry JK, Jazdzewski K, Robert H (2007) Catalogue of the Gammaridean and Corophiidean Amphipoda (Crustacea) of the Southern Ocean, with distribution and ecological data. In: De Broyer C (ed.). Census of Antarctic Marine Life: Synopsis of the Amphipoda of the Southern Ocean. I. Bulletin de l'Institut royal des Sciences naturelles de Belgique, Biologie. 77 suppl 1:1–325

De Broyer C, Koubbi P, Griffiths HJ, Raymond B, Udekem d'Acoz C d' et al (eds) (2014) Biogeographic Atlas of the Southern Ocean. Scientific Committee on Antarctic Research, Cambridge, p 510

Delord K, Gasco N, Barbraud C, Weimerskirch H (2009) Multivariate effects on seabird bycatch in the legal Patagonian toothfish longline fishery around Crozet and Kerguelen Islands. Polar Biol 33:367–378

Delyamure SL (1955) Helminth fauna of marine mammals in the light of their ecology and phylogeny. Izv. Acad. Sci. USSR, Moscow, p 517 de Meeus T, Renaud F (2002) Parasites within the new phylogeny of eukaryotes. Trends Parasito 18:247–251

Diaz JI, Fusaro B, Vidal V, González-Acuña D, Costa ES, Dewar M, Gray R, Power M, Miller G, Blyton M, Vanstreels R, Barbosa A (2017) Chapter 9 Macroparasites in Antarctic Penguins. In: Biodiversity and evolution of parasitic life in the Southern Ocean. Klimpel S, Kuhn T, Mehlhorn H (Eds). Parasitol Res Monogr 9:183–204. https://doi.org/10.1007/978-3-319-46343-8_4

Dobson A, Lafferty KD, Kuris AM, Hechinger RF, Jetz W (2008) Homage to Linnaeus: how many parasites? How many hosts? Proc Natl Acad Sci USA 105:11482–11489

Duhamel G et al (2014) Chapter 7. Biogeographic patterns of fish. In: De Broyer C, Koubbi P, Griffiths HJ, Raymond B, Udekem d'Acoz C d' et al (eds) Biogeographic Atlas of the Southern Ocean. Scientific Committee on Antarctic Research, Cambridge, pp 328–362

Eastman JT (1993) Antarctic fish biology: evolution in a unique environment. Academic, San Diego

Eastman JT (2000) Antarctic notothenioid fishes as subjects for research in evolutionary biology. Antarct Sci. https://doi.org/10.1017/s0954102000000341

Eiras JC, Saraiva A, Cruz C (2014) Synopsis of the species of Kudoa Meglitsch, 1947 (Myxozoa: Myxosporea: Multivalvulida). Syst Parasitol 87:153–180. https://doi.org/10.1007/s11230-013-9461-4

Faltýnková A, Georgieva S, Kostadinova A, Bray RA (2017) Chapter 5. Biodiversity and evolution of Digeneans of fishes in the Southern Ocean. In: Biodiversity and evolution of parasitic life in the Southern Ocean. Klimpel S, Kuhn T, Mehlhorn H (Eds). Parasitol Res Monogr 9:49–75. https://doi.org/10.1007/978-3-319-46343-8_4

Felix JR (2013) Reported incidences of parasitic infections in marine mammals from 1892 to 1978. Zea E-Books 20:149. https://digitalcommons.unl.edu/zeabook/20

Fonseca VG, Carvalho GR, Sung W, Johnson HF, Power DM, Neill SP, Packer M, Blaxter ML, Lambshead PJD, Thomas WK, Creer S (2010) Second-generation environmental sequencing unmasks marine metazoan biodiversity. Nat Commun 1(98). https://doi.org/10.1038/ncomms1095

Froese R, Pauly D (eds) (2019) FishBase. World Wide Web electronic publication. www.fishbase.org, version (12/2019).

Furuya H (2008) Redescription of *Dicyemennea nouveli* (Phylum: Dicyemida) from *Enteroctopus dofleini* (Mollusca: Cephalopoda: Octopoda). J. Parasitol 94(5):1064–1070

Furuya H, Hochberg FG (2002) New species of *Dicyemennea* (phylum: Dicyemida) in deep-water *Graneledone* (Mollusca: Cephalopoda: Octopoda) from the Antarctic. J. Parasitol 88(2):330–336. https://doi.org/10.1645/0022-3395(2002)088[0330:NSODPD]2.0.CO;2

Gaevskaya AV (1982) The discovering of the trematode metacercarias in mysids of the South Georgia Island. Nauchnye Doklady Vysshey Shkoly, Biologicheskie Nauki 8:27–29 (In Russian)

Gaevskaya AV, Rodyuk GN (1988) Ecological characteristics of the parasite fauna of the South Atlantic grenadier (*Macrourus carinatus* Gunther). Biol Sci Sci Rep Higher Schools, 21–25 (In Russian)

Gaevskaya AV, Rodyuk GN, Parukhin AM (1990) Features and formation of the parasite fauna of the patagonian toothfish *Dissostichus eleginoides*. Biol Sea 4:23–28 (In Russian)

Golemansky V (2011) Coccidian parasites (Apicomplexa) of penguins (*Pygoscelis* spp.) from Livingston Island and King George Island, Antarctica. Pol Polar Res 32:263–268

Gómez-Gutiérrez J, Morales-Ávila JR (2016) Parasites and Diseases. In: Siegel V (ed) Biology and ecology of Antarctic Krill. Advances in Polar Ecology. Springer, Cham. https://doi.org/10.1007/978-3-319-29279-3_10

Gordeev II (2015) Prevalence, geographical distribution and host specificity of parasitic copepod *Lophoura szidati* Stadler, 1978 (Copepoda: Sphyriidae) on grenadiers (*Macrourus* spp.) in the Antarctic. Inverteb Zool 12(2):207–212

Gordeev II, Sokolov SG (2016a) Helminths and the feeding habits of the marbled moray cod *Muraenolepis marmorata* Günther, 1880 (Gadiformes, Muraenolepididae) in the Ross Sea (Southern Ocean). Polar Biol DOI. https://doi.org/10.1007/s00300-016-2055-2

Gordeev II, Sokolov SG (2016b) Parasites of Antarctic toothfish *Dissostichus mawsoni* (Norman, 1937) (Perciformes, Nototheniidae). Polar Res. https://doi.org/10.3402/polar.v35.29364

Gravier C (1912a) Sur un Crustace parasite d'un Polynoidien de l'antarctique sudamericaine (*Selioides tardus* n. sp.). Bull Mus Paris 1912:63–67

Gravier C (1912b) Sur un type nouveau de Crustace parasite d'un Serpulien de l'antarctique sud-Americaine (*Bactropus* nov. g. *cystopomati* nov. sp.). Bull Mus Paris 1912:67–71

Gravier C (1912c) Sur un nouveau genre de Crustace parasite d'un Syllidien de l'antarctique sud-Americaine (*Thylacoides* nov. g. *sarsi* nov. sp.). Bull Mus Paris 1912:71–74

Gravier C (1912d) Sur une Copepode (*Zanclopus antarcticus* nov. sp.) parasite d'un Cephalodiscus recueilli par la seconde Expedition antarctique francaise et sur l'evolution du genre Zanclopus Calman. Bull Mus Paris 1912:240–245

Gravier C (1914) *Isidicola antarctica* Gravier. Crustace parasite de quelques Isidae de l'Antarctique sud-Americaine. C R hebd Séanc Acad Sci Paris 158:354–356

Grygier MJ (1981) A representative of the genus *Dendrogaster* new record (Cirripedia, Ascothoracica) parasitic in an Antarctic starfish *Acodontaster conspicuous*. Antarct Res Ser 32:1–15

Grygier MJ (1987) Antarctic records of asteroid-infesting Ascothoracida (Crustacea), including a new genus of Ctenosculidae. Proceed Biol Soc Washing 100:700–712

Grygier MJ, Fratt DB (1984) The ascothoracid crustacean *Ascothorax gigas*: Redescription, larval development, and notes on its infestation of the Antarctic ophiuroid *Ophionotus victoriae*. Antarctic Res Ser 41:43–58

Grygier MJ, Salvat MB (1984) *Dendrogaster argentinensis*, new species, a South American sea-star parasite (Crustacea: Ascothoracida). Proceed Biol Soc Washing 97:43–48

Heegard P (1951) Antarctic parasitic copepods and an ascothoracid cirriped from brittle – stars. Vidensk Meddr dansk naturh Foren 113:117–190

Ho J-s, Kim Il-H, Nagasawa K, Saruwatari T (2003) *Paeonocanthus antarcticensis* (Hewitt, 1965): A rare copepod parasite of bathypelagic fish, *Bathylagus antarcticus* Günther, from the Antarctic Ocean. Zool Sci 20:1315–1321

Hochberg FG (1990) Diseases of Cephalopoda. Diseases caused by protistans and mesozoans. In: Diseases of marine animals, Kinne O (Ed.). Biologische Anstalt Helgoland, Hamburg, Germany 3:47–202

Hochberg FG, Short RB (1970) *Dicyemennea littlei* sp. n. and *Dicyema benthoctopi* sp. n.: Dicyemid Mesozoa from *Benthoctopus magellanicus*. Trans Amer Microsc Soc 89(2):216–224

Janussen D, Downey RV (2014) Chapter 5.5 Porifera. In: De Broyer C, Koubbi P, Griffiths HJ, Raymond B, Udekem d'Acoz C d' et al (eds) Biogeographic Atlas of the Southern Ocean. Scientific Committee on Antarctic Research, Cambridge, pp 94–102

Jeżewski W, Zdzitowiecki K, Laskowski Z (2014) Digenea in nothothenioid fish in the Beagle Channel (Magellanic sub-region, sub-Antarctica). Acta Parasitol 59(1):42–49. https://doi.org/10.2478/s11686-014-0208-4

Kennicutt MC, Chown SL, Cassano JJ et al (2015) A roadmap for Antarctic and Southern Ocean science for the next two decades and beyond. Antarct Sci 27:3–18

Klapper R, Münster J, Kochmann J, Klimpel S, Kuhn T (2017) Chapter 4. Biodiversity and host specificity of Monogenea in Antarctic Fish Species. In: Biodiversity and evolution of parasitic life in the Southern Ocean. Klimpel S, Kuhn T, Mehlhorn H (eds). Parasitol Res Monogr 9:33–47. https://doi.org/10.1007/978-3-319-46343-8_4

Klimpel S, Kuhn T, Mehlhorn H (eds) (2017) Biodiversity and evolution of parasitic life in the Southern Ocean. Parasitol Res Monogr:226. https://doi.org/10.1007/978-3-319-46343-8

Klimpel S, Palm HW (2011) Anisakid nematode (Ascaridoidea) life cycles and distribution: increasing zoonotic potential in the time of climate change? In: Progress in parasitology. Springer, Berlin, pp 201–222

Kock KH (1992) Antarctic fish and fisheries. Cambridge University Press, Cambridge, 359 p

Kononenko AF (2009) Distribution of parasitic isopods suborder Flabellifera and suborder Gnathiidea in different regions of the Atlantic Ocean. In: X Congress of the Hydrobiological Soc. Russ. Acad. Sci. Vladivostok: Dalnauka, p 203 (In Russian)

Kovaleva AA, Rodyuk GN, Grudnev MA (2002) Mixosporidia (Cnidospora: Myxosporea) of Antarctic fishes. Parasitologiay 36(6):502–513 (In Russian)

Laskowski Z, Rocka A (2014) Molecular identification larvae of *Onchobothrium antarcticum* Cestoda: Tetraphyllidea) from marbled rockcod, *Notothenia rossii*, in Admiralty Bay (King George Island, Antarctica). Acta Parasitl 59:767–772

Laskowski Z, Zdzitowiecki K (2017) Chapter 8. Acanthocephalans in Sub-Antarctic. In: Biodiversity and evolution of parasitic life in the Southern Ocean. Klimpel S, Kuhn T, Mehlhorn H (Eds). Parasitol Res Monogr 9:77–107. https://doi.org/10.1007/978-3-319-46343-8_4

Laskowski Z, Rocka A, Zdzitowiecki K (2005) New data on the occurrence of internal parasitic worms in the *Gymnodraco acuticeps* and *Cygnodraco mawsoni* (Bathydraconidae) fish in the Ross Sea, Antarctica. Pol Polar Res 26(1):37–40

Laskowski Z, Rocka A, Zdzitowiecki K, Ozouf-Costaz C (2007) Occurrence of endoparasitic worms in dusky notothen, *Trematomus newnesi* (Actinopterygii Nototheniidae), at Adélie Land, Antarctica. Pol Polar Res 28:37–42

Leonardi MS, Bobinac M, Negrete J (2015) Redescription of *Antarctophthirus lobodontis* (Anoplura: Echinophthiriidae) from the crabeater seal and identification key for Antarctic lice. Polar Biol. https://doi.org/10.1007/s00300-015-1822-9

Leonardi MS, Poljak S, Carlini P, Galliari J, Bobinac M, Santos M, Márquez ME, Negrete J (2014) *Antarctophthirus carlinii* (Anoplura: Echinophthiriidae), a new species from the Weddell seal *Leptonychotes weddelli*. Parasitol Res 113:3947–3951. https://doi.org/10.1007/s00436-014-4058-7

López-González PJ, Bresciani J, Conradi BM (2006) New genus, three new species and new records of Herpyllobiidae Hansen, 1892 (Crustacea, Copepoda), parasites of polychaetes from Antarctica. Scientia Marina 70(2):243–259

Lützen J, Jones B (1976) Two new species of *Herpyllobius* parasitic Copepoda from New Zealand and the Antarctic. N Z Jl Mar Freshw Res 10:371–374

Lyadov VN (1981) Zoogeographic characteristics of fish helminths in the Antarctic zone of the World Ocean. In: Symposium on parasitology and pathology of marine organisms, pp 58–61 (In Russian)

Lyadov VN (1985) Zoogeographical characteristics of the helminths of fishes from the Antarctic zone of the World Ocean. NOAA Technical Rep NMFS 25:41–43

Lyadov VN, Parukhin AM, Mironova AV (1981) Helminth fauna of commercial fish of the family Chaenichthidae from the Kerguelen Islands region. Zool J 60(1):142–144 (In Russian)

Maltsev VN (2000) New data on four species of diphyllobothriids parasitizing pinnipeds in Antarctica. Proceedings of the Southern Scientific Research Institute of Marine Fisheries and Oceanography 45:93–105 (In Russian)

Marcogliese DJ (2004) Parasites: small players with crucial roles in the ecological theater. EcoHealth 1:151–164. https://doi.org/10.1007/s10393-004-0028-3

Margolis L (1955) Notes on the morphology, taxonomy and synonymy of several species of whale-lice (Cyamidae: Amphipoda). J Fish Res Board Canada 12(1):121–133

Mattiucci S, Paoletti M, Cipriani P, Webb SC, Timi JT, Nascetti G (2017) Chapter 7 Inventorying Biodiversity of Anisakid Nematodes from the Austral Region: A Hotspot of Genetic Diversity? In: Biodiversity and evolution of parasitic life in the Southern Ocean. Klimpel S, Kuhn T, Mehlhorn H (Eds). Parasitol Res Monogr 9:109–140. https://doi.org/10.1007/978-3-319-46343-8_4

Mehlhorn B, Mehlhorn H (2017) Chapter 10 Lice on Seals in the Antarctic Waters and Lice in Temperate Climates. In: Biodiversity and evolution of parasitic life in the Southern Ocean. Klimpel S, Kuhn T, Mehlhorn H (Eds). Parasitol Res Monogr 9:205–215. https://doi.org/10.1007/978-3-319-46343-8_4

Moser M, Cowen RK (1991) The effects of periodic eutrophication on parasitism and stock identification of *Trematomus bernacchii* (Pisces: Nototheniidae) in McMurdo Sound, Antarctica. J Parasitol 77:551–556

Mouritsen KN, Poulin R (2002) Parasitism, community structure and biodiversity in intertidal ecosystems. Parasitology 124:S101–S117

Oğuz MC, Tepe Y, Belk MC, Heckmann RA, Aslan B, Gürgen M, Bray RA, Akgül Ü (2015) Metazoan Parasites of Antarctic Fishes. Turkiye Parazitol Derg 39:174–178. https://doi.org/10.5152/tpd.2015.3661

Palm HW (2004) The Trypanorhyncha Diesing, 1863. PKSPL-IPB Press, Bogor, 710 pp

Palm HW, Reimann N, Spindler M, Plötz J (1998) The role of the rock cod *Notothenia coriiceps* Richardson, 1844 in the life-cycle of Antarctic parasites. Polar Biol 19:399–406

Parker E, Jones CD, Arana PM et al (2020) Infestation dynamics between parasitic Antarctic fish leeches (Piscicolidae) and their crocodile icefish hosts (Channichthyidae). Polar Biol 43:665–677. https://doi.org/10.1007/s00300-020-02670-x

Parukhin AM (1986) Features of the helminth fauna of commercial notothenium fish in the Subantarctic sector of the Indian Ocean. Bul Zool 3:6–10 (In Russian)

Parukhin AM, Lyadov VN (1981) Parasite fauna of notothenioid fishes (Notothenioidei) from the waters of the Atlantic and Indian oceans. Bul Zool 3:90–94 (In Russian)

Parukhin AM, Lyadov VN (1982) Helminth fauna of commercial fish of the Nototheniidae family of the Kerguelen subregion. Ecology of the Sea 10:49–56 (In Russian)

Peirce MA, Prince PA (1980) *Hepatozoon albatrossi* sp. nov. (Eucoccida: Hepatozoidae) from *Diornedea* spp. in the Antarctic. J Nat Hist 14:447–452. https://doi.org/10.1080/00222938000770381

Platonova TA, Potin VV (1972) On new genera *Harpagonchoides* (Nematoda, Chromadorida, Harpagonchidae, fam. n.) living on the parapodia and gills of the Antarctic polychaetes *Aglaophamus* Kinberg and *Hemipodus* Quatrifages. Issled Fauny Morei 19:81–87

Poulin R, Morand S (2000) Diversity of parasites. The Quarterly review of biology 75:277–293

Poulin R, Morand S (2004) Parasite biodiversity. Smithsonian Books, Washington, DC, p 216

Poulin R, Mouritsen KN (2006) Climate change, parasitism and the structure of intertidal ecosystems. J Helmint 80:183–191. https://doi.org/10.1079/JOH2006341

Rennie J, Reid A (1912) XXII. – The Cestoda of the Scottish National Antarctic Expedition. Trans Royal Soc Edinburgh 48:441–453. https://doi.org/10.1017/S0080456800002945

Rocka A (2003) Cestodes of the Antarctic fishes. Pol Polar Res 24:261–276

Rocka A (2006) Helminths of Antarctic fishes: life cycle biology, specificity and geographical distribution. Acta Parasitol 51:26–35

Rocka A (2017) Chapter 6. Cestodes and Nematodes of Antarctic fishes and birds. In: Biodiversity and evolution of parasitic life in the Southern Ocean. Klimpel S, Kuhn T, Mehlhorn H (Eds). Parasitol Res Monogr 9:77–107. https://doi.org/10.1007/978-3-319-46343-8_4

Rocka A (2019) Cestodes of the genus *Tetrabothrius* (Tetrabothridea: Tetrabothriidae) from Antarctic flying seabirds. Annals of Parasit 65(4):423–426. https://doi.org/10.17420/ap6504.230

Rocka A, Zdzitowiecki K (1998) Cestodes in fishes of the Weddell Sea. Acta Parasitol 43:64–70

Rodyuk GN (1988) Parasite fauna of teleost fishes of the Southwest Atlantic and adjacent Antarctic waters. PhD. dissertation. Moscow State University, p 503 (In Russian)

Rokicki J, Skora KE (1987) The dynamics of occurrence of *Eubrachiella antarctica* (Quidor, 1906) on *Notothenia gibberifrons* Lonnberg, 1905. Wiad Parazytol 32:511–515

Rokicki J, Zdzitowiecki K (1991) Dynamics of *Eubrachiella antarctica* (Quidor, 1906) (Copepoda) occurrence in *Notothenia rossii marmorata* (Fischer, 1885). Acta Ichthyol Piscat 21(2):45–52

Ropert-Coudert Y, Hindell MA, Phillips RA, Charrassin J-B, Trudelle L, Raymond B (2014) Chapter 8. Biogeographic patterns of birds and mammals. In: De Broyér C, Koubbi P, Griffiths HJ, Raymond B, Udekem d'Acoz C d' et al (eds) Biogeographic Atlas of the Southern Ocean. Scientific Committee on Antarctic Research, Cambridge, pp 364–387

Rubtsov RA (1977) A new genus and species of parasitic nematode, *Ananus asteroideus* (Nematoda, Marimermithidae), from the asteroid *Diplopteraster peregrinator*. Bull Mus Hist Nat Paris (Zoologie) 345:1113–1116

Santoro M, Mattiucci S, Cipriani P, Bellisario B, Romanelli F, Cimmaruta R, Nascetti G (2014) Parasite Communities of Icefish (*Chionodraco hamatus*) in the Ross Sea (Antarctica): influence of the Host Sex on the Helminth Infracommunity Structure. PLoS ONE 9(2):e88876. https://doi.org/10.1371/journal.pone.0088876

Sawyer RT (1972) A New Species of "Tentacled" Marine Fish Leech Parasitic on Notothenia from the Subantarctic Marion and Crozet Islands. Hydrobiologia 40(3):345–354

Sawyer RT, White MG (1969) A new genus and species of marine leech, *Glyptonobdella antarctica*, from an Antarctic isopod. Bull Br Antarct Surv 22:1–14

Schultz GA (1980) *Arcturocheres gaussicola* n. sp. (Cabiropsidae), parasite in *Antarcturus gaussianus* Vanhoffen (Arcturidae) from Antarctica (Isopoda). Crustaceana 39:153–156

Shultz RS, Gvozdev EV (1972) Fundamentals of general helminthology, 2. Ed. "Science", Moscow: 1–515 (In Russian)

Skryabin AS (1972) Larvae of cestodes of the genus *Phyllobothrium* van Beneden, 1850 (Tetraphyllidea), parasitizing whales and other marine animals. Parasitologiay 6(5):426–434 (In Russian)

Short RB, Hochberg FG (1970) A new species of *Dicyemennea* (Mesozoa, Dicyemidae) from near Antarctic Peninsula. J Parasitol 56:517–522

Stryukov AA (2000) *Corynosoma erignathi* (Acanthocephala, Polymorphidae) – a parasite of the seal *Erignathus barbatus nauticus*. Bull Zool 14:9–18 (In Russian)

Stryukov AA (2001) Parasite of the southern elephant seal from the Pacific sector of Antarctica. Sci Notes Tauride National University 14(53):68–74 (In Russian)

Stryukov AA (2002) About the scraper *Corynasoma arctocephali* (Acanthocephala, Polymorphidae) – a parasite of the leopard sea *Hydrurga leptonyx* from the Pacific sector of Antarctica. Bul Zool 36(4):71–77 (In Russian)

Studer A, Thieltges DW, Poulin R (2010) Parasites and global warming: net effects of temperature on an intertidal host–parasite system. Mar Ecol Prog Ser 415:11–22

Takahashi K, Kobayashi M, Kawaguchi S, Saigusa J, Tanimura A, Fukuchi M, Toda T (2008) Circumpolar occurrence of eugregarinid protozoan *Cephaloidophora pacifica* associated with Antarctic krill, *Euphausia superba*. Antarctic Sci 20(5):437–440. https://doi.org/10.1017/S0954102008001016

Targett TE (1981) Trophic ecology and structure of coastal Antarctic fish communities. Mar Ecol Prog Ser 4:243–263

Thomas F, Renaud F, Guégan J-F (2005) Parasitism and ecosystems. Oxford University Press, Oxford

Utevsky AYu (1995) A new genus of Antarctic leeches (Himdinea, Piscicolidae). Bul Zool 5–6:3–12 (In Russian)

Utevsky AY (2005) An identification key to Antarctic fish leeches (Hirudinea: Piscicolidae). Ukrains'kij Antark tychnij Zhurnal 3:135–144

Utevsky AY (2007) Antarctic Piscicolid Leeches. Bonner Zoologische Monographien 54:80

Utevsky A, Gordeev I (2015) New tentacled leech *Ceratobdella quadricornuta* n. g., n. sp. (Hirudinida: Piscicolidae) parasitic on the starry skate *Raja georgiana* Norman from the Scotia Sea, Antarctica. Syst Parasitol 91:203–210. https://doi.org/10.1007/s11230-015-9570-3

Utevsky A, Utevsky S (2018) New Antarctic deep-sea weird leech (Hirudinida: Piscicolidae): morphological features and phylogenetic relationships. Syst Parasitol. https://doi.org/10.1007/s11230-018-9816-y

Venekey V, Gheller PF, Kandratavicius N, Cunha BP, Vilas-Boas AC, Fonseca G, Maria TF (2019) The state of the art of Chromadoridae (Nematoda, Chromadorida): a historical review, diagnoses and comments about valid and dubious genera and a list of valid species. Zootaxa 4578(1):1. https://doi.org/10.11646/zootaxa.4578.1.1

Wallis JR, Smith, AJR, Kawaguchi, SO (2017) Discovery of gregarine parasitism in some Southern Ocean krill (Euphausiacea) and the salp *Salpa thompsoni*. Polar Biol: https://doi.org/10.1007/s00300-017-2089-0

Walter T, Palm HW, Piepiorka S, Rückert S (2002) Parasites of the Antarctic rattail *Macrourus whitsoni* (Regan, 1913) (Macrouridae, Gadiformes). Polar Biol 25:633–640

Wojciechowska A (1990a) *Onchobothrium antarcticum* sp. n. (Tetraphyllidea) from *Bathyraja eatonii* (Günther, 1876) and a plerocercoid from Notothenioidea (South Shetlands, Antarctic). Acta Parasitol Pol 35:113–117

Wojciechowska A (1990b) *Pseudanthobothrium shetlandicum* sp. n. and *P. notogeorgianum* sp. n. (Tetraphyllidea) from rays in the regions of the South Shetlands and South Georgia (Antarctic). Acta Parasitol Pol 35:181–186

Wojciechowska A (1991a) *Pseudanthobothrium shetlandicum* sp. n. and *P. notogeorgianum* sp. n. (Tetraphyllidea) from rays in the regions of the South Shetlands and South Georgia (Antarctic). Acta Parasitol Pol 35:181–186

Wojciechowska A (1991b) Some tetraphyllidean and diphyllidean cestodes from Antarctic batoid fishes. Acta Parasitol Pol 36:69–74

Wojciechowska A (1993a) The tetraphyllidean and tetrabothriid cercoids from Antarctic bony fishes. I. Morphology, Identification with adult forms. Acta Parasitol 38:15–22

Wojciechowska A (1993b) The tetraphyllidean and tetrabothriid cercoids from Antarctic bony fishes. II. Occurrence of cercoids in various fish species. Acta Parasitol 38:113–118

Wojciechowska A, Zdzitowiecki K, Pisano E, Vacchi M (1994) The tetraphyllidean cercoids from bony fishes occurring in the Ross Sea (Antarctic). Acta Parasitol 39:13–15

Wojciechowska A, Pisano E, Zdzitowiecki K (1995) Cestodes in fishes at the Heard Island (Subantarctic). Pol Polar Res 16:205–212

Yurakhno MV (1990) Pinniped helminths of the World Ocean (taxonomy, fauna, ecology, zoogeography, coevolution with hosts). Dr. Biol. Sciences. dissertation. Simferopol, p 530 (in Russian)

Yurakhno MV, Maltsev VN (1997) Infestation by tape helminths of real seals of the Antarctic. Parasitologiay 31(1):81–89 (in Russian)

Zdzitowiecki K (1991) Occurrence of digeneans in open sea fishes off the South Shetland Islands and South Georgia, and a list of fish digeneans in the Antarctic. Pol Polar Res 12:55–72

Zdzitowiecki K (1997) Antarctic Digenea, parasites of fish. Koeltz Scientific Books, Königstein

Zdzitowiecki K (2001a) New data on the occurrence of fish endoparasitic worms off Adelie Land, Antarctica. Pol Polar Res 22:159–165

Zdzitowiecki K (2001b) Acanthocephala occurring in intermediate hosts, amphipods, in Admiralty Bay (South Shetland Islands, Antarctica). Acta Parasitol 46:202–207

Zdzitowiecki K, Laskowski Z (2004) Helminths of an Antarctic fish, *Notothenia coriiceps*, from the Vernadsky Station (Western Antarctica) in comparison with Admiralty Bay (South Shetland Islands). Helminthologia 41:201–207

Zdzitowiecki K, Zadrozny T (1999) Endoparasitic worms of *Harpagifer antarcticus* off the South Shetland Islands (Antarctic). Acta Parasitol 44:125–130

Zdzitowiecki K, White MG, Rocka A (1997) Digenean, monogenean and cestode infection of inshore fish at the South Orkney Islands. Acta Parasitol 42:18–22

Zdzitowiecki K, Rocka A, Pisano E, Ozouf-Costaz C (1998) A list of fish parasitic worms collected off Adelie Land (Antarctica). Acta Parasitol 43:71–74

Chapter 24
Spatial Distribution, Species Composition, and Number of Seabirds in the Argentine Basin, Drake Passage, East of Antarctic Peninsula, and Powell Basin in January–March 2020

Sergey P. Kharitonov, Alexander L. Mischenko, Nikolai B. Konyukhov, Alexander E. Dmitriyev, Andrey V. Tretyakov, Gleb Yu. Pilipenko, Svetlana M. Artemyeva, and Matvey S. Mamayev

Abstract Results of observations of seabird occurrence during cruise 79 of the R/V "*Akademik Mstislav Keldysh*" in January, February, and early March 2020 are presented. We recorded 35,851 seabirds, 20,116 of them were identified to the species level, and a total of 43 species were recorded. The observed occurrence of seabirds in the ocean was correlated to a number of environmental parameters: in January, almost all species preferred "good weather", that is, higher air temperatures, lower humidity and, in some species, high air pressure. In February, a number of species maintained these correlations, but several species, in particular, Black-browed Albatross, Antarctic Skua, Cape Petrel, and Southern Fulmar revealed correlations opposite to the cited environmental parameters. They were mainly encountered during low air temperature and high humidity. This probably indicates that the birds found adequate foraging areas under these meteorological conditions. Several species (e.g., Antarctic Petrel, Antarctic Skua) were observed within regions with specific sea depths. The occurrence of most birds from January to February varied in the regions with different depths. Unlike marine mammals, several seabird species formed uniform distribution of occurrence along the vessel route. The Black-browed

S. P. Kharitonov (✉) · A. L. Mischenko · N. B. Konyukhov · A. E. Dmitriyev
A. V. Tretyakov · M. S. Mamayev
Severtsov Institute of Ecology and Evolution, Russian Academy of Sciences, Moscow, Russia
e-mail: serpkh@gmail.com

G. Yu. Pilipenko
Faculty of Geography, Lomonosov Moscow State University, Moscow, Russia

S. M. Artemyeva
Zoological Museum of the Lomonosov Moscow State University, Moscow, Russia

E. G. Morozov et al. (eds.), *Antarctic Peninsula Region of the Southern Ocean*, Advances in Polar Ecology 6, https://doi.org/10.1007/978-3-030-78927-5_24

Albatross and the Southern Giant Petrel maintained this distribution pattern during both months of the study. The mathematical evaluation of distributions together with a map analysis points to the competition between species or, possibly, in addition, intraspecies competition (probably for food) among these two species.

Keywords Antarctic birds · Seabirds · Albatrosses · Penguins · Petrels · Seabird distribution · Factors affecting the occurrence

24.1 Introduction

Antarctic studies are important for monitoring this very productive region of the Earth. The composition and distribution of species of marine mammals and sea birds might change over the years, so the fluctuating process should be monitored. Cruise 79 of the R/V "*Akademik Mstislav Keldysh*" of the Russian Academy of Sciences to the southern part of the Atlantic took place in January–March, 2020. The goal of the study was to investigate the current state of the oceanographic environment and the Antarctic biota.

Data on seabirds and marine mammals were collected from January 9 to March 3, 2020. This article presents the data and preliminary results of the occurrence of seabirds in this area. The cruise consisted of several stages. The first stage was the voyage from Montevideo to the entrance of the Beagle Channel. The second and third stages consisted of two trips to the Antarctic waters with an intermediate visit to Ushuaia. The total length of the scientific voyage was approximately 14,000 km.

It is known that the distribution of seabirds in Antarctica depends on the abundance of plankton (krill), but to a lesser extent than in whales (Joiris and Dochy 2013; Orgeira et al. 2015). Herewith, we have revealed a number of regularities in bird distribution in connection with several parameters of the environment, partly confirming previous studies and partly describing new regularities.

Detailed mechanisms of the distribution of seabirds in the Antarctic are particularly important at the present time of climate change, which largely concerns many regions of the Earth, as well as the Antarctic. Possibly, birds are more dependent on weather conditions than marine mammals. It is also possible that birds may be less reliant on the depths of the sea than mammals. The general feature of the distribution of seabirds at sea is still poorly investigated. Some articles contain information on the density of seabirds, for example, Joiris and Dochy (2013), but little information on the density of seabirds is particularly available in Antarctic waters. Evaluation of bird density within the area from a moving vessel (e.g., varying levels of visibility at different parts of the route) is complicated. On the other hand, data on the amount and composition of plankton have not always been available. For this purpose, we determined the type of distribution of the species along the vessel's route as along the line and tried to collect information on the various species from their type of distribution.

24.2 Materials and Methods

The data have been collected during three parts of the vessel voyage: (1) 09–15 January 2020, Montevideo–entrance to the Beagle Channel, near 2500 km; (2) 16 January–6 February, entrance to the Beagle Channel–Drake Passage–South Shetland Islands–tip of the Antarctic Peninsula–South Orkney Islands–back to Antarctic Peninsula–Drake Passage-Ushuaia, near 5500 km; (3) 08 February–03 March: Ushuaia–Drake Passage–tip of the Antarctic Peninsula–Powell Basin-back to the Antarctic Peninsula–Drake Passage–Ushuaia, more than 6000 km (Fig. 24.1).

Observations were made throughout all daylight hours from the broad direction-finder deck by two observers positioned there; watches lasted for standard 4 h or

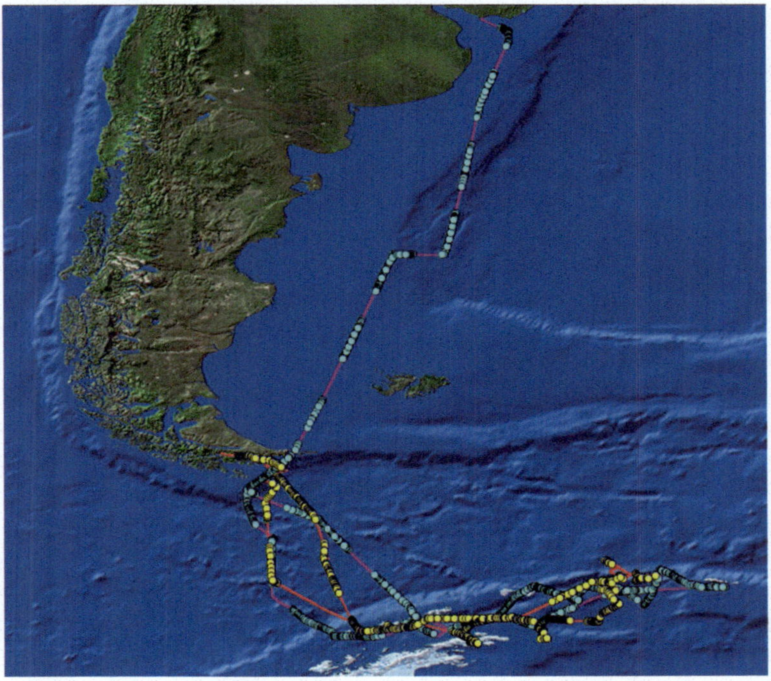

Fig. 24.1 Map of the R/V "*Akademik Mstislav Keldysh*" trips to Antarctica. Purple line is the trip on 9 January–6 February, Montevideo–entrance of the Beagle Channel–Drake Passage–top of the Antarctic Peninsula–South Orkney Islands–back to Antarctic Peninsula–Drake Passage–Ushuaia. Red line is the trip on 08 February–03 March: Ushuaia–Drake Passage–top of the Antarctic Peninsula–Powell Basin–back to Antarctic Peninsula–Drake Passage–Ushuaia. Blue-green dots are seabirds observed during January–early February, yellow dots are seabirds seen in February–early Mach. Empty parts of the tracks are related to nighttime

less in the morning or in the evening. Observers were equipped with high-power photo cameras and telephoto lenses. Most of the encountered seabirds were photographed. Species, number of individuals in the groups, behavior, ice characteristics, and general weather parameters were recorded. Large birds (like albatrosses) were detectable at distances up to 500 m from the ship, middle-size birds (like Southern Fulmar, Cape Petrel, etc.) were usually visible up to 300 m, and smaller birds, like prions or Storm Petrels, were normally visible up to 150 m. Birds on ice might be detected at a distance of up to 1 km from the ship. The route of the vessel was tracked by three GPS navigators; therefore, coordinates of each encounter with biological objects in the sea are available.

Depths along the route were determined by the Kongsberg EA600 12 kHz single-beam hydrographic echo sounder. However, it was not always possible to determine the depth of such exact points. The echo sounder did not work at "stations" (points where the ship stopped to collect sea samples) and during very rough seas. In these cases, we calculated the nearest point where the depth was measured. Usually, this value was less than 3 km from the real point. In two cases, the distance was approximately 100 km. This was along the Beagle Channel where the depth was defined only at its entrance, and with the flat area of about the same depth of the Powel Basin (see map in Fig. 24.1) where the depth was defined at several points. However, the Beagle Channel is shallow along its full extent, so it is not bias to consider one measurement from the entrance point. The flat area in the Powell Basin has about the same depth over its extent, so remote measurements can be accepted there as well. For these reasons, all depth measurements are considered as relevant for the analysis. The use of an echo sounder on the route from Montevideo to Ushuaia is prohibited; therefore, depths from the open-access database https://www.gebco.net/data_and_products/gridded_bathymetry_data/ were used. Although this is not as accurate as echo sounder data, it is acceptable for our analysis.

Common weather parameters were permanently recorded every 5 min by the weather station Davis Vantage Pro 2, managed by the scientific team of the V.I. Il'ichev Pacific Oceanological Institute Far Eastern Branch Russian Academy of Sciences.

During the entire described trip, we had 6179 encounters with seabirds, with a total of 35,851 birds detected (Table 24.1).

For the data analysis, we used commonly accepted factor analysis and correlation analysis via the Statistica-12, StatSoft Inc. program. In addition, for the comparison of averages, we used t- Bailey criterion (Plokhinskiy 1978) instead of the commonly accepted criteria (t-Student or U-Mann-Whitney). This criterion is parametric and, at the same time, does not require normal distribution of data. It is therefore applicable in all situations where the Mann-Whitney criterion is necessary and has the same statistical power as the Student criterion.

In order to obtain information on the form of distribution of species that have been abundant in the field, we used the method of assessment of distribution of objects along the line (Kharitonov 2005) derived from the Clark-Evans Nearest Neighbor Method (Clark and Evans 1954). As the observations were not performed in dark time, for the distribution evaluation, we subtracted night pieces from the

Table 24.1 Seabird species and systematic groups (if the species was not defined) during the Antarctic trip of the R/V "*Akademik Mstislav Keldysh*"

	Part of the voyage and dates					
	Montevideo–entrance to the Beagle Channel, 09–15 January 2020		Entrance to the Beagle Channel–Drake Passage–top of the Antarctic Peninsula–South Orkney Islands–back to Antarctic Peninsula–Drake Passage–Ushuaia, 16 January–6 February		Ushuaia–Drake Passage–top of the Antarctic Peninsula–Powell Basin–back to Antarctic Peninsula–Drake Passage–Ushuaia, 08 February–03 March	
Species	Number of encounters	Number of birds	Number of encounters	Number of birds	Number of encounters	Number of birds
Gentoo Penguin *Pygoscelis papua*	–	–	36	312	13	346
Adelie Penguin *Pygoscelis adeliae*	–	–	33	381	20	79
Chinstrap Penguin *Pygoscelis antarcticus*	–	–	359	3248	86	1389
Penguin Pygoscelis sp.	–	–	41	5355	17	61
Southern Rockhopper Penguin *Eudyptes chrysocome*	2	2	3	3	–	–
Magellanic Penguin *Spheniscus magellanicus*	–	–	1	1	9	28
Wilson's Storm Petrel *Oceanites oceanicus*	26	54	382	910	228	624
Black-bellied Storm Petrel *Fregetta tropica*	–	–	78	98	49	57
Storm Petrel Oceanitidae sp.	–	–	8	18	7	8
Wandering Albatross *Diomedea exulans*	37	49	117	138	72	87

(continued)

Table 24.1 (continued)

Species	Part of the voyage and dates					
	Montevideo–entrance to the Beagle Channel, 09–15 January 2020		Entrance to the Beagle Channel–Drake Passage–top of the Antarctic Peninsula–South Orkney Islands–back to Antarctic Peninsula–Drake Passage–Ushuaia, 16 January–6 February		Ushuaia–Drake Passage–top of the Antarctic Peninsula–Powell Basin–back to Antarctic Peninsula–Drake Passage–Ushuaia, 08 February–03 March	
	Number of encounters	Number of birds	Number of encounters	Number of birds	Number of encounters	Number of birds
Southern Royal Albatross *Diomedea epomophora*	9	20	9	24	31	33
Northern Royal Albatross *Diomedea sanfordi*	3	3	–	–	–	–
Albatross Diomedeidae sp.	3	7	–	–	8	8
Light-mantled Albatross *Phoebetria palpebrata*	8	13	42	47	32	35
Black-browed Albatross *Thalassarche melanophris*	200	396	366	800	279	648
Atlantic Yellow-nosed Albatross *Thalassarche chlororhynchos*	37	47	–	–	–	–
Grey-headed Albatross *Thalassarche chrysostoma*	–	–	30	32	32	33
Southern Giant Petrel *Macronectes giganteus*	87	213	364	1040	169	337
Northern Giant Petrel *Macronectes halli*	36	52	31	44	172	256
Giant Petrel Macronectes sp.	15	210	18	182	61	131

(continued)

Table 24.1 (continued)

Species	Part of the voyage and dates					
	Montevideo–entrance to the Beagle Channel, 09–15 January 2020		Entrance to the Beagle Channel–Drake Passage–top of the Antarctic Peninsula–South Orkney Islands–back to Antarctic Peninsula–Drake Passage–Ushuaia, 16 January–6 February		Ushuaia–Drake Passage–top of the Antarctic Peninsula–Powell Basin–back to Antarctic Peninsula–Drake Passage–Ushuaia, 08 February–03 March	
	Number of encounters	Number of birds	Number of encounters	Number of birds	Number of encounters	Number of birds
Southern Fulmar *Fulmarus glacialoides*	–	–	305	1941	250	2843
Cape Petrel *Daption capense*	1	11	242	1695	200	2429
Antarctic Petrel *Thalassoica antarctica*	–	–	21	30	24	578
Snow Petrel *Pagodroma nivea*	–	–	31	105	64	67
Blue Petrel *Halobaena caerulea*	1	1	–	–	–	–
Prion Pachyptila sp.	42	173	195	332	171	962
Great-winged Petrel *Pterodroma macroptera*	–	–	–	–	3	4
Atlantic Petrel *Pterodroma incerta*	8	9	2	2	–	–
Soft-plumaged Petrel *Pterodroma mollis*	79	133	4	4	2	2
White-chinned Petrel *Procellaria aequinoctialis*	102	132	124	178	113	147
Sooty Shearwater *Ardenna grisea*	9	10	35	203	20	54
Great Shearwater *Ardenna gravis*	32	54	–	–	3	4

(continued)

Table 24.1 (continued)

Species	Part of the voyage and dates					
	Montevideo–entrance to the Beagle Channel, 09–15 January 2020		Entrance to the Beagle Channel–Drake Passage–top of the Antarctic Peninsula–South Orkney Islands–back to Antarctic Peninsula–Drake Passage–Ushuaia, 16 January–6 February		Ushuaia–Drake Passage–top of the Antarctic Peninsula–Powell Basin–back to Antarctic Peninsula–Drake Passage–Ushuaia, 08 February–03 March	
	Number of encounters	Number of birds	Number of encounters	Number of birds	Number of encounters	Number of birds
Petrel Procellariidae sp.	6	9	12	13	13	21
Common Diving Petrel *Pelecanoides urinatrix*	–	–	1	10	14	25
Imperial Shag *Leucocarbo atriceps*	1	1	17	50	61	5029
Antarctic Shag *Leucocarbo bransfieldensis*	–	–	3	52	–	–
Anhinga *Anhinga anhinga*	–	–	–	–	2	2
Snowy Sheathbill *Chionis albus*	–	–	5	8	3	3
Phalarope Phalaropus sp.	1	4	–	–	–	–
Dolphin Gull *Leucophaeus scoresbii*	1	1	–	–	–	–
Kelp Gull *Larus dominicanus*	–	–	3	7	3	3
South American Tern *Sterna hirundinacea*	–	–	3	103	8	153
Common Tern *Sterna hirundo*	1	1	–	–	–	–
Arctic Tern *Sterna paradisaea*	–	–	4	22	24	63
Antarctic Tern *Sterna vittata*	–	–	14	26	14	15
Tern Sterna sp.	–	–	1	20	12	33

(continued)

Table 24.1 (continued)

	Part of the voyage and dates					
	Montevideo–entrance to the Beagle Channel, 09–15 January 2020		Entrance to the Beagle Channel–Drake Passage–top of the Antarctic Peninsula–South Orkney Islands–back to Antarctic Peninsula–Drake Passage–Ushuaia, 16 January–6 February		Ushuaia–Drake Passage–top of the Antarctic Peninsula–Powell Basin–back to Antarctic Peninsula–Drake Passage–Ushuaia, 08 February–03 March	
Species	Number of encounters	Number of birds	Number of encounters	Number of birds	Number of encounters	Number of birds
South Polar Skua *Stercorarius maccormicki*	–	–	–	–	2	2
Chilean Skua *Stercorarius chilensis*	1	1	–	–	–	–
Antarctic Skua *Stercorarius antarcticus*	2	3	102	114	62	90
Parasitic Jaeger *Stercorarius parasiticus*	3	3	–	–	–	–
Skua Stercorarius sp.	1	1	–	–	–	–
Total	754	1613	3043	17549	2382	16689

lines of the vessel route. Therefore, the submethod of interrupted lines was used (Kharitonov 2005).

24.3 Results

24.3.1 Species Composition and General Occurrence

In contrast to marine mammals, which were not found in all areas on the way (article on marine mammals, this issue), seabirds were seen almost anywhere along the vessel's path (Fig. 24.1). Birds, again opposite to mammals, are more difficult to identify to the species. Of the total number of 35,851 (44%) birds, 15,735 were identified only at the level of the genus and 20,116 (56%) at the level of the species. In total, we recorded 42 recognizable seabird species (Table 24.1). We really saw two species of prions, but it was very difficult to tell the difference between them under the available conditions (i.e., far from breeding colonies). Therefore, we considered and studied this systematic group – prions – together as a species. This adds

1467 birds as "identified"; therefore, it is possible to state that 21,584 (60%) birds have been identified to the species level, so the number of the analyzed species increases to 43. Some information about bird species content and recording is available in the work of Konyukhov et al. (2020). In penguins, Southern Rockhopper and Magellanic Penguins were seen only off the Patagonia and southeastern end of the South American continent (Table 24.1). Wilson's Storm Petrel, Northern and Southern Giant Petrels, Wandering, Light-mantled, and Black-browed Albatrosses were recorded everywhere (Fig. 24.2). Northern and Southern Giant Petrels, prions and White-chinned Petrel, displayed similar pattern of occurrence. Numerous Atlantic Yellow-nosed Albatrosses were reported only to the east of South America, in the northern half of the journey from Montevideo to the Beagle Channel. In most cases, Royal Albatrosses were identified as superspecies themselves, mainly Southern Royal Albatrosses, but three birds were identified as Northern Royal Albatross north of the Beagle Channel (Fig. 24.2).

A number of species were seen both in the Argentine Basin and Drake Passage with the tendency to occur in the northern part of the passage: Sooty and Great Shearwaters (Fig. 24.2). Soft-plumaged Petrels were recorded in the middle of the Drake Passage, in addition to the Argentine Basin (Fig. 24.2). Two species from different systematic groups (Cape Petrel and Antarctic Skua) were seen in the Argentine Basin only rarely and in small numbers (Table 24.1), a large number of groups and individuals were present in the Drake Passage and Antarctic waters (Fig. 24.3). Gentoo, Adelie, and Chinstrap Penguins, as well as any penguins of the genus Pygoscelis were seen in the Antarctic and, in small amount, in the Drake Passage; however, Adelies were found only in the Antarctic (Fig. 24.3). Black-bellied Storm Petrels were observed only in Antarctic waters and the Drake Passage, as well as Grey-headed Albatrosses (Fig. 24.3). Antarctic Petrels, Southern Fulmars, and Antarctic Terns were observed mostly in the Antarctic and a few birds in the southern part of the Drake Passage (Fig. 24.3). Snow Petrel and Snowy Sheathbill were seen only in the Antarctic with just one Snow Petrel encounter north of the South Shetland Islands. Common Diving Petrels were recorded in the middle of the Drake Passage only (Fig.24.3). A group of ten of these birds was noticed in mid-January in the middle of the Drake Passage. On February 11, only one bird was met there; however, on February 28, a total of 22 diving petrels were recorded along the part of the track of 154 km. Other species or groups, for example, Chilean Skua, Phalarope, etc., were occasionally observed in small amounts (Table 24.1).

24.3.2 Factors Affecting the Occurrence of the Different Species of Seabirds

The primary question in such studies is what factors affect the occurrence of birds that can be seen from the ship along the route. There are two categories of parameters: natural parameters (sea depth, ambient temperature, etc.) and conditions that

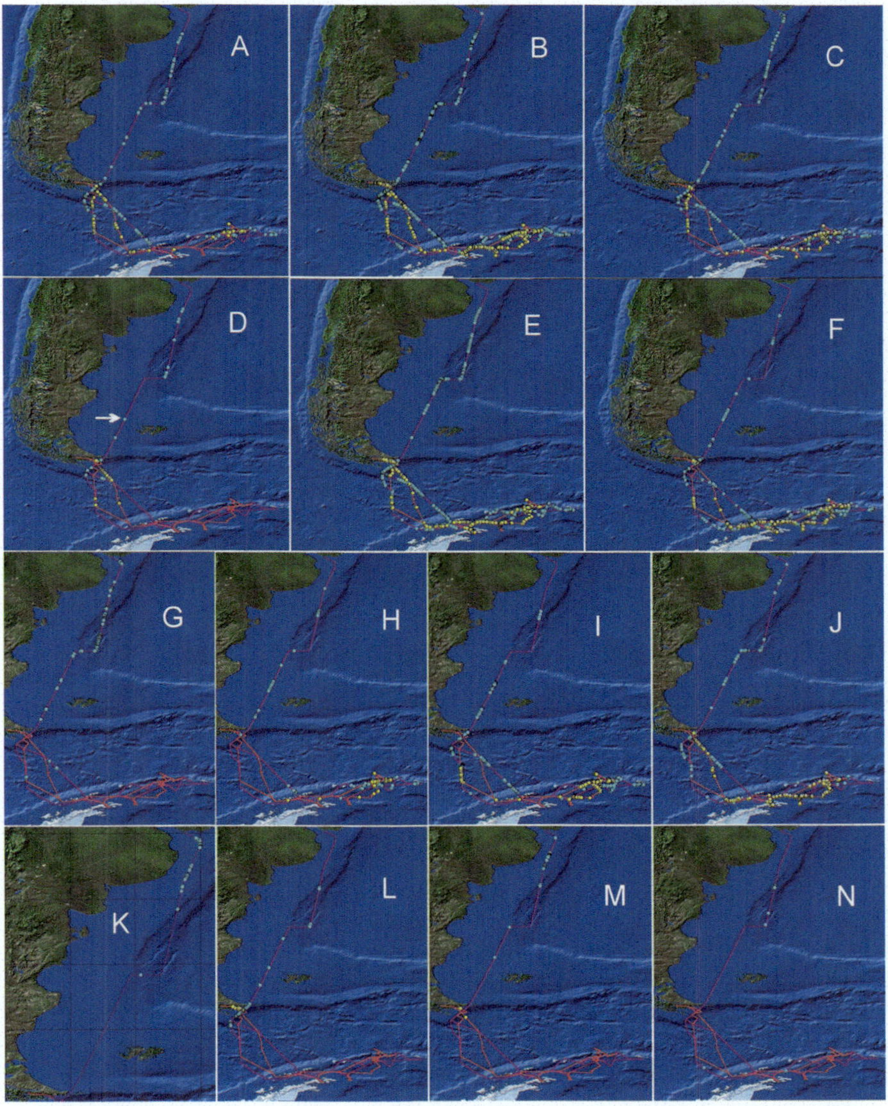

Fig. 24.2 Map of the sea bird encounters during second part of the vessel trip (first trip to Antarctic). (**a**) Wandering Albatross, (**b**) Black-browed Albatross, (**c**) White-chinned Petrel, (**d**) Southern and Northern (arrow) Royal Albatrosses, (**e**) Southern Giant Petrel, (**f**) Wilson's Storm Petrel, (**g**) Soft-plumaged Petrel, (**s**) Light-mantled Albatross, (**i**) Prions, (**j**) Northern Giant Petrel, (**k**) Yellow-nosed Albatross, (**l**) Sooty Shearwater, (**m**) Great Shearwater, (**n**) Atlantic Petrel. Purple and red lines represent Antarctic trips (Fig. 24.1)

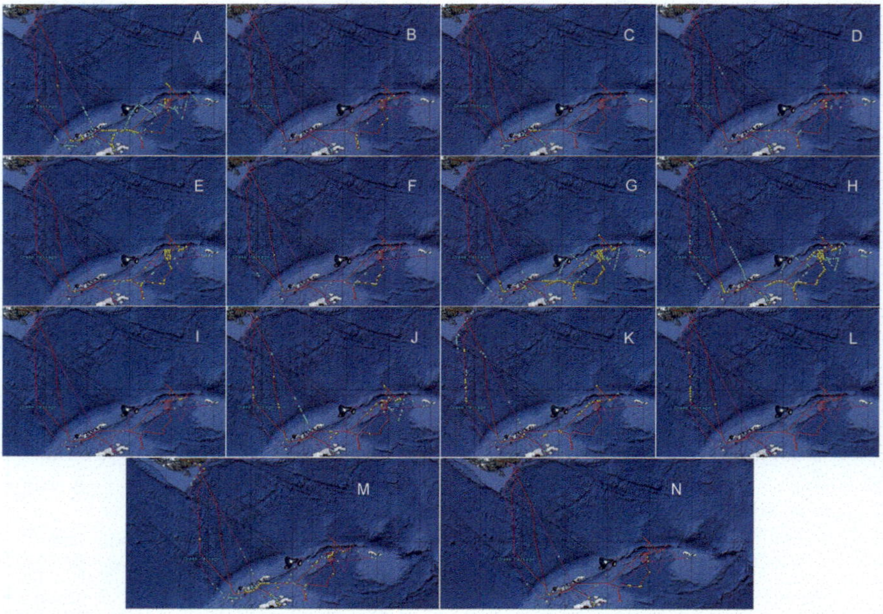

Fig. 24.3 Map of the sea bird encounters during third part of the vessel trip (second trip to Antarctic). (**a**) Chinstrap Penguin, (**b**) Adelie Penguin, (**c**) Gentoo Penguin, (**d**) Penguin Pygoscelis sp., (**e**) Snow Petrel, (**f**) Antarctic Petrel, (**g**) Southern Fulmar, (**h**) Cape Petrel, (**i**) Snowy Sheathbill, (**j**) Black-bellied Storm Petrel, (**k**) Grey-headed Albatross, (**l**) Common Diving Petrel, (**m**) Antarctic Skua, (**n**) Antarctic Tern. Purple and red lines represent two trips in the Antarctic (Fig. 24.1)

offer possibilities to detect birds (visibility for observers, vessel speed, etc.). Factor analysis in statistical methods is a good method for knowing the major variables and the main parameters of these factors. In factor analysis, we used the standard procedure for calculating the two principal components. The first component represents the variability of occurrence of the requested subject, for example, all birds in the sea or a separate species. This really is the answer to our main question. The second component represents the variability in the "speed" (how fast) of the occurrence change under the considered parameters. It is generally accepted that the influence of a parameter is considered important if the factor loading of the parameter is greater than 0.7 (Jeffers 1978).

Since large-size seabirds are detectable mainly at distances up to 500 m from the ship in any available visibility conditions, such human-ability parameters like visibility and ship speed (up to 10 knots, usually less) are not taken into account. Surprisingly, a group of birds has been identified with significantly less potential than single birds. It was possibly related to the size of the birds: large birds (e.g., albatrosses) were better identified and were generally single. Smaller birds (e.g., Southern Fulmars) formed flocks more often, but these birds were less detectable. Negative but little influence was noticed in bird fixation. This is understandable because birds were detected more likely in front of the ship. Since the effect of these two parameters was not important, we removed them from further consideration. Finally, the following parameters were considered: calendar date, time of the day, latitude, longitude, depth, ambient air temperature, humidity, wind speed, wind gusts, wind chill factor, air pressure. The main environmental parameters (factors) that correlated with the occurrence of birds as a whole systematic group were geographic coordinates, air temperature, humidity, wind chill factor; however, the latter parameter is usually closely related to the air temperature. Wind speed and wind gusts were of less importance.

Surprisingly, in the various stages of the vessel cruise, the listed parameters might be alternatively related to the bird occurrence. Occurrence of birds as whole systematic groups in the first stage (from Montevideo to the Beagle Channel) was positively related to latitude and longitude (the number of birds increased to the north and to the east), air temperature and wind chill factor (the greater both parameters the more birds were seen), but negatively related to humidity. Wind speed and gusts was negatively related to change in occurrence of birds. In the second stage (Ushuaia–Antarctic–Ushuaia, January and early February), the number of essential factors diminished; however, the vector of relation in the first principal component remained unchanged: the further to the north, the greater the occurrence of birds, which positively related to ambient temperature and wind chill factor. Wind speed and gusts at this stage was positively related to change in occurrence. In the third stage (second time Ushuaia–Antarctic–Ushuaia, February–early March), three main parameters changed their sign: relations between occurrence and latitude, air temperature and wind chill factor became negative (the lower temperature the more birds), relation to humidity became positive. Essential positive relation to longitude (more birds to the east, Figs. 24.2 and 24.3) appeared again. Calendar dates strongly related to speed of changing occurrence: the later the dates, the faster was the change in seabird occurrence.

As far as separate species are concerned, it was appropriate to analyze the species that produced larger samples and were met in a larger area and time, that is, species that were recorded in all three stages of the vessel voyage or during both trips to Antarctica. Several species like Black-browed Albatross, Antarctic Skua (data analyzed in the second and third stages of the voyage), Cape Petrel (second and third stages), and Southern Fulmar generally followed the tendency of all birds (Figs. 24.2 and 24.3). Surely, any particular species might be related to some additional factors; for example, occurrence of Black-browed Albatross in the second stage was positively related to air pressure, in the third stage, this relation was negative, which was related to the change in air temperature.

Several species, for example, Wandering and Royal Albatrosses, both Giant Petrels, and White-chinned Petrels, did not follow the general birds' tendency: they maintained positive relation with the air temperature along the entire route; however, the occurrence of all species did not essentially relate to humidity. The occurrence of Light-mantled Albatross at the first stage of the voyage was positively related to coordinates, almost essentially positively related to air temperature, and negatively related to humidity and air pressure. During the second stage, the occurrence of this species positively related to wind speed and gusts (more birds in windy conditions), which was totally unexpected. Change in occurrence was essentially related only to latitude, positively. During the third stage, the occurrence of this species was positively related only to both coordinates and calendar date (the later the season, the greater the occurrence of these albatrosses). Change in occurrence was negatively related to air temperature and chill factor. Wilson's Storm petrel occurrence was positively related to latitude and air pressure during all stages of the voyage. Air temperature was positively related only at the second stage. Occurrence of prions at the first stage was positively related to coordinates sea depth and air pressure. During the second stage, their relation to longitude became negative (more birds to the west), air temperature and chill factor became more important, and positively related. On the third stage, the same relation was positive to longitude, positive to latitude (more birds to the north), and positive to ambient temperature. Wind speed and gusts were positively related to change in occurrence.

In the three Antarctic species of Penguins, we noted the following relations with environmental parameters. The occurrence of Chinstrap Penguins during the second stage was positively related to coordinates (0.64 to latitude and 0.58 to longitude, the farther east and north, the more birds occurred) and also positively to calendar date. In addition, surprisingly, a positive relation to depth was revealed. Change in occurrence was in accordance with wind speed and gusts, positively, and chill factor, negatively. Adelie Penguin occurrence in January related to coordinates, positively (more birds to the north and to the east), and also positively to the time of the day: more birds were met in the afternoon. Besides that, positive relation to air pressure was also observed. Change in occurrence was related to wind speed, gusts, and chill factor. On the third stage, we observed a relation to the same parameters as in Chinstrap Penguins (essentially to coordinates), including positive relation to depths. The latter relation appeared in the two penguin species in February. In January, Gentoo Penguin occurrence was related to calendar date, longitude, air temperature, and humidity, all positively. The change in occurrence was related only to wind conditions. In February, relations to date, coordinates, and pressure continued, but negative relations to air temperature, humidity, and wind conditions appeared. Sea depth was not essential. Change in occurrence was related to wind speed positively, but to chill factor negatively.

24.3.3 Seabird Distribution in Concordance with the Sea Depth

We need a basis for measuring the correlation coefficient, that is, a collection of depths more or less relevant to the depth of the vessel's route. Herewith, we can only depend on the points of the route where the depths are known. At these points, we encountered biological objects: marine mammals and sea birds. Theoretically, it is possible to use either of them as a kernel; however, in mammals, the number of points of observations is lesser than in birds: in stages two and three of the vessel's route where correlation coefficient could be calculated, we have 690 and 452 points, respectively. At the same time, in birds, we have 3165 and 2382 points of encounters, respectively (see article on marine mammals, this issue).

As the number of all bird observation points is much greater than in mammals, herewith we rely on the depth measurements in birds' encounter points. We will use the Pearson's correlation coefficient (denoted as r), since the same type of coefficients is used in the factor analysis. Analysis will be performed for the second and third parts of the vessel route, because variability of the data is greater; it is possible to compare observations in January and February in the Antarctic.

Correlation between basic birds' sets of the second and third stages is significant and quite noticeable, but not high: $r = 0.62$, $n = 30$, $P < 0.001$ (Fig. 24.4). Within each stage, species have very variable correlation coefficients. Several species have relatively high correlations within each stage. Depths of the regions where Chinstrap Penguin were seen in January correlate with the set of depths of the stage very close to 1: $r = 0.93$, $n = 30$, $P < 0.001$; in February, the correlation is slightly reduced: $r = 0.73$, $n = 30$, $P < 0.001$. White-chinned Petrel has $r = 0.72$ and 0.83 in these months, respectively, which is significant. Depths of the regions where Cape petrels were seen highly correlated with the ocean depths: $r = 0.89$ and

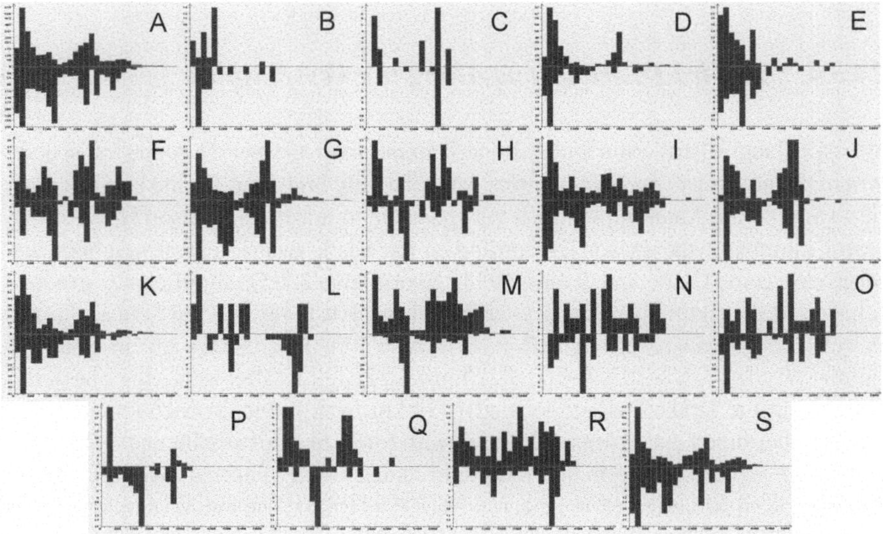

Fig. 24.4 Depths distribution along two vessel routes in Antarctic. Upper histograms are depths along the trip mid-January–early February, lower histograms – February–early Mach. (**a**) all depths along the route, (**b**) depths of Adelie Penguin, (**c**) Antarctic Petrel, (**d**) Chinstrap Penguin, (**e**) Antarctic Skua, (**f**) White-chinned Petrel, (**g**) Cape Petrel, (**h**) Black-bellied Storm-Petrel, (**i**) Black-browed Albatross, (**j**) Southern Fulmar, (**k**) Wilson's Storm Petrel, (**l**) Southern Royal Albatross, (**m**) Prions, (**n**) Grey-headed Albatross, (**o**) Northern Giant Petrel, (**p**) Snow Petrel, (**q**) Light-mantled Albatross, (**r**) Wandering Albatross, (**s**) Southern Giant Petrel. X axis – depths in meters, Y axes – number of encounters of the species

0.84. Black-browed Albatross, Antarctic Fulmar, Wilson's Storm Petrel, Snow Petrels, and Southern Giant Petrel had similar characteristic of correlations. The encounter depths of Adelie Penguin, Royal, and Wandering Albatrosses depths in both months were weakly correlated with the stage depth sets. Other species occupied intermediary positions in the series of correlation coefficients; meanwhile, some species, for example, Northern Giant Petrel, Grey-headed Albatross, might have low correlation in January, but high in February, while Antarctic Petrel had opposite regularity.

It is interesting to note that quite a number of species have high correlation coefficients between depths in different parts of the voyage; at the same time, even the species has low correlation within a stage. Such example displays Antarctic Petrel, which had high correlation in January, low in February, but quite high ($r = 0.73$, $n = 30$, $P < 0.001$) between January and February. A similar feature is characteristic of Chinstrap Penguin, Antarctic Skua, and Wilson's Storm Petrel. In some species, such correlation is very low, like in Adelie Penguin, Black-bellied Storm-Petrel, Royal Albatross, Snow Petrel, and Prions. Within the remaining species with great amount of data, Black-browed Albatross, Southern Fulmar, and Southern Giant Petrel have correlation coefficient that is slightly higher than 0.5. Cape Petrel, Northern Giant Petrel, and Wandering Albatross have coefficients between 0.4 and 0.5. White-chinned Petrel and Grey-headed Albatross have coefficient between 0.3 and 0.4.

24.3.4 Seabird Distributions Along the Vessel Route

We calculated distribution of bird encounters along the vessel route for species where the number of such encounters was relatively high, and for the two last parts of voyage to the Antarctic where it was possible to make comparison between different months of the season. According to the Clark-Evans nearest neighbor distance criterion (Clark and Evans 1954; Kharitonov 2005), significantly grouped (clustered) distribution in January–early February was revealed in White-chinned Petrel (Clark-Evans coefficient $R = 0.72$, $N = 124$, $P < 0.002$) and in Southern Fulmar ($R = 0.87$, $N = 305$, $P < 0.03$). Chinstrap Penguins after removing one outlying case in the Drake Passage located 1434 km from the main encounter set also displayed grouped distribution, although with lower level of significance ($R = 0.91$, $N = 358$, $P = 0.08$). Uniform distribution in January–early February was revealed in Black-browed Albatross ($R = 1.14$, $N = 366$, $P = 0.006$). The entire sample of Cape Petrel also showed uniform distribution but with low level of significance. If we remove outlying group of 7 encounters in the north, the significance of the uniform distribution becomes acceptable ($R = 1.21$, $N = 235$, $P = 0.002$). Southern Giant Petrel displayed uniform distribution with low significance ($R = 1.08$, $N = 364$, $P = 0.1$). Several species showed the tendency for grouped distribution (R considerably less than 1), but insignificant probably due to the relatively little samples. Examples are Antarctic Petrel ($N = 20$ after removing the outlying case), Northern Giant Petrel ($N = 31$), and Snow Petrel ($N = 31$). Tendency for uniform distribution, but with smaller samples was noticed in Grey-headed Albatross ($R = 1.29$, $N = 30$, $P = 0.12$). Other species displayed random distribution of encounters: either R is very close to 1 or the samples are great. These species are Antarctic Skua, Black-bellied Petrel, Wilson's Storm Petrel, prions (after removing one outlying case), Light-mantled, and Wandering Albatrosses.

In February–early March, grouped distribution was found in Antarctic Petrel ($R = 0.44$, $N = 24$, $P < 0.02$), Black-bellied Storm Petrel ($R = 0.51$, $N = 49$, $P < 0.002$), and prions ($R = 0.80$, $N = 143$, $P < 0.02$). Uniform distribution in February–early March was revealed in White-chinned Petrel ($R = 1.22$, $N = 113$, $P = 0.02$), Black-browed Albatross ($R = 1.13$, $N = 279$, $P = 0.03$), and Southern Giant Petrel ($R = 1.17$, $N = 168$, $P < 0.03$). Insignificant tendency for grouped distribution was found in Royal Albatrosses, while tendency for uniform distribution was observed in Adelie Penguin ($N = 19$) and Grey-headed Albatross ($N = 32$). Random distribution was revealed in Chinstrap Penguin (after removing the outlying case in Drake Passage located 1330 km from the main encounter set), Antarctic Skua, Cape Petrel (after removing the outlying group of 8 encounters near Beagle Channel located in 470 km (excluding trip during nights) from the Antarctic encounter set), Southern Fulmar, Wilson's Storm Petrel, and Northern Giant Petrel (after removing the outlying case located in 1866 km from the main encounter set). Mean linear densities are presented in Table 24.2.

Table 24.2 Mean linear densities of seabird encounters per 100 km

	Period of 2020 in the Antarctic	
Species	16 January–6 February	8 February–3 March
Adelie Penguin	–	3.84
Chinstrap Penguin	16.2	2.53
Wilson's Storm Petrel	10.2	7.2
Black-bellied Storm Petrel	7.07	0.61
Wandering Albatross	4.5	1.6
Royal Albatrosses	–	1.57
Light-mantled Albatross	0.71	1.07
Black-browed albatross	10.6	7.4
Grey-headed Albatross	2.9	1.5
Southern Giant Petrel	9.5	3.85
Northern Giant Petrel	0.68	7.2
Southern Fulmar	13.9	7.23
Cape Petrel	6.6	5.5
Antarctic Petrel	6.6	11.0
Snow Petrel	1.73	1.4
Prions	2.4	4.9
White-chinned Petrel	2.79	2.43
Antarctic Skua	3.6	1.28

– indicates small sample for calculation

24.4 Discussion

24.4.1 Factors Affecting the Occurrence of Different Species of Seabirds

A number of species (Northern and Southern Giant Petrels, Black-browed Albatross, Southern Fulmar, Sooty Shearwater, Prions, Light-mantled and Wandering Albatrosses, White-chinned Petrel, Wilson's Storm Petrel, Antarctic Petrel, Chinstrap Penguin, Antarctic Skua) displayed similar areas of occurrence in January–early March (Figs. 24.2 and 24.3). Other species displaced their areas of presence totally or partly. In February, the proportion of Royal Albatrosses increased in the northern part of the Drake Passage (Fig. 24.2), and they were mainly adult birds. This change might just be related to the change in the events of breeding season. On the other hand, occurrence of Common Diving Petrels definitely fluctuated with the season. Small amounts of these birds were seen in January–mid February. In late February, we recorded noticeable flights of this species. The increase in the number of Common Diving Petrels in the Drake Passage in this time might likely be related to their dispersion after the breeding season.

The distribution of Adelie Penguins from January to February displaced to the east (Fig. 24.3). Grey-headed Albatrosses did not change their distribution in the Drake Passage, but to the south of the line of islands, they displaced closer to the Powell Basin (Fig. 24.3). Black-bellied Storm Petrels did not change their location in the Drake Passage, but in February, they spread more over the Powell Basin (Fig. 24.3). Gentoo Penguin in February was observed in very restricted area in comparison to their occurrence in January (Fig. 24.3).

Snowy Sheathbills were more spread in January, but in February, these birds were encountered only near the Hesperides Trough (Fig. 24.3). Also, Snow Petrels moved more to the northern part of Powell Basin, close to the Hesperides Trough (Fig. 24.3). Such change in their occurrence might be related to the abundance of krill. The reason for this is that, in this very area, krill was the purest, with the lowest concentration of salps (S.H. Kobylyanskiy, personal communication). In addition to the enumerated species, some other birds (White-chinned Petrel, Light-mantled Albatross, Prions) formed noticeable concentrations there (Figs. 24.2 and 24.3). Seemingly, all these birds prefer pure krill. Antarctic Skuas, probably, had the same tendency: the birds were encountered in high numbers near shores and near the Hesperides Trough (Fig. 24.3); however, the latter area might attract this species with the predatory behavior as a source of potential bird prey.

Some previous studies highlighted the importance of environmental factors structuring the distribution of seabirds across gradients of ocean productivity and water mass properties (Serratosa et al. 2020; Veit 1995). Factor analysis showed that, generally, birds prefer good weather: higher air temperature (among the characteristic ones of the region), lower humidity, and, in some species, high air pressure. Typical examples are Antarctic bird species such as Wandering and Royal Albatrosses, both Giant Petrels, White-chinned Petrels, and Prions. Other numerous

Antarctic species like Cape Petrel and Southern Fulmar occurred at low temperatures and high humidity. Interestingly, some species, for example, Black-browed Albatross, changed their relation to air temperature and humidity alternatively: in February, they "prefer" low temperatures and high humidity. This occurred in spite of the fact that the mean air temperature in the second and third parts of the voyage of the vessel, including trips in the Beagle Channel, was precisely +3 °C. Such change in the vector of the relation to the environmental parameters, most likely, is caused by a much more powerful factor, presumably food. Satellite tracking of Black-browed Albatrosses also showed that this species prefers warm air, but they often use cold Antarctic areas for foraging (Wakefield et al. 2011). Young Black-browed Albatrosses, nevertheless, concentrated in the northern, warmer, part of the study area. Young Wandering Albatrosses were more often seen in the northern areas of the Drake Passage, as well.

24.4.2 Seabird Distribution in Concordance with the Sea Depth

High correlation between the depths of the regions where particular species were seen in different months might provide evidence about the preference of definite sets of depths, and those sea depths for some reason were important for even the species found in different months in different areas of the ocean. Sets of depths could be similar, whereas mean depths of encounters are different in different months. For instance, Antarctic Petrel that had high correlation of occurrence with depths in January and February, probably, displayed that this species definitely preferred to occur over specific depths at any time (Fig. 24.4). The presence of icebergs is very important for this species, since Antarctic Petrels use them as roosting platforms where they spend 60–80% of time (Delord et al. 2020). Therefore, we think that this species does not demand definite depths, but the depths where they can find icebergs. The position of the icebergs slightly changed from January to February: in February, more icebergs were seen in the Powell Basin. Therefore, the mean depths (M) where Antarctic Petrels were recorded in February were greater: $M = 1676 \pm 239.9$ m in January ($N = 21$) and 2280 ± 171.4 m in February ($N = 24$).

Mean depths of the regions where species had high correlations in January and February were recorded were as follows. Chinstrap Penguin were seen everywhere in the Antarctic and southern part of the Drake Passage (Fig. 24.3), with depths in January $M = 1257 \pm 62.3$ m ($N = 359$), and in February considerably less $M = 891.7 \pm 100.6$ m ($N = 86$); nevertheless, this species had the tendency to occur over smaller depths (Fig. 24.4). Antarctic Skuas in both months were encountered practically in the same areas that offered high correlation coefficient and almost the same depths ($M = 834 \pm 83.9$, $N = 102$ and 899 ± 92.6 m, $N = 62$, respectively). Encountered everywhere, Wilson's Storm Petrel had high correlation with depths and very close mean depths for the months ($M = 1418 \pm 62.0$, $N = 382$ and

1390 ± 69.4 m, $N = 228$). Prions are known for their extensive feeding area (Cherel et al. 2002; Quillfeldt et al. 2014). In our study, they displayed low correlation with depths where they were encountered and deep-sea areas of their presence ($M = 2445 \pm 94.4$, $N = 194$ and 1390 ± 69.4 m, $N = 228$), with noticeable concentration at the Hesperides Trough in February (Fig. 24.2).

Adelie Penguin was quite remarkable among the birds with low correlation coefficient. In January, the observers saw mostly adult birds not far from the colonies at the tip of the Antarctic Peninsula and nearby islands (Fig. 24.3). It is known that during breeding time, the adult birds find food at distances up to 120 km from the colonies (Kerry et al. 1997). In February, observers recorded mostly young birds on ice pieces: the period of dispersion of youngsters commenced, and many penguins appeared far from the breeding colonies. Therefore, in February, these penguins were encountered over deeper waters than in January: M = 999 + 213.2 m (N = 20) and M = 666 + 79.1 m (N = 33), respectively; the first datum varies more.

Royal Albatrosses had practically no correlation with depths between months and great differences in mean depths in January and February. Very little and even negative correlation with depths was revealed in Light-mantled Albatrosses: in different months, these birds occurred over very different depths (Fig. 24.4). In both months, these Albatrosses occurred near the Hesperides Trough; however, in February, their concentration there was greater than in January (Fig. 24.2). Such variations of depths where Royal and Light-mantled Albatrosses have been seen most likely suggests that sea depth is not important for these species.

Widely distributed Black-browed Albatrosses and Southern Giant Petrels (Fig. 24.2) had quite moderate correlation coefficients, with almost similar depths distributions (Fig. 24.4) and quite close mean depths, although Southern Giant Petrels were found over slightly lower depths: $M = 1823 \pm 72.1$, $N = 366$ and 1720 ± 82.9 m, $N = 279$, respectively, by months in Black-browed Albatrosses and $M = 1686 \pm 69.0$, $N = 364$ and 1528 ± 92.8 m, $N = 168$, respectively by months in Southern Giant Petrels.

24.4.3 Spatial and Linear Distributions of Seabirds in the Sea

Type of spatial or linear distribution is a good tool that allows us to understand additional features of the species population structure and prosperity. A mathematical pattern of the spatial distribution is the characteristic of biological interactions between objects. Clustered distribution of objects indicates that the objects follow some environmental features or social attraction (Odum 1983). Even (uniform) distribution of the biological objects usually indicates saturation of a given area and, very likely, competition between individuals.

In the majority of the seabird species, linear distributions in the Antarctic in January and February were generally similar to each other. Several species changed their distribution by months, they were: White-chinned Petrel (from clustered to uniform), Black-bellied Storm-Petrel (from random to clustered), and Southern Fulmar (from clustered to random).

In our study, the most likely reason for clustered distribution was food abundance. However, clustered distribution in the Antarctic Petrel, most likely, was caused by social reasons, since these birds were often encountered in groups. In several species with small samples, random distribution could be caused by the presence of small amount of data, for example, Adelie Penguin, Southern Royal Albatross, Northern Giant Petrel, and Snow Petrel in January, Light-mantled Albatross.

Two species, Black-browed Albatross and Southern Giant Petrel, displayed uniform distributions along the vessel route during both months of the study. It means that, during austral summer, the surveyed area in Antarctic was saturated with individuals of each species. Probably, there was intraspecies competition between birds, but it is difficult to tell definitely from our data. However, intense intraspecies competition for food in Black-browed Albatross was shown in other studies (Cherel et al. 2002; Wakefield et al. 2011). In our study, types of distribution of both species together with map analysis revealed the competition between these particular species. Looking at the species maps (Fig. 24.2), one can notice that south of the easternmost islands of the South Shetland Islands and in the middle of the Drake Passage in early February, these species replace each other. This replacement together with the revealed saturation of the area by these species pointed to the competition between them. In both months, Black-browed Albatross yielded some areas near the South Shetland Islands, while Southern Giant Petrel yielded areas in the Drake Passage and the Hesperides Trough. It is likely that in January, this competition was not so high (birds had close linear densities of encounters: 10.6 and 9.5 birds per 100 km for Black-brow and Southern, respectively). In February, the competition between these species was more intense: linear density of Black-browed Albatross decreased to 7.4 encounters per 100 km, in the case of Southern Giant Petrel, the decrease was twice, that is, 3.85 encounters per 100 km. Gaps of encounters in both species were more prominent during February (Fig. 24.2).

24.5 Conclusions

We present preliminary results on seabird occurrence during cruise 79 of the R/V "*Akademik Mstislav Keldysh*" to the Antarctic. However, even the preliminary results allowed us to make conclusions. In January, February, and early March, we recorded 35,851 seabirds, 20,116 of them were identified to the species level, and a total of 43 species were recorded. Surprisingly, groups of birds were detected with slightly lower possibility than single birds. This might be related to the size of birds: large birds more often were singles. The observed occurrence of seabirds in the ocean related to a number of environmental parameters: in January, almost all species preferred "good weather": higher air temperature, lower humidity, and, in some species, high air pressure. In February, a number of species maintained these relations, but several species, in particular, Black-browed Albatross, Antarctic Skua, Cape Petrel, and Southern Fulmar, had contrary relations to environmental parameters. They were mainly encountered during low air temperature and high humidity. This probably indicates that they found relevant foraging areas in the existing air conditions.

We expected that the depth of the sea is not as important for birds compared to whales. Nevertheless, the results show that depth might also influence bird occurrence. We found that several species (e.g., Antarctic Petrel, Antarctic Skua) were attached to definite depths. Most birds changed the depths where they occurred over from January to February.

Unlike marine mammals (see article in this issue), several seabird species formed uniform distribution of encounters along the vessel route. The Black-browed Albatross and the Southern Giant Petrel maintained this distribution pattern during both months of the study. The mathematical evaluation of distributions together with a map analysis points to the competition between or, possibly, in addition, intraspecies competition among these two species. They probably were restricted in food sources.

Acknowledgments We would like to express our gratitude to echo sounder managers Dmitry I. Frey and Alexander V. Gavrikov, Shirshov Institute of Oceanology, for offering us the detailed data on depths. Many thanks to the "Mercury detachment": Maryina E.M., Rudykh Ya.I., head-Kalinchuk V.V. of V.I. Ilyichev Pacific oceanological Institute Far Eastern Branch of RAS, who managed weather station Davis Vantage Pro 2 and offered us the weather data.

References

Cherel Y, Weimerskirch H, Trouve C (2002) Dietary evidence for spatial foraging segregation in sympatric albatrosses (*Diomedea* spp.) rearing chicks at Iles Nuageuses, Kerguelen. Mar Biol 141:1117–1129

Clark PJ, Evans FC (1954) Distance to nearest neighbor as a measure of spatial relationships in populations. Ecology 35:445–453

Delord K, Kato A, Tarroux A, Orgeret F, Cotté C, Ropert-Coudert Y, Cherell Y, Descamps S (2020) Antarctic petrels 'on the ice rocks': wintering strategy of an Antarctic seabird. R Soc Open Sci 7(4). https://doi.org/10.1098/rsos.191429

Jeffers JNR (1978) An introduction to systems analysis with ecological applications. University Park Press, Baltimore, 198 p

Joiris CR, Dochy O (2013) A major autumn feeding ground for fin whales, southern fulmars and grey-headed albatrosses around the South Shetland Islands, Antarctica. Polar Biol 36:1649–1658

Kerry KR, Clarke J, Eberhard S, Gardner H, Lawless RM, Trémont R, Wienecke B, Corsolini S, Focardi S, Franchi E, Rodary D, Thomson R (1997) The foraging range of Adélie penguins – implications for CEMP and interactions with the krill fishery. CCAMLR Sci 4:75–87

Kharitonov SP (2005) A "nearest neighbour distance" method for the quantitative evaluation of the distribution of biological objects in the plane or line. Vestnik of the Nizhny Novgorid University. Ser Biol 1(9):213–221. [in Russian]

Konyukhov NB, Mischenko AL, Kharitonov SP, Dmitriev AE, Artemyeva SM, Mamayev MS, Pilipenko GY, Tretyakov AV (2020) Notes on birds of the Atlantic sector of the Southern Ocean, Scotia Sea and Drake Passage. Rus J Ornithol 29(Express-issue 1945):3049–3082

Odum EP (1983) Basic ecology, vol 2. Sounders College Publishing, New York/London/ Toronto, 376 p

Orgeira JL, Alderete MC, Jimenez YG, Gonzalez JC (2015) Long-term study of at-sea distribution of seabirds and marine mammals in the Scotia Sea, Antarctica. Adv Polar Sci 26:158–167

Plokhinskiy NA (1978) Mathematical methods in biology. Moscow State University, Moscow. 265 p [in Russian]

Quillfeldt P, Phillips RA, Marx M, Masello J (2014) Colony attendance and at-sea distribution of thin-billed prions during the early breeding season. J Avian Biol 45:315–324

Serratosa J, Hyrenbach KD, Miranda-Urbina D, Portflitt-Toro M, Luna N, Luna-Jorquera G (2020) Environmental drivers of seabird at-sea distribution in the Eastern South Pacific Ocean: assemblage composition across a longitudinal productivity gradient. Front Mar Sci 6:838. https://doi.org/10.3389/fmars.2019.00838

Veit RR (1995) Pelagic communities of seabirds in the South Atlantic Ocean. Ibis 137:1–10

Wakefield ED, Phillips RA, Trathan PN, Arata J, Gales R, Huin N, Robertson G, Waugh SM, Weimerskirch H, Matthiopoulos J (2011) Habitat preference, accessibility, and competition limit the global distribution of breeding Black-browed Albatrosses. Ecol Monogr 81:141–167

Chapter 25
Spatial Distribution, Species Composition, and Number of Marine Mammals in the Argentine Basin, Drake Passage, East of Antarctic Peninsula, and Powell Basin in January–March 2020

Sergey P. Kharitonov, Andrey V. Tretyakov, Alexander L. Mischenko, Nikolai B. Konyukhov, Svetlana M. Artemyeva, Gleb Yu. Pilipenko, Matvey S. Mamayev, and Alexander E. Dmitriyev

Abstract Results of observations of marine mammals' occurrence during cruise 79 of the R/V "Akademik Mstislav Keldysh" in January, February, and early March 2020 are presented. We recorded 2140 marine mammals, 1990 of them were identified to the species level, and a total of 20 species were recorded. During the second and third parts of the trip, the most numerous marine mammals were: Antarctic fur seal, Humpback and Fin whales. It was revealed that the observed occurrence of animals in the ocean is related to a number of environmental parameters (general weather parameters, ice characteristics) and weakly related to the parameters of observation: visibility, ship speed and the difference in detection degree between one animal and groups of animals. The occurrence of the three most numerous animals positively correlated with the depths distribution along the ship route. This suggests that sea depth is important for these animals. Spatial distributions of Antarctic fur seals in January and February in the studied Antarctic sea area were generally similar to each other; distribution of Humpback and Fin whales varied

S. P. Kharitonov (✉) · A. V. Tretyakov · A. L. Mischenko · N. B. Konyukhov
M. S. Mamayev · A. E. Dmitriyev
A.N. Severtsov Institute of Ecology and Evolution, Russian Academy of Sciences, Moscow, Russia
e-mail: serpkh@gmail.com; atretyakovnn@mail.ru; almovs@mail.ru; konyukhov@gmail.com; trofim@riseup.net; zzu@inbox.ru

S. M. Artemyeva
Zoological Museum of the Lomonosov Moscow State University, Moscow, Russia
e-mail: artemieva.svetlana@gmail.com

G. Yu. Pilipenko
Faculty of Geography, Lomonosov Moscow State University, Moscow, Russia
e-mail: limitedkoxd@gmail.com

E. G. Morozov et al. (eds.), *Antarctic Peninsula Region of the Southern Ocean*, Advances in Polar Ecology 6, https://doi.org/10.1007/978-3-030-78927-5_25

from January to February: a large number of both species in February started to be concentrated in the area south of the South Shetland Islands. The type of distribution of Fin whales suggests that this species is not restricted to food supply. No competition for food has been noticed in this species and in the Humpback whales.

Keywords Marine mammals · Whales · Dolphins · Fur seals

Data on mammals were collected during cruise 79 of the R/V "Akademik Mstislav Keldysh" in the southern part of the Atlantic from January to March 2020. We recorded 2140 marine mammals, 1990 of them were identified to the species level, and a total of 20 species were recorded. The most numerous marine mammals were Antarctic fur seals and humpback and fin whales. It was revealed that the observed occurrence of animals in the ocean is related to a number of environmental parameters (general weather parameters, ice characteristics) and weakly related to the parameters of observation: visibility, ship speed, and difference in detection degree between one animal and groups of animals. The occurrence of the three most numerous animals positively correlated with the depth along the ship route. Spatial distributions of Antarctic fur seals in January and February were generally similar to each other; the distribution of humpback and fin whales varied between January and February: a large number of both species in February started to be concentrated in the area south of the South Shetland Islands. The type of distribution of fin whales suggests that this species is not restricted to food supplies. No competition for food has been noticed in this species and in the humpback whales.

25.1 Introduction

Antarctic studies are important for monitoring this very productive region of the Earth. The composition and distribution of species of marine mammals and seabirds might be changing over the years; the process of change should be monitored. Cruise 79 of the R/V "Akademik Mstislav Keldysh" of the Russian Academy of Sciences to the South Atlantic and Southern Ocean was held in January–March, 2020. The goal of the study was to investigate the current state of the oceanographic environment and Antarctic biota.

This article presents the data and preliminary results on the occurrence of marine mammals in this area. The cruise included several stages. Data on mammals and seabirds were collected from January 9 to March 3, 2020. The first stage was the voyage from Montevideo to the entrance of the Beagle Channel. The second and third stages consisted of two trips to Antarctic waters with an intermediate visit to Ushuaia. The total length of the scientific voyage considered here was approximately 14,000 km.

It is well known that the distribution of whales in Antarctica depends on the distribution of food (mostly crustaceans), as well as on the local sea depth (Burkhardt and Lanfredi 2012; Friedlaender et al. 2006; Herr et al. 2016). We revealed a number of regularities in the distribution of mammals correlated to many environmental parameters, partly confirming the previous studies and partly detecting new regularities. The regularity that spatial distribution of whales depends to a large degree on the position of the plankton is rational and demonstrated in a number of studies. This implies that, in general, the distribution of plankton could be evaluated from the distribution of whales over the sea basins. However, plankton is typically not homogeneous, and its quality varies in different areas of the sea. It is understood that the phase of penetration of Antarctic salps, *Salpa thompsoni*, into waters south of 60° S has occurred since the 1970s, creating a problem for Antarctic krill *Euphausia superba* trophic ability and declining krill content (Minkina et al. 1999). There is an extensive field for studies on how the composition and quality of plankton affect the species composition and distribution of whales and other marine mammals.

While there are studies on the correlation of marine mammals, mainly whales, with the depths of the sea, new data on this topic are included, not only on whales but also on the Antarctic fur seal. The general characteristic of the distribution of marine mammals is still poorly understood. There are numerous publications on the density of mammals at sea, e.g., Joiris and Dochy (2013) and Viquerat and Herr (2017), but little information on the density of mammals is available, particularly in Antarctic waters. The evaluation of the density of marine mammals within the area from a moving ship (e.g., different visibility at different parts of the route) is complicated. On the other hand, details on the volume and composition of plankton have not always been known. For this purpose, we determine the type of distribution of the species along the vessel's route as along the line and try to collect information on the species from their type of distribution.

25.2 Materials and Methods

Data on mammals were collected during several parts of the vessel voyage: (1) January 09–15, 2020, Montevideo–entrance to the Beagle Channel, near 2500 km; (2) January 16–February 6, entrance to the Beagle Channel–Drake Passage–tip of the Antarctic Peninsula–South Orkney Islands–back to Antarctic Peninsula-Drake Passage–Ushuaia (Fig. 25.1), near 5500 km; (3) February 08–March 03, Ushuaia–Drake Passage–tip of the Antarctic Peninsula–Powell Basin–back to Antarctic Peninsula–Drake Passage–Ushuaia (Fig. 25.1), more than 6000 km.

Observations were made throughout all day light hours from the broad direction-finder deck by two observers on the deck; watches lasted for standard 4 h or less in the morning or in the evening. The observers were fitted with high-power photo cameras and lenses. Nearly every sea mammal encountered was photographed. Species, number of individuals in the groups, behavior of mammals, ice characteristics, and general weather parameters were recorded. The route of the vessel was

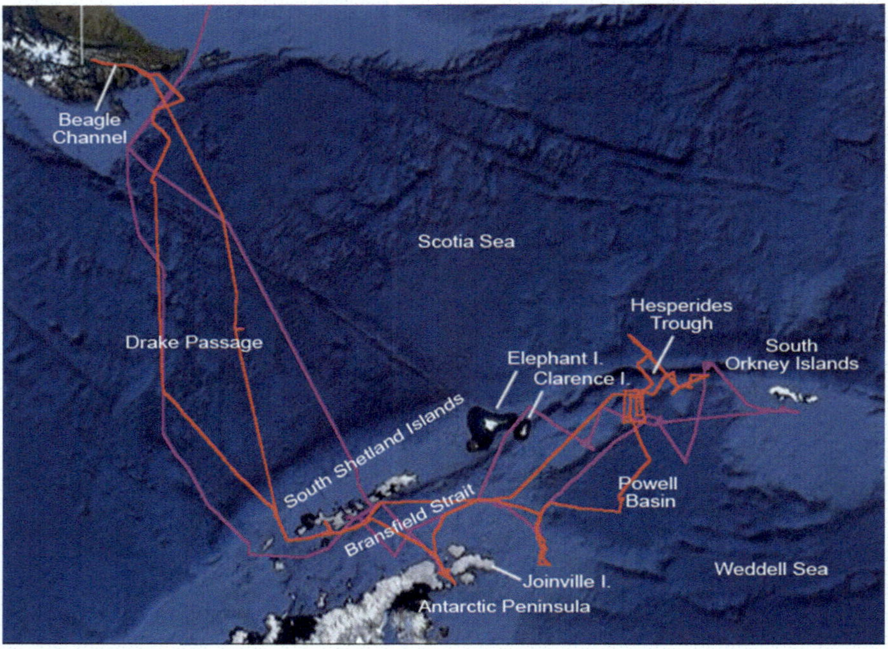

Fig. 25.1 Map of the R/V "Akademik Mstislav Keldysh" cruise to the Antarctic. Purple line is the trip on January 16–February 6, from the entrance to the Beagle Channel–Drake Passage–tip of the Antarctic Peninsula–South Orkney Islands–back to the Antarctic Peninsula–Drake Passage–Ushuaia. The red line is the trip on February 08–March 03: Ushuaia–Drake Passage–tip of the Antarctic Peninsula–Powell Basin–back to Antarctic Peninsula–Drake Passage–Ushuaia

tracked by three GPS navigators; thus, coordinates were available for each observation of a biological object in the sea.

Depths along the route were determined by the Kongsberg EA600 12 kHz single-beam hydrographic echo sounder. However, it was not always possible to determine the depth of such exact points. The echo sounder did not work at "stations" (points where the ship was not moving while collecting different kinds of sea samples) and in very stormy weather. In these cases, we calculated the nearest point where the depth was defined. Usually, this value was less than 3 km from the real point. In two cases, the distance was approximately 100 km. This was along the Beagle Channel where the depth was determined only at its entrance and over the flat area at about the same depth in the Powel Basin (see map in Fig. 25.1) where the depth was defined at several points. However, the Beagle Channel is shallow all over, so it is not bias to consider just one measurement at the entrance point. The flat area in the Powell Basin is almost of the same depth; hence measurements can be accepted there as well. All depth measurements were considered as relevant for the analysis.

The depths along the section from Montevideo to Ushuaia were from the GEBCO database: https://www.gebco.net/data_and_products/gridded_bathymetry_data/. Although this is not as accurate as echo sounder data, it is acceptable for our analysis.

Common weather parameters were permanently recorded every 5 min by the weather station Davis Vantage Pro2 managed by the scientific team of the V.I. Il'ichev Pacific Oceanological Institute, Far Eastern Branch, Russian Academy of Sciences.

The main analysis of the distribution of marine mammals was performed in the second and third parts of the voyage, i.e., within Antarctic waters. There were less observations during the second stage, mostly for the occurrence of Southern Right whale, *Eubalaena australis* (Fig. 25.4a). During the entire trip, we encountered marine mammals 1188 times; a total of 2140 animals were detected (Table 25.1).

Commonly accepted factor analysis and correlation analysis using the program Statistica-12 from StatSoft were used for the data analysis. In addition, the averages were compared using the t-Bailey criterion (Plokhinskiy 1978) instead of commonly accepted criteria (t-Student or U-Mann-Whitney). This criterion is parametric and, at the same time, does not require normal distribution of data. It is therefore applicable in all situations where the Mann-Whitney criterion is necessary and has the same statistical power as the Student criterion. In order to obtain information on the type of distribution of species that have been abundant in the field, we used the method of assessment of distribution of objects along the vessel's route (Kharitonov 2005) derived from the Clark-Evans nearest neighbor method (Clark and Evans 1954). Because the observations were not performed at night, we subtracted night periods from the lines of the vessel route for the distribution evaluation. Therefore, the sub-method of interrupted lines was used (Kharitonov 2005).

25.3 Results

25.3.1 Species Composition

Most of the mammals were identified to the species. A total of 20 marine mammal species were recorded (Table 25.1). A majority of the species were recorded in two areas: in the Beagle Channel and surroundings, and a large number of them were recorded south of 60° S, in the Antarctic (Figs. 25.2a and 25.3a). The Drake Passage had a quite poor mammal population. South American sea lions were seen only within the Beagle Channel. Long-finned pilot whales were numerous from Montevideo to the Beagle Channel and were detected twice on the southeastern edge of the South American continental shelf. Sei whales were concentrated at the edge of the continental shelf, at a distance of approximately 50–70 km southeast of the Beagle Channel entrance, in early February. In early March, these whales were noticed in the shallow waters at the entrance of the Beagle Channel, not far from their location in early February. Dusky and Peale's dolphins in the Antarctic regions of the trip displayed a similar occurrence pattern. Hourglass dolphin and Cuvier's beaked whales were observed only in the Drake Passage, including its southernmost part.

Table 25.1 Marine mammal species and systematic groups (if the species was not defined) during the Antarctic cruise of the "Akademik Mstislav Keldysh"

Species	Part of the voyage and dates					
	Montevideo–entrance to the Beagle Channel, January 09–15, 2020		Entrance to the Beagle Channel–Drake Passage–tip of the Antarctic Peninsula–South Orkney Islands–back to Antarctic Peninsula–Drake Passage–Ushuaia, January 16–February 6		Ushuaia–Drake Passage–tip of the Antarctic Peninsula–Powell Basin–back to Antarctic Peninsula–Drake Passage–Ushuaia, February 8–March 3	
	Number of encounters	Number of animals	Number of encounters	Number of animals	Number of encounters	Number of animals
Antarctic fur seal *Arctocephalus gazella*	1	1	315	396	109	169
South American sea lion *Otaria flavescens*	–	–	–	–	4	5
Crabeater seal *Lobodon carcinophaga*	–	–	2	2	–	–
Weddell seal *Leptonychotes weddellii*	–	–	6	10	–	–
Ross seal *Ommatophoca rossii*	–	–	1	1	–	–
Leopard seal *Hydrurga leptonyx*	–	–	4	4	1	1
Seal sp.	–	–	2	3	1	1
Southern right whale *Eubalaena australis*	32	51	–	–	2	2
Antarctic minke whale *Balaenoptera bonaerensis*	–	–	–	–	6	6
Humpback whale *Megaptera novaeangliae*	–	–	132	207	66	133
Fin whale *Balaenoptera physalus*	–	–	203	296	166	387
Sei whale *Balaenoptera borealis*	–	–	2	20	6	12

(continued)

Table 25.1 (continued)

	Part of the voyage and dates					
	Montevideo–entrance to the Beagle Channel, January 09–15, 2020		Entrance to the Beagle Channel–Drake Passage–tip of the Antarctic Peninsula–South Orkney Islands–back to Antarctic Peninsula–Drake Passage–Ushuaia, January 16–February 6		Ushuaia–Drake Passage–tip of the Antarctic Peninsula–Powell Basin–back to Antarctic Peninsula–Drake Passage–Ushuaia, February 8–March 3	
Species	Number of encounters	Number of animals	Number of encounters	Number of animals	Number of encounters	Number of animals
Blue whale *Balaenoptera musculus*	–	–	1	1	2	2
Baleen whale	–	–	5	6	70	130
Long-finned pilot whale *Globicephala melas*	2	62	2	30	–	–
Killer whale *Orcinus orca*	–	–	2	32	3	22
Southern bottlenose whale *Hyperoodon planifrons*	–	–	1	1	–	–
Cuvier's beaked whale *Ziphius cavirostris*	–	–	1	2	–	–
Toothed whale.	–	–	–	–	1	1
Dusky dolphin *Lagenorhynchus obscurus*	2	14	3	7	11	53
Peale's dolphin *Lagenorhynchus australis*	5	13	2	10	1	5
Southern right whale dolphin *Lissodelphis peronii*	2	7	–	–	–	–
Hourglass dolphin *Lagenorhynchus cruciger*	–	–	6	23	1	2
Dolphin sp.	2	7	1	1	1	1
Total	**46**	**155**	**690**	**1052**	**452**	**933**

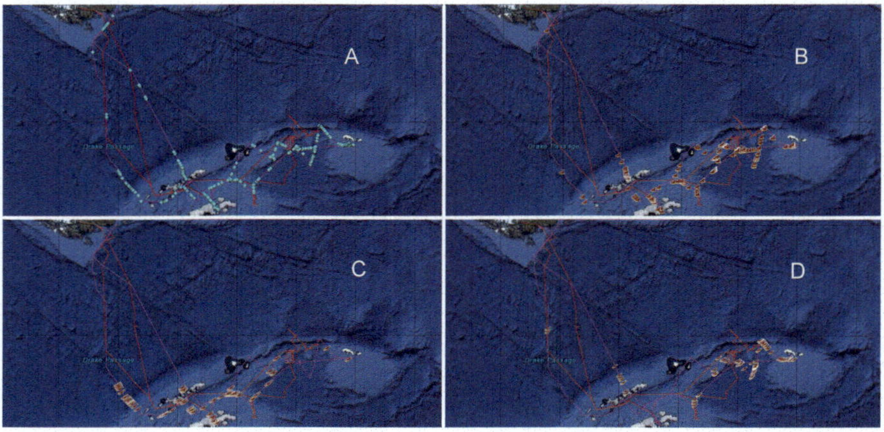

Fig. 25.2 Map of the marine mammal encounters in the second part of the trip (first trip to Antarctic). (**a**) all mammals (blue-green rhombuses); (**b**) Antarctic fur seals (symbols "eared seal"); (**c**) humpback whales (symbols "baleen whale"); (**d**) fin whales (symbols "baleen whale"). Purple and red lines represent trips in Antarctic (Fig. 25.1)

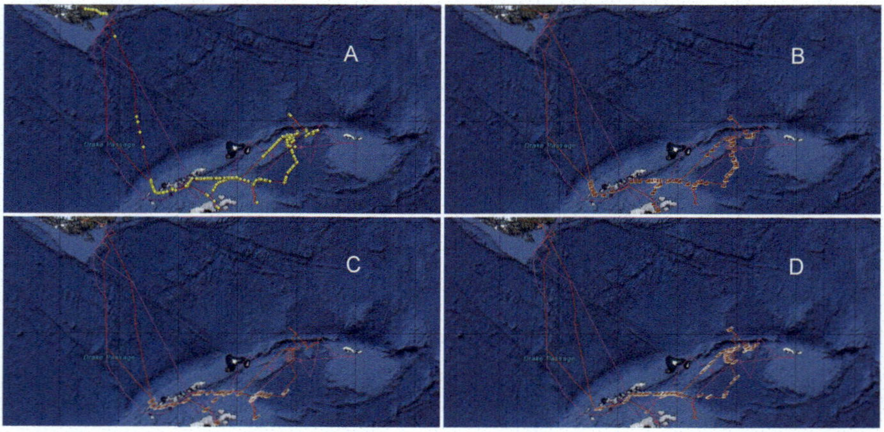

Fig. 25.3 Map of the marine mammals encountered during the third part of the trip (second trip to the Antarctic). (**a**) all mammals (blue-green rhombuses); (**b**) Antarctic fur seals (symbols "eared seal"); (**c**) humpback whales (symbols "baleen whale"); (**d**) fin whales (symbols "baleen whale"). Purple and red lines represent two trips in Antarctic (Fig. 25.1)

Killer whales were recorded both in the Drake Passage and further south in the Antarctic. Singles of the Antarctic minke whales were recorded everywhere during the third part of the trip: in the Beagle Channel, Drake Passage, and Antarctic. Blue whales and southern bottlenose whale were seen only in the Powell Basin. Crabeater, leopard, Weddell, and Ross seals were seen only in the Antarctic waters.

During the second and third part of the trip, the most numerous marine mammals were Antarctic fur seals, humpback, and fin whales (Table 25.1 and Figs. 25.2,

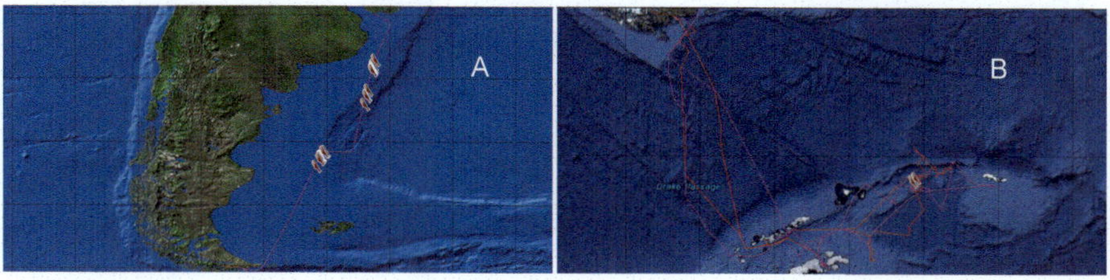

Fig. 25.4 Encounters of the southern right whales (symbols "baleen whale"). (**a**) in the first part of the trip (purple line); (**b**) in the third part of the trip (second trip to the Antarctic). The purple and red lines represent routes of the trips in the Antarctic (Fig. 25.1)

25.3). Only single fur seal was recorded on the way from Montevideo to the Beagle Channel; neither humpback nor fin whales were observed there. All these species were encountered in the Beagle Channel, in the Drake Passage, and Antarctic waters. The latter area was the location where all three species were commonly detected. Antarctic fur seals had the same area of distribution both in January and February (Figs. 25.2b and 25.3b), whereas humpbacks were more often found in January and February; they generally moved to the west clearing the eastern part of their previously occupied area (Figs. 25.2c and 25.3c). The filling pattern of the Antarctic area in fin whales looked like one complementary to humpback whales: in January, these whales in Antarctic waters were located in the eastern part of the study area (South Orkney Islands and Powell Basin); in February they occupied a much broader area increasing their presence in the west (Bransfield Strait).

Southern right whales were numerous in January; two such whales were detected in the middle of February in the Powell Basin next to the Hesperides Trough (Fig. 25.4).

25.3.2 *Factors Affecting the Occurrence of the Various Species of Marine Mammals*

The primary question in studies like this is what factors affect the occurrence of mammals that can be seen from a moving route. In this manner, there are two types of parameters: natural parameters (sea depth, ambient temperature, etc.) and conditions that offer possibilities to detect mammals (visibility to observers, vessel speed, etc.). Factor analysis in statistical method is a good method for finding the major variables and the main parameters of these factors. In the factor analysis, the standard procedure is used for calculating two principal components. The first component represents the variability of occurrence of the requested subject, e.g., all mammals in the sea or a specific species. This really is the answer to our main question. The second component represents the variability in the "speed" (how quickly) of change in the occurrence under the considered parameters. It is generally accepted that the influence of any parameter is considered important if the factor loading of the parameter is greater than 0.7 (Jeffers 1978).

For the preliminary steps of the factor analysis, the following human-ability parameters were included: visibility on the sea surface in meters, ship speed (up to 10 knots, usually less), and the difference in detection degree between one animal and groups of animals. To our surprise, all the parameters mentioned showed very little or no major effect on the occurrence of the animal. The most interesting issue was that the probability of detecting a single animal or a group was about the same. We therefore omitted these three parameters from further consideration. Finally, the following parameters were considered: calendar date, time, latitude, longitude, depth, ambient air temperature, humidity, wind speed, wind gusts, wind chill factor, and air pressure.

During the second trip (Drake Passage and Antarctic area) for all mammal species together, the two mentioned principal components gathered nearly half of the total variability of parameters. The essential factors responsible for the occurrence of any animal were only wind speed and wind gusts: more animals were recorded (occurred) when the wind speed was high. Regarding the rate of the occurrence change, the essential loadings were latitude (the further to the north the quicker the change in animal occurrences), ambient temperature, and wind chill factor: the greater the temperature the quicker the change in the number of animals around the vessel.

As far as different species are concerned, it makes sense to examine the species with larger occurrence. There are three species: humpback whale, fin whale, and Antarctic fur seal (Table 25.1). Regarding humpback whales they were wind speed, gusts, and negative wind chill factor. The rate of occurrence change was related to longitude (the further to the west, the greater the change in the occurrence of these whales) and humidity (the more the humidity, the quicker the change in the occurrence). Regarding fin whales, the most highly correlated parameters of occurrence were humidity, wind speed, and wind gusts. All these parameters had a positive effect. Occurrence change correlated with latitude. This dependence is not clearly seen: in January within the Antarctic waters, the further to the north, the greater the occurrence of fin whales (Fig. 25.2). During the second stage of the trip, a very interesting and very complicated factor and parameter compositions were observed related to Antarctic fur seal. The abundance (occurrence) of these numerous animals mostly correlated positively with the wind speed and, accordingly, with the wind gusts. However, the situation was shifting faster in the western and northern directions; thus, in January, the frequency of this seal species was the most stable in the eastern part of the Antarctic region under study (Fig. 25.2). The pattern of occurrence changing with depth revealed a quite unusual interaction. Although the correlation coefficient of the depths where we encountered Antarctic fur seals with the depths on the vessel route is high (see Sect. 25.3), the change in occurrence of this species in January is negatively correlated with the depth.

During part three of the trip, the two mentioned principal components gathered more than half of the total variability of parameters for all mammal species together. The essential factors loading for the occurrence of any animal was longitude (the farther east, the greater the number of animals seen; Fig. 25.3 confirms this regularity), ambient temperature, and the wind chill factor. Regarding the speed of changing occurrence, only the calendar date had an essential negative influence on this component: the greater the date, the lesser the speed of change in animal occurrence. The occurrence of humpback whales negatively correlated with humidity and wind gusts (with regard to speed, as well, but in distinction to gusts, the influence of the wind speed was slightly less than 0.7), and, although not expected, it positively correlated with the calendar date. The rate of change in occurrence correlated with longitude (in contrast to stage two of the trip, the farther east, the greater the change in occurrence of these whales) and negatively correlated with air temperature and wind chill factor; the latter two are logical. In February, the most stable fin whale concentration was observed in the Powell Basin (Fig. 25.3). Therefore, one of the most positively correlated parameters is longitude. The other related parameters are wind speed and gusts, positively related, and wind chill factor, negatively related. Change in occurrence essentially and positively correlates with calendar dates. It is noticeable that practically the same regularities occurred for the systematic group "baleen whales" where the species were not satisfactorily identified. This makes it possible to conclude that these whales were probably fin whales. In the Antarctic, fur seal factor loadings are not as complicated as in January: the occurrence correlates with longitude, positively (clearly seen in Fig. 25.3), and negatively with ambient temperature and wind chill factor. Depth correlated with the occurrence but does not reach the required threshold (just 0.5). Change in occurrence increases with calendar date.

Regarding the southern right whale, the correct amount of data was collected only from Montevideo to the entrance of the Beagle Channel. It turned out that many parameters correlated with the occurrence of this species: calendar date (negatively), time of day (more often seen in the afternoon), latitude and longitude (the further to the north and east, the greater the amount of these whales), ambient temperature and wind speed and wind gust factors (positively), and humidity (negatively). Change in occurrence positively correlated with wind chill factor which is complementary to the air temperature correlation with occurrence and strongly negative with depth.

25.3.3 Marine Mammal Distribution Correlated with the Sea Depth

We need a basis for measuring the correlation coefficient, i.e., a collection of depths more or less relevant to the depth of the vessel's path. Herewith, we can only depend on the points of the route where the depths are known. These points are encounters with biological objects: marine mammals and seabirds. Theoretically, it is possible to use either of them as a kernel, i.e., to calculate correlation coefficients of depths at encounters of each particular mammal species with all depths over where mammals were found. However, regarding mammals, the number of observations is much less than that of birds: during stages two and three of the vessel routes where a correlation coefficient was calculated, there are 690 and 452 points, respectively. At the same time, in birds, there are 3165 and 2382 points of occurrence, respectively (see article on seabirds, this issue).

The first procedure is to calculate the correlation coefficient between these two possible basics. We will use the Pearson's correlation coefficient (denoted as r) since the factor analysis uses the same type of coefficient. In stage two, the correlation between depths of mammals' and birds' encounter points is high and significant: $r = 0.90$, $n = 30$, $P < 0.001$. In stage three, the correlation between depths of mammal encounter points and bird encounter points is even greater: $r = 0.96$, $n = 30$, $P < 0.001$. This means that it does not matter what set we use as a basis for the calculation and mammals' sets are also valid despite fewer points than birds' sets.

Correlations between basic sets of stages two and three are significant, but not so high: in mammals' sets $r = 0.46$, $n = 30$, $P < 0.01$, in birds' sets $r = 0.62$, $n = 30$, $P < 0.001$. Correlation between depths of the three marine mammal species also shows low values: depths of humpback whale encounter points of stages two and three of the trip correlated as $r = 0.65$, $n = 30$, $P < 0.001$; in fin whales $r = 0.20$, $n = 30$, insignificant; in Antarctic fur seal $r = 0.36$, $n = 30$, $P = 0.05$ (Fig. 25.5). At the same time, correlations between the available depths and depths of the three species occurrence within each stage are much greater and positive. In stage two, the correlation coefficients between all mammals' depth set and depths of humpback whales are $r = 0.69$, $n = 30$, $P < 0.001$; fin whales $r = 0.90$, $n = 30$, $P < 0.001$; and Antarctic fur seal $r = 0.96$, $n = 30$, $P < 0.001$; in stage three the correlation coefficient between all mammals' set and depths of humpback whales is $r = 0.82$, $n = 30$, $P < 0.001$; fin whales $r = 0.93$, $n = 30$, $P < 0.001$; and Antarctic fur seal $r = 0.84$, $n = 30$, $P < 0.001$. These results show that all these three species prefer deep areas; however different depths are preferable for different species (Fig. 25.5).

Regarding southern right whale, if we consider all encounters with this species in stages one and three, the correlation between all available depths and depths of this whale encounters is negative: $r = -0.51$, $n = 21$, $P = 0.02$. If we increase the sample by using not only encounters but count the total number of individuals seen at these encounters, the correlation remains roughly the same: $r = -0.49$, $n = 21$, $P < 0.02$. This species prefers shallow waters (Fig. 25.6).

25.3.4 Marine Mammal Distributions Along the Vessel's Route

According to the Clark-Evans nearest neighbor distance criterion (Clark and Evans 1954; Kharitonov 2005), humpback distribution along the vessel route was significantly clustered (grouped) in both parts of the voyage in the Antarctic: in the second part (January), the Clark-Evans coefficient was (R) = 0.66, $N = 132$, $P < 0.001$; in the third part (February), the value was $R = 0.71$, $N = 66$, $P < 0.001$. Mean linear densities of encounters within those parts of the trip where humpback whales were

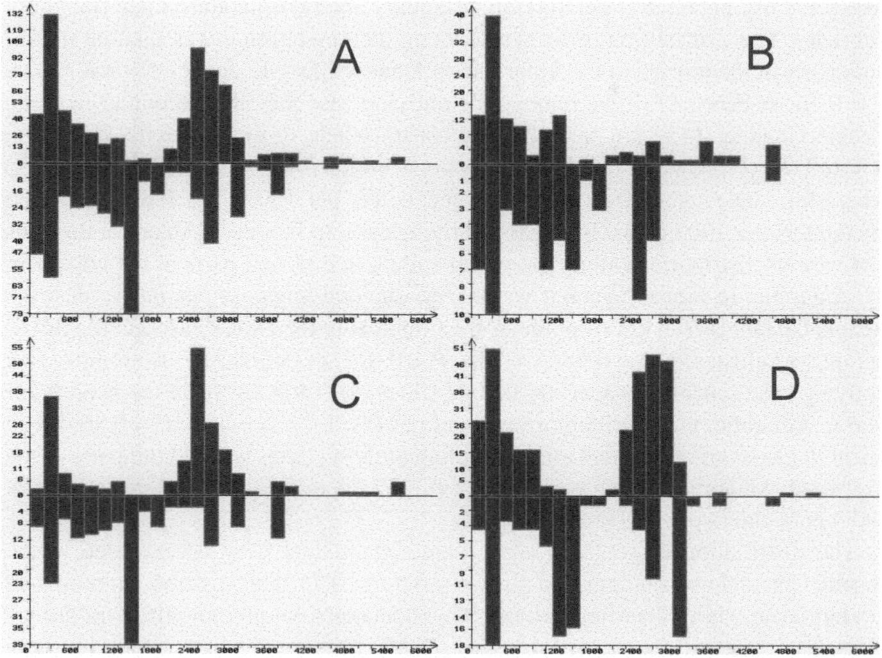

Fig. 25.5 Depth distribution along two routes of the vessel in the Antarctic. Upper histograms are depths along the route mid-January–early February; lower histograms February–early Mach. (**a**) all depths along the route; (**b**) depths of humpback whales; (**c**) depths of fin whales; (**d**) depths of Antarctic fur seals. X axis, depths in meters; Y axes, numbers of encounters with the species

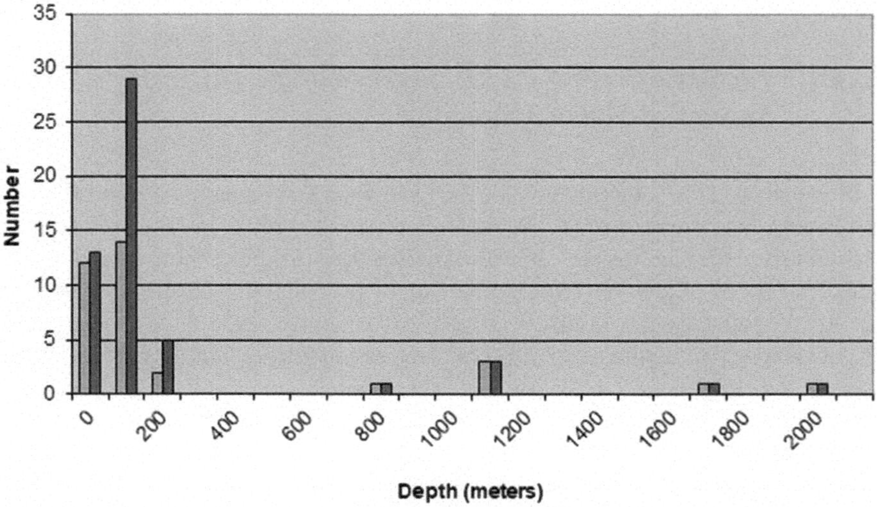

Fig. 25.6 Depth distribution in southern right whale encounters. Light gray histogram is the number of encounters; dark gray histogram is the number of whales

seen were five encounters per 100 km in January and two encounters per 100 km in February. The distribution of fin whale along the vessel route was random during both parts of the voyage in the Antarctic: in January Clark-Evans $R = 0.95$, $N = 203$, $P = 0.46$; in February, if we remove the outlying case, i.e., the encounter near the Beagle Channel 1452 km from the nearest fin whale in the Antarctic, $R = 0.91$, $N = 165$, $P = 0.23$. Mean linear densities of encounters within those parts of the trip where fin whales were seen were 17.2 encounters per 100 km in January and 6.5 encounters per 100 km, excluding the outlying case, in February. Antarctic fur seals had various distributions along the vessel's route in different parts of the voyage in the Antarctic: in January, even if we remove one outlying case (encounter near the Beagle Channel), 790 km from the nearest fur seal in the Drake Passage, the distribution was clustered: $R = 0.95$, $N = 314$, $P = 0.012$; in February, if we remove one outlying case (encounter near the Beagle Channel), 1468 km from the nearest fur seal in Antarctic, the distribution was random, $R = 1.09$, $N = 108$, $P = 0.36$. Mean linear densities of encounters after excluding outlying cases within both parts of the trip where Antarctic fur seal were seen were 15.2 encounters per 100 km in January and 4 encounters per 100 km in February.

The distribution of southern right whale encounters from Montevideo to the Beagle Channel was random: $R = 1.06$, $N = 32$, $P = 0.75$. This distribution remained random even while recording not only the encounters but all individuals in groups at these encounters: $R = 0.94$, $N = 51$, $P = 0.67$. The linear density of the southern right whale encounters was five encounters per 100 km or eight whales per 100 km along the part of the trip where these whales were seen.

25.4 Discussion

25.4.1 Factors Affecting the Occurrence of the Different Species of Marine Mammals

Factor analysis shows that in three of the most abundant Antarctic species (humpbacks, fin whales, and Antarctic fur seals) in different months, in January when the second part of the trip was performed and in February (the third part), animal occurrence and speed of change in occurrence might vary not only for different species but for the same species in different months. Generally, these three species were present at lower air temperatures and chill factors and greater humidity. This is clear because the closer we get to the Antarctic such animals are usually more numerous. These parameters might serve as the environmental determinants of the distribution of these whales, because air temperature, surely, affects sea surface temperature and the latter affects whale spreading (Santora et al. 2014). In contrast to humpbacks and fin whales, the southern right whale definitely prefers high temperature and low humidity. In this manner, their appearance in February in cold Antarctic waters was the outrange event, which rarely occurred (Bannister et al. 1999).

Regarding fin whales, change in occurrence positively correlates with latitude. The latter feature indicates a rare regularity: in January within the Antarctic, the further to the north, the greater the amount of fin whales. It seems they prefer the long basins that stretch along the longitude and south of the chain of islands: from the South Shetland Islands via Powell Basin to the South Orkney Islands, not among but close to the floating ice (icebergs). It is known that fin whales prefer offshore waters far from land. Our data show that they avoid mainland but do not avoid islands: there is a number of fin whale encounters very close to the South Shetland Islands (Fig. 25.2, 25.3). It is known that fin whales while feeding on large aggregations of Antarctic krill near Elephant Island stick to the shallow waters (<300 m) (Burkhardt and Lanfredi 2012).

We notice that change in the occurrence of the Antarctic fur seal in January negatively correlated with depth, whereas in February correlations with depths in occurrence and its changing are not essential. To our mind, this may be linked to such a feature: in February, the frequency of this species increased and decreased with depth approximately twice, while in January it increased noticeably only once at a depth of around 2000 m but decreased twice with an increase in depth (Fig. 25.5). This is likely to negatively correlate with the change in speed of occurrence with depth.

Interesting observations were made when, on February 22–24, the ship moved slowly between 60° 30'S and 48° 45'W. On February 22–23 in the morning, there were many fin whales, and on February 23, almost no whales were observed there. The next day, on February 24, many fin whales were found in this area again. This change in the fin whale occurrence assumes that availability of food is not compulsory since fin whales are consumers of this food. It is a known phenomenon that consumers might leave a rich foraging area looking for other foraging areas (review in Kharitonov (1983)). On the other hand, it is also assumed that the presence of consumers does not compulsorily points to concentration of food there.

In February, southern right whales were seen near the Hesperides Trough close to the area of depths exceeding 5000 m (Fig. 25.4). This implies that, in this very area, the krill is very pure, with the lowest concentration of salps (Kobylyanskiy, personal communication). In addition, several bird species, e.g., white-chinned petrel, light-mantled albatrosses, and snow petrels, also prefer pure krill. Precisely in this location, these bird species formed the greatest concentration (article on seabirds, this issue).

25.4.2 Sea Mammal Distribution Related to the Sea Depth

Correlation between the depths of the occurrence of fin and humpback whales in January is $r = 0.43$, $n = 30$, $P < 0.02$; in February it is greater: $r = 0.67$, $n = 30$, $P < 0.001$. This is understandable since in February fin whales become more abundant south of the South Shetland Islands, the main area occupied by humpback both in January and February. The mean depth of the humpback occurrence in January is

M = 1217.4 ± 108.1 m, N = 132; in February the value is very close M = 1270.6 ± 114.5 m N = 66. Regarding fin whales in January, it is M = 2120.6 ± 86.2 m N = 203; in February the value is significantly shallower M = 1503.3 ± 77.6 m N = 166; as expected, it is closer to the depths of humpback whales for the same reason: in this month, fin whales occupied a considerable part of the humpback area. The mean depth of the occurrence of fin whales encountered in January is significantly greater than in February, which is another confirmation that in these months these whales generally occupy different areas. Such a change in the depths between months is understandable since in February this whale performed a mass "intrusion" into the main humpback area (south of the South Shetland Islands). If we join samples of fin whales and baleen whales (this is possible in February), the mean depth of occurrence becomes M = 1501.8 ± 64.1 m N = 236, which is very close to fin whales only. This is the second argument that unidentified baleen whales have actually been fin whales.

After 2 months of observations in the same region of the sea, it can be reported that the distribution of the fin whales essentially matched the distribution of the depths, while this whale does not like shallow depths. Humpback whales, unlike fin whales, do not like deeper waters. Probably, this feature is related to different diving abilities of these species: humpback whales during foraging dive less than 150 m (Friedlaender et al. 2009), while fin whales normally dive deeper (fin whale 2020; Orgeira et al. 2015). This disparity probably means that the humpback whales have more trouble feeding in deeper waters.

The mean depth of the Antarctic fur seal occurrence in January was M = 1726.5 ± 64.9 m N = 315; in February, it was M = 1589.6 ± 99.5 m N = 109, greater but not significant. However, the Antarctic fur seals are widely spread in the Antarctic waters. Southern right whales prefer shallower waters (Fig. 25.6), but to reach rich foraging areas, particularly for "good food" (pure krill), they are capable of covering large areas with great depths, tolerating low temperatures and high humidity.

25.4.3 Spatial and Linear Distributions of Sea Mammals

Spatial distributions of Antarctic fur seals in January and February in the study site are generally similar to each other. However, among the most abundant whales, humpbacks and fin whales, the general spatial distribution changed from January to February. In January, humpbacks were both in the west and in the east of the study region. By February, not many humpbacks moved from the eastern part of the study site and concentrated in the western part of the study site. At the same time, fin whales from the east, where they were observed in January, spread throughout the entire region, and their concentration in its western part was even greater than in the eastern part. Previous recent studies have recorded midsummer concentrations of fin whales around the South Orkney Islands (Viquerat and Herr 2017). Concentrations of this whale near the Southern Shetland Islands were observed in the southern fall

(March–April), with concentrations of fin wales appearing to indicate pre-migrating feed aggregation (Joiris and Dochy 2013). Definitely, depths were not a reason: although humpbacks in February concentrated in the areas with depths below 2700 m (excluding one whale at the depth 4700 m), in January, quite a few of these whales were observed over depths up to 4700 m. It is known that the main food of humpbacks is Antarctic krill. However, the purest krill (with a low proportion of salps) in January and February occurred especially in the east. The western area contained (1) krill with greater proportion of salps and (2) the juvenile krill occurring in the western part and to the west of the Powell Basin (untreated data of the invertebrate branch of the vessel scientific team). Therefore, humpback whales need either salps or juvenile krill or both as food sources. Jarrod Santora and colleagues (Santora et al. 2010) revealed that humpback whales really prefer to feed on juvenile krill, whereas fin whales feed mostly on adult plankton (Santora et al. 2010). If this is true, the role of salps is still questionable: do humpback whales really need salps or they just do not "pay attention" to them? On the other hand, we should not exclude the possibility that humpbacks have been forced to move from the east by fin whales, which probably do not tolerate the presence of humpbacks. We assume this, because in our observations fin whales and humpbacks in February were seen when moving to the west of the Powell Basin mostly in the mono-species patches; there was no considerable interspecies mixing (although, some studies have shown that humpback and fin whales, when intermixing, display horizontal niche partitioning during mutual presence in the region) (Herr et al. 2016).

The type of spatial or linear distribution is a good tool for understanding the additional features of the population structure and prosperity of the species. A mathematical pattern of the spatial distribution is the characteristic of biological interactions between objects. Clustered distribution in humpback whales in January and February might indicate following some environmental features or social attraction (Odum 1983). The type of prey concentration seems the most likely reason for this. In January, Antarctic fur seals had clustered distribution with large clusters of these animals located near the tip of the Antarctic Peninsula and neighboring islands, near the South Shetland and South Orkney Islands, i.e., not far from land. By February, they were more widely spread, and their distribution became random. The change in distribution by months is likely connected with the breeding cycle of this seal: in January they are still not very far from the rookeries, while in February, they begin to migrate (Dykyy and Peklo 2012).

Even (uniform) distribution of the biological objects usually indicates saturation of a given area and even competition between individuals. In this study, no uniform distributions of marine mammals were seen; therefore no competition was noticed. Random distribution of fin whales both in January and February, in cases of a large data set, suggests that the area is not saturated with this species and the available amount of food is enough for whales currently present in the area.

Southern right whale distribution was random. However, there are insufficient data on this species to make a definite explanation. Random distribution could be a result solely of the small amount of data.

25.5 Conclusions

This article represents preliminary results on the occurrence of marine mammals during cruise 79 of the R/V "Akademik Mstislav Keldysh" to the Antarctic. However, even the preliminary results allowed us to make several conclusions. In January, February, and the first 3 days of March, we recorded 2140 marine mammals, 1990 of which were identified to the species level, and a total of 20 species were recorded. It has been revealed that the observed occurrence of animals in the ocean is related to a number of environmental parameters and weakly related to human-ability parameters like visibility, ship speed, and the difference in detection degree between one animal and groups of animals.

The occurrence of the three most numerous animals positively correlated with the depth distribution along the ship route; correlations were greater within each stage of the trip than between stages. This suggests that sea depth is important for these animals. Spatial distributions of Antarctic fur seals in January and February in the studied Antarctic sea area were generally similar to each other; distribution of humpback and fin whales varied from January to February: a large number of both species in February concentrated in the area south of the South Shetland Islands. The type of distribution of fin whales suggests that this species is not restricted by food supply. No competition for food has been noticed in this species and in the humpback whales. The group distribution of the latter might be caused not only by food distribution but also other unknown reasons.

Acknowledgments We would like to express our gratitude to echo sounder managers Dmitry I. Frey and Alexander V. Gavrikov, Shirshov Institute of Oceanology RAS for offering us the detailed data on depths. Many thanks to the "Mercury detachment": Maryina E.M., Rudykh Ya.I., head-Kalinchuk V.V. of V.I. Il'ichev Pacific Oceanological Institute, Far Eastern Branch of RAS, who managed weather station Davis Vantage Pro2 and offered us the weather data.

References

Bannister JL, Pastene LA, Burnell LA (1999) First record of movement of a southern right whale (Eubalaena australis) between warm water breeding grounds and the Antarctic Ocean, south of 60 degrees South. Mar Mamm Sci 15(4):1337–1342

Burkhardt E, Lanfredi C (2012) Fall feeding aggregations of fin whales off Elephant Island (Antarctica). Proceeding of 64th meeting of the International Whaling Commission (IWC). SC-64-SH9 6 p

Clark PJ, Evans FC (1954) Distance to nearest neighbor as a measure of spatial relationships in populations. Ecology 35(4):445–453

Dykyy IV, Peklo AM (2012) Seals of the Argentine islands (Antarctica). Proc Zool Mus 43:104–116

Fin whale. International Whaling Commission (IWC, 2020). https://iwc.int/fin-whale

Friedlaender AS, Halpin PN, Qian SS, Lawson GL, Wiebe PH, Thiele D, Read AJ (2006) Whale distribution in relation to prey abundance and oceanographic processes in shelf waters of the Western Antarctic Peninsula. Mar Ecol Prog Ser 317:297–310

Friedlaender AS, Lawson GL, Halpin PN (2009) Evidence of resource partitioning between humpback and minke whales around the western Antarctic Peninsula. Mar Mamm Sci 25(2):402–415

Herr H, Viquerat S, Siegel V, Kock KH, Dorchel B, Huneke WGC, Bracher A, Schroder M, Gutt J (2016) Horizontal niche partitioning of humpback and fin whales around the West Antarctic Peninsula: evidence from a concurrent whale and krill survey. Polar Biol 39:799–818

Jeffers JNR (1978) An introduction to systems analysis, with ecological applications. University Park Press, Baltimore, 198 p

Joiris CR, Dochy O (2013) A major autumn feeding ground for fin whales, southern fulmars and grey-headed albatrosses around the South Shetland Islands, Antarctica. Polar Biol 36:1649–1658

Kharitonov SP (1983) Black-headed gulls (Larus ridibundus) foraging pattern. Zool Zhurnal 62(5):748–754

Kharitonov SP (2005) A "nearest neighbour distance" method for the quantitative evaluation of the distribution of biological objects in the plane or line. Vestnik of the Nizhny Novgorod University. Series Biology 1(9):213–221

Minkina NI, Samyshev EZ, Chmyr VD, Seregin SA (1999) Relative evaluation of assimilation of primary production by krill, salps and bacterioplankton in Atlantic Sector of Antarctic (ASA) under the condition of mass development of gelatinous animals. Abstracts of the 2nd Int. Symp. on Krill (23–27 Aug. 1999.- Santa Cruz, USA, 1999), pp 33–35

Odum EP (1983) Basic ecology, vol 2. Sounders College Publishing, New York-London-Toronto, 376 p

Orgeira JL, Alderete MC, Jimenez YG, Gonzalez JC (2015) Long-term study of at-sea distribution of seabirds and marine mammals in the Scotia Sea, Antarctica. Adv Polar Sci 26(2):158–167

Plokhinskiy NA (1978) Mathematical methods in biology. Moscow State University, Moscow, 265 p

Santora JA, Reiss CS, Loeb VJ, Veit RR (2010) Spatial association between hotspots of baleen whales and demographic patterns of Antarctic krill *Euphausia superba* suggests size-dependent predation. Mar Ecol Prog Ser 405:255–269

Santora JA, Schroeder ID, Loeb VJ (2014) Spatial assessment of fin whale hotspots and their association with krill within an important Antarctic feeding and fishing ground. Mar Biol 61(10):2293–2305

Viquerat S, Herr H (2017) Mid-summer abundance estimates of fin whales *Balaenoptera physalus* around the South Orkney Islands and Elephant Island. Endanger Species Res 32:515–524. https://doi.org/10.3354/esr00832

Chapter 26
Meat in the Ocean: How Much and Who Is to Blame?

Sergey P. Kharitonov, Andrey V. Tretyakov, and Alexander L. Mischenko

Abstract Cruise 79 of the R/V "*Akademik Mstislav Keldysh*" of the Russian Academy of Sciences in the Atlantic sector of the Southern Ocean took place from January to March 2020. The data were collected on 13.02–26.02.2020 when pieces of bird meat were discovered on the surface of the sea along the vessel's route. Observations were performed on all days from dawn to dusk, from the broad direction-finder deck by one or two observers. We encountered remains of birds (carcasses, skins, pieces of meat). On some days, the sea surface was covered by these remains. It was mostly the remains of dead penguins, or sometimes whole bodies. In February, these were mostly juvenile Adélie Penguins, since they leave their colonies in this month, and high mortality of the young generation is known in birds and other animals. However, despite the fact that the absolute number of dead penguins is high, the proportion of them compared to the total population is quite low. Antarctic Fur Seals are suggested to be the main cause of the mortality of the young Adélie Penguins, and not Leopard Seals or Killer Whales. It looks like, the penguins are not just an occasional, but really an essential part of the Antarctic Fur Seal diet, especially in the period when starving males leave breeding grounds and commence migration.

Keywords Marine mammals · Seabirds · Predation · Adelie Penguin · Pygoscelis adeliae · Antarctic Fur Seal · Arctocephalus gazella

---

S. P. Kharitonov (✉) · A. V. Tretyakov · A. L. Mischenko
A.N. Severtsov Institute of Ecology and Evolution, Russian Academy of Sciences, Moscow, Russia
e-mail: serpkh@gmail.com

E. G. Morozov et al. (eds.), *Antarctic Peninsula Region of the Southern Ocean*, Advances in Polar Ecology 6, https://doi.org/10.1007/978-3-030-78927-5_26

26.1 Introduction

The Antarctic marine biota is very rich in plankton organisms that serve as the basis of the food web for the upper levels of food chains. On the other hand, some higher-level planktivorous animals comprise a part of forage source for other higher-level animals: penguins are known as food for Leopard Seals, *Hydrurga leptonyx*, and, occasionally, for Killer Whales, *Orcinus orca,* and Antarctic Fur Seals, *Arctocephalus gazella* (Shirihai 2008). Until now, it is not known how many penguins these predators can consume; the importance of penguins as food is also not known. Our observations from the research vessel in Antarctic waters in February 2020 may shed light on these ecological features. Cruise 79 of the research vessel "*Akademik Mstislav Keldysh*" of the Russian Academy of Sciences was conducted in the southern part of Atlantic in January–March 2020. The goal of the research was multiple: to study the current situation in the oceanographic environment and Antarctic biota. Data on seabirds and marine mammals were collected on January 9–March 3, 2020. Twice on 16.01–06.02.2020 and 09.02–03.03.2020, the vessel made trips to the Antarctic waters visiting approximately the same area (see articles on marine mammals and seabirds, this issue). Each trip was approximately 6000 km long.

The main task of the zoology team of the expedition was to study the distribution of marine mammals and seabirds. However, beginning on February 13 in the Antarctic (south of the Southern Shetland Islands), we noticed pieces of birds and bird flesh (meat) around the vessel. Since then, these pieces were counted, their distribution studied, and efforts made to understand why such amounts of meat were on the surface of the Antarctic waters.

26.2 Materials and Methods

The data were collected on 13.02–26.02.2020 after pieces of bird meat were found on the surface of the sea along the vessel's route. Observations were performed every day from dawn to dusk, from the broad direction-finder deck by one or two observers. Observers were equipped with high-power photo cameras and telephoto lenses.

This observation and resultant survey of meat in the ocean was supplementary work, which was additional to the main task of the expedition. For this reason, the former observation could not be completed, because our attention was focused on marine mammals or seabirds whenever they appeared in our field of vision. In any case, while recording the meat pieces, they were counted within the 100-m-wide strip along both sides of the ship.

The route of the vessel was tracked by three GPS navigation devices; therefore, coordinates of each encounter with biological objects in the sea are available. To obtain knowledge on the type of distribution of the species that were abundant in the area, the method of evaluation of distribution of objects along the line (Kharitonov

2005) derived from the Clark-Evans nearest neighbor method (Clark and Evans 1954) was used. Since the observations were not performed at night, the night periods were subtracted from the lines of the vessel route for distribution evaluation. The submethod of interrupted lines was used (Kharitonov 2005).

26.3 Results

During the first Antarctic trip (see articles on marine mammals and birds, this issue) of 09.01–06.02.2020, the observers did not see bird remains in the ocean (Nikolay Konyukhov and other observes, personal communication). We encountered remains of birds (carcasses, skins, pieces of meat) on February 13–26, 2020 (Fig. 26.1). On some days, the sea surface was covered by these remains. Most remains belonged to small penguins. A total of 74 meat objects were recorded, from small pieces to whole penguins. About half of the objects were big enough to be recognized as penguin. Only once was it possible to identify the species, that is, juvenile Adelie Penguin *Pygoscelis adeliae*. Efforts were made to evaluate the number of dead penguins that produced such a noticeable number of remains. We considered the following remains of penguins: (1) carcasses of penguins, whether destroyed or not; (2) large pieces of skin or flesh at a distance from each other. If we saw a set of small pieces very close to each other (several meters), we considered this set as belonging to one bird. Consequently, 68 "whole" penguins were counted along the route in areas where remains were observed. Six of them were full penguin carcasses without visible injury.

It was surprising that only eight pieces (12% of the total) were being eaten by various types of birds, the remaining 60 were unattended. The recorded species recorded eating meat pieces were: Southern Giant Petrels *Macronectes giganteus* (Fig. 26.2), Wilson's Storm Petrels *Oceanites oceanicus* (Fig. 26.2), Black-bellied

Fig. 26.1 Examples of penguin remains (meat objects) in the sea

Fig. 26.2 Photos of bird behavior while eating and after touching the meat pieces. Upper row of pictures and left picture in the second row displays how Southern Giant Petrel and Wilson's Storm Petrel consume remains of penguins. Right picture in the second row shows Southern Fulmars eating a piece of meat. The four pictures below show the situation when Wandering Albatross after touching the piece of meat did not continue eating, but immediately swam away and started to wash its bill (see text for details)

Storm Petrels *Fregetta tropica*, Wandering Albatrosses *Diomedea exulans* (Fig. 26.2), Light-mantled Albatrosses *Phoebetria palpebrata*, White-chinned Petrels *Procellaria aequinoctialis*, Southern Fulmars *Fulmarus glacialoides* (Fig. 26.2), and Cape Petrels *Daption capense*.

Areas where meat pieces were recorded were generally located east of the Antarctic Peninsula and neighboring islands. This is understandable, because most of currents are directed to the east and northeast within the Bransfield Strait (Gordon et al. 2000; Morozov 2007). According to the Clark-Evans criterion (1954) with the modification for lines (Kharitonov 2005), the distribution of encounters with meat objects in that part of the sea (where these pieces were found) was random (Clark-Evans R-criterion = 1.097 $N = 30$, $P = 0.6$). Line for the observation on one side of the vessel was 100 m, and on both sides, it was 200 m (0.2 km) in total. The distance where meat pieces could be surveyed was 2322 km along the vessel's route. This distance excludes nighttime parts of the route (night parts of the route were already subtracted). The area where we detected 68 penguins comprised $0.2 \times 2322 = 464.4$ sq. km. Since the evaluation was very approximate, we drew more or less arbitrary three sites where meat pieces were recorded (Fig. 26.3). The total area of these sites is about 30,000 sq. km. Therefore, extrapolating the number of dead penguins over the entire area where they were encountered, we get (30,000 km^2/464.4 sq.km) × 68 penguins = 4393 dead or eaten penguins.

26.4 Discussion

During the 14-day period (13–26 February), extrapolation of our survey of penguin meat and skin pieces showed the presence of approximately 4393 eaten or dead penguins within a sea surface area of 30,000 square km. However, this amount is highly underestimated. It should be considered that our observations were not continuous, since the process was often interrupted and our attention switched to marine mammals and seabirds and because fixation of the two latter was our main task. We believe that the area of 30,000 square km actually contained at least 10,000 eaten or dead penguins.

Now there are two questions to be answered: What is the source of such large amounts of penguins and who eats so many penguins? Gentoo penguins, *Pygoscelis papua,* are not so abundant in this area (Baylis et al. 2013, see article on seabirds in this issue). Young Gentoos leave their colonies in March (Shirihai 2008). Young Chinstrap Penguins, *Pygoscelis antarctica,* also leave their colonies in March (Shirihai 2008). The tip of the Antarctic Peninsula and the South Shetland Islands accommodate large colonies of Chinstrap Penguins, hundreds of thousands or even up to one million birds (Trivelpiece and Trivelpiece 2013). Adult Chinstrap Penguins were numerous in the study site in January and February (article on seabirds, this issue). Since in January, the observers did not see any remains of penguins, it implies that adult chinstraps were mostly able to escape the predators that produced "meat in the ocean" in February the same as in January. Regarding Adélie Penguin, the situation was quite different. In January, these penguins were scarce in the study site; only the adult birds were seen by the observers; whereas in February, Adélie Penguins were much more abundant, and there were juvenile birds (article on seabirds, this issue). In addition, it is known that juvenile Adélies leave their colonies in February. This species is numerous on the Antarctic Peninsula (colonies are located in multiple parts of the peninsula) and on the islands (Shirihai 2008). The counted numbers there exceed half a million breeding pairs (Lynch and La Rue 2014). In recent years, a large colony (1.5 million individuals) was found on Danger Island, southeast of Joinville Island (Borowicz et al. 2018). Hence, this area contains more than two million Adélie Penguins who can produce more than one million offspring. In our observations, currents were directed to the study site exactly from these regions of the Antarctic Peninsula and nearby islands, as well as from the colonies near the South Shetland Islands (Fig. 26.3). Therefore, it is most likely that the main source of meat in the ocean that we recorded were juvenile Adélie Penguins.

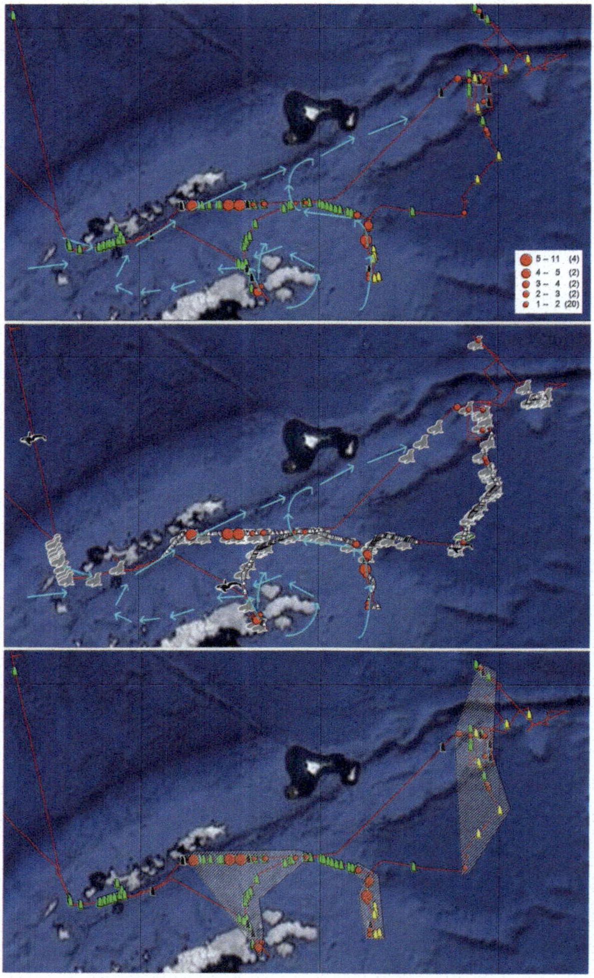

Fig. 26.3 Maps of pieces of meat distribution along the vessel's route. Red line is the trip on 08 February–03 March: Ushuaia-Drake Passage–top of the Antarctic Peninsula–Powell Basin–back to Antarctic Peninsula–Drake Passage–Ushuaia. Symbols "green penguin" mean surveyed alive Chinstrap Penguins (see article on seabirds, this issue), "yellow penguins" represent encountered alive Adélie Penguins, "black penguin" symbols mean unidentified penguins of the genus *Pygoscelis*, blue-green penguin symbols mean Gentoo Penguins. Blue-green arrows represent ocean currents (redrawn from Gordon et al. 2000; Morozov 2007). Red circles represent grades of pieces of meat counts during 2-h observations in areas where meat had been encountered (the legend in the right bottom corner of the top map): the largest circles denote 5–11 pieces of meat during 2 h, secondary large ones are 3–4 pieces, thirdly sized circles denote two pieces, the four grades of the circle size represent one piece per two hours. Symbols "killer whale" mean encounters with schools of Killer Whales. Symbol "seal" means the only observation of a Leopard Seal. Symbols "eared seal" mean encounters with Antarctic Fur Seals. White triangles mark the areas where icebergs were observed. Dashed sites are areas where meat pieces were recorded

Since most of the penguin carcasses observed were already destroyed and there were large amounts of floating meat pieces, we assume that some predators might be responsible for this situation. Few carcasses were intact; possibly, some dead birds in colonies were washed out by storm waves and transported by currents. In any case, birds that fed on meat objects were the secondary consumers. Surprisingly, these birds utilized only a small part of the available meat. It is important to emphasize here that some pieces of meat were not edible by the birds: they tried pecking them, but soon stopped doing it. As an example, we described our observation on 24.02.2020 at 12:03: a piece of meat together with skin of a penguin floated near the drifting vessel. A Wandering Albatross approached the piece, pecked it two times, then swam away, and washed its bill (see Fig. 26.2). After that, a group of ten Cape Petrels landed nearby, and one of them pecked the meat several times and swam away. Presumably, this piece of meat was rotten compared to some other pieces of meat. However, most of the pieces looked fresh and we cannot explain why most of them were unattended.

Subregions were identified in the study site with greater or lesser concentration of meat, but generally, they were randomly distributed. This implies that there are no external factors that influence the dispersion concentration of these pieces of meat.

We can consider Leopard Seals, Killer Whales, and Antarctic Fur Seals as possible predators of Penguins. In February, a Leopard Seal was recorded only once; groups of Killer Whales were recorded three times, but only one time in the area where meat was encountered (Fig. 26.3). The area where Antarctic Fur Seals were encountered strongly overlapped with the area of meat (Fig. 26.3). Therefore, Antarctic Fur Seals seem the most likely predator of penguins, especially if they are young and unexperienced. It is known that the main food of the Antarctic Fur Seal is krill, fish, and squid (Doidge and Croxall 1985); these seals can eat penguins; however, this is only occasional (Shirihai 2008). Some studies have shown that Antarctic Fur Seals hunt King Penguins, *Aptenodytes patagonicus,* to eat them (Hofmeyer and Bester 1993); however, King Penguins are considerably larger than Adélie or Chinstrap Penguins. Young fur seals play with penguins of various species, inflicting injuries on them, which may be fatal sometimes (Hofmeyer and Bester 1993). Haddad et al. (2015) also remind that Antarctic Fur Seals eat King Penguins; however, these penguins can be actually raped by Fur Seals (http://www.bbc.com/earth/story/20141117-why-seals-have-sex-with-penguins).

The period when we observed floating meat coincided also with the leaving of this territory by young Antarctic Fur Seal males in February and the commencement of northward migration (Dykyy and Peklo 2012; article on marine mammals, this issue). Territorial males do not feed during breeding season; young males that hoped to conquer territory were also starving. Surely, during this period, fur seals need to restore their supply of fat. It is obvious that the penguins become food for them during this period. This case, together with some other observations (e.g., hunting on King Penguins), suggests that penguins are not only occasional, but rather important food for Antarctic Fur Seals, at least in February.

Concerning the proportion between the existing number of Adélie Penguins (Chinstrap Penguins might also be consumed, but in small amounts), 10,000 youngsters cannot be compared to the millions of available birds. Antarctic Fur Seals do not significantly harm penguin populations. Seals, however, eat thousands of penguins; therefore, this kind of food is not only occasional.

It is interesting to know how open sea antipredator behavior occurs in seabirds and, in particular, in penguins, that is, how these birds react to danger. Some observations during other Antarctic trips near Cuverville Island (not far from our study site) show that, while a Leopard Seal eats a penguin, other penguins are not anxious and swim around nearby (Eugene G. Morozov, personal communication). However, this happens when the Leopard Seal has already killed a penguin and eats it. Many observations showed that seabirds and penguins fear predators. We can conclude this from the group behavior of some bird species. When several birds rest or even sleep on the sea surface, one of them is always alert (Fig. 26.4). No alert birds were observed in cases where a flock of birds (e.g., Giant Petrels) rest on tall icebergs. In resting flocks of penguins, alert birds were always observed: in flocks of several

Fig. 26.4 Alert birds while other members of flock rest. Upper photo: Black-browed Albatrosses on sea surface, center: group of young Adélie Penguins on the ice piece, bottom photo: two young Adélie Penguins on a small piece of ice

birds, at least one was alert; if the group consisted of only two penguins, one was alert anyway (Fig. 26.4). One penguin was never observed; it was always alert.

26.5 Conclusions

Our observations have shown that in some months of the year (e.g., February), we can see huge amounts of pieces of meat on the sea surface in the Antarctic. It is mostly the remains of dead or killed penguins, or sometimes whole bodies. In February, these remains consist of juvenile Adelie Penguins, since they leave their colonies in this month, and high mortality of the young generation is known for birds and other animals. However, despite the fact that the absolute number of dead penguins is high, the proportion compared to the total population is quite low. Antarctic Fur Seals are suggested to be the main cause of the mortality of the young Adélie Penguins, and not Leopard Seals or Killer Whales. It looks like, penguins are not just an occasional, but really an essential part of the Antarctic Fur Seal diet, especially in the period when starving males leave breeding grounds and commence migration.

Our results are preliminary. This is the first communication where the Antarctic Fur Seal is considered to be the main predator of penguins, at least during the period when young penguins leave their colonies. Our data is not so conclusive; we can only suggest a hypothesis. The purpose of this communication is to attract attention to the subject of "meat in the ocean." This subject is related not only to penguin and fur seal biology, but includes recent studies of Antarctic biological resources. More complete and deep studies of this resource are needed.

References

Baylis AMM, Crofts S, Wolfaardt AC (2013) Population trends of Gentoo Penguins Pygoscelis papua breeding at the Falkland Islands. Mar Ornithol 41:1–5

Borowicz A, McDowall P, Youngflesh C, Sayre-McCord T, Clucas G, Herman R, Forrest S, Rider M, Schwaller M, Hart T, Jenouvrier S, Polito MJ, Singh H, Lynch HJ (2018) Multi-modal survey of Adélie penguin mega-colonies reveals the Danger Islands as a seabird hotspot. Sci Rep 8(3926):1–9

Clark PJ, Evans FC (1954) Distance to nearest neighbor as a measure of spatial relationships in populations. Ecology 35:445–453

Doidge DW, Croxall JP (1985) Diet and Energy Budget of the Antarctic Fur Seal, Arctocephalus gazella, at South Georgia. In: Siegfried WR, Condy PR, Laws RM (eds) Antarctic nutrient cycles and food webs. Springer, Berlin/Heidelberg

Dykyy IV, Peklo AM (2012) Seals of the Argentine islands (Antarctica). Proc Zool Mus 43:104–116. [in Russian]

Gordon AL, Mensch M, Dong Z, Smethie WM Jr, de Bettencourt J (2000) Deep and bottom water of the Bransfield Strait eastern and central basins. J Geophys Res 105(C5):11,337–11,346

Haddad WA, Reisinger RR, Scott T, Bester MN, de Bruyn PJN (2015) Multiple occurrences of king penguin (*Aptenodytes patagonicus*) sexual harassment by Antarctic fur seals (*Arctocephalus gazella*). Polar Biol 38:741–746

Hofmeyer G, Bester MN (1993) Predation on king penguins by Antarctic fur seals. S Afr J Antarct Res 23:71–74

Kharitonov SP (2005) A "nearest neighbour distance" method for the quantitative evaluation of the distribution of biological objects in the plane or line. Vestnik of the Nizhny Novgorod University. Ser Biol 1(9):213–221. [in Russian]

Lynch HJ, LaRue MA (2014) First global census of the Adélie penguin. Auk 131:457–466

Morozov EG (2007) Currents in the Bransfield Strait. Dokl Earth Sci 415(6):984–986

Shirihai H (2008) The complete guide to Antarctic wildlife. In: Birds and marine mammals of the Antarctic Continent and the Southern Ocean. Princeton University Press, Princeton/Oxford, 546 p

Trivelpiece W, Trivelpiece S (2013) The chinstrap penguin. In: Boersma PD, Borboroglu PG (eds) Biology and conservation of the world's penguins. University of Washington Press, Seattle

Chapter 27
Macro- and Mesozooplankton in the Powell Basin (Antarctica): Species Composition and Distribution of Abundance and Biomass in February 2020

Vladimir A. Yakovenko, Vassily A. Spiridonov, Konstantin M. Gorbatenko, Nickolai V. Shadrin, Ernest Z. Samyshev, and Natalia I. Minkina

Abstract Macro- and mesozooplankton were studied in the Powell Basin in February 2020. Species composition, abundance and biomass distribution, the population size and sex structure of krill, salps, and mesozooplankton were analyzed. Cluster analysis allowed us to distinguish two areas differing in the ratio of salps, krill, and mesozooplankton. The maximum zooplankton biomass (1285.96 mg/m^3) was recorded in the central deep-water area and in the Antarctic Sound due to high abundance of salps, contributing up to 98.5% and 98.7% of the total zooplankton abundance, in both areas, respectively. The maximum euphausiid biomass (208.29 mg/ m^3) was found in the western part of the study site, where *Euphausia superba* dominated. According to the ICS index, the significance of *Salpa thompsoni* in zooplankton was on average 2.35 times higher than *E. superba*. The concentration of oxygen, temperature, and turbidity affected zooplankton structure. Principal component analysis (PCA) indicated that the first two principal components, related to transparency, oxygen, salinity, and temperature, explained 93.4% of the total variance in the zooplankton structure. Results of the correlation and principal component analysis revealed the key role of oxygen and temperature in the development of zooplankton in the Powell Basin.

Keywords Antarctica · Krill · Powell Basin · Salps · Zooplankton structure

V. A. Yakovenko (✉) · N. V. Shadrin · E. Z. Samyshev · N. I. Minkina
A. O. Kovalevsky Institute of Biology of the Southern Seas, Russian Academy of Sciences, Sevastopol, Russia

V. A. Spiridonov
Shirshov Institute of Oceanology, Russian Academy of Sciences, Moscow, Russia

K. M. Gorbatenko
Pacific Branch of VNIRO (TINRO), Vladivostok, Russia

E. G. Morozov et al. (eds.), *Antarctic Peninsula Region of the Southern Ocean*, Advances in Polar Ecology 6, https://doi.org/10.1007/978-3-030-78927-5_27

27.1 Introduction

The Atlantic sector of Antarctica is the area of the greatest abundance of Antarctic krill (*Euphausia superba* Dana, 1851) (Mackintosh 1972; Murphy et al. 2007). Antarctic krill, as well as other zooplankton species, feeds on phytoplankton. *Euphausia superba* is the keystone species in the Southern Ocean ecosystem, and plays a crucial role in the Antarctic food web providing an important source of food for enormous populations of seabirds, penguins, and marine mammals (Loeb et al. 1997; Ross and Quetin 1995; Schofield et al. 2017; Smith et al. 2008; Steinberg et al. 2015). The ability to reproduce at a fast rate allows *E. superba* to create an immense biomass varying from 61 million tons to 600 million tons (Atkinson et al. 2019; McQuillan 1962; Nemoto 1972; Samyshev 2002). Furthermore, krill contribute to the biological pump by production of faecal pellets (Mauchline and Fisher 1969). The highest concentrations of krill were found in the northwestern part of the Antarctic Peninsula as well as in the Scotia and Weddell seas where the eastward-flowing Antarctic Circumpolar Current (ACC) and waters from the Weddell–Scotia Confluence resulted in a strong advective flow, intense eddy activity, and mixing (Hewitt et al. 2004; Murphy et al. 2007). The Weddell Sea is a marginal sea of the Southern Ocean that is situated along the coast of Antarctica south of the Atlantic Ocean. The continental shelf along the eastern coast of the Antarctic Peninsula and the Antarctic continent covers about a quarter of the total Weddell Sea area. The ecosystem of the Weddell Sea deserves special study due to many aspects. The Weddell Sea is characterized by an extremely cold environment, the biota lives here most of the year in the presence of ice, and this sea is very important for biodiversity conservation in the Southern Ocean (Teschke et al. 2020). The Weddell Sea is a region of high phytoplankton productivity due to the mixing of micronutrients in the surface waters. The upper layer of cyclonic outflow of the Weddell Sea gyre over the shelf regions becomes enriched in sediment-derived nutrients. The northerly outflow of the Weddell Sea shelf water affects the deep-water Powell Basin located in the northeastern part of the Weddell Sea. The Powell Basin is isolated from the Weddell Sea by seamounts with the associated Weddell Front and characterized by cyclonic circulation (Heywood et al. 2004). A high abundance of *E. superba* was reported in the Powell Basin (Kasatkina et al. 2004; Siegel et al. 2004; Siegel and Watkins 2016); therefore, it is likely that krill juveniles from the western part of the Weddell Sea accumulate here.

In addition to the regular seasonal and interannual fluctuations in krill abundance associated with hydrological features in various years, there has been a decrease in krill catches up to 80% in recent decades in various regions of Antarctica, observed together with a significant increase of salp populations (*Salpa thompsoni* Foxton, 1961) and (*Ihlea racovitzai,* Van Beneden and De Selys Longchamp, 1913). The main cause of the decline in krill stock and the increase in salp abundance was the increase in the water temperature as a result of global warming and the retreat of Antarctic ice from the tip of the Antarctic Peninsula, because ice algae are important for krill recruitment.

It should be noted that the summer water temperature near the Antarctic Peninsula increased by 1.3 °C during a period of 50 years (Meredith and King 2005). Even the deep-water layers of the Weddell Sea warmed up by 0.032 °C over a decade (Whitehouse et al. 2008). Thus, considering the commercial and trophic importance of krill, it is important to study the abundance and biomass of krill, salps, and mesozooplankton in the Powell Basin that allows us to verify the hypothesis of the transport of juvenile krill from the Weddell Gyre to the Powell Basin.

27.2 Materials and Methods

In the Powell Basin area, the samples were collected in January–February 2020 during cruise 79 of the R/V "*Akademik Mstislav Keldysh*" (AMK) (Fig. 27.1). In a network of stations located in the Powell Basin (Fig. 27.1), the abundance, species composition, and biomass of macro- and mesozooplankton were analyzed using a double Bongo net (mouth area 0.6 m^2, mesh size 0.300 mm) hauled vertically between a depth of 200 m (or less in shallower stations) and the surface. The volume of water filtered was measured with sensing device placed at the mouth of the net. The samples were mixed and homogenized in a container, and then concentrated and fixed in 4% formalin in sea water (final concentration) for abundance, taxonomic, and biomass studies.

Biological analysis of krill was carried out after fixing the samples with formalin and reaching a constant weight of crustaceans based on the formalin concentration. Biological analysis of krill specimens included measurement of their individual length, mass, sex determination, and maturity stage (the degree of development of external and internal genitals). The length of krill crustaceans was measured with an

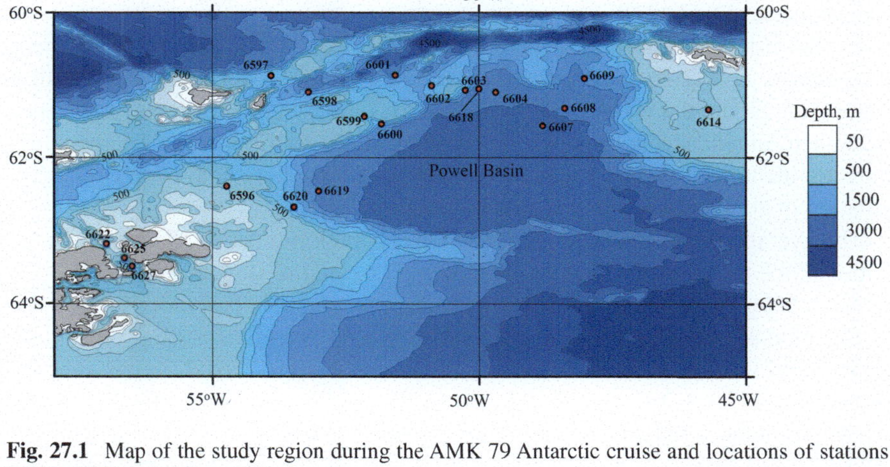

Fig. 27.1 Map of the study region during the AMK 79 Antarctic cruise and locations of stations (6597–6627) in the Powell Basin

accuracy of 1 mm from the outer edge of the orbit to the end of the telson. The individual mass of krill was determined by weighing the crustaceans dried with filter paper, on a torsion balance (WT-1000) (SC-CCAMLR 2010). The maturity stages of krill were determined according to the scale proposed by Makarov and Denys (1980).

Each individual salp was measured and weighted. Zooplankton samples were fixed with 4% formaldehyde and then identified under binocular MBS-9 (LOMO, Saint Petersburg, Russia) using a Bogorov chamber.

The dominant species were identified by the index of coenotic significance (ICS) according to formula (27.1):

$$\mathrm{ICS} = p\sqrt{B}, \tag{27.1}$$

where *p* is species occurrence (%), and *B* is species average biomass (g/m^3), respectively. This index indicates the significance of a given species in the community (Mordukhay–Boltovskoy 1975).

Data processing and statistical analyses were performed using the MS Excel 2010 program and STATISTICA 12.0 (StatSoft) software package. The PCA analysis, multiple regression analyses, and correlation coefficients between abundance and biomass and hydrological and hydrochemical parameters were calculated with the standard MS Excel 2010 program. Cluster analysis was conducted in the STATISTICA 12.0 (StatSoft) software package. This analysis was applied to visualize similarities in the physicochemical parameters as well as macro- and mesozooplankton species between the stations.

27.3 Results

In the Powell Basin, zooplankton taxa consisted of 7 phylums, 10 classes, 14 orders, and 34 species (Table 27.1).

In the order Euphausiacea, there were five species, but only three of them (*E. superba, E. frigida, E. crystallorophias*) belong to the designation of «krill». Among the phyla, the greatest number of species belonged to Arthropoda (17 species), then Annelida (four species), the phylums Mollusca, Chordata, Chaetognatha, Cnidaria included 3 species each and the phylum Ctenophora one species. The distribution of zooplankton biomass was not uniform.

The average biomass varied from 0 mg/m^3 at station 6620 to 2777.6 mg/m^3 at station 6599; the average value was 760.36 ± 777.1 mg/m^3 (Fig. 27.2).

Note. 6596 – 6627 are the designations of the stations of observations.

The biomass of salps was the maximal compared to krill as well as other zooplankton groups, but it varied to the greatest extent from 0–2761.2 mg/m^3, making on average 737.3 ± 767.5 mg/m^3. Between the stations, the percentage of *Salpa thompsoni* in the zooplankton community varied from 0% to 98.47 %. The biomass of krill varied from 0 to 496.3 mg/m^3, on average 40.6 ± 110.1 mg/m^3, while the

Table 27.1 List of collected species in the Powell Basin

Phylum Arthropoda	**Phylum** Chordata
Class Malacostraca	**Class** Thaliacea
Order Euphausiacea	**Order** Salpida
Species *Euphausia superba* Dana, 1851	**Species** *Salpa thompsoni* Foxton, 1961
E. frigida Hansen, 1911	*Ihlea racovitzai* Van Beneden and De Selys Long, 1913
E. crystallorophias Holt and Tattersall, 1906	**Phylum** Chordata
Thysanoessa macrura G.O. Sars, 1883	**Class** Appendicularia
Th. vicina Hansen, 1911	**Order** Colepata
Phylum Arthropoda	**Species** *Oikopleura vanhoeffeni* Lohmann, 1896
Class Hexanauplia	**Phylum** Chaetognatha
Order Calanoida	**Class** Sagittoidea
Species *Rhincalanus gigas* Brady, 1883	**Order** Aphragmophora
Calanoides acutus Giesbrecht, 1902	**Species** *Sagitta maxima* Conant, 1896
Calanus propinquus Brady, 1883	*Sagitta gazellae* Ritter-Záhony, 1909
Calanus simillimus Giesbrecht, 1902	**Phylum** Chaetognatha
Metridia gerlachei Giesbrecht, 1902	**Class** Sagittoidea
Paraeuchaeta antarctica Giesbrecht, 1902	**Order** Phragmophora
Gaidius sp. Giesbrecht, 1895	**Species** *Eukrohnia hamata* Möbius, 1875
Phylum Arthropoda	**Phylum** Cnidaria
Class Malacostraca	**Class** Hydrozoa
Order Amphipoda	**Order** Siphonophorae
Species *Themisto gaudichaudii* Guérin, 1825	**Species** *Dimophyes arctica* Chun, 1897
Primno macropa Guérin-Méneville, 1836	*Diphyes chamissonis* Huxley, 1859
Cyllopus magellanicus Dana, 1853	*D. dispar* Chamisso & Eysenhardt, 1821
Scina antarctica Wagler, 1926	**Phylum** Ctenophora
Phylum Arthropoda	**Class** Nuda
Class Malacostraca	**Order** Beroida
Order Mysidacea	**Species** *Beroe* sp. Muller, 1776
Phylum Arthropoda	**Phylum** Annelida
Class Ostracoda	**Class** Polychaeta
Order Halocyprida	**Order** Phyllodocida
Species *Conchoecia* Dana, 1849	**Species** *Sagitella* sp. Wagner, 1872
Phylum Mollusca	*Tomopteris pacifica* Izuka, 1914
Class Gastropoda	*Pelagobia longicirrata* Greeff, 1879
Order Pteropoda	**Phylum** Annelida
Species *Limacina helicina* Phipps, 1774	**Class** Polychaeta
Clio antarctica Dall, 1908	**Order** Sabellida
Spongiobranchaea australis d'Orbigny, 1836	Species *Hydroides* sp. Gunnerus, 1768

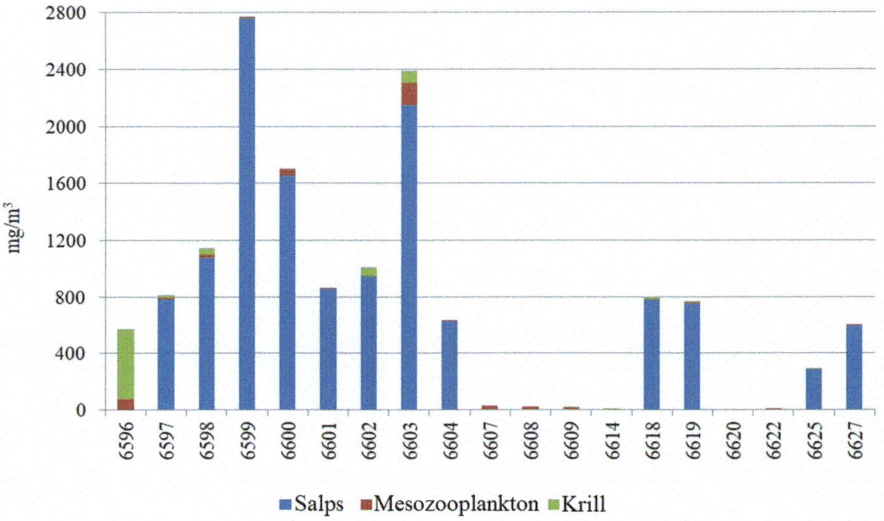

Fig. 27.2 Biomass of salp (*S. thompsoni, I. racovitzae*), krill (*E. superba, E. frigida*), and other groups of zooplankton (mesozooplankton) in the Powell Basin.

biomass of mesozooplankton varied from 0 to 158.4 mg/m^3, on average 21.2 ± 36.9 mg/m^3. The maximum biomass was observed in the central deep-water region (stations 6597–6604 and 6618–6619), as well as at stations 6625–6627, where salps were absolutely dominant, amounting to 97.4% in biomass.

The distribution of zooplankton abundance varied greatly among the study sites (Fig. 27.3).

Note. 6596–6627 are the designations of the stations of observations.

At other stations, krill or mesozooplankton prevailed, with maximal biomass up to 496.3 mg/m^3 in the western part of the study site where juvenile krill developed significantly. The share of krill in the total zooplankton abundance was only 3.1%, whereas the proportion of salps was 39.5% and the proportion of mesozooplankton was 57.4%. According to ICS, the species *S. thompsoni* dominates and its ICS exceeds 2.35 times that of *E. superba* (Fig. 27.4).

Due to the considerable dominance of salps, the ICS distribution of the species was very uneven.

The size of species *Euphausia frigida, E. crystallorophias,* and *Thysanoessa macrura*, which are Antarctic circumpolar neritic species, varied from 10 to 20 mm. The species Thysanoessa *vicina* was encountered occasionally and was represented by crustaceans ranging in size from 12 to 15 mm. Of the copepods, the *Calanoides acutus* species had the greatest weight according to the ICS; species *Calanus propinguus, Rhincalanus gigas*, and *Metridia gerlachei* were also found. In the Powell Basin, the biomass of copepods was low: from 0.38 to 2.24 mg/m^3 and composed from 0.03% to 9.47% of zooplankton biomass. Siphonophores were represented by three species, of which *Dimophyes antarctica* and *Diphyes chamissonis* ranked in

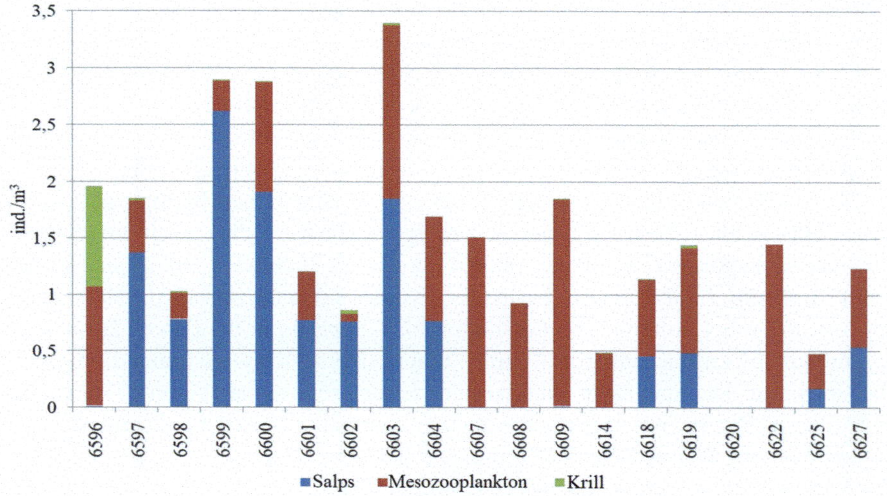

Fig. 27.3 Abundance of salp (*S. thompsoni, I. racovitzae*), krill (*E. superba, E. frigida*), and other groups of zooplankton (mesozooplankton) in the Powell Basin.

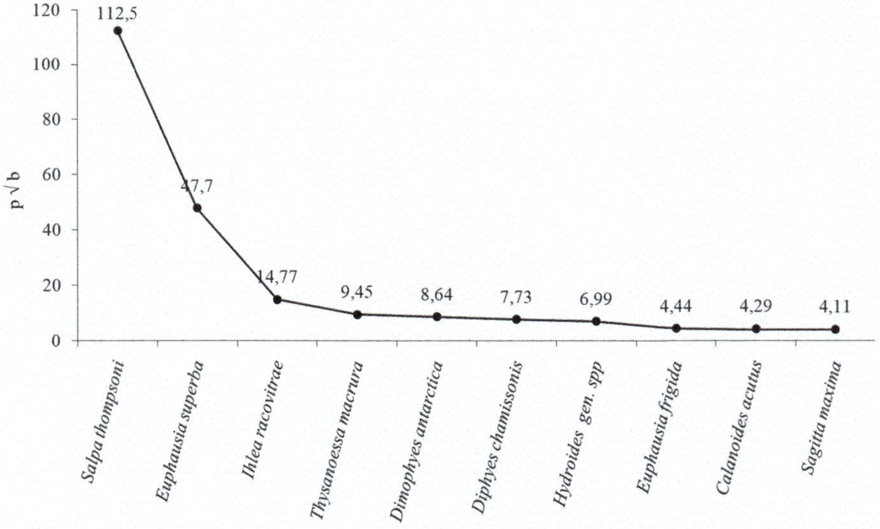

Fig. 27.4 Structure of zooplanktocenosis in the Powell Basin according to the ICS index

the fourth and fifth places, respectively, according to the ICS. Chaetognaths, which are predators, mainly on copepods or euphausiids larvae, were represented by three species: *Sagitta maxima, Sagitta gazellae,* and *Eukrochia hamata*. The species *Sagitta maxima* ranked tenth according to the ICS, but the biomass of sagitta was relatively low, from 0.08 to 1.23 mg/m^3, and their proportion ranged from 0.01% to 5.19%. The biomass of hyperiids being another group of predators, was low, from 0 to 0.38 mg/m^3. Hyperiids were represented by four species: *Themisto gaudichaudii,* which was the most abundant, as well as *Primno macropa, Cyllopus magellanicus,* and *Scina antarctica.* It is likely that low occurrence and abundance of the species was associated with the low abundance of copepods and euphausiids.

The size of species *Salpa thompsoni* varied from 5 to 120 mm, with a prevalence of specimens from 30 to 40 mm (25.4%) and from 40 to 50 mm (26.0%) (Fig. 27.5).

Many salps were confined to deep-water stations due to their biotopic allocation in a 1500-m thick layer of water according to literature data (Voronina 1998). Another salp species *I. racovitrae* ranked third in the ICS index and its size varied from 2 to 6 mm. *I. racovitrae* is a salp species confined to the surface (0–50 m deep) layer of the water column. The population of *E. superba* in the Powell Basin was represented by crustaceans from 10 to 53 mm in size, which were found throughout the study site. The maximum abundance of *E. superba* was at station 6596, where juveniles of 20–30 mm predominated; they were transported here, probably by waters from the Weddell–Scotia Confluence. The size of species *E. superba* varied from 10 to 53 mm, with prevalence of specimens from 40 to 50 mm (35.2%); those from 30 to 40 mm (33.5%) predominated in the size structure of krill of the Powell Basin (Fig. 27.6).

The correlation between mesozooplankton abundance and turbidity was high (Table 27.2).

Krill biomass positively correlated significantly with oxygen content and negatively with temperature (Table 27.2). The abundance of zooplankton, primarily copepods, increases with a decrease in the number of salps and krill, which feed not only on phytoplankton, but also on small zooplankton (Loeb et al. 1997). Salps consume phytoplankton competing with euphausiids. This explains the low inverse correlation between the abundance of salps and krill, which was $r = -0.23$, but, however, was not significant. Multiple regression analysis of the krill, salps, and mesozooplankton indices and oceanographic parameters revealed a significant cumulative effect of temperature, oxygen, salinity, and turbidity on krill abundance (*p*-value $<$ 0.001) and biomass (p-value $<$ 0.001). The oxygen level exerted a

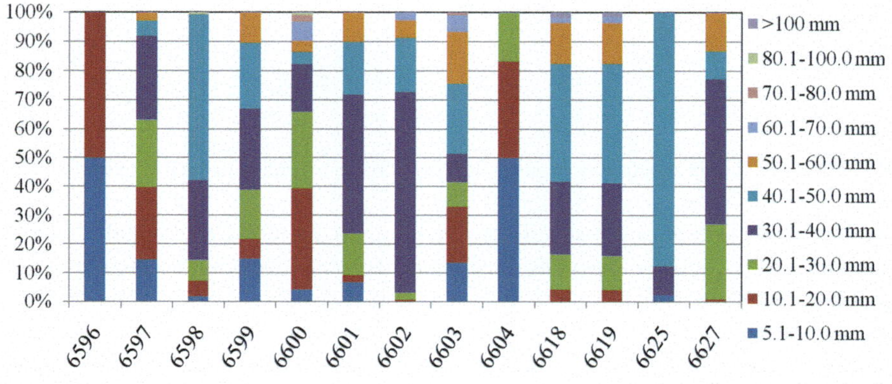

Fig. 27.5 Size structure of *S. thompsoni* in the Powell Basin (6596–6627 are station numbers)

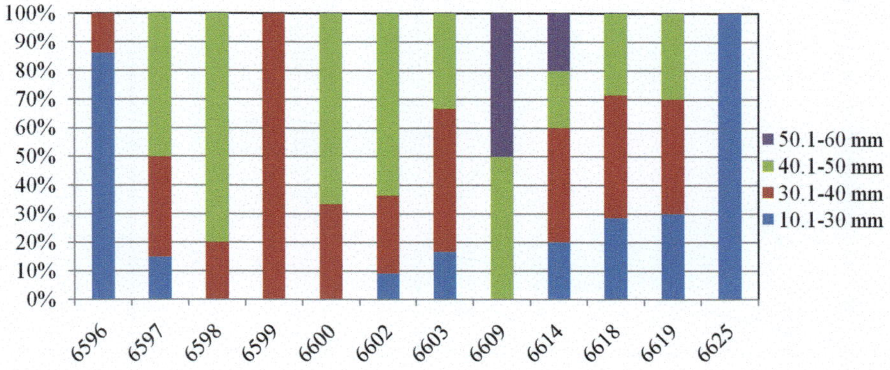

Fig. 27.6 Size structure of krill (*Euphausia superba*) in the Powell Basin.

Table 27.2 Correlation coefficient matrix for the main oceanographic parameters and quantitative indicators of krill, salps, and mesozooplankton in the Powell Basin (significant values are in bold)

	Temperature	Salinity	Oxygen	Turbidity
Krill abundance	−0,331	0,302	0,392	−0,065
Krill biomass	**−0,402**	0,378	**0,418**	−0,085
Salp abundance	−0,254	0,044	0,152	−0,162
Salp biomass	−0,328	0,107	0,173	−0,158
Mesozooplankton abundance	0,333	0,137	−0,309	**0,662**
Mesozooplankton biomass	−0,282	0,319	0,187	0,028

significant stimulatory effect (p-value ≤ 0.05) on both abundance and biomass of krill and salps. Thus, temperature decreased the combined effect of oxygen, turbidity, and salinity, which could be attributed to the delayed effect of temperature on the quantitative indicators of krill, salps, and mesozooplankton of the Powell Basin compared with the influence of oxygen, which determines the level of metabolism. Thus, oxygen content plays a leading role in the determination of krill development, especially in the conditions of the Powell Basin.

The PCA was used to assess the most important factors explaining the variation in the environmental data set (Fig. 27.7). The PCA is an indirect ordination technique for obtaining a low-dimensional representation of multivariate data, such that the data can be presented visually in a two-dimensional PCA correlation biplot. The PCA of the physicochemical variables and the abundance and biomass of krill, salps, and mesozooplankton through the first two components explained 93.4% of the total variance.

The PCA confirmed the stimulatory role of oxygen and salinity and the inhibitory effect of temperature and turbidity on both abundance and biomass of krill and salps. In addition, salinity was the main parameter (with a loading of 85.4%) in the second principal component and explained 24.4% of the total variance in the data. Nevertheless, turbidity had a positive effect on mesozooplankton abundance, which was confirmed by the correlation analysis. Thus, PCA showed that, of the

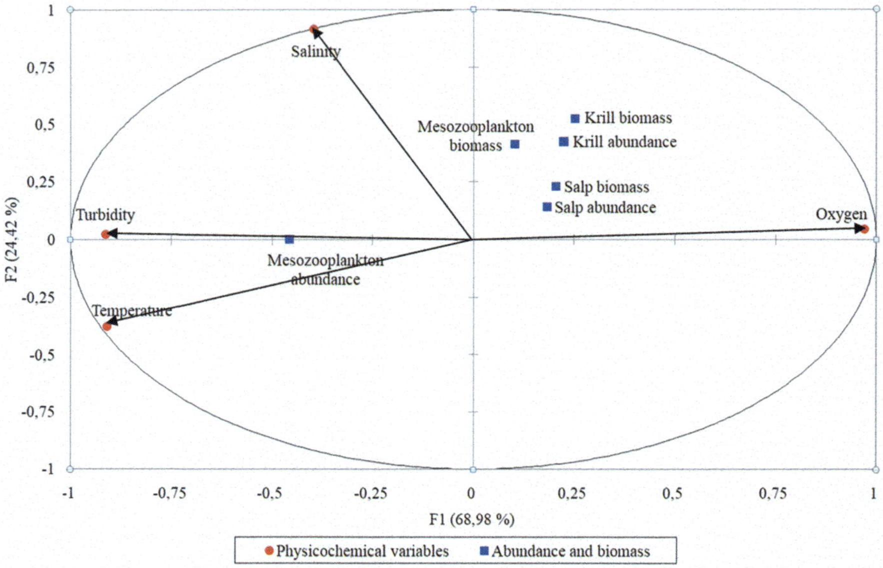

Fig. 27.7 Correlation biplot diagram based on the principal component analysis (PCA) of the physicochemical variables and the abundance and biomass of krill, salps, and mesozooplankton in the Powell Basin

physicochemical parameters, only temperature was negatively correlated with abundance and biomass of krill, salps, and mesozooplankton in the Powell Basin.

Cluster analysis taking into account all physicochemical and biotic parameters did not show complete coincidence of physicochemical parameters on the one hand and quantitative indicators of zooplankton on the other. (Fig. 27.8). Cluster analysis conducted on the basis of the parameters of hydrobionts showed similarity among two groups of stations. The first cluster included stations 6607, 6608, 6609, 6614, 6620, and 6596, whereas the second cluster included stations 6597, 6618, 6601, 6602, 6598, 6604, 6600, 6603, and 6599. The stations of the first cluster had lower total biomass but higher quantitative characteristics of krill and mesozooplankton compared to the second cluster due to high salp development at the stations of the latter.

27.4 Conclusions

Two subareas differing in the ratio of salps, krill, and mesozooplankton exist in the study site based on the cluster analysis. The maximum biomass of zooplankton (1285.96 mg/m^3) was recorded in the central deep-water area and in the Antarctic Sound due to salps. Results of the correlation and principal component analysis showed the key role of oxygen and temperature in the development of zooplankton

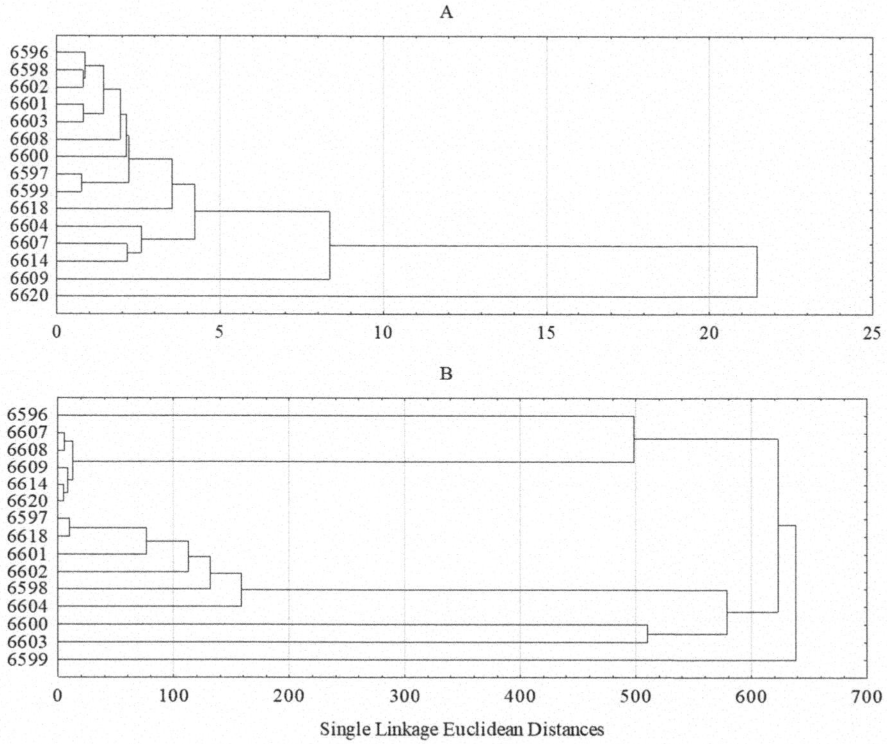

Fig. 27.8 Cluster analysis of physicochemical variables (a) as well as abundance and biomass of krill, salps, and mesozooplankton (b) in the Powell Basin in January–February 2020

in the Powell Basin. To fully understand the "salps–krill" relationships, wider and deeper studies are needed.

Acknowledgments This research was conducted in the framework of the state assignments of A.O. Kovalevsky Institute of Biology of the Southern Seas of the RAS (№ AAAA-A19-119100290162-0 and № AAAA-A19-119100790153-3), Shirshov Institute of Oceanology of the RAS (№ 0128-2019-0008).

References

Atkinson A, Hill S, Pakhomov EA, Siegel V, Reiss CR, Loeb VJ, Steinberg DK, Schmidt K, Tarling GA, Gerrish L, Sailley SF (2019) Krill (*Euphausia superba*) distribution contracts southward during rapid regional warming. Nat Clim Chang 9:142–147

Hewitt RP, Watkins J, Naganobu M, Sushin V, others (2004) Biomass of Antarctic krill in the Scotia Sea in January/February 2000 and its use in revising an estimate of precautionary yield. Deep-Sea Res II 51:1215–1236

Heywood KJ, Naveira Garabato AC, Stevens P, Muench RD (2004) On the fate of the Antarctic Slope Front and the origin of the Weddell Front. J Geophys Res 109:C06021
Kasatkina SM, Goss C, Emery JH, Takao Y, Litvinov FF, Malyshko AP, Shnar VN, Berezhinsky OA (2004) A comparison of net and acoustic estimates of krill density in the Scotia Sea during the CCAMLR 2000 Survey. Deep-Sea Res II 51:1289–1300
Loeb V, Siegel V, Holm-Hansen O, Hewitt R, Fraser W, Trivelpiece W, Trivelpiece S (1997) Effects of sea-ice extent and krill or salp dominance on the Antarctic food web. Nature 387:897–900
Mackintosh NA (1972) Life cycle of Antarctic krill in relation to ice and water conditions. Discov Rep 36:1–94
Makarov RR, Denys CJ (1980) Stages of sexual maturity of *Euphausia superba* Dana. BIOMASS Handbook 11:11 p
Mauchline J, Fisher LR (1969) The biology of euphausiids. Adv Mar Biol 7:1–454
McQuillan H (1962) The Antarctic krill. West Fish 63(4)
Meredith MP, King JC (2005) Rapid climate change in the ocean west of the Antarctic Peninsula during the second half of the 20th century. Geophys Res Lett 32(19):604–609
Mordukhay–Boltovskoy FD (1975) Metodika izucheniya biotsenozov vnutrennikh vodoyemov. Nauka, Moscow. 240 p. (in Russian)
Murphy EJ, Watkins JL, Trathan PN, Reid K, others (2007) Spatial and temporal operation of the Scotia Sea ecosystem: a review of large-scale links in a krill centred food web. Philos Trans R Soc Sect B Biol Sci 362:113–148
Nemoto T (1972) History of research into the food and feeding of euphausiids. Proce R Soc Edinb Sect B Biol Sci 73:259–265
Ross R, Quetin L (1995) Distribution of Antarctic krill and dominant zooplankton west of the Antarctic peninsula. Antarct Res Ser 70:199–217
Samyshev EZ (2002) Antarctic krill and the structure of planktonic community in its distribution area. ECOSEA Publ. House, Sevastopol, 168p
SC-CCAMLR (2010) Report of the twenty-ninth meeting of the Scientific Committee (SC-CCAMLR-XXIX), Hobart, 426 p
Schofield O, Saba K, Coleman K, Carvalho F, Couto N, Ducklow H, Finkel Z, Irwin A, Kahl A, Miles T, Montes-Hugo M, Stammerjohn S, Waite N (2017) Decadal variability in coastal phytoplankton community composition in a changing West Antarctic Peninsula. Deep-Sea Res I 124:42–54
Siegel V, Watkins JL (2016) Distribution, biomass and demography of Antarctic Krill, Euphausia superb. In: Biology and ecology of Antarctic krill, Advances in Polar Ecology, vol 1, pp 21–100
Siegel V, Kawaguchi S, Ward P, Litvinov F, Sushin V, Loeb V, Watkins J (2004) Krill demography and large-scale distribution in the southwest Atlantic during January/February 2000. Deep-Sea Res II 51:1253–1273
Smith RC, Martinson DG, Stammerjohn SE, Iannuzzi RA, Ireson K (2008) Bellingshausen and western Antarctic Peninsula region: pigment biomass and sea-ice spatial/temporal distributions and interannual variability. Deep-Sea Res II 55:1949–1963
Steinberg DK, Ruck KE, Gleiber MR, Garzio LM, Cope JS, Bernard KS, Stammerjohn SE, Schofield OME, Quetin LB, Ross RM (2015) Long-term (1993–2013) changes in macrozooplankton off the Western Antarctic Peninsula. Deep-Sea Res I Oceanogr Res Pap 101:54–70
Teschke K, Pehlke H, Siegel V, Bornemann H, Knust R, Brey T (2020) An integrated data compilation for the development of a marine protected area in the Weddell Sea. Earth Syst Sci Data 12:1003–1023
Voronina NM (1998) Comparative abundance and distribution of major filter-feeders in the Antarctic pelagic zone. J Mar Syst 17:375–390
Whitehouse MJ, Meredith MP, Rothery P, Atkinson A, Ward P, Korb RE (2008) Rapid warming of the ocean around South Georgia, Southern Ocean, during the 20th century: forcings, characteristics, and implications for lower trophic levels. Deep-Sea Res I 55:1218–1228

Chapter 28
Application of Autonomous Underwater Vehicles for Research of Ecosystems in the Southern Ocean

Alexander Yu. Konoplin, Alexey I. Borovik, Denis N. Mikhailov, Yuriy V. Vaulin, Alexander F. Scherbatyuk, Alexey A. Boreiko, Roman A. Babaev, Dmitriy A. Bolovin, and Dmitriy I. Tregubenko

Abstract This paper considers a technology based on an autonomous unmanned underwater vehicle (AUV) MMT-3000, which was deployed in the integrated surveys of Antarctic ecosystems and deep-sea biological resources in the southern seas during the expedition onboard the R/V "Akademik Mstislav Keldysh" (cruise #79) in 2020. The proposed technology included also mission planning, software development, AUV operation, and organization of submersions and interactions with the mothership's crew. The report addresses issues such as upgrading the AUV MMT-3000 to provide visual evaluation of zooplankton density in the water column and distribution of benthic animals, as well as equipping this AUV with a set of sensors to measure the hydrophysical and hydrochemical characteristics of the environment. The main results obtained in the course of the deep-sea missions are presented.

Keywords Autonomous unmanned underwater vehicle · Deep-sea research · Marine expedition · Monitoring of marine benthic ecosystems · Navigation system · Software

A. Yu. Konoplin (✉) · A. I. Borovik · D. N. Mikhailov · Y. V. Vaulin · A. F. Scherbatyuk
A. A. Boreiko · R. A. Babaev · D. A. Bolovin · D. I. Tregubenko
Institute of Marine Technology Problems, Far Eastern Branch, Russian Academy of Sciences, Vladivostok, Russia
e-mail: konoplin@marine.febras.ru

E. G. Morozov et al. (eds.), *Antarctic Peninsula Region of the Southern Ocean*, Advances in Polar Ecology 6, https://doi.org/10.1007/978-3-030-78927-5_28

28.1 Introduction

Investigations into the deep-sea environment of the ocean using unmanned underwater vehicles of various types have been increasingly developed in recent years. The particularly urgent issues that the Antarctic deep-sea research commonly addresses are the impact of the current climate trends on the natural ecosystems of the Southern Ocean, modern biological productivity, and the structure and spatial organization of pelagic and benthic ecosystems. This provides crucial data to estimate the potential allowable consumption of biological resources and assess other forms of human impacts (Morozov et al. 2020).

To date, contactless monitoring of marine benthic ecosystems using unmanned remotely operated underwater vehicles (ROVs) and autonomous underwater vehicles (AUVs) has become one of the most promising approaches to deep-sea research. In the conditions of seamounts and strong currents, various research operations are successfully performed using a Comanche 18, a work-class ROV towed via a depressor unit (Filaretov et al. 2017, 2019) that couples the rigid wire rope of the winch with the flexible communication cable of the vehicle. When surveying large areas or long routes, these ROVs need the mothership to be moved in the proper direction, which is not always possible in heavy ice conditions. The use of vehicles of this class in integrated polar expeditions is limited by the strict requirements to the mothership, as well as by the high cost of transportation and loading of the multiton winch and other equipment of the robotic system onboard.

Furthermore, the extreme conditions of the Antarctic seas, with their great depths, rapid currents, complex bottom topography, and drifting icebergs and sea ice, substantially restrict the use of ROVs connected to the mothership by a flexible communication cable. Being exposed to negative hydrodynamic effects in the surrounding aquatic environment, this cable not only hinders ROV's movements (Filaretov et al. 2019), but can also be damaged around the propellers of the main propulsion and transverse thrusters of a support vessel.

Russian and foreign researchers have already accumulated substantial experience in deploying AUVs during Arctic and Antarctic expeditions (Morozov et al. 2020; Dowdeswell et al. 2003; McPhail 2007; Inzartsev et al. 2010). Even when operated under ice, AUVs can carry sets of sensors that provide a continuous representation of the physical and hydrochemical parameters of the environment. These vehicles can efficiently be used for both mapping bottom biocenoses and assessing abundance of aquatic organisms without taking them out of their habitat.

To conduct integrated biological studies (along with hydrophysical, hydrochemical, and other surveys) in the expedition aboard the R/V *Akademik Mstislav Keldysh* (cruise #79) in the Atlantic sector of the Antarctic, it was proposed to use an AUV MMT-3000 (Gornak et al. 2006; Borovick et al. 2018) designed by the Institute of Marine Technology Problems, Far Eastern Branch, Russian Academy of Sciences (IMTP FEB RAS).

To efficiently fulfill the research objectives of the expedition, the AUV MMT-3000 was upgraded, equipped with a set of sensors for measuring the environmental parameters, and set up with auxiliary equipment. A technology for performing research AUV operations, which includes deep-sea mission planning, software development, AUV operations, as well as organization of submersions and interactions with the motherships' crew, was designed and tested.

28.1.1 Upgrading the AUV MMT-3000

The AUV MMT-3000 was designed to conduct observation and search at depths of up to 3000 m (Gornak et al. 2006). The vehicle is normally equipped with an onboard integrated navigation system (Dubrovin et al. 2018), including a GPS unit, a depth sensor, a Doppler velocity log, a compass, an echo-sounding system, and a hydroacoustic navigation and communication system (HANCS), which makes possible not only real-time tracking of AUV motions, but also controlling the motion parameters (direction, speed, distance to the bottom, and depth) via telecommands sent by the operator from the above-water control post.

To carry out field studies, the AUV MMT-3000 was additionally equipped with FLCDRT-926 and FLNTU-665 fluorescence sensors, as well as with a CTD-NV-2406 profiler. To provide free water flow through these sensors during vehicle motion, they were fixed using a frame mounted on the upper part of the AUV hull parallel to its longitudinal axis (Fig. 28.1). A power battery and a single-board computer that logs data from the sensors were enclosed in a separate sealed container also fixed to the vehicle hull. When the AUV is above water, this control computer connects to the main AUV network via a USB-WiFi modem to synchronize time with the onboard computer of the vehicle and to upload the logged data to the operator's computer.

The mounted set of autonomous oceanographic sensors makes possible composing a table of measurements of the following marine environment parameters: concentration of dissolved organic matter (CDOM) and chlorophyll, conductivity, water temperature and turbidity, and sound speed. All measurements were matched to the AUV navigation data, including its current coordinates, heading, list angle, trim angle, depth, and speed.

During the preparatory works for operation in the Antarctic zone, a number of measures were taken to upgrade the navigation system of the AUV MMT-3000 to be deployed in a high-latitude region, including the one near the South Magnetic Pole, in the areas with low magnetic field strength, or in areas with a high degree of magnetic declination variability. In such areas, the error-of-heading measurements with the onboard MTI X-Sense magnetic compass increase and, thus, a decision was made to install a VG-035 fiber-optic gyroscope on the AUV to correct the magnetic heading. Off the Antarctic coast, the AUV was launched in the region with approximate coordinates 62° S, 54° W characterized by large magnetic declination (8.31° E), which was taken into account prior to launching the AUV.

The necessary accuracy of coordinate measurements aboard the AUV was provided by the use of the HANCS with an Evologics ultrashort base line (USBL). A

Fig. 28.1 The AUV MMT-3000 with sensors mounted on its hull aboard the R/V *Akademik Mstislav Keldysh*

S2C M 18/34 modem was installed on the AUV for data exchange via a hydroacoustic communication channel. The support vessel had an USBL S2C 18/34 modem with an integrated USBL HANCS antenna. When the AUV was traveling underwater, its coordinates were measured with the USBL HANCS on the mothership every 2–4 s in the real-time mode and transmitted to the vehicle via an acoustic channel for further correction of the vehicle's dead-reckoning system. A special feature of the USBL HANCS installed on the AUV MMT-3000 is that it includes an Applanix POS/MV V5 satellite inertial measuring system for determining the spatial orientation of the support vessel. This solution, as compared to measuring vessel's heading with magnetic compasses or inertial navigation systems, has certain advantages in Arctic or Antarctic conditions, because it provides a high accuracy of heading measurements (0.1°) irrespective of the geographical latitude of survey area.

The AUV MMT-3000 was also equipped with a digital photo system based on a Prosilica GC1380 camera with a short-pulse illumination source. The system makes possible taking high-quality images with a resolution of 2 MP and a frequency of up to one frame per two seconds. The time of all images taken with the camera is synchronized with the general AUV onboard time by the software. This makes it possible to obtain data of any accompanying parameters of vehicle's motion—coordinates, altitude above the bottom, depth, speed, etc. for each frame using the

IMTPLooker data postprocessing program. During the expedition, the IMTPLooker was upgraded to display a ruler in the photo frame in order to estimate the geometric sizes of underwater objects. In addition, a streaming image-processing mode was set up to upload the selected images with a specified downsampling and to display the information selected by the user in each photograph.

To obtain data on the bottom topography and bottom sediment characteristics, the AUV MMT-3000 normally includes a side-scan sonar (SSS) and an acoustic profiler designed by IMTP FEB RAS. The SSS has an operating frequency of 470 kHz and a viewing band of 70 m onboard, which provides a resolution with a 5 cm distance and 0.5° bearing. The acoustic profiler makes possible scanning the bottom to a depth of 40 m.

To increase the maneuverability of the AUV and provide its motion in the photography mode at a low speed (from 0.1 m/s) near the bottom (up to 0.5 m), a unit with a vertical transverse thruster having a thrust force of at least 7 kgf was designed. This unit was installed in the central part of the vehicle (Fig. 28.1), with the longitudinal axis of the thruster oriented perpendicular to the AUV longitudinal axis. The thruster makes possible taking high-quality photos without blurring, thanks to a more accurate stabilization of the altitude above the bottom, which makes it possible to identify and measure the size of benthic animals in the images taken.

A comprehensive upgrade of the AUV software with a few additional modules, developed on the platform of the RCE software, was performed prior to the expedition. Drivers for the photosystem and the set of autonomous oceanographic sensors were written; the dynamic model of the AUV was improved taking into account the new transverse thruster unit, and the high-level mission-planning system (Bagnitckii et al. 2011) was adjusted to the hardware and software of the AUV MMT-3000.

To provide efficient interaction with the mothership's crew, a navigation software interface for the ship's bridge was developed. This simple and intuitive web interface clearly displays the relative positions and the tracks of the AUV and the vessel, as well as all the parameters of vehicle motions.

Moreover, a number of other operations to install auxiliary equipment for the robotic system were performed as follows:

- Installing a container to store the laboratory equipment
- Preparing the hull structure, mechanical elements, and outboard AUV cables
- Making a pole for mounting a USBL HANCS antenna
- Setting up the control and communication post
- Making a beacon for the calibration of USBL HANCS
- Preparing a kit of spare parts, tools, and accessories of the AUV system

The sea-based tests of the AUV MMT-3000 in Patrokl Bay, Primorsky Krai, Russia, confirmed the operability of all the systems of the vehicle and its readiness for the research fieldwork.

28.1.2 Specifics of Undersea Operations

To fulfill the goals of the scientific research using the AUV, it is necessary to address a number of important issues such as surveying large deep-sea areas. Furthermore, this requires high-quality photography, SSS scanning, profiling, and measuring the parameters of the surrounding aquatic environment along vehicle's routes that will be conducted in the conditions of rough terrain and strong near-bottom currents.

Since 2007, the AUV MMT-3000 has been used in various deep-sea operations such as, in particular, hydrographic surveys for the project to lay undersea communication cables in the Sea of Okhotsk (Boreiko et al. 2019). Based on the accumulated experience, we propose an AUV-based technology for solving research problems that includes undersea operation planning, AUV operation, organization of submersions, and interactions with the mothership's crew.

Before launching the AUV, the study area is surveyed with the vessel's echo sounder; based on its measurements, a refined depth map for the selected polygons is composed, and the AUV mission (route and motion modes) is designed taking into account the objectives set forth. The route of the support vessel is planned in a way to keep the AUV always within the range of the USBL HANCS during the mission. The operability of all the AUV subsystems and the above-water control system is tested. The test consists of a simulation of the vehicle's mission that the operator runs to check whether all the AUV mechanisms and equipment respond properly to the mission commands and whether any failures of software and hardware might occur.

After the vessel arrives at the specified point of the study site, if weather conditions permit, the AUV is safely lowered, while the vessel is dynamically positioned at this point in a mode that can compensate for the wind- and current-induced drift (Fig. 28.2). Assistants on a motor boat manually pull the vehicle to a safe distance away in order to avoid its being driven to the blades of the main and transverse thrusters of the vessel. After that, the operator sends the command to launch the mission via a radio channel, and the AUV starts submersion.

Based on the navigation data, the graphical interface displays the tracks of motions of the support vessel and the AUV, as well as the telemetry data transmitted via the hydroacoustic communication system. This gives us the opportunity to set target points and indicate the location of underwater objects by inputting geographical coordinates or using the marks previously made in the loaded map of the survey area. This graphical interface is displayed not only on the screen of the AUV operator, but also on the onboard computer of the vessel's watch officer, which provides timely adjustments of the vessel's course in the real-time mode. Prior to sending the AUV along the planned route, the operator provides the mothership's crew with all necessary commands to move the vessel in a specified direction for a specified distance so that the AUV does not go beyond the range of USBL HANCS.

For mapping the benthic biocenoses and assessing the abundance of aquatic organisms without taking them out from their habitat, it was proposed to design the AUV missions using the method of similar ROV-based operations conducted in the Sea of Japan, Sea of Okhotsk, Bering Sea, and in the Pacific Ocean (Filaretov et al.

Fig. 28.2 Launching the AUV in the Bransfield Strait

2017, 2019; Galkin et al. 2017a, b). Quantification of objects during submersions requires performing photographic transects with a straight-lined motion of the AUV at a distance of about 1 m from the bottom and at a constant speed of approximately 0.5 m/s. The preliminary experimental surveys in Patrokl Bay showed that this method provides adequate information about the composition and distribution of animals larger than 1 cm.

For SSS scanning and obtaining data from the acoustic profiler and the echo-sounding system, the AUV should move at a speed of approximately 1.5 m/s and a distance of 7 m from the bottom. The obtained data can be used for mapping the survey area and determining the sediment composition and roughness of bottom topography. For the research studies of the integrated expedition, decisions about feasibility of deploying a Sigsbee trawl within the study site, as well as about dredging and use of grab samplers, should be made on the basis of this data.

Furthermore, the AUV equipment with a set of sensors for measuring the parameters of the surrounding aquatic environment makes it possible to record the hydrophysical and hydrochemical characteristics of the water along the AUV route and also during the submersion and ascent of the vehicle.

It should be noted that, simultaneously with an AUV mission, other outboard research operations, such as bottom grab sampling or dipping an oceanographic rosette sampler, may also be performed on the support vessel kept at a specified position.

After completing the mission, the AUV surfaces and transmits the coordinates of its location to the operator's post via a radio channel. After being found, the vehicle is towed by a motor boat to the vessel and lifted aboard.

28.1.3 Results of Surveys Conducted during the Expedition

The technological solutions presented here were applied in the study of the biodiversity of benthic communities in extreme conditions of the southern seas during the expedition aboard the R/V *Akademik Mstislav Keldysh* (cruise #79). Prior to the beginning of works, the AUV oceanographic sensors and USBL HANCS were calibrated. A test mission was also run to ballast the AUV in the mode of hovering at depths of 0.5 and 5 m. During this mission, the stability of the vehicle and the operability of the main and transverse thrusters were evaluated. Based on the results of the evaluation, the AUV was ballasted for operation in the Antarctic conditions.

Below, we consider the results of a submersion of the AUV MMT-3000 at a station with coordinates 61.43° S, 52.22° W and a depth of 570 m as an example of the work mission. During this mission, the AUV made two straight-lined parallel 1-km runs (transects) spaced 100 m apart. Figure 28.3 shows the tracks of the vehicle on a 3D model of bottom topography built using the data from the onboard echo sounder. While moving down these transects, the vehicle performed profiling and SSS scanning of the bottom topography. Then, moving along a different specified course at a height of 1 m above the bottom, it also took a series of photographs of the sea bottom. The images of aquatic organisms taken along the photographic transect made it possible to assess the pattern of their distribution. The software provides cross-linking the photographic images taken and estimating the area of the bottom surface in each image. Based on the data collected (Fig. 28.4), the benthic team of the expedition made a decision to carry out a trawl haul (with a Sigsbee trawl) along the AUV track. The image received from the SSS shows a strange object (similar to a rod) with a height of 3.16 m. According to the profilogram, the sediments to a depth of 10–15 m consist mainly of dense sandstone with a small amount of pebble that has a small admixture of silt on the surface.

The parameters of the aquatic environment were also measured with a set of oceanographic sensors (CTD, CDOM, Chlorophyll-a, and turbidity sensor) during the AUV missions.

As an example, describing the evaluation of the bottom topography roughness, Fig. 28.5 shows a segment of the AUV motion in the mode of hovering at a specified altitude above the bottom using the transverse thrusters. Within this 187-m segment, elevation variations did not exceed 3 m.

Figure 28.6 shows the results of an autonomous operation of the AUV onboard integrated navigation system. In the first half of the mission, there was a period of temporary loss of communication with the support vessel, during which the coordinates of the vehicle were not corrected by USBL HANCS data. Within this segment, the error of dead reckoning by the AUV onboard navigation system can be

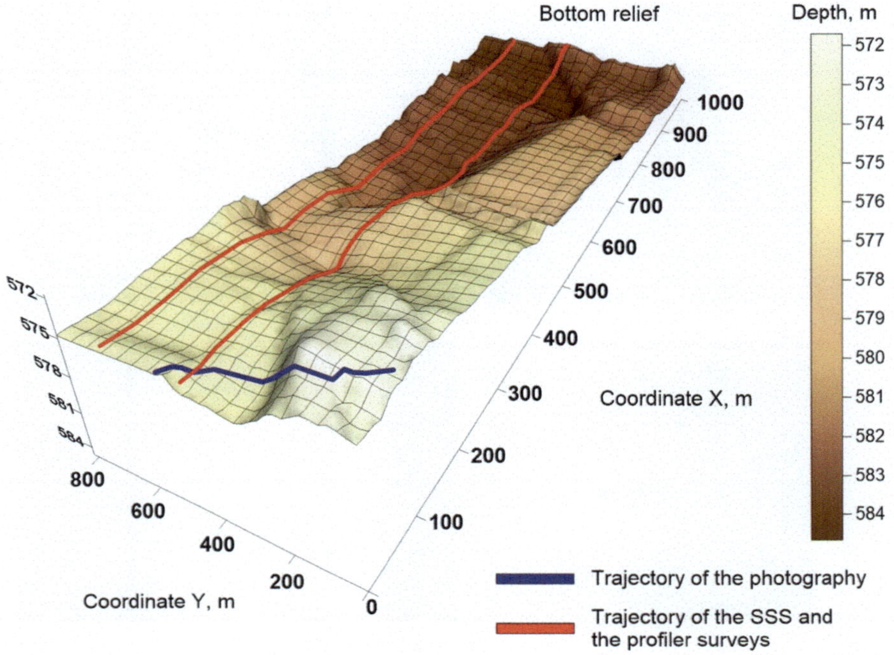

Fig. 28.3 A 3D model of bottom topography inferred from the AUV echo sounder data

estimated. The duration of the autonomous operation of this system without corrections from USBL HANCS lasted for 17.5 min. During this time, the accumulated error of coordinate determination was equal to 25 m, which constitutes 3% of the distance travelled. These values demonstrate the satisfactory operation of the AUV onboard navigation system in the fully autonomous mode under the current conditions. The loss of communication between the vessel and the AUV probably occurred due to some of the vessel's maneuvers and the insufficient depth of the USBL HANCS antenna installed on the external submersible pole and vertically fixed to the vessel's hull side.

28.2 Conclusions

Following are the main results of the deployment of the AUV MMT-3000 during the expedition aboard the R/V *Akademik Mstislav Keldysh* (cruise #79) in the Atlantic sector of the Antarctic.

The technological solutions implemented for upgrading the vehicle and in its preliminary tests were used and improved in the expedition to carry out a wide range of research works in harsh Antarctic conditions. The marine biology, hydrophysics, and hydrochemistry studies, as well as the survey of the topography and structure of the sea floor, were of primary importance.

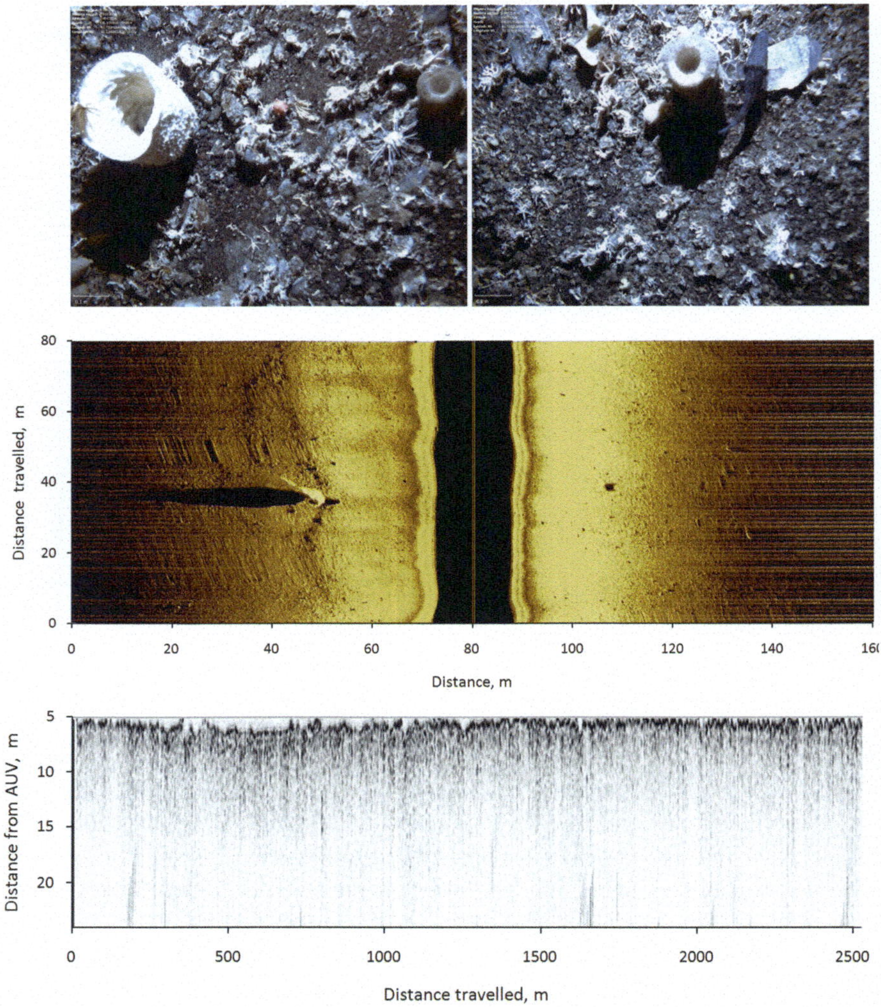

Fig. 28.4 Photos of benthic organisms at a depth of ~570 m at 61.4325° S, 52.2258° W (top panel); image of a vertical object like a rod in the bottom (middle panel); profile of sediments below the AUV to a depth of ~20 m below the bottom (bottom panel)

The technological results achieved are as follows:

- The AUV research equipment provides visual evaluation of the distribution of benthic animals in coastal ecosystems of the Antarctic and deep-sea biological resources in the southern seas.
- The efficiency of the software and special algorithms for assessing the distribution and surveying benthic bioresources over a large area, as well as for estimating the zooplankton density in the water column, has been confirmed.

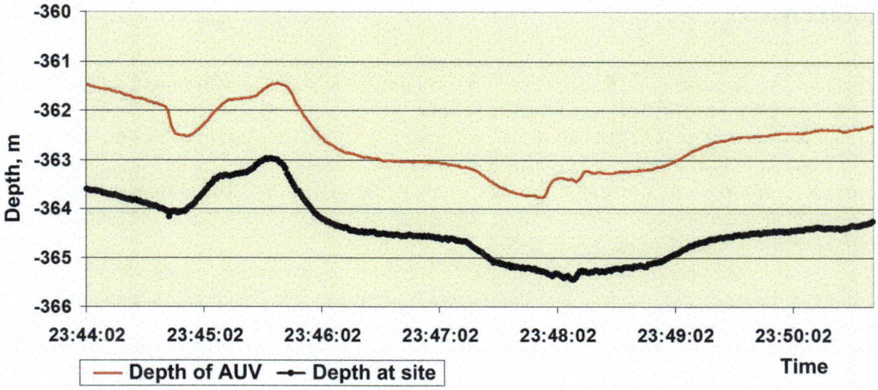

Fig. 28.5 A segment of AUV movement in the mode of hovering at a specified altitude above bottom. The figure shows a graph ~187 m long

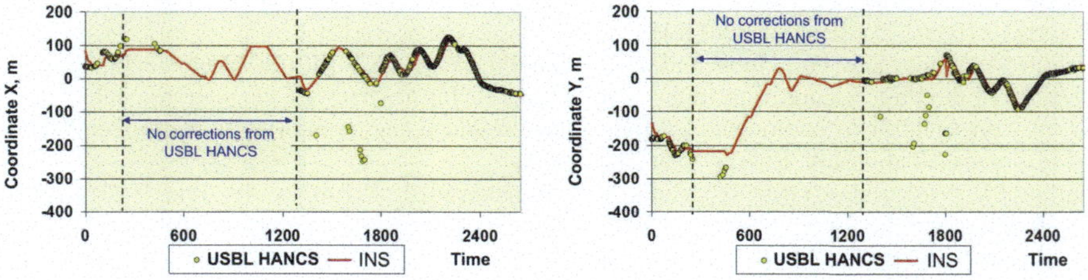

Fig. 28.6 Coordinates of the AUV based on the data of USBL HANCS and the onboard integrated navigation system (INS)

- The set of hydrophysical and hydrochemical sensors provides assessment of the state of the environment.
- For a general environmental assessment, there is an opportunity to measure the chlorophyll-a concentration in the upper water layers, which determines the conditions for primary production in the Antarctic ecosystems.

Acknowledgments The authors express their sincere gratitude to the crew of the R/V *Akademik Mstislav Keldysh* for the highly professional support of the AUV launches. The authors also thank all members of the Institute of Marine Technology Problems, Far Eastern Branch, Russian Academy of Sciences, who participated in the preparatory works, upgrading, and preliminary tests of the AUV MMT-3000.

Funding The improvement of the operator's post software and the algorithms for processing data from the onboard sensors were implemented with the financial support of the Grant of the Ministry of Science and Education, Russian Federation **13.1902.21.0012** «Fundamental Problems of Study and Conservation of Deep-Sea Ecosystems in Potentially Ore-Bearing Areas of the Northwestern Pacific» (agreement №**075-15-2020-796**).

References

Bagnitckii A, Inzartsev A, Senin R. (2011) Facilities of AUV search missions planning. OCEANS'11 – MTS/IEEE Kona, Program Book

Boreiko AA, Kushnerik AA, Mikhailov DN, Scherbatyuk AF (2019) Current experience for usage of some AUV developed in IMTP FEB RAS. Extreme Robotics 1(1):226–232

Borovick A, Inzartsev A, Kamomiy A, Scherbatyuk A, Sporyshev M (2018) Some results of the ecological investigation in Zolotoy Rog Bay using AUV MMT 3000. 2018 OCEANS MTS/IEEE Kobe Techno-Oceans, OCEANS

Dowdeswell JA, Evans J, Mugford R et al (2003) Autonomous underwater vehicles and investigations of the ice-ocean interface in Antarctic and Arctic waters. J Glaciol 54(187):661–672

Dubrovin F, Scherbatyuk A, Vaulin Y (2018) Some results of operation for the AUV MMT 3000 mobile navigation system on long and deep-water trajectories. 2018 OCEANS – MTS/IEEE Kobe Techno-Oceans, OCEANS

Filaretov VF, Konoplin NYu, Konoplin AYu (2017) Approach to creation of information control system of underwater vehicles. In: Proceedings of 2017 international conference on industrial engineering, applications and manufacturing (ICIEAM), S.-Petersburg, Russia, May 16–19

Filaretov VF, Konoplin AYu, Konoplin NYu (2019) Development of intellectual support system for ROV Operators. IOP Conf Ser Earth Environ Sci 272:032101. https://doi.org/10.1088/1755-1315/272/3/032101

Galkin SV, Vinogradov GM, Tabachnik KR, Konoplin AY, Ivin VV (2017a) Biological investigations of the Piip Volcano with ROV "Comanche 18". In: Abstracts of the international conference on scientific technological developments of research and monitoring of marine biological resources. Vladivostok, Russia, May 22–24, p 41

Galkin SV, Vinogradov GM, Tabachnik KR, Rybakova EI, Konoplin AY, Ivin VV (2017b) Megafauna of the Bering Sea slope based on observations and imaging from ROV "Comanche". Marine Imaging Workshop, Kiel, Germany, 2017

Gornak VE, Inzartsev AV, Lvov OY, Matvienko YV, Scherbatyuk AF (2006) MMT 3000 – Small AUV of New Series of IMTP FEB RAS. OCEANS 2006, Boston

Inzartsev AV, Kamornyi AV, Kiselev LV, Matvienko YuV, Pavin AM, Rylov NI, Rylov RN, Vaulin YuV (2010) The integrated navigation system of an autonomous underwater vehicle and the experience from its application in high Arctic latitudes. Gyroscopy and Navigation 1:107–112

McPhail S (2007) Autosub operations in the Arctic and the Antarctic. In: Proceedings of the international masterclass, 28–30 March 2006, pp 27–38

Morozov EG, Spiridonov VA, Molodtsova TN, Frey DI, Demidova TA, Flint MV (2020) Investigations of the ecosystem in the Atlantic sector of Antarctica (Cruise 79 of the R/V Akademik Mstislav Keldysh). Oceanology 60(5):721–723

Chapter 29
Heavy Metals and Anthropogenic Radionuclides in the Region of the Antarctic Peninsula

Artem A. Paraskiv, Natalia Yu. Mirzoeva, Nataliya N. Tereshchenko, Vladislav Yu. Proskurnin, Ilya G. Sidorov, Svetlana I. Arkhipova, and Eugene G. Morozov

Abstract The relevance of content monitoring of heavy metals and anthropogenic radionuclides in the Atlantic waters of the Antarctic region is determined by the need for a modern assessment of the aquatic environmental quality for making responsible decisions about the protection of marine living resources in this unique region of the World Ocean. The goal of the research is to obtain new data on the levels and spatial and depth distribution of the concentration of trace elements, mainly heavy metals, and anthropogenic radioisotopes in seawater. Seawater samples were taken during the Antarctic expedition of cruise 79 of the R/V "*Akademik Mstislav Keldysh*" at 21 stations in the Drake Passage, Bransfield Strait, Antarctic Sound, and Weddell and Scotia seas. Increased concentrations of some trace elements, including heavy metals in seawater, were found in some regions of the Atlantic sector of the Southern Ocean. According to international environmental legislation, such as the "Dutch sheets," some cases exceeding the MPC (maximum permissible concentration for short-term exposure) of Cd and Zn were found; concentrations of Cu, Pb, Cd, Zn, Se, and Co exceeded their target value (TV) at several stations (control levels for chronic exposure). Analysis of vertical distribution of trace elements and heavy metals in the water column showed that concentrations of Zn, Pb, Sb, Mo, Cd, V, and Ni in the surface layer of the deep-water station in the Powell Basin were higher than in the water near the Antarctic Peninsula. The content of elements such as Fe, Co, and Cu in the surface–water layer at these stations

A. A. Paraskiv (✉) · N. Yu. Mirzoeva · N. N. Tereshchenko · V. Yu. Proskurnin · I. G. Sidorov
S. I. Arkhipova
A. O. Kovalevsky Institute of Biology of the Southern Seas, Russian Academy of Sciences, Moscow, Russia
e-mail: paraskiv@ibss-ras.ru

E. G. Morozov
Shirshov Institute of Oceanology, Russian Academy of Sciences, Moscow, Russia

E. G. Morozov et al. (eds.), *Antarctic Peninsula Region of the Southern Ocean*, Advances in Polar Ecology 6, https://doi.org/10.1007/978-3-030-78927-5_29

was almost the same, and the Se concentration was five times higher in the surface waters of the coastal station in the Antarctic Sound. The concentration of trace elements and heavy metals in the water column of the Antarctic Sound coastal area was higher than in the deep-water part of the Weddell Sea. This is likely caused by erosion of elements of tectonic origin from the mainland as a result of ice cover melting. As for the ^{90}Sr and ^{137}Cs radioisotopes, the waters of the Southern Ocean belong to the cleanest areas of the World Ocean. Further study of the sources of trace elements and the features of their distribution in the seawaters of the Atlantic part of Antarctica is necessary to explain the ongoing processes and to ensure rational environmental management and environmentally sound conservancy in the Antarctic region.

Keywords Heavy metals · Surface and deep-water distribution · Anthropogenic radioisotopes ^{90}Sr and ^{137}Cs · Seawater · Atlantic sector of Antarctica

29.1 Introduction

Trace elements are one of the most important components of the marine environment, since they play an essential role in the functioning of living organisms (Casas et al. 2008; Mertz 2013). However, an excess of certain limits of concentration of these substances in organisms may lead to toxic effects (SS of RF 56219-2014; Rainbow 2002). It is generally accepted that V, Co, Ni, Cu, and Zn are considered moderately hazardous and Cd and Pb highly hazardous toxicants (Directory 2016). These chemical elements and a number of other heavy metals are included in the list of substances whose amounts are subject to regular monitoring in natural environments including oceanic and sea waters (Klenkin et al. 2007). Studying the levels and redistribution features of heavy metals is also relevant for the Antarctic region. According to international agreements (Antarctic Treaty 1959; CCAMLR 1982), limited human activities occur on the Antarctic continent and in the waters of the Southern Ocean. For example, the activities of scientific research expeditions and stations, as well as of ships extracting seafood and providing tourist activities are carried out here. Previous studies have shown that an increase in the concentration of heavy metals in the waters near Antarctica can be caused by natural processes (Ahn et al. 1996, 2004; Samyshev and Minkina 2019). These processes include, in particular, the erosion of elements of tectonic origin from the mainland as a result of increased melting of glaciers, which is caused by global warming in recent decades (Ahn et al. 1996, 2004; Samyshev and Minkina 2019; Sanchez et al. 2019), and upwelling of deep waters (Honda et al. 1987).

The relevance of monitoring of heavy metals in the Atlantic waters of Antarctica is determined, first of all, by the fact that they possess favorable conditions for the existence of stable clusters of Antarctic krill (*Euphausia superba* Dana, 1852).

Clusters with the highest biomass density in comparison with the other basins of the Southern Ocean were found earlier in this region (Byikova et al. 2004; Honda et al. 1987), and currently, this is a region of krill fishery. Many aquatic biota species, including krill, accumulate heavy metals up to concentrations that are many times higher than their concentrations in the water (Polikarpov and Egorov 1986; Honda et al. 1987; Mertz 2013). At the same time, they can both be exposed to toxic effects themselves, especially their most sensitive stages of ontogenesis (eggs, larvae, juveniles), and serve as a transmitting link within the transfer of toxic substances along the trophic chains up to human beings consuming seafood (Casas et al. 2008; Samyshev and Minkina 2019). Therefore, monitoring the concentration levels of heavy metals in the Southern Ocean waters is of significant scientific and practical interest.

Anthropogenic radionuclides entered the Atlantic Ocean primarily with global fallout from nuclear-weapon tests in the atmosphere, which were carried out mainly in the late 1950s and early 1960s (UNSCEAR 1982). The main anthropogenic dose-forming radioactive isotopes in the marine environment today are ^{90}Sr and ^{137}Cs. The necessity for large-scale radioecological research in the Atlantic Ocean is caused by various features of radioactive fallout in the Northern and Southern hemispheres after nuclear-weapon tests and accidents at nuclear facilities (UNSCEAR 2000; Aarkrog 2003; Gudiksen et al. 1989; Gulin and Stokozov 2005; Buesseler et al. 2017, 2018). The differences in the intensity of anthropogenic nuclear activity in the Northern and Southern hemispheres as well as features of planetary-scale atmospheric processes resulted in the properties of the fallout revealed by the fact that about three-quarters of the radioactivity inputs in the Atlantic Ocean were in the North Atlantic, and a quarter in the South Atlantic. In each hemisphere, the maximum fallout of radioactive substances was found at mid-latitudes, and the minimum at the equator and poles (Aarkrog 1994; Gulin and Stokozov 2005).

The first large-scale study of radionuclides distribution in the Atlantic Ocean was carried out in 1972–1973 under the Geochemical Ocean Exploration Program GEOSECS (Ostlund et al. 1993). Since the GEOSECS maritime expeditions, the distribution of artificial radionuclides, especially ^{137}Cs, has been extensively studied on an oceanic scale through several major programs in the Atlantic Ocean. These expeditions were the Swedish Antarctic Research Expedition SWEDARP in 1988–1989, which covered the regions of the North and South Atlantic (Holm et al. 1996), and the French expedition on the R/V "*Jeanne d'Arc*" across the Atlantic Ocean in 1992–1993 (Bourlat et al. 1996; Gulin and Stokozov 2005). These early measurements clearly showed distinct zones of ^{137}Cs latitudinal distribution with higher concentrations at mid-latitudes in both the Northern and Southern hemispheres, as was expected from the global fallout pattern. Comparison of the data collected during these expeditions made it possible to trace long-term changes in the surface concentrations of ^{137}Cs indicating a gradual decrease over the past decades due to radioactive decay and the physical mixing of water masses (Bourlat et al. 1996; Holm et al. 1996; Gulin and Stokozov 2005).

The latitudinal distribution of ^{137}Cs in the surface waters of the Atlantic and Western Antarctica was studied by the IBSS staff during the 7th Ukrainian Antarctic Expedition in January–May 2002 (Gulin et al. 2003; Gulin and Stokozov 2005). Concentrations of ^{137}Cs were also measured in the upper ice of the Woozle Hill coastal glacier, located near the Ukrainian Antarctic station "Akademik Vernadsky" (Western Antarctica, 65° 15′ S, 64° 16′ W). Comparison of these data with the results of previous expeditions along the same route (Holm et al. 1996; Bourlat et al. 1996) showed almost the same ^{137}Cs concentrations between 40° N and 20° S, which indicates a decrease in ^{137}Cs activity at these latitudes with the effective half-life of 10–15 years (on average 12.5 ± 2.1 years) (Gulin et al. 2003; Gulin and Stokozov 2005). This suggests a decrease in ^{137}Cs surface ocean water concentration at this latitude range to be controlled, in addition to the radioactive decay of ^{137}Cs (half-life is 30 years), by vertical mixing of the upper water masses. South of 20° S, concentrations of ^{137}Cs in surface waters decreased more rapidly due to the influence of less polluted Antarctic waters. At 50–60° S and off the Antarctic coast, the activity of ^{137}Cs in 2002 was similar to the activity measured during the SWEDARP and Jeanne d'Arc expeditions, which was likely to be caused by an additional input of ^{137}Cs into these waters due to ice melting from the coastal glaciers (Gulin et al. 2003; Gulin and Stokozov 2005).

Thus, the goal of this study was to obtain new data on the levels and spatial distribution of the concentration of heavy metals and the main anthropogenic radionuclides ^{90}Sr and ^{137}Cs in seawater in order to assess the current quality of natural complexes in the Atlantic sector of the Antarctic.

29.2 Materials and Research Methods

29.2.1 Study Sites and Materials

Sampling seawater was carried out for determining the content of dissolved forms of trace elements, including heavy metals, and radioactive isotopes ^{90}Sr and ^{137}Cs. Samples were taken in the Atlantic sector of Antarctica at 22 stations in the Drake Passage, Bransfield Strait, and Antarctic Sound separating the Antarctic Peninsula from the Joinville Islands group, and in the Weddell and Scotia seas (Fig. 29.1). Coordinates of sampling locations and characteristics of the aquatic environment (temperature, salinity) at sampling stations are shown in Table 29.1.

For the determination of trace elements, including heavy metals, 21 samples of surface (0–5 m) seawater were taken from the shipboard in the aquatic environment of the study site (Fig. 29.1 and Table 29.1). Samples were also taken to study the vertical distribution of heavy metals and trace elements. Station 8, located in the

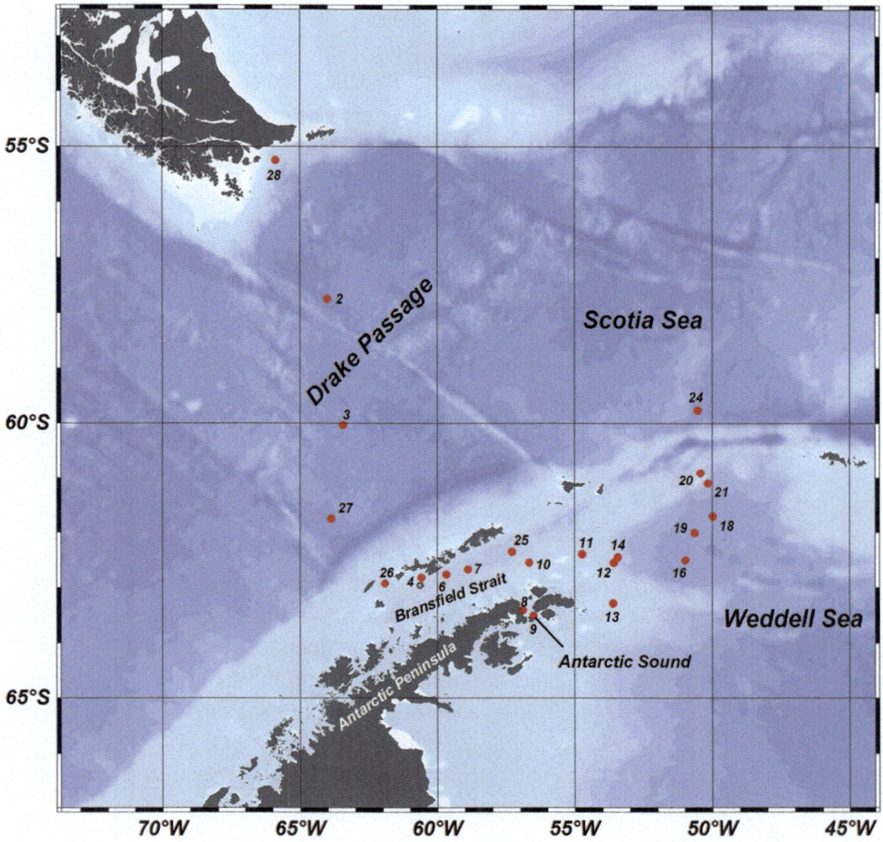

Fig. 29.1 Map of the study region and location of sampling stations

Antarctic Sound, the shallowest in the strait at a depth of 187 m, and station 21 at 2766 m water depth located in the northwestern part of the Weddell Sea, was chosen for the water column sampling. Water temperature and salinity (Table 29.1) were determined using the SBE 911 Plus CTD probe equipped with a cassette of 21 Niskin bottles (5 l each) for water column sampling.

Twelve samples of seawater for ^{90}Sr concentration measurement and 7 samples for ^{137}Cs were taken at six stations (3, 6, 8, 12, 16, 21) (Fig. 29.1). Surface water was sampled into plastic containers (the sample volume for ^{90}Sr analyses was 20 l and 100 l for ^{137}Cs) using the ship's flow-through system from an approximate depth of sampling at 6 m. At station 12, additional samples for ^{90}Sr analyses were taken from six depths: surface, 200 m, 400 m, 600 m, 800 m, and 1025 m. At station 16 besides the surface water, Antarctic Bottom Water was sampled from a depth of 3290 m.

Table 29.1 Metadata of sampling stations for determination of heavy metals

Number of station	Date of sampling	Coordinates		Water depth (m)	Temperature of water (°C)	Salinity (PSU)
		S	W			
1	2	3	4	5	6	7
2	10.02.2020	57° 44.4074	63° 59.9839	4130	8.62	33.92
3	11.02.2020	60° 01.5902	63° 25.3325	3778	5.73	33.63
4	12.02.2020	62° 48.3626	60° 33.7133	373	3.16	34.11
6	12.02.2020	62° 45.2536	59° 39.5351	548	2.78	34.11
7	12.02.2020	62° 39.3628	58° 52.8365	1574	4.19	34.14
8	13.02.2020	63° 24.4219	56° 53.5885	187	−0.91	34.28
9	13.02.2020	63° 30.5974	56° 29.7344	736	0.64	34.27
10	14.02.2020	62° 32.0215	56° 39.2660	345	1.64	34.04
11	14.02.2020	62° 23.0395	54° 43.0033	362	−0.72	34.21
12	15.02.2020	62° 32.2922	53° 33.7689	1030	−0.67	34.32
13	15.02.2020	63° 17.4354	53° 35.9791	362	0.25	33.88
14	16.02.2020	62° 26.0611	53° 25.7918	1383	−0.02	33.66
16	17.02.2020	62° 29.0724	50° 57.3809	3290	−0.42	33.27
18	18.02.2020	61° 42.2470	49° 58.2523	1743	0.26	32.74
19	19.02.2020	61° 59.7085	50° 37.6094	700	1.0	33.97
20	19.02.2020	60° 53.9809	50° 24.0082	940	0.89	34.03
21	20.02.2020	61° 05.5251	50° 08.3302	2766	0.31	32.99
24	24.02.2020	59° 45.2400	50° 30.7537	3787	2.94	34.12
25	26.02.2020	62° 20.7280	57° 16.5980	1331	3.22	34.12
26	27.02.2020	62° 55.1235	61° 54.2073	800	2.61	33.8
27	27.02.2020	61° 45.3571	63° 50.9526	3622	4.17	33.64
28	28.02.2020	55° 14.6157	65° 51.9639	1098	10.64	33.46

29.2.2 Research Methods

29.2.2.1 Heavy Metals and Trace Elements

Water samples for heavy metals and trace elements were treated on board by the extraction concentration and isolation technique in accordance with the Russian Federation regulating document (RD 52.10.243-92). The method is based on the carbon tetrachloride extraction of the diethyldithiocarbamate (DDC) complexes of the determined elements using sodium DDC (Na-DDC) as a complexing agent. The extraction step is followed by the destruction of extracted complexes with concentrated nitric acid and re-extraction of the elements to be determined into an aqueous solution. After the extraction step, most of the interfering macroelements remain in the primary aqueous phase and the re-extraction step leads to a concentration increase of extractable elements up to 100 times. The flowchart of the trace elements concentration in accordance with the technique governed by the regulating document (RD 52.10.243-92) was the following. The samples were vacuum-filtered through FMNC-0.45 membrane nitrocellulose filters with a nominal pore size of 0.45μm, acidified to pH 4 with concentrated nitric acid of the high-purity grade; then a 2% Na-DDC solution was added and the two-stage extraction with carbon tetrachloride was carried out. Re-extraction was carried out by sequential addition of 0.5 ml of concentrated nitric acid and 4.5 ml of deionized water. The obtained re-extracts were stored in a refrigerator until the end of the expedition and transported to the coastal laboratory for analysis. In accordance with the regulating document (RD 52.10.243-92), quantitative extraction using the described procedure was performed for Cu, Zn, Cd, and Pb. At the same time, Be, V, Fe, Co, Ni, Se, Mo, Sb, and Tl were also extracted using the same analysis scheme (Zolotov and Kuzmin 1971; Stary and Kratzer 1968; Yeh et al. 1980).

Measurement of trace elements (Be, V, Fe, Co, Ni, Cu, Zn, Se, Mo, Cd, Sb, Tl, and Pb) in re-extracts from seawater samples was carried out in the laboratory of the Center for Collective Use "Spectrometry and Chromatography" of the FRC IBSS on a mass spectrometer with inductively coupled plasma (ICP-MS) PlasmaQuant MS Elite (Analytik Jena AG) (SS 56219-2014; PlasmaQuant 2014). The ICP-MS unit was calibrated using a standard solution "Multi-element calibration standard IV-28, HNO3/HF" (Inorganic Ventures). The mass-spectrometer measurement mode included seven replicates of ten scans for each measured element with a duration from 10,000 to 100,000μs, depending on expected concentration and typical count rate intensity of each element. Processing and presentation of measurement results were carried out in accordance with SS 56219-2014 and RD 52.10.243-92. The detection limits (DLs) were individual for each determined element and depended on the response intensity and noise level of the mass spectrometer for each specific ratio of the ion mass to its charge. Thus, the lowest DL of 0.0003$\mu g \cdot l^{-1}$ was observed for Co, and the highest 0.5$\mu g \cdot l^{-1}$ for Fe. For the rest of the determined elements, the DLs ($\mu g \cdot l^{-1}$) were 0.001 for Sb and Tl; 0.002 for Cd; 0.005 for Be; 0.01 for V, Ni, Se, and Pb; 0.04 for Cu; 0.05 for Zn; and 0.1 for Mo. The measurement uncertainty for each element did not exceed 15% of the result for concentrations greater than 5 DL and varied from 15% to 50% for concentrations less than 5 DL.

29.2.2.2 ^{90}Sr Radioisotope

Determination of ^{90}Sr concentrations in seawater was carried out using a standard technique (Polikarpov et al. 2008). The preliminary treatment of water samples for the ^{90}Sr concentration determination implying the precipitation of strontium carbonates ($SrCO_3$) in them was carried out onboard directly after sampling. Each water sample was acidified with concentrated HCl (one ml of HCl per one l of sampled water), and a solution of stable carriers Sr^{+2} (1000 mg per sample) and Fe^{+3} (500 mg per sample) was added as well. After 2 h of holding with periodic stirring, 120 g of NH_4Cl was added to the water sample. After 3–4 h of periodic stirring, 200 ml of a 1.5 M sodium carbonate solution was added as a precipitant. The sample was thoroughly mixed and after 12–24 h of the strontium carbonate sediment settlement, the supernatant was decanted, and the suspension was filtered using a vacuum pump through a paper filter for slow filtration (Polikarpov et al. 2008). Further radioanalytical treatment was carried out in the radiochemical laboratory of the Department of Radiation and Chemical Biology (DRCB) of the FRC IBSS.

The Measurement Principle The measurement method is based on the radiometric determination of ^{90}Sr from the Cherenkov radiation of its daughter element ^{90}Y using a low-background liquid scintillation counter (LSC) LKB "QUANTULUS 1220" (Instrument description 1989). The lower limit of detectable activity (LLD) is $0.01 \div 0.04$ Bq•m^{-3}. The control over the correctness of the method and reliability of the results was carried out by means of constant international intercalibration under the auspices of the IAEA (Vienna, Austria) and the National Laboratory (RISOE, Denmark). The results were subjected to a common mathematical data processing (Polikarpov et al. 2008). The intercalibration data from measurements of reference samples and field parallel determinations between the IBSS and other institutions confirmed that the used methodological base made it possible to assess the contamination of the studied ecosystems with the long-lived radionuclide ^{90}Sr at the necessary and sufficient degree of reliability.

29.2.2.3 ^{137}Cs Radioisotope

Activity concentration of ^{137}Cs in seawater samples (100 l) was determined by a sorption technique using two series-connected adsorbers with subsequent measurement of ^{137}Cs content via measurements of its gamma-emitting daughter element ^{137m}Ba. Sampled water was first filtered through a foam polypropylene filter-cartridge with a nominal pore size of 0.5μm to remove suspended matter. Then, the filtrate passed through plastic 10-ml absorbers filled with particulate sorbent. This inorganic composite sorbent consists of a thin layer of mixed nickel-potassium ferrocyanide chemically deposited from water solution on delignificated wood powder as a carrier. The approximate composition of the sorptive layer is $K_{1.33}Ni_{1.33}[Fe(CN)_4]$ according to the sorbent documentation (Betenekov et al. 1976). The mass of the sorbent in each absorber is 2.5 g. The absorber is a column with sorbent of 7 cm height and column inner diameter of 1.5 cm. The filtered water passed through the absorbers forced by a peristaltic pump (Elpan 372.C, Poland) at a rate of 0.07 $l \cdot min^{-1}$. To reduce sorbent volume for its further adjustment to standard measurement geometry, they were reduced to ash in a muffle furnace at 400 °C. The ash-residue weight was about 0.8 g. Then the ashed sorbent was placed into standard beakers for gamma-spectrometric measurement. Content of ^{137}Cs in the samples was determined using a scintillation spectrometer with a detector based on the single crystal of NaI(Tl) (STC "RADEK", Russia) with close to 4π counting geometry (well type detector). The radioactive background and detection efficiency of ^{137}Cs (^{137m}Ba) radiation were taken into account while processing the measurement results. Sorption efficiency was assessed by the difference in ^{137}Cs activity on the first and the second absorbers. Knowing the ^{137}Cs activity on the first absorber, sorption efficiency, and sample volume, the ^{137}Cs activity concentration ($Bq \cdot m^{-3}$) in water was calculated (Miroshnichenko et al. 2019).

29.3 Results and Discussion

29.3.1 Heavy Metals and Trace Elements in Surface Waters

Among the list of studied heavy metals previously noted, the most toxic pollutants include copper, zinc, cadmium, and lead (Klenkin et al. 2007). The measurement results of their concentrations in surface waters are shown in Fig. 29.2.

Trace elements such as beryllium and thallium are also highly toxic substances (Directory 2016), but their concentrations in all samples were too low to exceed their detection limits. Hence, their concentrations did not exceed 0.005$\mu g \cdot l^{-1}$ for Be and 0.001$\mu g \cdot l^{-1}$ for Tl. Molybdenum, cobalt, and nickel are characterized as toxic substances, but they are also characterized by carcinogenic and mutagenic effects in living organisms. The results of these trace elements determination are shown in Fig. 29.3.

Among the studied trace elements are toxic substances such as selenium and vanadium, as well as hazardous substance antimony and less toxic iron. Concentration distribution of these chemical elements in surface waters is shown in Fig. 29.4.

Since the study sites belong to the Antarctic krill fishery area, the assessment of the ecological state of the Antarctic region was carried out by comparison of the data with the reference water quality values given in the "Dutch sheets" (Warmer and van Dokkum 2002). The values used were MRS (maximum permissible concentration of a pollutant considering short-term exposure) and target value (TV,

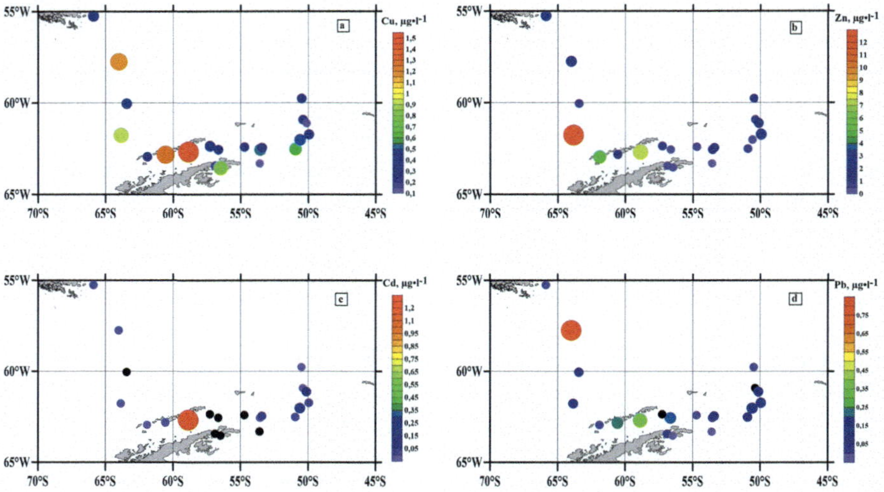

Fig. 29.2 Concentration of Cu (**a**), Zn (**b**), Cd (**c**), and Pb (**d**) in surface seawater in the study sites of the Antarctic region. Black dots show data below the detection limit

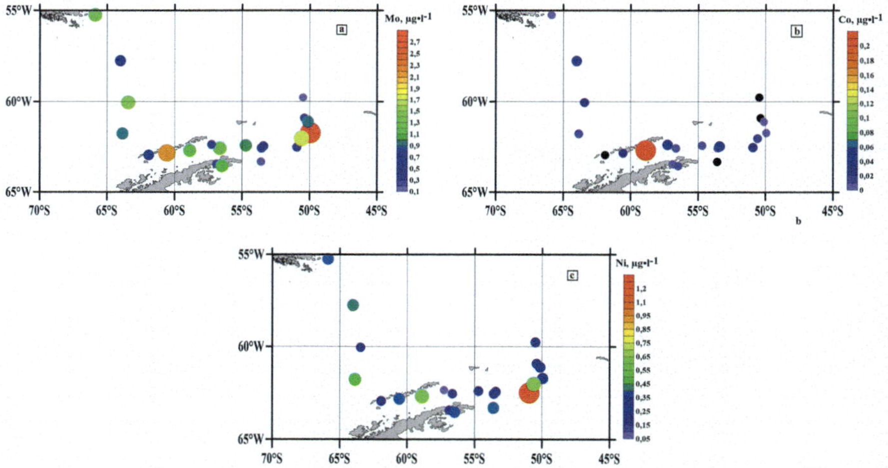

Fig. 29.3 Concentration of Mo (**a**), Co (**b**), and Ni (**c**) in surface seawater in the study sites of the Antarctic region. Black dots show data below the detection limit

reference level of pollutant concentration considering long-term (chronic) exposure) (Warmer and van Dokkum 2002). Comparison of the obtained levels with the background concentrations of the studied elements in surface seawaters (Crommentuijn et al. 1997) and in surface waters in the open ocean (Izrael and Tsyiban 2009) was also carried out.

Analysis of results (Figs. 29.2, 29.3 and 29.4) indicates relatively low levels of studied trace element concentrations in the surface seawater in the study site.

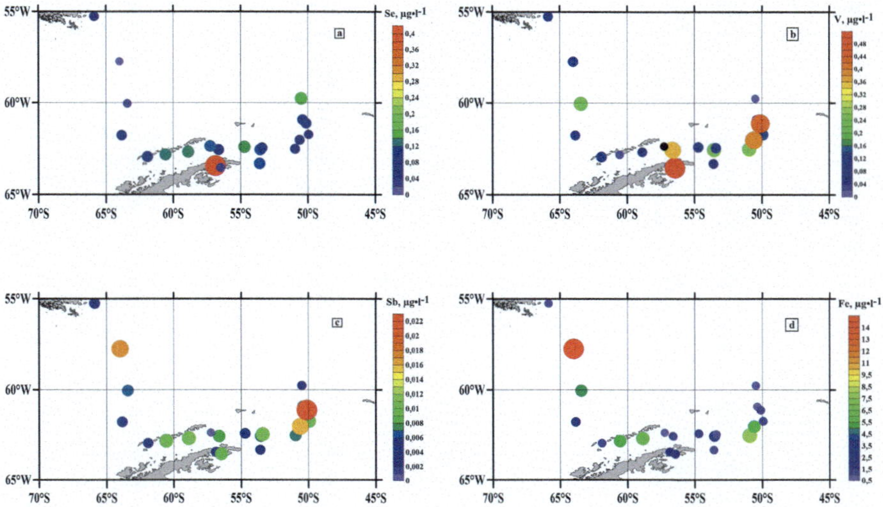

Fig. 29.4 Concentrations of microelements Se (a), V (b), Sb (c), and Fe (d) in surface seawater in the Antarctic region. Black dots show data below the detection limit

However, comparing the data with the concentrations of these trace elements in the waters of the open ocean (Izrael and Tsyiban 2009), it was noted that maximum concentrations of the majority of the studied chemical elements (Pb, Cd, Zn, Cu, Be, Mo, Co, and V) in the Antarctic region were higher than those, typical for the waters of the open ocean. This excess was as high as 2.3 times for Se and 211.3 times for Zn (Table 29.2).

According to the "Dutch sheets," the measured concentrations of Tl, Sb, Be, Ni, V, and Fe were significantly lower than MPC and TV (Warmer and van Dokkum 2002) (Table 29.2) at all stations. An excess of the MPC was observed for Cd at station 7 and for Zn at station 27 (Fig. 29.2), and an excess of TV was noted for Cu at 7 stations, for Zn at 3 stations, for Cd and Pb at 2 stations, for Se at 9 stations, and for Co at one station (Table 29.2 and Figs. 29.2, 29.3 and 29.4).

The significance of hydrological processes, primarily currents, in the distribution of increased concentrations of dissolved forms of chemical elements in Antarctic waters is proved by the similarity of the distribution of the studied trace elements in the marine environment at sampling stations (Figs. 29.2, 29.3 and 29.4). The coastal boundary current in the Bransfield Strait is an important factor of circulation. Together with the Antarctic Circumpolar Current (ACC) in the Drake Passage, it plays a significant role in the transport of waters eastward (Morozov 2007).

Table 29.2 Comparison of the measured values of the concentrations of chemical elements in surface ocean waters of the Antarctic area with their background concentrations and with reference water quality values given in the "Dutch sheets"

Chemical element	Concentration of elements in water ($\mu g \cdot l^{-1}$): Concentration range of the dissolved form of the elements measured in water	Background concentrations: C_{OPEN}[a] min/max	Background concentrations: C_{SWB}[b]	Reference water quality values for chemical elements: MPC[c]	Reference water quality values for chemical elements: TV[d]
Pb	0.01–0.90	0.001/0.034	0.02	11	0.3
Cd	<0.002–1.17	0.0001/0.120	0.025	0.4	0.08
Zn	0.39–12.26	0.003/0.058	0.35	9.4	2.9
Cu	0.10–1.49	–	0.25	1.5	0.5
Tl	<0.001	–	–	1.6	0.06
Be	<0.005	–	–	0.2	0.02
Mo	0.21–2.84	–	–	290	4.3
Co	0.0003–0.21	–	–	2.8	0.2
V	<0.01–0.67	1.02/1.78	–	4.3	0.9
Ni	0.05–1.22	0.116/0.70	–	5.1	3.3
Se	<0.01–0.419	0.04/0.18	–	5.3	0.09
Sb	<0.001–0.02		–	0.4	1.6
Fe	0.76–14.33	0.005/0.140	–	–	–

Notes:

[a]C_{OPEN} is the typical surface water concentration of an element in the waters of the open ocean (background) (Izrael and Tsyiban 2009)

[b]C_{SWB} is the background concentration of an element in surface sea waters (Crommentuijn et al. 1997)

[c]MPC is the maximum permissible concentration for the dissolved form of an element in surface water (with short-term exposure) (Warmer and van Dokkum 2002)

[d]TV is the target value of concentration for the dissolved form of an element conserved during chronic (long-term) exposure in surface water (Warmer and van Dokkum 2002)

29.3.2 The Depth Distribution of Heavy Metals and Trace Elements

To study the depth distribution of heavy metals and microelements, two stations were selected: a coastal one in the Antarctic Sound (station no. 8, depth 187 m) and a deep-water one in the northwestern part of the Weddell Sea (station no. 21, depth 2766 m). These stations differed in the thermohaline properties of the water column.

Thus, the water in the Antarctic Sound was characterized by small changes in temperature and salinity with depth (Fig. 29.5a). In the depths of 0–140 m, the water was colder (−0.93 – −0.95 °C) than at the same depths at a deep-water station (−0.93 °C – +0.29) (Fig. 29.5b). This difference is caused by a large amount of ice and icebergs that cool water to negative temperatures. In the deeper layers, water temperature decreased at both stations; in the bottom layer of the coastal station at a depth of 184 m, it was −1.06 °C. The water structure at station 21 significantly differed from the coastal region and was generally typical for the Weddell Sea (Klepikov 1963; Morozov et al. 2020). A warm surface layer with an apparent vertical temperature gradient was located at depths down to 140–150 m. Below, at depths of 150–200 m, there was a summer cold intermediate layer with temperatures below −1.00 °C—with the lowest temperature of −1.28 °C at a depth of about 180 m (Fig. 29.5b).

Further, there was a layer of warm deep water with temperatures of up to +0.20 °C at a depth of 550–600 m. Then, with increasing depth, periodic fluctuations of temperature in the range from −0.67 to +0.29 °C were observed. Below a depth of 1460 m, a zone of stable temperature decrease was recorded; the temperature decreased with depth to −0.51 °C in the bottom layer at a depth of 2766 m (Fig. 29.5b).

Figures 29.6 and 29.7 show the vertical distribution of the studied trace elements and heavy metals at station 8.

The first group of elements (Co, Ni, Cu, Sb, Zn, and Pb) was characterized by their relatively low concentrations in the surface water layer and by the presence of

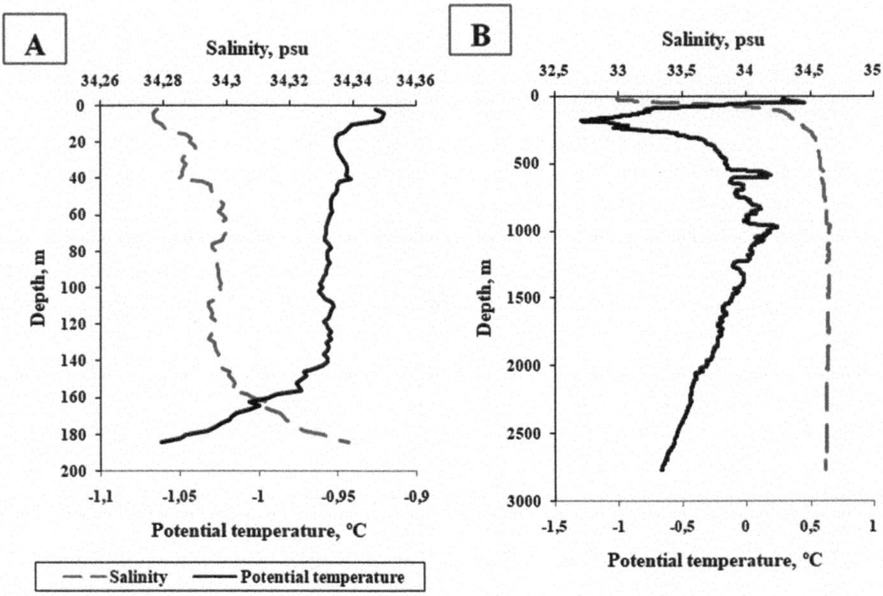

Fig. 29.5 Salinity and potential temperature at stations 8 (**a**) and 21 (**b**)

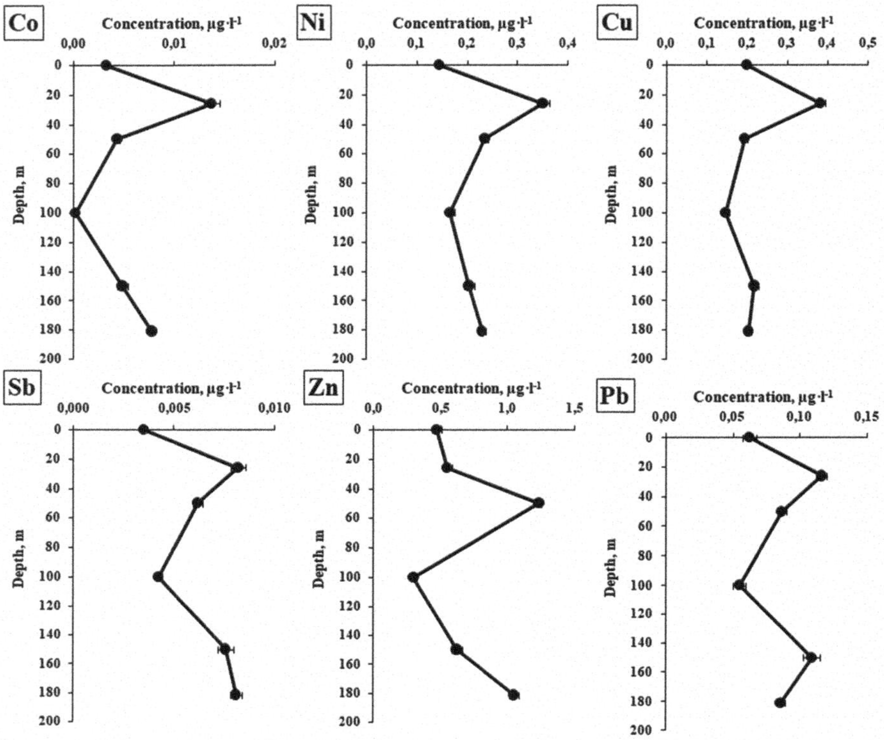

Fig. 29.6 Vertical distribution of Co, Ni, Cu, Sb, Zn, and Pb at station 8. The uncertainty is shown at the points where it exceeds the symbol size

subsurface concentration maxima at a depth of 20–50 m (Fig. 29.6). The second group of elements was characterized by an almost uniform increase (V, Mo) or a decrease (Fe, Se) in their concentration with depth. The exception was observed in the vertical profile of the Cd distribution with multidirectional fluctuations of its concentration with depth (Fig. 29.7).

Figures 29.8 and 29.9 show the vertical distribution of trace elements and heavy metals at station 21, located in the northwestern part of the Weddell Sea. In contrast to the coastal region, here the maximum concentrations of all elements were observed in the surface water layer. However, further analysis also showed that all the elements may be divided into two groups. The first group (Cu, Ni, Pb, Cd, Mo, and V) was characterized by a decrease in concentration with a decrease in water temperature down to the cold intermediate layer and their further increase with an increase in temperature in a layer of deep warm water (Fig. 29.8). Then, downward, the concentrations of the studied elements also decreased with decreasing water temperature.

Their maximum concentration of another group of elements (Fe, Co, Zn, Se, and Sb) was also found in the surface water layer. A decrease of their concentrations

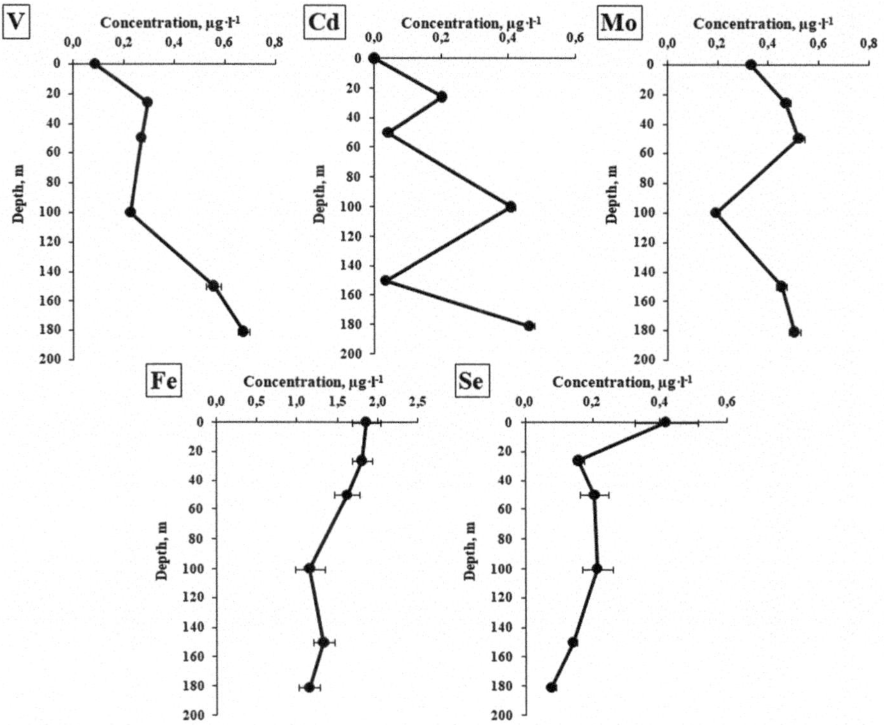

Fig. 29.7 Vertical distribution of V, Cd, Mo, Fe, and Se at station 8. The uncertainty is shown at the points where it exceeds the symbol size

with depth was observed. However, there was no further increase in the layer of deep warm water, noticed as typical for the elements of the first group (Fig. 29.9).

Comparative analysis of the content of trace elements and heavy metals in the water column 0–200 m at both stations showed that the content of Zn, Pb, Sb, Mo, Cd, V, and Ni was higher in the surface layer of deep-water station no. 21 (Fig. 29.10).

The content of elements such as Fe, Co, and Cu in the surface water layer at both stations was practically the same, and the Se concentration was five times higher in the surface layer of coastal station no. 8 (Fig. 29.11).

However, it can be seen from the plots shown in Figs. 29.10 and 29.11 that in the coastal waters of the Antarctic Sound, the concentrations of trace elements and heavy metals in the water column are higher than in the deep-water part of the Weddell Sea. This fact may serve as a confirmation of the previously made assumptions that an increase of the heavy metal concentrations in waters near Antarctica may be caused by natural processes, including the erosion of tectonic elements from the mainland as a result of ice-cover melting (Ahn et al. 1996, 2004; Samyshev and Minkina 2019).

At the same time, it is difficult to determine the main sources of trace elements input to the Antarctic waters by a single study. Their entry can be both the result of

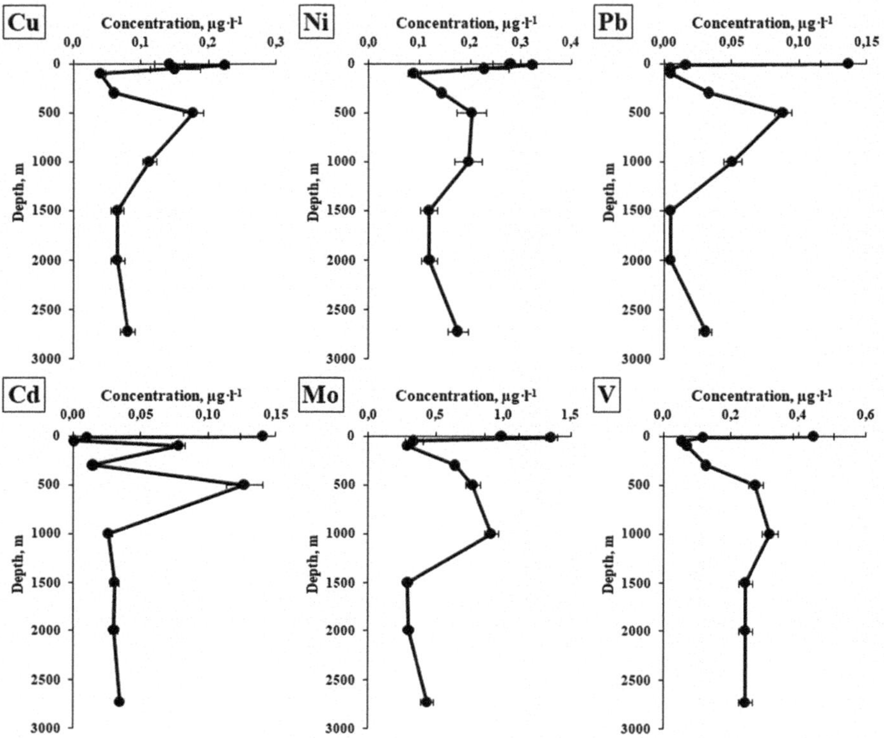

Fig. 29.8 Vertical distribution of Cu, Ni, Pb, Cd, Mo, and V at station 21. The uncertainty is shown at the points where it exceeds the symbol size

anthropogenic and natural processes on the mainland and in the water column, as well as in other regions of the planet, and these elements may reach the Antarctic region as a result of transboundary transport of substances (Klenkin et al. 2007, Directory 2016; Samyshev and Minkina 2019; Sanchez et al. 2019).

29.3.3 ^{90}Sr and ^{137}Cs Radioisotopes

At all six sampling stations (stations: 3, 6, 8, 14, 17, 21) (Fig. 29.1), the ^{90}Sr activity concentrations both in surface water and in the water column were below the detection limit (0.04 Bq·m^{-3}) of the applied technique (Polikarpov et al. 2008). This is likely to be caused by the fact that the atmospheric transport of ^{90}Sr from the areas of large-scale nuclear weapons testing, as well as from the nuclear accident regions, was insignificant, because this radioisotope is less volatile than, for example, ^{137}Cs (Appleby et al. 1999). Also, the radioactive decay of radionuclides and hydrological processes occurring in the Southern Ocean are the main mechanisms for the

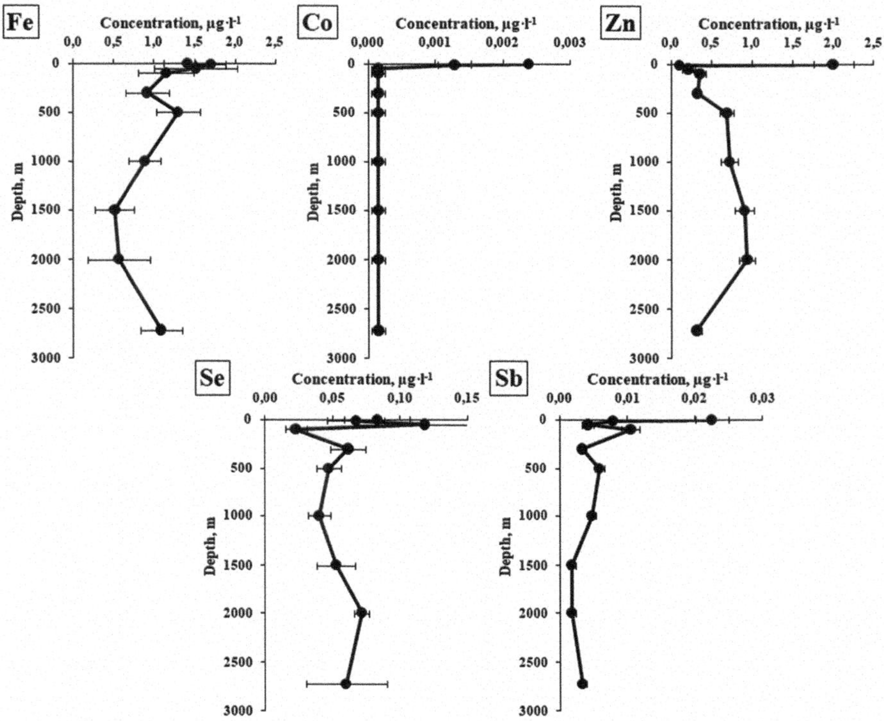

Fig. 29.9 Vertical distribution of Fe, Co, Zn, Se, and Se at station 21. The uncertainty is shown at the points where it exceeds the symbol size

elimination of various pollutants from the Antarctic aquatic environment up to complete self-purification of waters, as in the case with ^{90}Sr in the studied period.

Results of ^{137}Cs measurements showed that at three stations (3, 16, and 21), activity concentration of the radionuclide did not exceed the detection limit, while at stations 6, 8, and 12, it was 0.88, 0.21, and 0.17 Bq·m^{-3}, respectively. It is noteworthy that the stations with detectable activity concentration of ^{137}Cs (in contrast to stations 3, 6, and 21) were located near the Antarctic Peninsula and the lowest ^{137}Cs value was recorded at station 12, the most distant among them from the shore.

Analyses of the ^{137}Cs content dynamics in the surface waters of the South Atlantic showed that the effective half-life of the radionuclide concentration here is less than its radioactive half-life (30.17 years) and averaged about 12.5 years, with the lowest value about 8 years. This fact can be associated with the influence of the Falkland Current and active mixing of the surface water layer in stormy regions (Bourlat et al. 1996; Gulin and Stokozov 2005; Holm et al. 1996). At the same time, in the area adjacent to the Antarctic Peninsula, the activity concentration of ^{137}Cs has not changed significantly, despite the radioactive decay or mixing of water. According to various sources, it varied from 0.2 to 0.7 Bq·m^{-3} in recent decades (Bourlat et al. 1996; Gulin and Stokozov 2005), which must have been associated with the influx

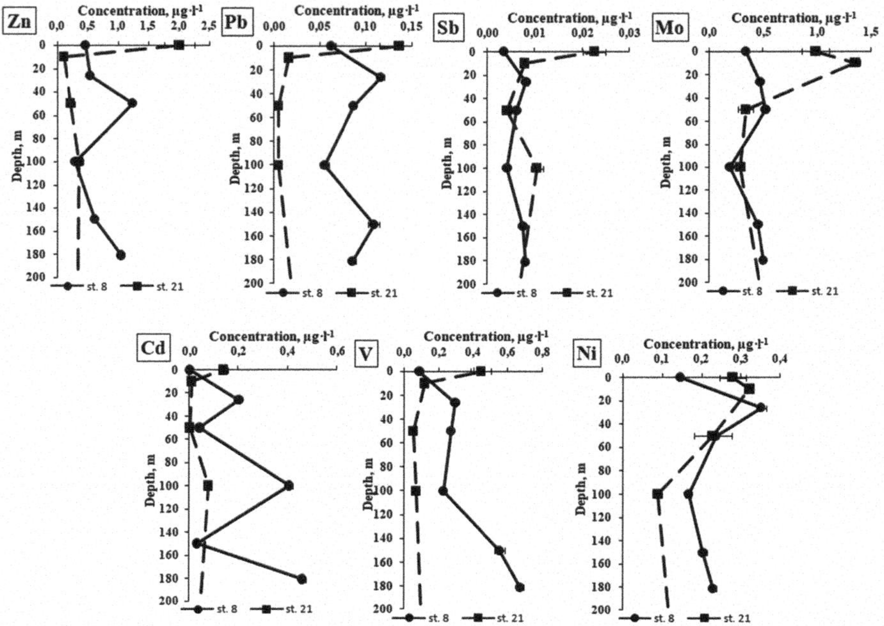

Fig. 29.10 Comparison of vertical distribution of Zn, Pb, Sb, Mo, Cd, V, and Ni at stations 8 and 21. The uncertainty is shown at the points where it exceeds the symbol size

of additional amounts of ^{137}Cs with melt water from adjacent continental glaciers, where the ^{137}Cs content was shown to be much higher than in the coastal Antarctic waters (Gulin and Stokozov 2005).

The presented results show that the values of ^{137}Cs activity concentration obtained in this study in the water areas of the Antarctic Peninsula generally fit into the reported range. Taking into account the data on the content of ^{137}Cs in the continental glacier (Gulin and Stokozov 2005), by 2020, its activity concentration in melt water should be about 1.84 Bq·m^{-3}, which is still sufficient to ensure a constant level of activity in the Antarctic coastal waters.

The lack of ^{137}Cs detectable activity at stations 3, 16, and 21 is obviously associated with the factors already mentioned: currents and active water exchange. Our measurements confirm previous results of low concentrations in the Drake Passage. In 2002, activity of the ^{137}Cs concentration in the central part of Drake Passage (60° S) was recorded as 0.57 Bq·m^{-3} (Gulin and Stokozov 2005).

In the same work, it was found that in the surface waters of the Drake Passage between 50° and 60° S, the effective half-life of ^{137}Cs was determined as 16 years; thus, by 2020, the activity concentration of ^{137}Cs in this region should have been 0.26 Bq·m^{-3}. The effective half-life of a radionuclide in surface waters implies not only the radioactive half-life of an isotope, but also a decrease in its concentration due to dilution, water-mixing, and sedimentation, which leads to self-purification of water masses. However, according to the SWEDARP expedition data in the time

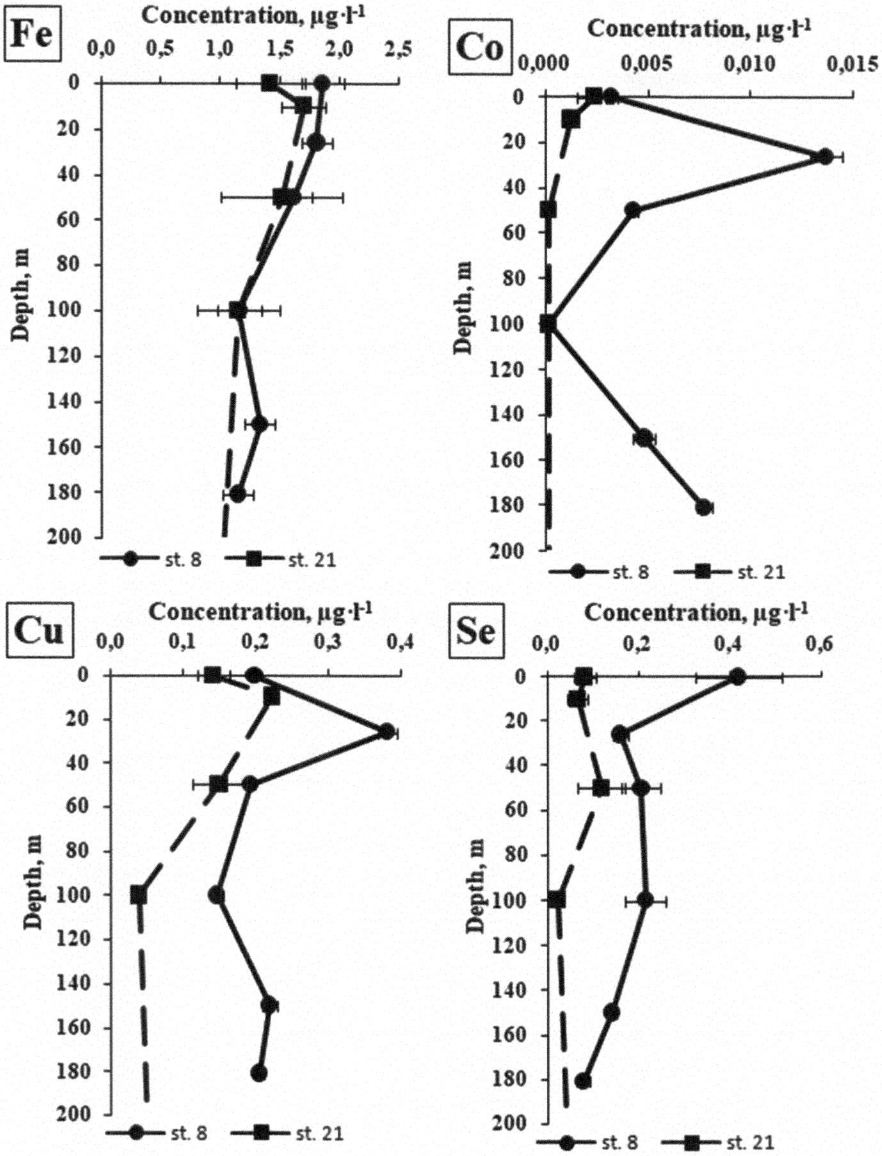

Fig. 29.11 Comparison of vertical distribution of Fe, Co, Cu, and Se at stations 8 and 21. The uncertainty is shown at the points where it exceeds the size of the icon

period 1988–1989, the ^{137}Cs activity concentration at the Drake Passage station was more than two times less than in 2002 (Bourlat et al. 1996; Gulin and Stokozov 2005). Such ambiguous results may be associated with high storm activity and sporadic vertical water mixing in this region. In general, in relation to ^{90}Sr and ^{137}Cs radioisotopes, the waters of the Southern Ocean belong to the cleanest of the World Ocean, which agrees with the data shown earlier (WOMARS 2005).

29.4 Conclusions

The new data on the levels and spatial distribution of 13 trace elements (Be, Se, Sb, Tl, V, Pb, Cd, Cu, Zn, Ni, Mo, Co, and Fe) including heavy metals was collected in 2020 in the Antarctic region.

According to the "Dutch sheets," concentrations of Cd and Zn exceeded their MPC values in single cases and at several stations, concentrations of Cu, Pb, Cd, Zn, Se, and Co exceeded their target values (TVs).

The observed levels of trace element concentrations in the seawater of the Atlantic sector of the Antarctic region indicate the presence of increased concentrations of certain heavy metals, despite the limited anthropogenic activity in this region of the Southern Ocean.

The increased concentrations of some trace elements in comparison with the established MPC and TV in the northern part of the Drake Passage are possibly associated with water runoff from industrial facilities in Chile, which have been mining and processing copper for more than 100 years, but at the same time, the influence of natural sources is also possible. The increased concentration of iron in the northern part of the Weddell Sea (Powell Basin) is most likely to be determined by the input of iron compounds from natural sources in the west of the Weddell Sea, which agrees with the earlier studies by Sanchez et al. 2019.

The ^{90}Sr activity concentration in all study sites of the region was below the detection limit, both in surface water and in the water column. The ^{137}Cs concentration was low (from 0.17 to 0.88 $Bq \cdot m^{-3}$) in the water of three stations close to the Antarctic Peninsula, while at other stations, it was below the detection level. This indicates that in relation to the ^{90}Sr and ^{137}Cs radioisotopes, the waters of the Southern Ocean belong to the cleanest of the World Ocean, as was stated before.

Acknowledgments The authors express their sincere gratitude to the Shirshov Institute of Oceanology, to the Atlantic Fleet Base, and to the crew of the R/V *"Akademik Mstislav Keldysh"* for the excellent organization and high-level performance of the sea expedition to the Atlantic sector of Antarctica. The authors are grateful to the staff of the Center for Collective Use of the Federal Research Center IBSS "Spectrometry and Chromatography," as well as to the leading engineer of the Department of Radiation and Chemical Biology N.V. Kravchenko for assisting in measurements for this study.

Funding The research was supported by the state assignment of the FRC IBSS "Comprehensive studies of the current state of the ecosystem of the Atlantic sector of Antarctica" (Registration number AAAA-A19-119100290162-0) and state assignment of the IO RAS 0128-2019-0008 "The assessment of the current state of natural complexes in the Atlantic sector of the Southern Ocean and their multi-periodical variability (ecosystems, bioproductivity, hydrophysics, hydro- and geochemistry" AAAA-A18-118051490130-3.

References

Aarkrog A (1994) Source terms and inventories of anthropogenic radionuclides. In: Holm E (ed) Radioecology. Lectures in environmental radioactivity. World Scientific Publishing, Lund, pp 21–38

Aarkrog A (2003) Input of anthropogenic radionuclides into the World Ocean. Deep Sea Res Part II Topic Stud Oceanogr 50(17–21):2597–2606. https://doi.org/10.1016/S0967-0645(03)00137-1

Ahn IY, Lee SH, Kim KT, Shim JH, Kim DY (1996) Baseline heavy metal concentrations in the Antarctic clam, Laternula elliptica in Maxwell Bay, King George Island, Antarctica. Mar Pollut Bull 32(8–9):592–598. https://doi.org/10.1016/0025-326X(95)00247-K

Ahn IY, Chung KH, Cho HJ (2004) Influence of glacial runoff on baseline metal accumulation in the Antarctic limpet *Nacella concinna* from King George Island. Mar Pollut Bull 49(1):119–127. https://doi.org/10.1016/j.marpolbul.2004.03.008

Antarctic Treaty (1959) United States Department of State. https://www.state.gov/antarctic-treaty/. Last accessed in November 2020

Appleby LJ, Dewell L, Mishara YK (1999) The ways of migration of artificial radionuclides in the Environment. In: Warner F, Harrison (eds) "Mir" Publ., Moscow, 512 p (in Russian)

Betenekov ND, Egorov YV, Popov VI, Puzako VD, Cheremukhin YG (1976) Method for producing a thin layer sorbent. Ural Polytechnic Institute. (in Russian)

Bourlat Y, Millies-Lacroix JC, Le Petit G, Bourguignon J (1996) ^{90}Sr, ^{137}Cs and ^{239,240}Pu in world ocean water samples collected from 1992 to 1994. In: Gueguenlat P, Germain P, Metivier H (eds) Radionuclides in the oceans: inputs and inventories. Editions de Physique, pp 75–93

Buesseler K, Dai M, Aoyama M, Benitez-Nelson C, Charmasson S, Higley K, Maderich V, Masqé P, Morris PJ, Oughton D, Smith JN (2017) Fukushima Daiichi-Derived radionuclides in the Ocean: transport, fate, and impacts. Annu Rev Mar Sci 9:173–203. https://doi.org/10.1146/annurev-marine-010816-060733

Buesseler KO, Charette MA, Pike SM, Henderson PB, Kipp LE (2018) Lingering radioactivity at the Bikini and Enewetak Atolls. Sci Tot Environ 621:1185–1198. https://doi.org/10.1016/j.scitotenv.2017.10.109

Byikova VM, Byikov VP, Krivosheina LI, Radakova TN, Groysman MY, Glazunov OI (2004) Main directions of integrated use of krill. Trudy TINRO 143:17–32. (in Russian)

Casas S, Gonzalez JL, Andral B, Cossa D (2008) Relation between metal concentration in water and metal content of marine mussels (*Mytilus galloprovincialis*): impact of physiology. Environ Toxicol Chem 27(7):1543–1552. https://doi.org/10.1897/07-418.1

Commission for the Conservation of Antarctic Marine Living Resources CCAMLR (1982). www.ccamlr.org

Crommentuijn T, Polder MD, van de Plassche EJ (1997) Maximum permissible concentration and negligible concentration for metals, taking background concentration into account. Report no. 601501 001. Netherlands, 260 p

Directory. Characteristics of pollutants (2016) Section II. For water objects. "Ecologia" Publ., Perm. 296 p (in Russian)

Gudiksen PH, Harvey TF, Lange R (1989) Chernobyl source term, atmospheric dispersion and dose estimation. Health Phys 57:697–705

Gulin SB, Stokozov NA (2005) ^{137}Cs concentrations in Atlantic and Western Antarctic surface waters: results of the 7th Ukrainian Antarctic Expedition, 2002. J Environ Radioact 83(1):1–7. https://doi.org/10.1016/j.jenvrad.2005.02.005

Gulin SB, Samyshev EZ, Stokozov NA, Sysoev AA (2003) Estimation of the mass accumulation rate from the sea surface layer of Bransfield Strait (West Antarctica) using thorium-234 as a natural radiotracer. Ukr Antarct J 1:30–36. (in Russian)

Holm E, Roos P, Josefsson D, Persson B (1996) Radioactivity from the North Pole to the Antarctic. In: Gueguenіat P, Germain P, Metivier H (eds) Radionuclides in the oceans: inputs and inventories. Editions de Physique, pp 59–74

Honda K, Yamamoto Y, Tatsukava R (1987) Distribution of heavy metals in Antarctic marine ecosystem. Proc NIPR Symp Polar Biol 1:184–197

Instrument Description «Quantulus 1220» (1989) Instrument manual LKB Wallac, Pharmacia Wallac Oy, Ontario, Canada, 350 p

Izrael YA, Tsyiban AV (2009) Anthropogenic ecology of the Ocean. “Flinta Nauka” Publ., Moscow. 519 p (in Russian)

Klenkin AA, Korpakova IG, Pavlenko LF, Temerdasheva ZA (2007) Sea of Azov Ecosystem: anthropogenic pollution. Krasnodar. 324 p (in Russian)

Klepikov VV (1963) Hydrology of the Weddell Sea. Proc Sov Antarct Exped 17:45–93. (in Russian)

Mertz W (2013) Trace elements in human and animal nutrition, 5th edn. Academic, Orlando. 499 p

Miroshnichenko ON, Paraskiv AA, Gulin SB (2019) Cesium-137 concentration in the surface waters of Eurasian seas: evidence from the expedition research of 2017. Geochem Int 57(12):1349–1354. https://doi.org/10.1134/S0016702919120073

Morozov EG (2007) Currents in Bransfield Strait. Doklady Earth Sci 415(6):984–986

Morozov EG, Frey DI, Polukhin AA, Krechik VA, Artemiev VA, Gavrikov AV, Kasian VV, Sapozhnikov PV, Gordeeva NV, Kobylyansky SG (2020) Mesoscale variability of the ocean in the northern part of the Weddell Sea. Oceanology 60(5):573–588. https://doi.org/10.1134/S0001437020050173

Ostlund HG, Craig H, Broecker WS, Spencer D (1993) Geosecs Atlantic. Pacific, and Indian Ocean Expeditions (National Science Foundation, Washington, DC. 1987). vol 7. SL Hautala and SC Riser. J Phys Oceanogr 23:1975

PlasmaQuant MS Elite (2014) Mass-spectrometers with inductively coupled plasma (ICP-MS). User’s manual. Analytik Jena AG, 143 p

Polikarpov GG, Egorov VN (1986) Marine dynamic radiochemoecology. “Energoatomizdat” Publ., Moscow. 176 p (in Russian)

Polikarpov GG, Egorov VN, Gulin SB, Stokozov NA, Lazorenko GE, Mirzoyeva NY, Tereshchenko NN, Tsitsugina VG, Kulebakina LG, Popovichev VN, Korotkov AA, Evtushenko DB, Zherko NV, Malahova LV (2008) Radioecological response of the Black Sea to the Chernobyl accident. “EKOSI-Gidrofizika” Publ., Sevastopol. 667 p (in Russian)

Rainbow PS (2002) Trace metal concentrations in aquatic invertebrates: why and so what? Environ Pollut 120(3):497–507. https://doi.org/10.1016/S0269-7491(02)00238-5

RD 52.10.243-92. Guide to the chemical analysis of marine waters. http://docs.cntd.ru/document/1200042955. (in Russian)

Samyshev EZ, Minkina NI (2019) Coastal ecosystem contamination by heavy metals as an indicator of climate change in Antarctica. J Comput Theor Nanosci 16(1):228–236. https://doi.org/10.1166/jctn.2019.7729

Sanchez N, Reiss CS, Holm-Hansen O, Hewes CD, Bizsel KC, Ardelan MV (2019) Weddell-Scotia confluence effect on the iron distribution in waters surrounding the South Shetland (Antarctic Peninsula) and South Orkney (Scotia Sea) Islands during the Austral Summer in 2007 and 2008. Front Mar Sci 6(771):1–16. https://doi.org/10.3389/fmars.2019.00771

SS of RF 56219-2014 State Standard of Russian Federation (2009) Water. Determination of 62 elements content by inductively coupled mass spectrometry technique. Standartinform, Moscow, 36 p (in Russian)

Stary J, Kratzer K (1968) Determination of extraction constants diethyldithiocarbamates. Analyt Chim Act 40:93–100. https://doi.org/10.1016/S0003-2670(00)86699-4

UNSCEAR (1982) Ionizing radiation: sources and biological effects. United Nations, New York, 773 p

UNSCEAR (2000) Sources and effects of ionizing radiation. Sources, United Nations Scientific Committee on the Effects of Atomic Radiation, vol I. Report to the General Assembly, with Scientific Annexes. United Nations, New York, 649 p

Warmer H, van Dokkum R (2002) Water pollution control in the Netherlands Policy and practice RIZA report 2002.009. Netherlands, 76 p

WOMARS (2005) Worldwide marine radioactivity studies. Radionuclide levels in oceans and seas. IAEA, Vienna, 287 p

Yeh SJ, Lo JM, Shen LH (1980) Determination of extraction constants for metal diethyldithiocarbamates and chloride-mixed metal diethyldithiocarbamates by substoichiometric extraction. Analyt Chem 52(3):528–531. https://doi.org/10.1021/ac50053a035

Zolotov YA, Kuzmin NM (1971) Extraction concentrating. "Khimia" Publ., Moscow. 272 p (in Russian)

The manufacturer's authorised representative in the EU is Springer Nature Customer Service Centre GmbH, Europaplatz 3, 69115 Heidelberg, Germany. If you have any concerns regarding our products, please contact ProductSafety@springernature.com

Printed and bound by CPI Group (UK) Ltd, Croydon, CR0 4YY
30/09/2026
02225248-0015